Solutions MANUAL

Merle C. Potter • David C. Wiggert • Midhat Hondzo

MICHIGAN STATE UNIVERSITY Purdue University

Mechanics of FLUIDS

SECOND EDITION

Merle C. Potter
David C. Wiggert

PRENTICE HALL Upper Saddle River, NJ 07458

Executive Editor: *Bill Stenquist*
Production Editor: *Dawn Blayer*
Manufacturing Buyer: *Julia Meehan*
Special Projects Manager: *Barbara A. Murray*
Supplement Cover Manager: *Paul Gourhan*
Supplement Cover Designer: *PM Workshop Inc.*

Simon & Schuster/A Viacom Company
Upper Saddle River, NJ 07458

Printed in the United States of America

10 9 8 7 6 5 4 3 2

ISBN 0-13-286593-9

Prentice-Hall International (UK) Limited, *London*
Prentice-Hall of Australia Pty. Limited, *Sydney*
Prentice-Hall Canada, Inc., *Toronto*
Prentice-Hall Hispanoamericana, S.A., *Mexico*
Prentice-Hall of India Private Limited, *New Delhi*
Prentice-Hall of Japan, Inc., *Tokyo*
Simon & Schuster Asia Pte. Ltd., *Singapore*
Editora Prentice-Hall do Brasil, Ltda., *Rio de Janeiro*

PREFACE

This manual provides the solutions to the problems in our book, MECHANICS OF FLUIDS. In many cases, the solutions are not as detailed as the examples in the book; they are intended to provide the primary steps in each solution. The discussion of a subtle point, should one exist in a particular problem, is left as a task for the instructor. On the other hand, some of the solutions are presented in detail, especially in Chapter 10 through 14, where many of them focus on applied fluid mechanics. In general, some knowledge of a problem may be needed to fully understand all of the steps presented. This manual is not intended to be a self-paced workbook —the instructor is critically needed to provide explanations, discussions, and illustrations of the myriad of phenomena encountered in the study of fluids.

The degree of difficulty and length of solution for each problem varies considerably. Some are relatively easy and others quite difficult. This allows for flexibility in assignments or in practice sessions. Typically, the easier problems are the first problems for a particular section. A number of the problems have a design aspect. A lesser number of them are true "open-ended" design, in as much as there is more than one correct solution. Those problems are listed in the table immediately following this preface.

The examples and problems have been carefully solved with the hope that errors have not been introduced. Even though extreme care is taken and problems are reworked, errors creep in. We would appreciate knowing about any errors that you may find. They can be eliminated in future printings. Suggestions for changes and improvements are also welcomed. We have class tested most of the chapters with good response from our students, but we are sure that there are improvements to be made.

East Lansing, Michigan

West Lafayette, IN

Merle C. Potter
David C. Wiggert
Midhat Hondzo

DESIGN-FOCUSSED PROBLEMS

Chapter	Problems
2	50, 53, 54, 73, 80, 82
3	66, 68, 69
4	42, 43, 48, 95, 96, 97, 122, 127, 139, 141, 146, 148, 149
5	41
6	32, 34, 52, 55, 57
7	9, 10, 14, 21, 33, 49, 59, 89, 90, 91, 92, 103, 108, 119
8	21, 26, 31, 39, 48, 58, 113
9	24
10	7, 9, 12, 13, 14, 15, 18, 26, 30, 31*, 34, 35, 39, 45, 55, 65*, 68*, 70, 71*, 72*
11	4, 5, 14, 16, 17, 19, 24E, 25E, 26E, 29, 36, 37, 38*, 39E*, 40*, 41E*, 42E*, 43, 45, 47*
12	6*, 7*, 9, 12, 13, 14, 15, 16, 17, 19, 20*, 21*, 23*, 26, 27*, 29*, 30, 31E*, 33*, 34, 38, 39, 40*, 42*, 43, 44E*, 45, 46, 47, 48, 49*, 50*
14	4, 5, 14

*These problems have a more extensive design orientation, some of which are "open-ended."

ONE

Basic Considerations

1.1 Conservation of mass — Mass — density
Newton's second law — Momentum — velocity
The first law of thermodynamics — internal energy — temperature

1.2 a) density = mass/volume = M / L^3
b) pressure = force/area = $F / L^2 = ML / T^2 L^2 = M / LT^2$
c) power = force × velocity = $F \times L / T = ML / T^2 \times L / T = ML^2 / T^3$
d) energy = force × distance = $ML / T^2 \times L = ML^2 / T^2$
e) mass flux = $\rho AV = M/L^3 \times L^2 \times L/T = M/T$
f) flow rate = $AV = L^2 \times L/T = L^3/T$

1.3 a) density = $\frac{M}{L^3}\frac{FT^2 / L}{L^3} = FT^2 / L^4$
b) pressure = F/L^2
c) power = $F \times$ velocity = $F \times L/T = FL/T$
d) energy = $F \times L = FL$
e) mass flux = $\frac{M}{T} = \frac{FT^2 / L}{T} = FT / L$
f) flow rate = $AV = L^2 \times L/T = L^3/T$

1.4 a) $L = C\,T^2$. $\quad \therefore [C] = L/T^2$
b) $F = C\,M$. $\quad \therefore [C] = F/M = ML/T^2 M = L/T^2$
c) $L^3/T = C\,L^2\,L^{2/3}$. $\quad \therefore [C] = L^3 / T \cdot L^2 \cdot L^{2/3} = L^{1/3} T$
Note: the slope S_o has no dimensions.

1.5 a) $\text{m} = C\,\text{s}^2$. $\quad \therefore [C] = \text{m/s}^2$
b) $\text{N} = C\,\text{kg}$. $\quad \therefore [C] = \text{N/kg} = \text{kg} \cdot \text{m/s}^2 \cdot \text{kg} = \text{m/s}^2$
c) $\text{m}^3/\text{s} = C\,\text{m}^2\,\text{m}^{2/3}$. $\quad \therefore [C] = \text{m}^3/\text{s} \cdot \text{m}^2 \cdot \text{m}^{2/3} = \text{m}^{1/3}/\text{s}$

1.6 a) pressure : $\text{N/m}^2 = \text{kg} \cdot \text{m/s}^2/\text{m}^2 = \text{kg/m} \cdot \text{s}^2$
b) energy: $\text{N} \cdot \text{m} = \text{kg} \cdot \text{m/s}^2 \times \text{m} = \text{kg} \cdot \text{m}^2/\text{s}^2$
c) power: $\text{N} \cdot \text{m/s} = \text{kg} \cdot \text{m}^2/\text{s}^3$
d) viscosity: $\text{N} \cdot \text{s/m}^2 = \frac{\text{kg} \cdot \text{m}}{\text{s}^2} \cdot \text{s} \frac{1}{\text{m}^2} = \text{kg} / \text{m} \cdot \text{s}$
e) heat flux: $\text{J/s} = \frac{\text{N} \cdot \text{m}}{\text{s}} = \frac{\text{kg} \cdot \text{m}}{\text{s}^2} \cdot \frac{\text{m}}{\text{s}} = \text{kg} \cdot \text{m}^2 / \text{s}^3$

f) specific heat: $\dfrac{\text{J}}{\text{kg}\cdot\text{K}} = \dfrac{\text{N}\cdot\text{m}}{\text{kg}\cdot\text{K}} = \dfrac{\text{kg}\cdot\text{m}}{\text{s}^2}\cdot\dfrac{\text{m}}{\text{kg}\cdot\text{K}} = \text{m}^2/\text{K}\cdot\text{s}^2$

1.7 $\text{kg}\dfrac{\text{m}}{\text{s}^2} + c\dfrac{\text{m}}{\text{s}} + k\text{m} = f$. Since all terms must have some dimensions (units) we require:

$[C]$ = kg/s, $[k]$ = kg/s^2 = $\text{N}\cdot\text{s}^2/\text{m}\cdot\text{s}^2 = \text{N/m}$, $[f] = \text{kg}\cdot\text{m}/\text{s}^2 = \text{N}$.

Note: we could express the units on C as $[C] = \text{kg/s} = \text{N}\cdot\text{s}^2/\text{m}\cdot\text{s} = \text{N}\cdot\text{s/m}$

1.8 a) 250 kN b) 572 G Pa c) 42 n Pa d) 17.6 cm^3
e) 1.2 cm^2 f) 76 mm^3

1.9 a) 1.25×10^8 N b) 3.21×10^{-5} s c) 6.7×10^8 Pa
d) 5.6×10^{-12} m^3 e) 5.2×10^{-2} m^2 f) 7.8×10^9 m^3

1.10 $\lambda = .225\dfrac{.06854m}{.00194\rho\times3.281^2d^2} = .738\dfrac{m}{\rho d^2}$

where m is in slugs, ρ in slug/ft^3 and d in feet. We used the conversions in the front cover.

1.11 a) 20 in/hr = $\dfrac{20}{12}\times.3048/3600 = 1.411\times10^{-4}$ m/s

b) 2000 rev/min = $2000\times2\pi/60 = 209.4$ rad/s

c) 50 Hp = $50\times745.7 = 37\,285$ W

d) 100 ft^3/min = $100\times.02832/60 = .0472$ m^3/s

e) 20 kip/ft^2 = $20\times4448/.0929 = 9.576\times10^5$ N/m^2

f) 4 slug/min = $4\times14.59/60 = .9727$ kg/s

g) 50 in-lb = $50\times\dfrac{0.3048}{12}\times4.448 = 5.649$ N·m

h) 500 kWh = $500\times1000\times3600 = 1.8\times10^9$ J

1.11E a) 20 cm/hr = $20\times\dfrac{.3937}{12}/3600 = 1.823\times10^{-4}$ ft/sec

b) 2000 rev/min = $2000\times2\pi/60 = 209.4$ rad/sec

c) 50 Hp = $50\times745.7\times.7376 = 27{,}500$ ft-lb/sec

d) 3 m^3/min = $3\times35.32/60 = 1.766$ ft^3/sec

e) 2000 N/cm^2 = $2000\times.2248\times2.54^2\times144 = 417{,}700$ lb/ft^2

f) 10 kg/min = $10\times.06854/60 = .01142$ slug/sec

g) 200 N·mm = $200\times.2248\times.1\times.3937/12 = .1475$ lb-ft

h) 500 kWh = $500\times1000\times3600\times.7376 = 1.3277\times10^9$ ft-lb

1.12 a) $F = ma = 10\times40 = \underline{400\text{ N}}$.

b) $F - W = ma$. $\therefore F = 10\times40 + 10\times9.81 = \underline{498.1\text{ N}}$.

c) $F - W\sin30° = ma$ $\therefore F = 10\times40 + 9.81\times0.5 = \underline{449\text{ N}}$.

1.13 $m = \frac{W}{g} = \frac{250}{9.81} = 25.48$ kg. $\therefore W_{moon} = 25.48 \times 1.6 = \underline{40.77\text{ N}}$.

1.13E $m = \frac{60}{32.2} = 1.863$. $\therefore W_{moon} = 1.863 \times 5.4 = \underline{10.06\text{ lb}}$.

1.14 a) $\lambda = .225\frac{m}{\rho d^2} = .225\frac{4.8 \times 10^{-26}}{.184 \times (3.7 \times 10^{-10})^2} = .43 \times 10^{-6}\text{ m}$ or $\underline{0.00043\text{ mm}}$

b) $\lambda = .225\frac{m}{\rho d^2} = .225\frac{4.8 \times 10^{-26}}{.00103 \times (3.7 \times 10^{-10})^2} = 7.7 \times 10^{-5}\text{ m}$ or $\underline{.077\text{ mm}}$

c) $\lambda = .225\frac{m}{\rho d^2} = .225\frac{4.8 \times 10^{-26}}{.00002 \times (3.7 \times 10^{-10})^2} = .0039\text{m}$ or $\underline{3.9\text{ mm}}$

1.15 a) 52.3 + 101.3 = 153.6 kPa
b) 52.3 + 89.85 = 142.2 kPa
c) 52.3 + 54.43 = 106.7 kPa
d) 52.3 + 26.49 = 78.8 kPa
e) 52.3 + 1.196 = 53.5 kPa

1.16 a) 101 - 31 = $\underline{70\text{ kPa abs}}$.

b) $760 - \frac{31}{101} \times 760 = \underline{527\text{ mm of Hg abs}}$.

c) $14.7 - \frac{31}{101} \times 14.7 = \underline{10.2\text{ psia}}$.

d) $34 - \frac{31}{101} \times 34 = \underline{23.6\text{ ft of } H_2O\text{ abs}}$.

e) $30 - \frac{31}{101} \times 30 = \underline{20.8\text{ in. of Hg abs}}$.

1.17 $p = p_o\, e^{-gz/RT} = 101\, e^{-9.81 \times 4000/287 \times (15 + 273)} = \underline{62.8\text{ kPa}}$
From Table B.3, at 4000 m/p = 61.6 kPa. The percent error is

$$\%\text{ error} = \frac{62.8 - 61.6}{61.6} \times 100 = \underline{1.95\ \%}.$$

1.18 a) $p = 35.65 + \frac{8320 - 8000}{10\,000 - 8000}(26.49 - 35.65) = \underline{34.18\text{ kPa}}$.

$T = 236.2 + \frac{8320 - 8000}{10\,000 - 8000}(223.3 - 236.2) = \underline{234.1\text{ K}}$.

b) $p = 35.65 + .16\,(26.49 - 35.65) + \frac{.16}{2}(.16\text{-}1)\,(19.4 - 2 \times 26.49 + 35.65) = \underline{34.04\text{ kPa}}$.

$T = 236.2 + .16(223.3 - 236.2) + \frac{.16}{2}(.16\text{-}1)\,(216.7 - 2 \times 223.3 + 236.2) = \underline{233.7\text{ K}}$.

Note: The results in (b) are more accurate than the results in (a). When we use a linear interpolation, we lose significant digits in the result.

1.18E a) $p = 973 + \dfrac{22{,}560 - 20{,}000}{25{,}000 - 20{,}000}(785 - 973) = \underline{877\text{ psf}}$

$T = -12.3 + \dfrac{22{,}560 - 20{,}000}{25{,}000 - 20{,}000}(-30.1 + 12.3) = \underline{-21.4°\text{F}}$

b) $p = 973 + .512\,(785 - 973) + \dfrac{.512}{2}(-.488)\,(628 - 2\times 785 + 973) = \underline{873\text{ psf}}$

$T = -12.3 + .512\,(-30.1 + 12.3) + \dfrac{.512}{2}(-.488)\,(-48 + 2\times 30.1 - 12.3) = -\underline{21.4°\text{F}}$

Note: See the note above.

1.19 $T = -48 + \dfrac{33{,}000 - 30{,}000}{35{,}000 - 30{,}000}(-65.8 + 48) = \underline{-59°\text{F}}$ or $(-59 - 32)\,\dfrac{5}{9} = \underline{-50.6°\text{C}}$

1.20 $p = \dfrac{F_n}{A} = \dfrac{26.5\cos 42°}{152 \times 10^{-4}} = 1296\text{ MN/m}^2 = \underline{1296\text{ MPa}}.$

1.21 $\left.\begin{aligned} F_n &= (120000)\times.2\times10^{-4} = 2.4\text{N} \\ F_t &= 20\times.2\times10^{-4} = .0004N \end{aligned}\right\}$ $F = \sqrt{F_n^2 + F_t^2} = \underline{2.400\text{ N}}.$ $\theta = \tan^{-1}\dfrac{.0004}{2.4} = \underline{.0095°}$

1.22 $\rho = \dfrac{m}{V} = \dfrac{0.1}{100\times10^{-6}} = \underline{1000\text{ kg/m}^3}.$ $\gamma = \rho g = 1000\times 9.81 = \underline{9810\text{ N/m}^3}.$

1.22E $\rho = \dfrac{m}{V} = \dfrac{0.2}{180/1728} = \underline{1.92\text{ slug/ft}^3}.$ $\gamma = \rho g = 1.92\times 32.2 = \underline{61.8\text{ lb/ft}^3}.$

1.23 $\rho = 1000 - (T-4)^2/180 = 1000 - (70-4)^2/180 = \underline{976\text{ kg/m}^3}$

$\gamma = 9800 - (T-4)^2/18 = 9800 - (70-4)^2/180 = \underline{9560\text{ N/m}^3}$

% error for $\rho = \dfrac{976 - 978}{978}\times 100 = \underline{-.20\%}$

% error for $\gamma = \dfrac{9560 - 978\times 9.81}{978\times 9.81}\times 100 = \underline{-.36\%}$

1.24 $S = 13.6 - .0024T = 13.6 - .0024\times 50 = 13.48.$ % error $= \dfrac{13.48 - 13.6}{13.6}\times 100 = \underline{-.88\%}$

1.25 a) $m = \frac{W}{g} = \frac{\gamma \, \forall}{g} = \frac{12\,400 \times 500 \times 10^{-6}}{9.81} = \underline{.632 \text{ kg}}$

b) $m = \frac{12\,400 \times 500 \times 10^{-6}}{9.77} = \underline{.635 \text{ kg}}$

c) $m = \frac{12\,400 \times 500 \times 10^{-6}}{9.83} = \underline{.631 \text{ kg}}$

1.26 $S = \frac{\rho}{\rho_{water}} = \frac{m / \forall}{\rho_{water}}$ $\quad 1.2 = \frac{200 / \forall}{1000}.$ $\quad \therefore \forall = \underline{.1667 \text{ m}^3}$

1.26E $S = \frac{\rho}{\rho_{water}} = \frac{m / \forall}{\rho_{water}}$ $\quad 1.2 = \frac{10 / \forall}{1.94}$ $\quad \therefore \forall = \underline{4.30 \text{ ft}^3}$

1.27 $\tau = \mu \left| \frac{du}{dr} \right| = 0.9 \times 10^{-3} \left[\frac{10(2 \times (0.01)}{0.01^2} \right] = \underline{1.8 \text{ Pa}}$

1.27E $\tau = \mu \left| \frac{du}{dr} \right| = 1.92 \times 10^{-5} \left[\frac{30(2 \times (1/12)}{(1/12)^2} \right] = \underline{0.014 \text{ psf}}$

1.28 $T = \text{force} \times \text{moment arm} = \tau \, 2\pi RL \times R = \mu \left| \frac{du}{dr} \right| 2\pi R^2 L = \mu \left(\frac{0.4}{R^2} + 1000 \right) 2\pi R^2 L.$

$$\therefore \mu = \frac{T}{\left(\frac{0.4}{R^2} + 1000 \right) 2\pi R^2 L} = \frac{0.0026}{\left(\frac{0.4}{(0.01)^2} + 1000 \right) 2\pi \times .01^2 \times 0.2} = \underline{0.414 \text{ N} \cdot \text{s/m}^2.}$$

1.29 Use Eq. 1.5.8: $T = \frac{2\pi R^3 \omega L \mu}{h} = \frac{2\pi \times .01^3 \times \frac{2000 \times 2\pi}{60} \times 1.2 \times .42}{.0003} = \underline{2.21 \text{ N} \cdot \text{m}}.$

$$Hp = \frac{T\omega}{746} = \frac{2.21 \times 209.4}{746} = \underline{0.62} \text{ Hp}$$

1.29E Use Eq. 1.5.8: $T = \frac{2\pi R^3 \omega L \mu}{h} = \frac{2\pi \times (.5/12)^3 \times \frac{2000 \times 2\pi}{60} \times 4 \times .006}{.01/12} = \underline{2.74 \text{ ft-lb}}.$

$$Hp = \frac{T\omega}{550} = \frac{2.74 \times 209.4}{550} = \underline{1.04} \text{ Hp}$$

1.30 $F_{belt} = \mu \frac{du}{dy} A = 1.31 \times 10^{-3} \frac{10}{.002} (.6 \times 4) = 15.7 \text{ N}.$ $\quad Hp = \frac{F \times V}{746} = \frac{15.7 \times 10}{746} = \underline{0.210} \text{ Hp}$

1.31 Assume a linear velocity so $\frac{du}{dy} = \frac{r\omega}{h}$. Due to the area element shown, $dT = dF \times r = \tau dA \times r = \mu \frac{du}{dy} 2\pi r\, dr \times r$.

Thus, $T = \int_0^R \frac{\mu\omega 2\pi}{h} r^3 dr = \frac{2\pi\mu\omega}{h}\frac{R^4}{4} = \frac{\pi \times 1.31\times 10^{-3} \times \frac{400\times 2\pi}{60} \times .125^4}{2\times .002}$ = $\underline{.0105 \text{ N·m}}$

1.31E From the solution above

$$T = \frac{2\pi\mu\omega}{h}\frac{R^4}{4} = \frac{\pi \times 2.36\times 10^{-5} \times \frac{400\times 2\pi}{60} \times (3/12)^4}{2\times .08/12} = \underline{91 \times 10^{-5} \text{ ft-lb}}.$$

1.32 $\tau = \mu\left|\frac{du}{dr}\right| = \mu\left[4\left(\frac{8r}{r_0^2}\right)\right] = 32\mu r / r_0^2$. $\quad \therefore \tau_{r=0} = 0$,

$\tau_{r=2.25} = 32 \times 1 \times 10^{-3} \times \frac{.25/100}{(.5/100)^2} = 3.2$ Pa. $\quad \tau_{r=.5} = 32 \times 1 \times 10^{-3} \times \frac{(.5/100)}{(.5/100)^2} = 6.4$ Pa

1.33 The velocity at a radius r is $r\omega$. The shear stress is $\tau = \mu\frac{\Delta u}{\Delta y}$. The torque is $dT = \tau r dA$ due to a differential element. We have

$$T = \int \tau r dA = \int_0^{.08} \mu \frac{r\omega - 0}{.0002} 2\pi r dx \qquad \omega = \frac{2000\times 2\pi}{60} = 209.4 \text{ rad/s}$$

where x is measured along the rotating surface. From the geometry $x = \sqrt{2}\, r$, so that

$$T = \int_0^{.08} 0.1 \frac{x/\sqrt{2}}{.0002} 2\pi \frac{x}{\sqrt{2}} \times 209.4 dx = \frac{1571}{3} \times 209.4(.08^3) = \underline{56.14 \text{ N·m}}$$

1.34 If $\tau = \mu\frac{du}{dy}$ = cons't and $\mu = Ae^{B/T} = Ae^{By/K} = Ae^{Cy}$, then

$$Ae^{Cy}\frac{du}{dy} = \text{cons't}. \qquad \therefore \frac{du}{dy} = De^{-Cy}.$$

Finally, $\int_o^u du = \int_o^y De^{-Cy} dy$ or $u(y) = -\frac{D}{C}e^{-Cy}\Big|_o^y = \underline{E(e^{-Cy} - 1)}$.

Where A, B, C, D, E, and K are constants.

1.35 $\left.\begin{array}{l}\mu = Ae^{B/T}, \quad .001 = Ae^{B/293} \\ .000357 = Ae^{B/353}\end{array}\right\}$ $\quad \therefore A = 2.334 \times 10^{-6}$, $B = 1776$.

$\mu_{40} = 2.334 \times 10^{-6} e^{1776/313} = \underline{6.80 \times 10^{-4} \text{ N}\cdot\text{s/m}^2}$

1.36 $m = \rho V$ Then $dm = \rho dV + V d\rho$. Assume mass to be constant in a volume subjected to a pressure increase; then $dm = 0$. $\therefore \rho dV = -V d\rho$, or $\frac{dV}{V} = -\frac{d\rho}{\rho}$.

1.37 $B = -\frac{V\,\Delta p}{\Delta V} = 2200\text{MPa}$. $\therefore \Delta V = \frac{-2\times 10}{2200} = \underline{-0.00909\text{ m}^3}$ or $\underline{-9090\text{ cm}^3}$

1.38 Use $c = 1450$ m/s. $L = c\Delta t = 1450 \times .62 = \underline{899\text{ m}}$

1.39 $\Delta p = -\frac{B\Delta V}{V} = -2100\,\frac{-1.3}{20} = \underline{136.5\text{ MPa}}$

1.40 a) $c = \sqrt{2110\times 10^6 / 1000} = \underline{1453\text{ m/s}}$ b) $\sqrt{2300\times 10^6 / 988} = \underline{1526\text{ m/s}}$
c) $\sqrt{2120\times 10^6 / 962} = \underline{1485\text{ m/s}}$

1.40E a) $\sqrt{294{,}000\times 144 / 1.94} = \underline{4670\text{ fps}}$ b) $\sqrt{327{,}000\times 144 / 1.93} = \underline{4940\text{ fps}}$
c) $\sqrt{308{,}000\times 144 / 1.87} = \underline{4870\text{ fps}}$

1.41 $\Delta V = 3.8\times 10^{-4}\times -20\times 1 = \underline{0.0076\text{ m}^3}$. $\Delta p = -B\,\frac{\Delta V}{V} = -2270\frac{-.0076}{1} = \underline{17.25\text{ MPa}}$

1.42 $p = \frac{2\sigma}{R} = \frac{2\times.0741}{5\times 10^{-6}} = 2.96\times 10^4$ Pa or $\underline{29.6\text{ kPa}}$
bubbles: $p = 4\sigma/R = \underline{59.3\text{ kPa}}$

1.43 From Table B.1 $\sigma = .0748$ N/m. $\therefore p = \frac{4\sigma}{R} = \frac{4\times.0748}{.001} = \underline{299\text{ Pa}}$

1.43E From Table B.1 $\sigma = .00504$ lb/ft. $\therefore p = \frac{4\sigma}{R} = \frac{4\times.00504}{1/(32\times 12)} = 7.74$ psf or $\underline{0.0538\text{ psi}}$

1.44 See Example 1.4: $h = \frac{4\sigma\cos\beta}{\rho g D} = \frac{4\times.0736\times.866}{1000\times 9.81\times.0002} = 0.130\text{ m}.$

1.45 $h = \frac{4\sigma\cos\beta}{\rho g D}$ from Example 1.4. $\therefore h = \frac{4\times.5\times\cos 130^\circ}{1000\times 13.6\times 9.81\times.002} = -4.82\times 10^{-3}\text{ m}$
or $\underline{-4.82\text{ mm}}$

1.45E See Example 1.4: $h = \frac{4\sigma\cos\beta}{\rho g D} = \frac{4\times.032\cos 130^\circ}{1.94\times 13.6\times 32.2\times.8/12} = -.00145$ ft. or $\underline{-0.0174\text{ in.}}$

1.46 force up = $\sigma \times L \times 2\cos\beta$ = force down = $\rho g h t L$. $\therefore h = \dfrac{2\sigma\cos\beta}{\rho g t}$.

1.47 Draw a free-body diagram:
The force must balance:

$$W = 2\sigma L \;\text{ or }\; \left(\frac{\pi d^2}{4}L\right)\rho g = 2\sigma L. \qquad \therefore d = \sqrt{\frac{8\sigma}{\pi\rho g}}$$

1.48 From the free-body diagram in No. 1.47, create a force balance. Is $\dfrac{\pi d^2}{4}\rho g < 2\sigma$?

$$\frac{\pi(.004)^2}{4}7850 \times 9.81 < 2 \times .0741$$

$$.968 < .1482 \qquad \therefore \text{No}$$

1.49

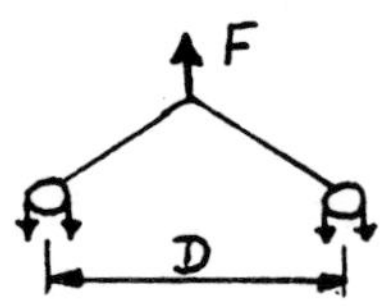

Each surface tension force = $\sigma \times \pi D$. There is a force on the outside and one on the inside of the ring.
$\therefore F = \underline{2\sigma\pi D}$ neglecting the weight of the ring.

1.50

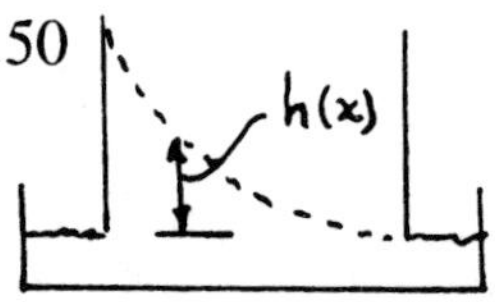

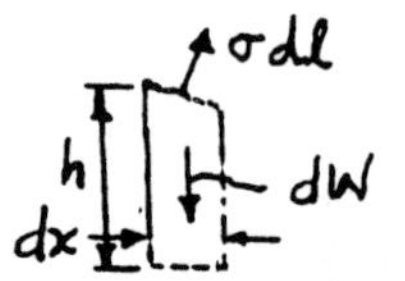

From the infinitesimal free-body shown:
$\sigma d\ell \cos\theta = \rho g h \alpha x dx$. $\cos\theta = \dfrac{dx}{d\ell}$.

$$\therefore h = \frac{\sigma d\ell dx / d\ell}{\rho g \alpha x dx} = \frac{\sigma}{\rho g \alpha x}.$$

We assumed small α so that the element thickness is αx.

1.51 The absolute pressure is p = -80 + 92 = 12 kPa. At 50°C water has a vapor pressure of 12.2 kPa; so T = 50°C is a maximum temperature. The water would "boil" above this temperature.

1.52 The engineer knew that water boils near the vapor pressure. At 82°C the vapor pressure from Table B.1 is 50.8 (by interpolation). By looking in Table B.3, the elevation that has a pressure of 50.8 kPa is interpolated to be <u>5500 m</u>.

1.53 At 40°C the vapor pressure from Table B.1 is <u>7.4 kPa</u>. This would be the minimum pressure that could be obtained since the water would vaporize below this pressure.

1.54 The absolute pressure is 92 - 77 = 15 kPa. If bubbles were observed to form at 15 kPa (this is boiling), the temperature from Table B.1 is interpolated, using vapor pressure, to be <u>53.8°C</u>.

1.54E The absolute pressure is 14.5 - 11.5 = 3.0 psia. If bubbles were observed to form at 3.0 psia (this is boiling), the temperature from Table B.1 is interpolated, using vapor pressure, to be <u>141°F</u>.

1.55 The inlet pressure to a pump cannot be less than 0 kPa absolute. Assuming atmospheric pressure to be 100 kPa, we have

$$10\,000 + 100 = 600\,x. \qquad \therefore x = \underline{16.83\text{ km}}.$$

1.56 $\rho = \dfrac{p}{RT} = \dfrac{101.3}{0.287\times(273+15)} = \underline{1.226\text{ kg/m}^3}.$ $\qquad \gamma = 1.226\times 9.81 = \underline{12.03\text{ N/m}^3}$

1.57 $\rho_{in} = \dfrac{85}{0.287\times 293} = \underline{1.01\text{ kg/m}^3}.$ $\qquad \rho_{out} = \dfrac{85}{0.287\times 248} = \underline{1.19\text{ kg/m}^3}.$

Yes. The heavier air outside tends to enter at the bottom and the lighter air tends to exit at the top. A circulation is set up and air moves from the outside in and the inside out — infiltration. This is the "chimney" effect.

1.58 $\rho = \dfrac{p}{RT} = \dfrac{5000}{.287\times 281} = \underline{62.0\text{ kg/m}^3}.$ $\qquad m = \rho V = 62\times 5 = \underline{310\text{ kg}}.$

1.58E $\rho = \dfrac{p}{RT} = \dfrac{750\times 144}{1716\times 470} = \underline{0.1339\text{ slug/ft}^3}.$ $\qquad m = \rho V = .1339\times 15 = \underline{2.01\text{ slug}}.$

1.59 $\rho = \dfrac{p}{RT} = \dfrac{20}{287\times 291} = \underline{2.39\times 10^{-4}\text{ kg/m}^3}.$ $\qquad m = \rho V = 2.39\times 10^{-4}\times 5$

$= \underline{1.197\times 10^{-3}\text{ kg}}.$

(Note: Since pressure is in Pa, R has units $\dfrac{\text{J}}{\text{kg}\cdot\text{K}}$)

1.60 $W = \dfrac{p}{RT} V\, g = \dfrac{100}{.287\times 293}\times(10\times 20\times 4)\times 9.81 = \underline{9333\text{ N}}$

1.61 We will assume that the tire's volume remains essentially unchanged (it's steel-belted and quite undeformable unless there are extremely high pressures). Therefore,

$$V_1 = V_2 \text{ or } \frac{m_1RT_1}{p_1} = \frac{m_2RT_2}{p_2}.$$

Since $m_1 = m_2$ we have

$$p_2 = p_1\frac{T_2}{T_1} = (240+100)\frac{65+273}{-20+273} = 454\text{ kPa or } \underline{354\text{ kPa gage}}$$

1.61E Assume the volume remains unchanged so that (using $m_1 = m_2$):

$$\frac{m_1RT_1}{p_1} = \frac{m_2RT_2}{p_2} \text{ or } p_2 = p_1\frac{T_2}{T_1} = (35+14.7)\frac{150+460}{-10+460} = 67.4\text{ psia}.$$

This is $\underline{52.7\text{ psi gage}}$.

1.62 The pressure holding up the mass is 100 kPa. Hence, using $pA = W$, we have

$$100\ 000 \times 1 = m \times 9.81. \qquad \therefore m = 10\ 200 \text{ kg}.$$

Finally, $m = \dfrac{p\mathcal{V}}{RT} = \dfrac{100 \times 4\pi r^3 / 3}{.287 \times 288} = 10\ 200.$ $\quad \therefore r = 12.6$ m or $\underline{d = 25.2 \text{ m}}$.

1.63 $0 = \Delta KE + \Delta PE = \dfrac{1}{2}mV^2 + mg(-3).$ $\quad \therefore V^2 = 6 \times 9.81.$ $\quad \therefore V = \underline{7.67 \text{ m/s}}$.

$0 = \dfrac{1}{2}mV^2 + mg(-6).$ $\quad \therefore V^2 = 12 \times 9.81.$ $\quad \therefore V = \underline{10.85 \text{ m/s}}$.

1.63E $0 = \Delta KE + \Delta PE = \dfrac{1}{2}mV^2 + mg(-10).$ $\quad \therefore V^2 = 20 \times 32.2.$ $\quad \therefore V = \underline{25.4 \text{ fps}}$.

$0 = \dfrac{1}{2}mV^2 + mg(-20).$ $\quad \therefore V^2 = 40 \times 32.2.$ $\quad \therefore V = \underline{35.9 \text{ fps}}$.

1.64 $W_{1-2} = \Delta KE.$ $\quad$ a) $200 \times 10 = \dfrac{1}{2} \times 15\left(V_f^2 - 10^2\right).$ $\quad \therefore V_f = \underline{19.15 \text{ m/s}}$.

b) $\displaystyle\int_0^{10} 20s\,ds = \frac{1}{2} \times 15\left(V_f^2 - 10^2\right).$

$$20 \times \frac{10^2}{2} = \frac{1}{2} \times 15\left(V_f^2 - 10^2\right). \qquad \therefore V_f = \underline{15.27 \text{ m/s}}.$$

c) $\displaystyle\int_0^{10} 200 \cos\frac{\pi s}{20} ds = \frac{1}{2} \times 15\left(V_f^2 - 10^2\right).$

$$\frac{20}{\pi} 200 \sin\frac{\pi}{2} = \frac{1}{2} \times 15\left(V_f^2 - 10^2\right). \qquad \therefore V_f = \underline{16.42 \text{ m/s}}.$$

1.65 $E_1 = E_2.$ $\quad \dfrac{1}{2} \times 10 \times 40^2 + .2\tilde{u}_1 = 0 + .2\tilde{u}_2$ $\quad \therefore \tilde{u}_2 - \tilde{u}_1 = \dfrac{10 \times 40^2}{2 \times .2}.$

$\Delta\tilde{u} = c_v \Delta T.$ $\quad \therefore \Delta T = \dfrac{10 \times 40^2}{2 \times .2 \times 717} = \underline{55.8°\text{C}}.$ (Find c_v in Table B.4.)

(Units: $\dfrac{\text{kg} \cdot \dfrac{\text{m}^2}{\text{s}^2}}{\text{kg} \cdot \dfrac{\text{J}}{\text{kg} \cdot {}^\circ C}} = \dfrac{\text{kg} \cdot \text{m}^2 \cdot {}^\circ\text{C}}{\text{N} \cdot \text{m} \cdot \text{s}^2} = \dfrac{\text{kg} \cdot \text{m}^2 \cdot {}^\circ\text{C}}{\text{kg} \cdot \dfrac{\text{m}}{\text{s}^2} \cdot \text{m} \cdot \text{s}^2} = {}^\circ\text{C}$)

Friction which would heat up the plunger/side walls.

1.66 $E_1 = E_2$. $\frac{1}{2}\times 1500\times\left(\frac{100\times 1000}{3600}\right)^2 = 1000\times 2000\times 10^{-6}\times 4180\Delta T$

$\frac{1}{2}mV^2 = m_{H_2O}\, c\Delta T$. The specific heat c is found in Table B.5. (Use c = 4180 J/kg·°C.)

The above gives ΔT = 69.2°C. (See Prob. 1.38 for a units check.)

1.67 $m_f h_f = m_{H_2O}\, c\Delta T$. $0.2\times 40\,000 = 100\times 4.18\,\Delta T$

(The specific heat c is found in Table B.5.) $\therefore \Delta T$ = 19.1 °C.

Note: We used kJ on the left and kJ on the right.

1.68 $W_{1-2} = \int p\,dV = \int \frac{mRT}{V}dV = mRT\int\frac{dV}{V} = mRT\ln\frac{V_2}{V_1} = mRT\ln\frac{p_1}{p_2}$

Since, if T = const, pV = const so that $p_1V_1 = p_2V_2$. Finally,

$W_{1-2} = 2\times .287\times 293\ln\frac{1}{2}$ = -116.6 kJ.

The first law states:

$Q - W = m\Delta u = mc_v\Delta T = 0.$ $\therefore Q = W$ = -116.6 kJ.

1.68E $W = \int p\,dV = \int\frac{mRT}{V}\,dV = mRT\ln\frac{V_2}{V_1} = mRT\ln\frac{p_1}{p_2}$

Since, if T = const, pV = const so that $p_1V_1 = p_2V_2$. Finally,

$W_{1-2} = \frac{4}{32.2}\times 1716\times 530\ln\frac{1}{2}$ = -78,310 ft-lb.

The first law states:

$Q - W = m\Delta u = mc_v\Delta T = 0.$ $\therefore Q = W$ = -78,310 ft-lb or -101 Btu.

1.69 If volume is fixed, the work is zero since the boundary does not move. Also, since $V = \frac{mRT}{p}$, $\frac{T_1}{p_1} = \frac{T_2}{p_2}$ so the temperature doubles if the pressure doubles. Hence, using Table B.4 and Eq. 1.7.17,

a) $Q = mc_v\Delta T = \frac{200\times 2}{.287\times 293}(1.004-.287)(2\times 293-293)$ = 999 kJ

b) $Q = mc_v\Delta T = \frac{200\times 2}{.287\times 373}(1.004-.287)(2\times 373-373)$ = 999 kJ

c) $Q = mc_v\Delta T = \frac{200\times 2}{.287\times 473}(1.004-.287)(2\times 473-473)$ = 999 kJ

1.70 $W = \int p d\mathcal{V} = p(\mathcal{V}_2 - \mathcal{V}_1)$. If p = const, $\frac{T_1}{\mathcal{V}_1} = \frac{T_2}{\mathcal{V}_2}$ so if $T_2 = 2T_1$,

then $\mathcal{V}_2 = 2\mathcal{V}_1$.

Hence, $W = p(2\mathcal{V}_1 - \mathcal{V}_1) = p\mathcal{V}_1 = mRT_1$.

a) $W = 2 \times .287 \times 333 = \underline{191 \text{ kJ}}$

b) $W = 2 \times .287 \times 423 = \underline{243 \text{ kJ}}$

c) $W = 2 \times .287 \times 473 = \underline{272 \text{ kJ}}$

1.71 $T_2 = T_1\left(\frac{p_2}{p_1}\right)^{\frac{k-1}{k}} = (273+20)\left(\frac{500}{5000}\right)^{\frac{1.4-1}{1.4}} = 151.8$ K or $\underline{-121.2°\text{C}}$.

1.72 We assume an isentropic process for the maximum pressure:

$$p_2 = p_1\left(\frac{T_2}{T_1}\right)^{\frac{k}{k-1}} = (150+100)\left(\frac{423}{293}\right)^{\frac{1.4}{1.4-1}} = 904 \text{ kPa abs or } \underline{804 \text{ kPa}} \text{ gage.}$$

Note: We assumed p_{atm} = 100 kPa since it was not given. Also, a measured pressure is assumed to be a gage pressure.

1.73 $p_2 = p_1\left(\frac{T_2}{T_1}\right)^{k/(k-1)} = 100\left(\frac{473}{293}\right)^{1.4/.4} = \underline{534.5 \text{ kPa}}$

$w = -\Delta u = -c_v(T_2 - T_1) = -(1.004 - .287)(473 - 293) = \underline{-129 \text{ kJ/kg}}$.

1.74 a) $c = \sqrt{kRT} = \sqrt{1.4 \times 287 \times 293} = 343.1$ m/s

b) $c = \sqrt{kRT} = \sqrt{1.287 \times 188.9 \times 293} = 266.9$ m/s

c) $c = \sqrt{kRT} = \sqrt{1.4 \times 296.8 \times 293} = 348.9$ m/s

d) $c = \sqrt{kRT} = \sqrt{1.4 \times 4124 \times 293} = 1301$ m/s

e) $c = \sqrt{kRT} = \sqrt{1.33 \times 461.5 \times 293} = 424.1$ m/s

We have used the units on R to be J/kg·K in the above.

1.75 At 10 000 m the speed of sound $c = \sqrt{kRT}$

$= \sqrt{1.4 \times 287 \times 223} = 299$ m/s.

At sea level, $c = \sqrt{kRT}$

$= \sqrt{1.4 \times 287 \times 288} = 340$ m/s.

$$\% \text{ decrease} = \frac{340 - 299}{340} \times 100 = \underline{12.06\ \%}.$$

1.76 a) $c = \sqrt{1.4 \times 287 \times 253} = 319$ m / s

$L = c\Delta t = 319 \times 8.32 = 2654$ m.

b) $c = \sqrt{1.4 \times 287 \times 293} = 343$ m / s

$L = c\Delta t = 343 \times 8.32 = 2854$ m.

c) $c = \sqrt{1.4 \times 287 \times 318} = 357$ m / s

$L = c\Delta t = 357 \times 8.32 = 2970$ m.

TWO

Fluid Statics

2.1 $\Sigma F_y = ma_y$: $p_y \Delta z - p\Delta s \sin\alpha = \rho \frac{\Delta y \Delta z}{2} a_y$

$\Sigma F_z = ma_z$: $p_z \Delta y - p\Delta s \cos\alpha = \rho \frac{\Delta y \Delta z}{2} a_z + \rho g \frac{\Delta y \Delta z}{2}$

Since $\Delta s \cos\alpha = \Delta y$ and $\Delta s \sin\alpha = \Delta z$, we have

$$p_y - p = \rho \frac{\Delta y}{2} a_y$$

$$p_z - p = \rho \frac{\Delta z}{2}(a_z + g)$$

Let $\Delta y \to 0$ and $\Delta z \to 0$:

$$\left.\begin{aligned} p_y - p &= 0 \\ p_z - p &= 0 \end{aligned}\right\} \qquad \therefore p_y = p_z = p.$$

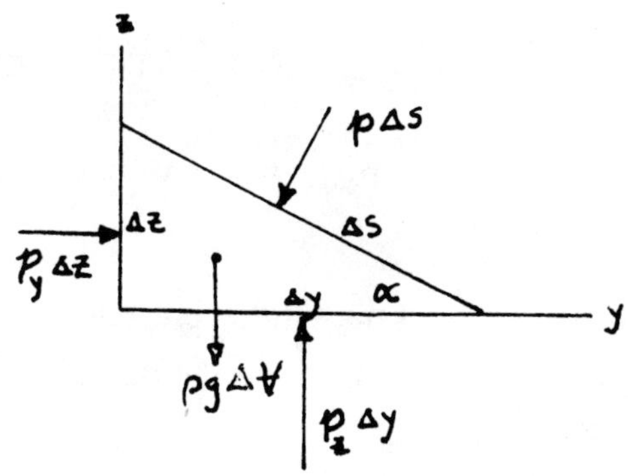

2.2 $p = \gamma h$.
a) $9810 \times 10 = 98\ 100$ Pa or 98.1 kPa
b) $(0.8 \times 9810) \times 10 = 78\ 480$ Pa or 78.5 kPa
c) $(13.6 \times 9810) \times 10 = 1\ 334\ 000$ Pa or 1334 kPa
d) $(1.59 \times 9810) \times 10 = 155\ 980$ Pa or 156.0 kPa
e) $(0.68 \times 9810) \times 10 = 66\ 710$ Pa or 66.7 kPa

2.3 $h = p/\gamma$.
a) $h = 250\ 000/9810 =$ 25.5 m
b) $h = 250\ 000/(0.8 \times 9810) =$ 31.9 m
c) $h = 250\ 000/(13.6 \times 9810) =$ 1.874 m
d) $h = 250\ 000/(1.59 \times 9810) =$ 16.0 m
e) $h = 250\ 000/(0.68 \times 9810) =$ 37.5 m

2.4 $S = \frac{p}{\gamma h} = \frac{136\ 000}{9810 \times 6} = 2.31$. $\rho = 2.31 \times 1000 = 2310\ \text{kg/m}^3$.

2.4E $S = \frac{p}{\gamma h} = \frac{20}{62.4 \times 20} = 2.31$. $\rho = 1.94 \times 2.31 = 4.48\ \text{slug/ft}^3$.

2.5 a) $p = \gamma h = 0.76 \times (13.6 \times 9810) = 9810\ h$. $\therefore h =$ 10.34 m.
b) $(13.6 \times 9810) \times 0.75 = 9810\ h$. $\therefore h =$ 10.2 m.
c) $(13.6 \times 9810) \times 0.01 = 9810\ h$. $\therefore h = 0.136$ m or 13.6 cm.

2.6 a) $p = \gamma_1 h_1 + \gamma_2 h_2 = 9810 \times .2 + (13.6 \times 9810) \times .02 = 4630$ Pa or $\underline{4.63 \text{ kPa}}$.
b) $9810 \times .052 + 15\,630 \times .026 = 916$ Pa or $\underline{0.916 \text{ kPa}}$.
c) $9016 \times 3 + 9810 \times 2 + (13.6 \times 9810) \times 0.1 = 60\,010$ Pa or $\underline{60.0 \text{ kPa}}$.

2.7 $\Delta p = \rho g h = 1.23 \times 9.81\,(-3000) = -36\,200$ Pa or $\underline{-36.2 \text{ kPa}}$.

2.7E $\Delta p = \rho g h = 0.0024 \times 32.2\,(-10{,}000) = -773$ psf or $\underline{-5.37 \text{ psi}}$.

2.8
$$\left.\begin{aligned} \Delta p_{outside} &= \rho_o g \Delta h = \frac{pg}{RT_o}\Delta h = \frac{100 \times 9.81}{.287 \times 253} \times 3 = 13.51 \text{ Pa} \\ \Delta p_{inside} &= \rho_i g \Delta h = \frac{pg}{RT_i}\Delta h = \frac{100 \times 9.81}{.287 \times 293} \times 3 = 11.67 \text{ Pa} \end{aligned}\right\} \Delta p_{base} = \underline{1.84 \text{ Pa}}$$

If no wind is present this Δp_{base} would produce a small infiltration since the higher pressure on the outside would force in the outside air in the bottom region (in cracks).

2.9 $dp = \rho g dh$ where $h = -z$. $S = 1.0 + h/100$ since $S(0) = 1$ and $S(10) = 1.1$.
$\rho = 1000$ S where $\rho_{water} = 1000$ kg/m^3. Then $dp = 1000\,(1 + h/100)\,g dh$. Integrate:

$$\int_0^p dp = \int_0^{10} 1000(1 + h/100) g dh$$

$$p = 1000 \times 9.81(10 + \frac{10^2}{2 \times 100}) = 103\,000 \text{ Pa or } \underline{103 \text{ kPa}}$$

Note: we could have used an average S: $S_{avg} = 1.05$, so that $\rho_{avg} = 1050$ kg/m^3.

2.10
$$\vec{\nabla} p = \frac{\partial p}{\partial x}\hat{i} + \frac{\partial p}{\partial y}\hat{j} + \frac{\partial p}{\partial z}\hat{k}$$
$$= -\rho a_x \hat{i} - \rho a_y \hat{j} - \rho a_z \hat{k} - \rho g \hat{k} = -\rho\left(a_x \hat{i} + a_y \hat{j} + a_z \hat{k}\right) - \rho g \hat{k}$$
$$= -\rho \vec{a} - \rho \vec{g}$$
$$\therefore \underline{\vec{\nabla} p = -\rho(\vec{a} + \vec{g})}$$

2.11 $p = p_{atm}[(T_o - \alpha z)/T_o]^{g/\alpha R}$
$= 100\,[(288 - .0065 \times 300)/288]^{9.81/.0065 \times 287} = \underline{96.49 \text{ kPa}}$

$$p = p_{atm} - \rho g h = 100 - \frac{100}{.287 \times 288} \times 9.81 \times 300/1000 = \underline{96.44 \text{ kPa}}$$

$$\% \text{ error} = \frac{96.44 - 96.49}{96.49} \times 100 = \underline{-.052\%}$$

The density variation can be ignored for heights of 300 m or less.

2.12 $\Delta p = p - p_o = p_{atm}\left(\frac{T_o - \alpha z}{T_o}\right)^{g/\alpha R} - p_{atm}$

$$= 100\left[\left(\frac{288-.0065\times 20}{288}\right)^{9.81/.0065\times 287} - 1\right] = -.237 \text{ Pa or } \underline{.000237 \text{ kPa}}$$

This change is very small and can usually be ignored.

2.13 Eq. 1.5.10 gives $2100\times 10^6 = \rho\frac{dp}{d\rho}$. But, $dp = \rho g dh$. Therefore,

$$\rho g dh = \frac{2100\times 10^6}{\rho}d\rho$$

$$\frac{d\rho}{\rho^2} = \frac{9.81}{2100\times 10^6}dh$$

Integrate, using $\rho_o = 1030$ kg/m^3:

$$\int_{1030}^{\rho}\frac{d\rho}{\rho^2} = \frac{9.81}{2100\times 10^6}\int_0^h dh$$

$$-\left(\frac{1}{\rho} - \frac{1}{1030}\right) = 4.67\times 10^{-9}\, h \qquad \text{or} \qquad \rho = \frac{1000}{.971 - 4.67\times 10^{-6}h}$$

Now,

$$p = \int_0^h \rho g dh = \int_0^h \frac{1000g}{.971 - 4.67\times 10^{-6}h}dh = \frac{1000g}{-4.67\times 10^{-6}}\ln(.971 - 4.67\times 10^{-6}h)\Big|_0^{5000}$$

$$= 5.11\times 10^7 \text{ Pa or } \underline{51.1 \text{ MPa}}$$

Assume ρ = const:

$$p = \rho g h = 1030\times 9.81\times 5000 = 5.05\times 10^7 \text{ Pa or } \underline{50.5 \text{ MPa}}$$

$$\%\ \text{error} = \frac{50.5 - 51.1}{51.1}\times 100 = -\underline{1.96\ \%}$$

2.14 Use the result of Example 2.2: $p = 101\, e^{-gz/RT}$.

a) $p = 101\, e^{-9.81\times 10\,000/287\times 273} = \underline{28.9 \text{ kPa}}$.

b) $p = 101\, e^{-9.81\times 10\,000/287\times 288} = \underline{30.8 \text{ kPa}}$.

c) $p = 101\, e^{-9.81\times 10\,000/287\times 258} = \underline{26.9 \text{ kPa}}$.

2.15 Use Eq. 2.4.8: $p = 101(1 - .0065z/288)^{\frac{9.81}{.0065\times 287}}$.

a) $z = 3000$. $\therefore p = \underline{69.9 \text{ kPa}}$.

b) $z = 6000$. $\therefore p = \underline{47.0 \text{ kPa}}$.

c) $z = 9000$. $\therefore p = \underline{30.6 \text{ kPa}}$.

d) $z = 11\,000$. $\therefore p = \underline{22.5 \text{ kPa}}$.

2.16 Use the result of Example 2.2: $p = p_o e^{-gz/RT}$

$\frac{1}{101\,000} = e^{-9.81z/287\times 253}$. $\quad \ln\frac{1}{101\,000} = -\frac{9.81z}{287\times 253}$. $\quad \therefore z = \underline{85\,290 \text{ m}}$.

2.16E Use the result of Example 2.2: $\frac{p}{p_o} = e^{-gz/RT}$.

$\ln \frac{p}{p_o} = -\frac{gz}{RT}$. $\ln \frac{.001}{14.7} = -\frac{32.2z}{1716 \times 455}$. $\therefore z = \underline{232{,}670 \text{ ft}}$.

2.17 $p = \gamma h = (13.6 \times 9810) \times 0.25 = 33\ 354$ Pa or $\underline{33.4 \text{ kPa}}$.

2.18 a) $p = \gamma h$. $450\ 000 = (13.6 \times 9810)$ h. $\therefore h = \underline{3.373 \text{ m}}$

b) $p + 11.78 \times 1.5 = (13.6 \times 9810)$ h. Use $p = 450\ 000$, then $h = \underline{3.373 \text{ m}}$

% error is $\underline{0.000\ \%}$.

2.19 Referring to Fig. 2.7a, the pressure in the pipe is $p = \rho gh$. If $p = 2400$ Pa, then

$2400 = \rho gh = \rho \times 9.81$ h.

a) $\rho = \frac{2400}{9.81 \times .36} = 680 \text{ kg/m}^3$. $\therefore$ gasoline

b) $\rho = \frac{2400}{9.81 \times .272} = 899 \text{ kg/m}^3$. $\therefore$ benzene

c) $\rho = \frac{2400}{9.81 \times .245} = 999 \text{ kg/m}^3$. $\therefore$ water

d) $\rho = \frac{2400}{9.81 \times .154} = 1589 \text{ kg/m}^3$. $\therefore$ carbon tetrachloride

2.20 Referring to Fig. 2.7a, the pressure is $p = \rho_w gh = \frac{1}{2}\rho_a V^2$. Then

$$V^2 = \frac{2\rho_w gh}{\rho_a}.$$

a) $V^2 = \frac{2 \times 1000 \times 9.81 \times .06}{1.23} = 957$. $\therefore V = \underline{30.9 \text{ m/s}}$

b) $V^2 = \frac{2 \times 1.94 \times 32.2 \times 3/12}{.00238} = 13{,}124$. $\therefore V = \underline{115 \text{ ft/sec}}$

c) $V^2 = \frac{2 \times 1000 \times 9.81 \times .1}{1.23} = 1595$. $\therefore V = \underline{39.9 \text{ m/s}}$

d) $V^2 = \frac{2 \times 1.94 \times 32.2 \times 5/12}{.00238} = 21{,}870$. $\therefore V = \underline{148 \text{ ft/sec}}$

2.21 See Fig. 2.7b: $p_1 + \gamma_1 h = \gamma_2 H$.

$p_1 = -0.86 \times 9810 \times 0.1 + 13.6 \times 9810 \times 0.197 = 25\ 440$ Pa or $\underline{25.4 \text{ kPa}}$.

2.21E See Fig. 2.7b: $p_1 = -\gamma_1 h + \gamma_2 H$.

$p_1 = -0.86 \times 62.4 \times \frac{5}{12} + 13.6 \times 62.4 \times \frac{9.5}{12} = 649.5$ psf or $\underline{4.51 \text{ psi}}$.

2.22 $p = p_o + \rho_1 g h_1 + \rho_2 g h_2 + \rho_3 g h_3 + \rho_4 g h_4$

$= 3200 + 917 \times 9.81 \times .2 + 1000 \times 9.81 \times .1 + 1258 \times 9.81 \times .15 + 1593 \times 9.81 \times .18$

$= 10\ 640$ Pa or 10.64 kPa

2.23 $p_1 - p_4 = (p_1 - p_2) + (p_2 - p_3) + (p_3 - p_4)$ (Use $\Delta p = \rho g \Delta h$)

$40\ 000 - 16\ 000 = 1000 \times 9.81(-.2) + 13\ 600 \times 9.81 \times H + 920 \times 9.81 \times .3$. $\therefore H = .1743$ m or 17.43 cm

2.24 $p_1 - p_4 = (p_1 - p_2) + (p_2 - p_3) + (p_3 - p_4)$ (Use $\Delta p = \rho g \Delta h$)

$p_o - p_w = 900 \times 9.81(-.05) + 13\ 600 \times 9.81(-.1) + 1000 \times 9.81 \times .15 = -12\ 300$ Pa or -12.3 kPa

2.25 $p_1 - p_5 = (p_1 - p_2) + (p_2 - p_3) + (p_3 - p_4) + (p_4 - p_5)$

$p_1 = 9810(-.02) + 13\ 600 \times 9.81(.-04) + 9810(-.02) + 13\ 600 \times 9.81 \times .16 = 15\ 620$ Pa or 15.62 kPa

2.26 $p_w + 9810 \times .15 - 13.6 \times 9810 \times .1 - .68 \times 9810 \times .2 + .86 \times 9810 \times .15 = p_o$.

$\therefore p_w - p_o = 11\ 940$ Pa or 11.94 kPa.

2.27 $p_w - 9810 \times .12 - .68 \times 9810 \times .1 + .86 \times 9810 \times .1 = p_o$.

With $p_w = 15\ 000$, $p_o = 14\ 000$ Pa or 14.0 kPa.

2.28 a) $p + 9810 \times 2 = 13.6 \times 9810 \times .1$. $\therefore p = -6278$ Pa or -6.28 kPa.

b) $p + 9810 \times .8 = 13.6 \times 9810 \times .2$. $\therefore p = 18\ 835$ Pa or 18.84 kPa.

c) $p + 62.4 \times 6 = 13.6 \times 62.4 \times 4/12$. $\therefore p = -91.5$ psf or -0.635 psi.

d) $p + 62.4 \times 2 = 13.6 \times 62.4 \times 8/12$. $\therefore p = 441$ psf or 3.06 psi.

2.29 $p - 9810 \times 4 + 13.6 \times 9810 \times .16 = 0$. $\therefore p = 17\ 890$ Pa or 17.89 kPa.

2.30 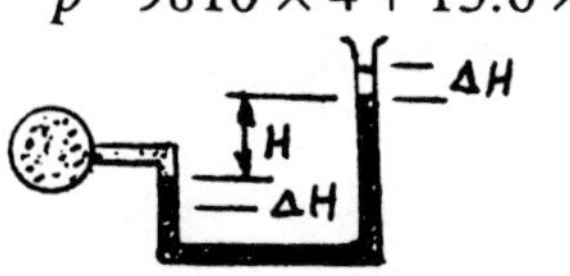

$8200 + 9810 \times .25 = 1.59 \times 9810 \times H$. $\therefore H = 0.683$ m

$H_{new} = .683 + .273 = .956$ m. $\Delta H = \dfrac{.273}{2} = .1365$.

$p + 9810\,(.25 + .1365) = 1.59 \times 9810 \times .956$. $\therefore p = 11\ 120$ Pa or 11.12 kPa.

2.31 $p + 9810 \times .05 + 1.59 \times 9810 \times .07 - .8 \times 9810 \times .1 = 13.6 \times 9810 \times .05$.

$\therefore p = 5873$ Pa or 5.87 kPa.

Note: In our solutions we retain 3 significant nos. in the answers, usually. In most problems a material property is used, i.e., $S = 1.59$. This is only 3 sig. nos! $\therefore$ keep only 3!

2.32 Before pressure is applied the air column on the right is 1.4 m high. After pressure is applied, it is (1.4 - $H/2$) m high. For an isothermal process $p_1 V_1 = p_2 V_2$, using absolute pressures. Thus,

$$101 \times 1.4\,A = p_2\ (1.4 - \frac{H}{2})A \text{ or } \quad p_2 = \frac{141.4}{1.4 - H/2}.$$

From a pressure balance on the manometer (pressures in Pa):

$$200\ 000 + 101\ 000 = 13.6 \times 9810\,H + \frac{141{,}400}{1.4 - H/2},$$

or $H^2 - 5.056\,H + 4.197 = 0$. $\therefore H = 4.01$ or $\underline{1.047 \text{ m}}$.

2.32E Before pressure is applied the air column on the right is 48" high. After pressure is applied, it is (4 - $H/2$) ft high. For an isothermal process $p_1 V_1 = p_2 V_2$, using absolute pressures. Thus,

$$14.7 \times 144 \times 4A = p_2\ (4 - H/2)A \quad \text{or} \quad p_2 = \frac{8467}{4 - H/2}.$$

From a pressure balance on the manometer (pressures in psf):

$$30 \times 144 + 14.7 \times 144 = 13.6 \times 62.4\,H + \frac{8467}{4 - H/2},$$

or $H^2 - 15.59\,H + 40.73 = 0$. $\therefore H = 12.27$ or $\underline{3.32 \text{ ft}}$.

2.33 a) $p_1 - p_5 = (p_1 - p_2) + (p_2 - p_3) + (p_3 - p_4) + (p_4 - p_5)$

4000=9800(.16-.22) + 15 600(.10-.16) + 133 400H + 15 600(.07-H). $\therefore H$ = .0376 m or $\underline{3.76 \text{ cm}}$

b) .6×144=62.4(-2/12)+99.5(-2/12)+849H + 99.5(2.5/12 - H). $\therefore H$ = .1236 ft or $\underline{1.483 \text{ in.}}$

2.34a)

$$\frac{\Delta H}{\Delta p_1} = \frac{2D^2/d^2}{-\gamma_1 + 2\gamma_2 + 2(\gamma_3 - \gamma_2)D^2/d^2} = \frac{2(.1/.005)^2}{-9800 + 2 \times 15\ 600 + 2(133\ 400 - 15\ 600)(.1/.005)^2}$$

$$= 8.487 \times 10H^{-6}$$

$$\therefore \Delta H = 8.487 \times 10^{-6} \times 400 \quad = .0034 \text{ m} \quad \text{or} \quad 3.4 \text{ mm}$$

b) $$\Delta H = \frac{2(4/.2)^2}{-62.4 + 2 \times 99.5 + 2(849 - 99.5)(4/.2)^2} .06 \times 144 = .01153 \text{ ft} \quad \text{or} \quad \underline{.138 \text{ in.}}$$

2.35 $p_1 - p_4 = (p_1 - p_2) + (p_2 - p_3) + (p_3 - p_4)$ (p_{oil} = 14.0 kPa from No. 2.27)

15 500 - 14 000 = 9800(.12+Δz) + 680(.1-2Δz) + 860(-.1-Δz). ∴Δz = .0451 m or $\underline{4.51 \text{ cm}}$

2.36 a) p_{air} = -6250 + 625 = -5620 Pa.

-5620 + 9800(2+Δz) - 13 600 × 9.81(.1+2Δz) = 0. ∴Δz = .0025.

∴h = .1 + 2Δz = .15 m or $\underline{15 \text{ cm}}$

b) p_{air} = 18 800 + 1880 = 20 680 Pa.

20 680 + 9800(.8+Δz) - 13 600 × 9.81(.2+2Δz) = 0. ∴Δz = .00715 m

∴h = .2+ 2Δz = .214 or $\underline{21.4 \text{ cm}}$

c) p_{air} = -91.5 + 9.15 = -82.4 psf.

-82.4 + 62.4(6+Δz) - 13.6 × 62.4(4/12+2Δz) = 0. ∴Δz = .00558 ft.

∴h = 4/12 + 2 (.00558) = .3445 ft or $\underline{4.13 \text{ in.}}$

d) p_{air} = 441 + 44.1 = 485 psf

485 + 62.4(2+Δz) - 13.6 × 62.4(8/12+2Δz) = 0. ∴Δz = .0267 ft.

∴h = 8/12 + 2 (.0267) = .7205 ft or $\underline{8.65 \text{ in.}}$

2.37 $F = \gamma \bar{h} A = 9810 \times 10 \times \pi \times .3^2/4 = \underline{6934\ \text{N}}$.

2.38 a) $F = p_c A = 9800 \times 2 \times 4^2 = 313\ 600$ N or $\underline{313.6\ \text{kN}}$

b) $F = p_c A = 9800 \times 1 \times (2 \times 4) = 78\ 400$ N or $\underline{78.4\ \text{kN}}$

c) $F = p_c A = 9800 \times 1 \times 2 \times 4 \times \sqrt{2} = 110\ 900$ N or $\underline{110.9\ \text{kN}}$

d) $F = p_c A = 9800 \times 1 \times 2 \times 4/.866 = 90\ 500$ N or $\underline{90.5\ \text{kN}}$

2.39 For saturated ground, the force on the bottom tending to lift the vault is
$F = p_c A = 9800 \times 1.5 \times (2 \times 1) = 29\ 400$ N.
The weight of the vault is approximately

$$W = \rho g V_{walls} = 2400 \times 9.81\ [2(2\times1.5\times.1) + 2(2\times1\times.1) + 2(.8\times1.3\times.1)] = 28\ 400\ \text{N}.$$

The vault will tend to rise out of the ground.

2.40 $F = p_c A = 6660 \times 2 \times \pi \times 2^2 = 167\ 400$ N or $\underline{167.4\ \text{kN}}$
Find γ in Table B.5 in the Appendix.

2.41 a) $F = p_c A = 9800\ (10\text{-}2.828/3)\ (2.828\times2/2) = 251\ 000$ N or $\underline{251.0\ \text{kN}}$
where the height of the triangle is $(3^2\text{-}1^2)^{1/2} = 2.828$ m.

b) $F = p_c A = 9800 \times 10\ (2.828\times2/2) = 277\ 100$ N or $\underline{277.1\ \text{kN}}$

c) $F = p_c A = 9800\ (10\text{-}2.828\times.866/3)\ (2.828\times2/2) = 254\ 500$ N or $\underline{254.5\ \text{kN}}$

2.42 a) $F = \gamma \bar{h} A = 9810 \times (10 - \frac{4}{3}) \times 6 = 510\ 100$ N or $\underline{510.1\ \text{kN}}$.

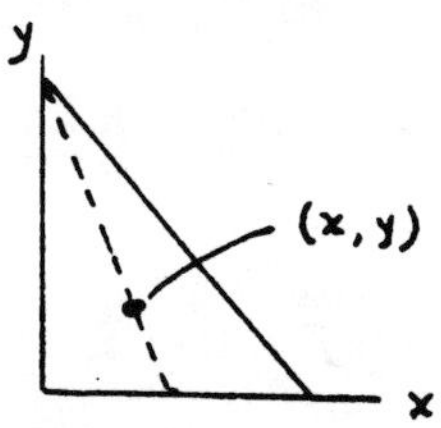

$$y_p = \bar{y} + \frac{\bar{I}}{A\bar{y}} = 8.667 + \frac{3 \times 4^3/36}{6 \times 8.667} = 8.770\ \text{m}.$$

$y = 10 - 8.770 = 1.230$ m.

$$\frac{4}{2.770} = \frac{1.5}{x} \qquad \therefore x = 1.039\ \text{m. The force acts at } \underline{(1.039, 1.230)\ \text{m}}.$$

b) $F = \gamma \bar{h} A = 9810 \times 10 \times 6 = 588\ 600$ N or $\underline{588.6\ \text{kN}}$.
It is located at the centroid of the area: $\underline{(1.001, 1.333)\ \text{m}}$.

$$y = 1.333. \qquad \frac{4}{2.667} = \frac{1.5}{x} \qquad \therefore x = 1.001.$$

c) $F = \gamma \bar{h} A = 9810 \times (10 - 1.333\times.707) \times 6 = 533\ 100$ N or $\underline{533.1\ \text{kN}}$.

$$y_p = 12.81 + \frac{3 \times 4^3/36}{6 \times 12.81} = 12.88\ \text{m}.$$

$$y = 14.14 - 12.88 = 1.26\ \text{m} \qquad \frac{4}{2.74} = \frac{1.5}{x} \qquad \therefore x = 1.027 \qquad \underline{(1.027, 1.26)\ \text{m}}.$$

2.42E a) $F = \gamma \bar{h} A = 62.4 \times 27.33 \times 24 = \underline{40{,}930 \text{ lb}}$.

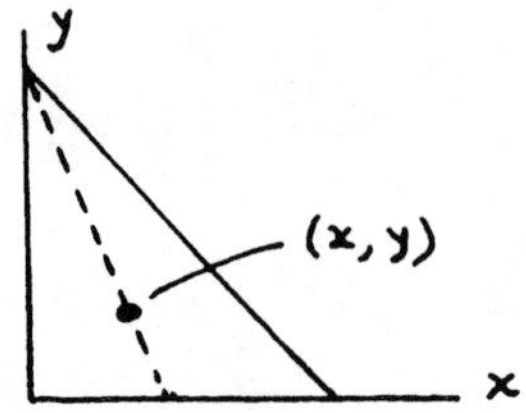

$$y_p = 27.33 + \frac{6 \times 8^3 / 36}{27.33 \times 24} = 27.46'. \therefore y = 30 - 27.46 = 2.54'.$$

$8/5.46 = 3/x$. $\therefore x = 2.05'$. $\underline{(2.05, 2.54) \text{ ft}}$.

b) $F = 62.4 \times 30 \times 24 = \underline{44{,}930 \text{ lb}}$. The centroid is the center of pressure.

$y = 2.667'$. $8/5.333 = 3/x$. $\therefore x = 2.000'$ $\underline{(2.000, 2.667) \text{ ft}}$.

c) $F = 62.4\,(30 - 2.667 \times .707) \times 24 = \underline{42{,}100 \text{ lb}}$.

$$y_p = 39.77 + \frac{6 \times 8^3 / 36}{39.77 \times 24} = 39.86'. \qquad y = 42.43 - 39.86 = 2.57'$$

$8/5.43 = 3/x$. $\therefore x = 2.04'$. $\underline{(2.04, 2.57) \text{ ft}}$.

2.43 a) $F = \gamma \bar{h} A = 9810 \times 6 \times \pi 2^2 = 739\,700 \text{ N}$ or $\underline{739.7 \text{ kN}}$.

$$y_p = \bar{y} + \frac{\bar{I}}{A\bar{y}} = 6 + \frac{\pi \times 2^4 / 4}{4\pi \times 6} = 6.167 \text{ m}. \qquad \therefore (x,y)_p = \underline{(0, -0.167) \text{ m}}$$

b) $F = \gamma \bar{h} A = 9810 \times 6 \times 2\pi = 369\,800 \text{ N}$ or $\underline{369.8 \text{ kN}}$.

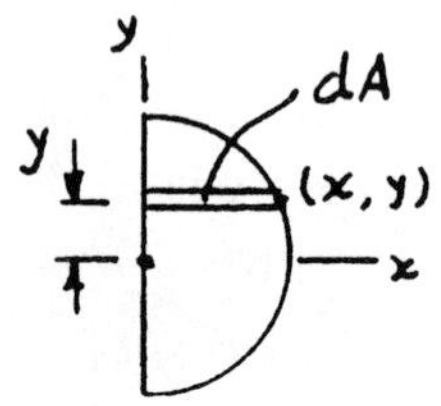

$$y_p = 6 + \frac{\pi \times 2^4 / 8}{2\pi \times 6} = 6.167 \text{ m}. \qquad x^2 + y^2 = 4$$

$$x_p F = \int \frac{x}{2} p\,dA = \frac{\gamma}{2} \int_{-2}^{2} x(6-y)x\,dy = \frac{\gamma}{2} \int_{-2}^{2} (4-y^2)(6-y)\,dy.$$

$$\therefore x_p \gamma 6 \times 2\pi = \frac{\gamma}{2} \int_{-2}^{2} (24 - 4y - 6y^2 + y^3)\,dy = 32\gamma. \qquad \therefore x_p = 0.8488 \text{ m}$$

$\therefore (x,y)_p = \underline{(0.8488, -0.167) \text{ m}}$

c) $F = 9810 \times (4 + 4/3) \times 6 = 313\,900 \text{ N}$ or $\underline{313.9 \text{ kN}}$.

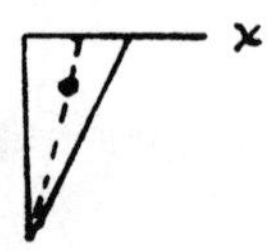

$$y_p = 5.333 + \frac{3 \times 4^3 / 36}{5.333 \times 6} = 5.500 \text{ m}. \qquad \therefore y = -1.5 \text{ m}$$

$4/2.5 = 1.5/x$. $\therefore x = 0.9375$. $\therefore (x,y)_p = \underline{(0.9375, -1.5)}$

d) $F = 9810 \times (4 + \frac{2}{3} \times 4 \sin 36.9°) \times 6 = \underline{330\,000 \text{ N}}$

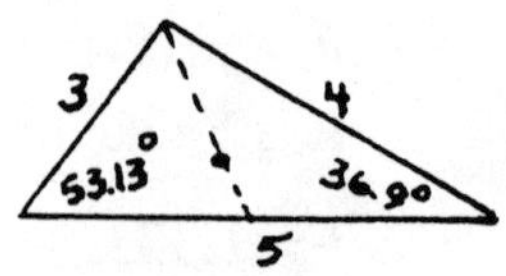

$$y_p = 5.6 + \frac{5 \times 2.4^3 / 36}{6 \times 5.6} = 5.657 \text{ m}. \qquad \therefore y = 0.343 \text{ m}$$

$3 \cos 53.13° = 1.8$. $2.5 - 1.8 = .7$ $2.4/2.057 = .7/x_1$ $\therefore x_1 = 0.6$.

$x = 1.8 + 0.6 = 2.4$. $\therefore (x,y)_p = \underline{(2.4, 0.343) \text{ m}}$.

2.44 $F = \gamma \bar{h} A = 9810 \times 3.5 \times (2 \times 3) = 622\,900 \text{ N}$.

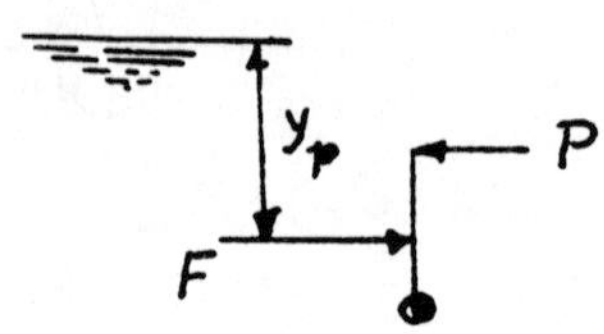

$$y_p = \bar{y} + \frac{\bar{I}}{A\bar{y}} = 3.5 + \frac{2 \times 3^3 / 12}{2 \times 3 \times 3.5} = 3.714.$$

$(5 - 3.714)\,622.9 = 3P$. $\therefore P = \underline{207.6 \text{ kN}}$

2.44E $F = \gamma \bar{h} A = 62.4 \times 11 \times (6 \times 10) = 41{,}180$ lb.

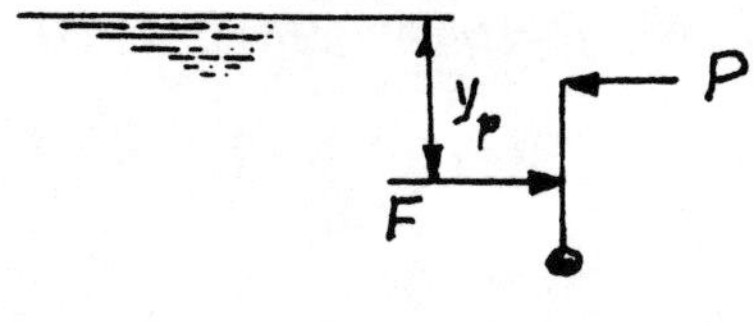

$$y_p = \bar{y} + \frac{\bar{I}}{\bar{y}A} = 11 + \frac{6 \times 10^3 / 12}{11 \times 60} = 11.758'.$$

$(16 - 11.758)\ 41{,}180 = 10P. \quad \therefore P = \underline{17{,}470 \text{ lb}}.$

2.45 $F = \gamma \bar{h} A = 9810 \times 6 \times 20 = 1.777 \times 10^6$ N, or 1177 kN.

$$y_p = \bar{y} + \frac{\bar{I}}{A\bar{y}} = 7.5 + \frac{4 \times 5^3 / 12}{7.5 \times 20} = 7.778 \text{ m}.$$

$(10 - 7.778)\ 1177 = 5\,P. \qquad \therefore P = \underline{523 \text{ kN}}.$

2.46 a) $F = \gamma \bar{h} A = 9810 \times 8 \times 20 = 1.570 \times 10^6 \text{ N} = 1570$ kN.

$$y_p = \bar{y} + \frac{\bar{I}}{A\bar{y}} = 10 + \frac{4 \times 5^3 / 12}{10 \times 20} = 10.208 \text{ m}.$$

$(12.5 - 10.208)\ 1570 = 5\,P. \qquad \therefore P = \underline{720 \text{ kN}}.$

b) $F = \gamma \bar{h} A = 9810 \times 10 \times 20 = 1962$ kN.

$$y_p = \bar{y} + \frac{\bar{I}}{A\bar{y}} = 12.5 + \frac{4 \times 5^3 / 12}{12.5 \times 20} = 12.667 \text{ m}.$$

$(15 - 12.667)\ 1962 = 5\,P. \quad \therefore P = \underline{915 \text{ kN}}.$

c) $F = \gamma \bar{h} A = 9810 \times 12 \times 20 = 2.354 \times 10^6$ N, or 2354 kN.

$$y_p = \bar{y} + \frac{\bar{I}}{A\bar{y}} = 15 + \frac{4 \times 5^3 / 12}{15 \times 20} = 15.139 \text{ m}.$$

$(17.5 - 15.139)\ 2354 = 5\,P. \qquad \therefore P = \underline{1112 \text{ kN}}.$

2.47 $y_p = \bar{y} + \dfrac{\bar{I}}{A\bar{y}} = \dfrac{H}{2} + \dfrac{bH^3/12}{bH \times H/2} = \dfrac{H}{2} + \dfrac{H}{6} = \dfrac{2}{3}H.$ y_p is measured from the surface.

$\therefore$ From the bottom, $H - y_p = H - \dfrac{2}{3}H = \dfrac{1}{3}H.$

Note: this result is independent of the angle α, so it is true for a vertical area or a sloped area.

2.48 a) $F = \gamma \bar{h} A = 9810 \times 1 \times \sin 40^\circ \times 15 = 94.6$ kN.

$94.6 \times \dfrac{2}{3} = 4\,P \quad \Rightarrow \quad P = \underline{15.8 \text{ kN}}$

b) $F = \gamma \bar{h} A = 9810 \times 2 \times \sin 40^\circ \times 15 = 189.2$ kN.

$189.2 \times \dfrac{4}{3} = 6\,P \quad \Rightarrow \quad P = \underline{42.04 \text{ kN}}$

c) $F = \gamma \bar{h} A = 9810 \times 1.607 \times 15 = 236\ 500$ N or 236.5 kN.

$236.5 \times \frac{5}{3} = 7P. \quad \therefore P = \underline{56.3 \text{ kN}}$

2.49 $h = \sqrt{1.2^2 - .4^2} = 1.1314$ m. $A = 1.2 \times 1.1314 + .4 \times 1.1314 = 1.8102$ m^2

Use 2 forces: $F_1 = \gamma h_c A_1 = 9800 \times .5657 \times (1.2 \times 1.1314) = 7527$ N

$$F_2 = \gamma h_c A_2 = 9800 \times \frac{1.1314}{3} \times (.4 \times 1.1314) = 1673 \text{ N}$$

$$y_{p1} = \frac{2}{3}(1.1314). \qquad y_{p2} = \bar{y} + \frac{\bar{I}_2}{A_2 \bar{y}} = \frac{1.1314}{3} + \frac{.4 \times 1.1314^3 / 36}{.4 \times 1.1314 / 2 \times 1.1314 / 3} = 0.5657 \text{ m}$$

$$\Sigma M_{hinge} = 0: \quad 7527 \times \frac{1.1314}{3} + 1673 \times (1.1314 - 0.5657) - 1.1314P = 0. \qquad \therefore \underline{P = 3346 \text{ N}}.$$

2.50 To open, the resultant force must be just above the hinge, i.e., y_p must be just less than h. Let $y_p = h$, the condition when the gate is about to open.

$$\bar{y} = (h+H)/3, \ A = (h+H)^2, \ \bar{I} = [2(h+H)](h+H)^3/36$$

$$\therefore y_p = \frac{h+H}{3} + \frac{2(h+H)^4/36}{(h+H)^2(h+H)/3} = \frac{h+H}{3} + \frac{h+H}{6} = \frac{h+H}{2}$$

a) $h = \frac{h+H}{2}$. $\qquad \therefore h = H = \underline{0.9 \text{ m}}$

b) $h = H = \underline{1.2 \text{ m}}$

c) $h = H = \underline{1.5 \text{ m}}$

2.51 The gate is about to open when the center of pressure is at the hinge.

a) $y_p = 1.2 + H = (1.8/2 + H) + \frac{b \times 1.8^3/12}{(.9+H)1.8b}$. $\qquad \therefore H = \underline{0}$.

b) $y_p = 1.2 + H = (2.0/2 + H) + \frac{b \times 2^3/12}{(1+H)2b}$. $\qquad \therefore H = \underline{0.6667 \text{ m}}$.

c) $y_p = 1.2 + H = (2.2/2 + H) + \frac{b \times 2.2^3/12}{(1.1+H)2.2b}$. $\qquad \therefore H = \underline{2.933 \text{ m}}$.

2.52 $F_1 = \gamma \frac{H}{2} \times bH = \frac{1}{2}\gamma b H^2$

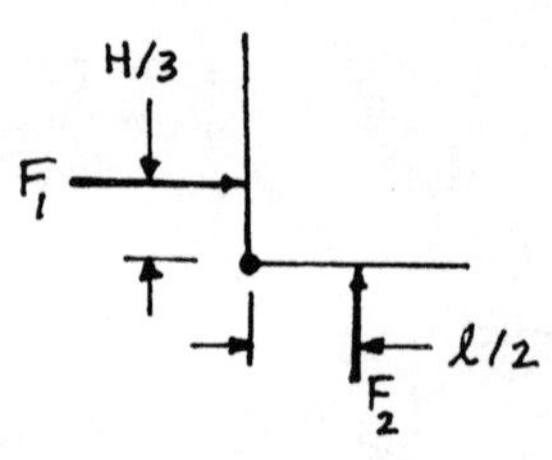

$F_2 = \gamma H \times \ell b = \gamma b \ell H$

$\frac{1}{2}\gamma b H^2 \times \frac{H}{3} = \gamma b \ell H \times \frac{\ell}{2}$. $\qquad \therefore H = \sqrt{3}\ell$

a) $H = \sqrt{3} \times 2 = \underline{3.464 \text{ m}}$ $\quad$ b) $H = \underline{1.732 \text{ m}}$ $\quad$ c) $H = \underline{10.39'}$. $\quad$ d) $H = \underline{5.196'}$

2.53 The dam will topple if the moment about "O" of F_1 and F_3 exceeds the restoring moment of W & F_2.

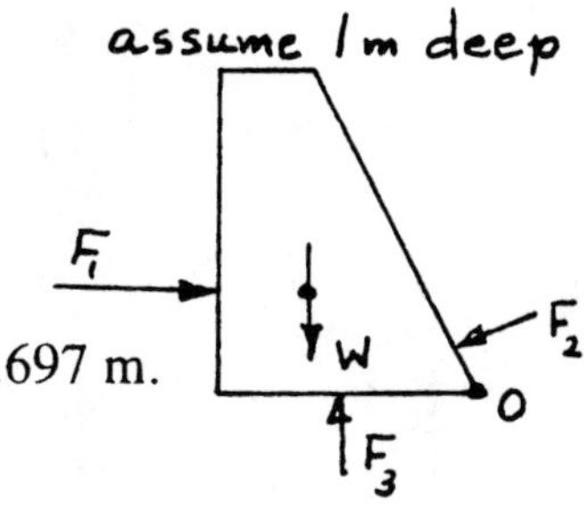

a) $W = (2.4 \times 9810)(6 \times 50 + 24 \times 50/2) = 21.19 \times 10^6$ N

$$d_w = \frac{300 \times 27 + 600 \times 16}{300 + 600} = 19.67 \text{ m.}$$

$F_2 = 9810 \times 5 \times 11.09 = 0.544 \times 10^6$ N. $\quad d_2 = \frac{11.09}{3} = 3.697$ m.

$F_1 = 9810 \times \frac{45}{2} \times 45 = 9.933 \times 10^6$ N. $\quad d_1 = 15$ m.

$F_3 = 9810 \times \frac{45+10}{2} \times 30 = 8.093 \times 10^6$ N. $\quad d_3 = \frac{2.943 \times 15 + 5.150 \times 20}{2.943 + 5.150} = 18.18$ m.

$$\left.\begin{aligned} Wd_w + F_2 d_2 &= 418.8 \times 10^6 \text{ N}\cdot\text{m} \\ F_1 d_1 + F_3 d_3 &= 296.1 \times 10^6 \text{ N}\cdot\text{m} \end{aligned}\right\} \therefore \underline{\text{will not topple}}.$$

b) $W = (2.4 \times 9810)(6 \times 65 + 65 \times 12) = 27.55 \times 10^6$ N.

$$d_w = \frac{390 \times 27 + 780 \times 16}{390 + 780} = 19.67 \text{ m.}$$

$F_2 \cong 0.54 \times 10^6$ N. $\quad d_2 \cong 3.70$ m.

$F_1 = 9810 \times 30 \times 60 = 17.66 \times 10^6$ N. $\quad d_1 = 20$ m.

$F_3 = 9810 \times \frac{60+10}{2} \times 30 = 10.3 \times 10^6$ N. $\quad d_3 = \frac{2.943 \times 15 + 7.358 \times 20}{2.943 + 7.358} = 18.57$ m.

$$\left.\begin{aligned} Wd_w + F_2 d_2 &= 543.9 \times 10^6 \text{ N}\cdot\text{m} \\ F_1 d_1 + F_3 d_3 &= 544.5 \times 10^6 \text{ N}\cdot\text{m} \end{aligned}\right\} \therefore \underline{\text{it will topple}}.$$

c) Since it will topple for $H = 60$, it certainly will topple if $H = 75$ m.

2.54 The dam will topple if there is a net clockwise moment about "O".

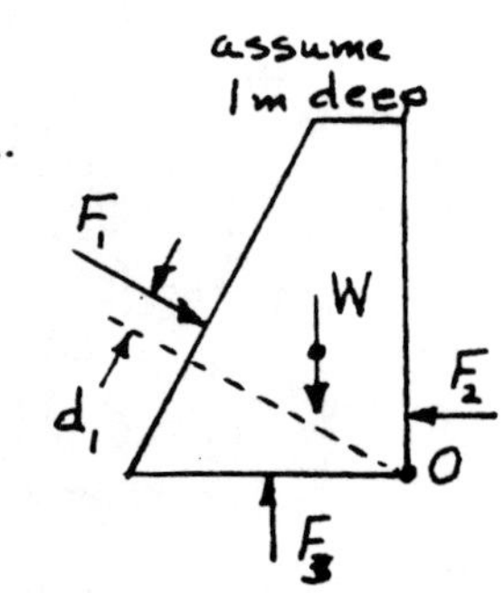

a) $W = (2.4 \times 9810)(6 \times 48 + 48 \times 12) = 20.34 \times 10^6$ N.

$$d_w = \frac{288 \times 3 + 576 \times 14}{6 \times 48 + 48 \times 12} = 10.33 \text{ m.}$$

$F_2 = 9810 \times 5 \times 10 = 0.49 \times 10^6$ N. $\quad d_2 = 3.33$ m.

$F_1 = 9810 \times \frac{45}{2} \times 50.31 = 11.10 \times 10^6$ N. $\quad d_1 = \frac{50.31}{3} - 30 \sin 26.57^\circ = 3.354$ m.

$F_3 = 9810 \times \frac{45+10}{2} \times 30 = 8.093 \times 10^6$ N. $\quad d_3 = \frac{2.943 \times 15 + 5.15 \times 20}{2.943 + 5.150} = 18.18$ m.

$\Sigma M = (8.093 \times 18.18 + 11.1 \times 3.354 - 20.34 \times 10.33 - .49 \times 3.33) \times 10^6 = -27.4 \times 10^6$ N·m.

$\therefore$ There is a ccw moment of -27.4×10^6 N·m. $\therefore$ it will not topple.

b) $W = (2.4 \times 9810)(6 \times 63 + 63 \times 12) = 26.7 \times 10^6$ N. $d_w = \dfrac{378 \times 3 + 756 \times 14}{378 + 756} = 10.33.$

$F_2 = 0.49 \times 10^6$ N. $\quad d_2 = 3.33$ m.

$F_1 = 9810 \times 30 \times 64.21 = 18.9 \times 10^6$ N. $\quad d_1 = \dfrac{64.21}{3} - 30\sin 20.85^\circ = 10.73$ m.

$F_3 = 9810 \times \dfrac{60+10}{2} \times 30 = 10.3 \times 10^6$ N. $\quad d_3 = \dfrac{2.943 \times 15 + 7.358 \times 20}{2.943 + 7.358} = 18.57$ m.

$\Sigma M = (10.3 \times 18.57 + 18.9 \times 10.73 - 26.7 \times 10.33 - .49 \times 3.33) \times 10^6 = 116.6 \times 10^6$ N·m.

$\therefore$ There is a cw moment of 116.6×10^6 N·m. $\therefore$ it will topple.

c) Since it will topple for $H = 60$ m, it certainly will topple if $H = 75$ m.

2.54E a) $W = W_1 + W_2$. $W_1 = (6 \times 43 \times 1) \times 62.4 \times 2.4 = 38{,}640$ lb

$W_2 = (24 \times 43 / 2) \times 62.4 \times 2.4 = 77{,}280$ lb

$F_1 = 62.4 \times 20 \times (40 \times 1) = 49{,}920$ lb. $W_3 = (40 \times 22.33 / 2) \times 62.4 = 27{,}870$ lb

$F_2 = 62.4 \times 5 \times (10 \times 1) = 3120$ lb $d_2 = 3.33$ ft

$F_{p1} = 62.4 \times 10 \times (30 \times 1) = 18{,}720$ lb. $F_{p2} = 62.4 \times 30 \times (30 \times 1)/2 = 28{,}080$ lb

$$\therefore F_3 = F_{p1} + F_{p2} = 62.4 \times \left(\frac{40+10}{2}\right) \times (30 \times 1) = 46{,}800 \text{ lb}$$

$F_1 = 62.4 \times 20 \times (45.81 \times 1) = 57{,}170$ lb

$d_1 = \dfrac{45.81}{3} - 30\sin 29.17^\circ = 0.649$ m

$\Sigma M_{corner} = 0$:

$$57{,}170 \times 0.649 + 18{,}720 \times 15 + 28{,}080 \times 20 - 3120 \times \frac{10}{3}$$

$$-38{,}640 \times 3 - 77{,}280(6+8) \overset{?}{=} 0. \qquad -328{,}740 < 0. \qquad \therefore \text{won't tip.}$$

b) $W_1 = 6 \times 63 \times 62.4 \times 2.4 = 56{,}610$ lb. $W_2 = (24 \times 63/2) \times 62.4 \times 2.4 = 113{,}220$ lb.

$F_1 = 62.4 \times 30 \times 60 = 112{,}300$ lb. $\quad W_3 = (60 \times 22.86 / 2) \times 62.4 = 42{,}790$ lb.

$F_2 = 62.4 \times 5 \times 10 = 3120$ lb

$F_{p1} = 62.4 \times 10 \times 30 = 18{,}720$ lb. $\quad F_{p2} = 62.4 \times 50 \times 30 / 2 = 46{,}800$ lb.

$\Sigma M_{corner} = 0$:

$$112{,}300 \times 20 - 42{,}790\left(30 - \frac{22.86}{3}\right) + 18{,}720 \times 15 + 46{,}800 \times 20 - 3120 \times \frac{10}{3}$$

$$-56{,}600 \times 3 - 113{,}220 \times 14 \overset{?}{=} 0 \qquad 739{,}900 > 0. \qquad \therefore \text{will tip.}$$

c) $W_1 = 6 \times 83 \times 62.4 \times 2.4 = 74{,}580$ lb. $W_2 = (24 \times 83/2) \times 62.4 \times 2.4 = 149,200$ lb.

$F_1 = 62.4 \times 40 \times 80 = 199{,}680$ lb. $\quad F_2 = 62.4 \times 5 \times 10 = 3120$ lb

$W_3 = (80 \times 23.13 / 2) \times 62.4 = 57{,}730$ lb.

$F_{p1} = 62.4 \times 10 \times 30 = 18{,}720$ lb. $\quad F_{p2} = 62.4 \times 70 \times 30 / 2 = 65{,}520$ lb.

$$\Sigma M_{corner} = 0:$$

$$199{,}680 \times \frac{80}{3} - 57{,}730\left(30 - \frac{23.13}{3}\right) + 18{,}720 \times 15 + 65{,}520 \times 20 - 3120 \times \frac{10}{3}$$

$$-74{,}580 \times 3 - 149{,}200 \times 14 \overset{?}{=} 0$$

$$4{,}593{,}000 > 0. \qquad \therefore \underline{\text{will tip.}}$$

2.55

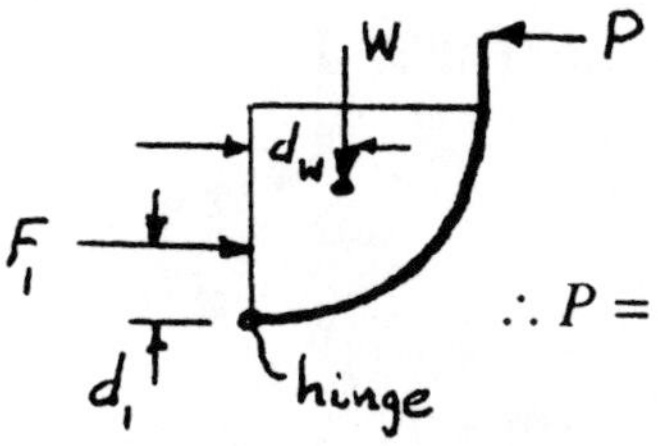

$$\Sigma M_{hinge} = 0.$$

$$2.5P - d_w \times W - d_l \times F_l = 0.$$

$$\therefore P = \frac{1}{2.5}\left[\frac{2}{3} \times 9800 \times 1 \times 8 + \frac{4 \times 2}{3\pi} \times 9800 \times \frac{\pi \times 2^2}{4} \times 4\right] = \underline{62\ 700\ \text{N}}$$

Note: This calculation is simpler than that of Example 2.7.

2.56 Since all infinitesimal pressure forces pass thru the center, we can place the resultant forces at the center. Since the vertical components pass thru the bottom point, they produce no moment about that point. Hence, consider only horizontal forces.

$$(F_H)_{H_2O} = 9810 \times 2 \times (4 \times 10) = 784\ 800\text{N}$$

$$(\text{F}_\text{H})_{oil} = 0.86 \times 9810 \times 1 \times 20 = 168\ 700\text{N}$$

$$\Sigma M:\ 2P = 784.8 \times 2 - 168.7 \times 2. \qquad \therefore \underline{P = 616.1\ \text{kN}}.$$

2.57 Place the resultant force $\vec{F}_H + \vec{F}_V$ at the center of the circular arc. $\vec{F}_H$ passes thru the hinge showing that $P = F_V$.

a) $P = F_V = 9810(6 \times 2 \times 4 + \pi \times 4) = 594\ 200$ N or $\underline{594.2\ \text{kN}}$.

b) $P = F_V = 62.4\ (20 \times 6 \times 12 + 9\pi \times 12) = \underline{111{,}000\ \text{lb}}$.

2.58 Place the resultant $\vec{F}_H + \vec{F}_V$ at the circular arc center. $\vec{F}_H$ passes thru the hinge so that $P = F_V$. Use the water that could be contained above the gate; it produces the same pressure distribution and hence the same F_V.

$P = F_V = 9810\ (6 \times 3 \times 4 + 9\pi) = 983\ 700$ N or $\underline{983.7\ \text{kN}}$.

2.59 Place the resultant $\vec{F}_H + \vec{F}_V$ at the center. $\vec{F}_V$ passes thru the hinge

$2 \times (9810 \times 1 \times 10) = 2.8$ P. $\qquad \therefore P = 70\ 070$ N or $\underline{70.07\ \text{kN}}$.

2.60 The incremental pressure forces on the circular quarter arc pass through the hinge so that no moment is produced by such forces. Moments about the hinge gives:

$$3\,P = 0.9\ W = 0.9 \times 400. \qquad \therefore \underline{P = 120\ \text{N}}.$$

2.61 The resultant $\vec{F}_H + \vec{F}_V$ of the unknown liquid acts thru the center of the circular arc. $\vec{F}_V$ passes thru the hinge. Thus we use only $(F_H)_{oil}$. Assume 1 m wide.

a) ΣM: $\frac{R}{3}\left(9810\frac{R}{2}R\right)+\frac{4R}{3\pi}\left(9800S\frac{\pi R^2}{4}\right)=R\left(\gamma_x\frac{R}{2}R\right)$. $\therefore \gamma_x = \underline{4580 \text{ N/m}^3}$

b) ΣM: $\frac{R}{3}\left(62.4\frac{R}{2}R\right)+\frac{4R}{3\pi}\left(62.4S\frac{\pi R^2}{4}\right)=R\left(\gamma_x\frac{R}{2}R\right)$. $\therefore \gamma_x = \underline{29.1 \text{ lb/ft}^3}$

2.62 The force of the water is only vertical $(F_V)_w$, acting thru the center. The force of the oil can also be positioned at the center.

a) $P=(F_H)_o=(0.8\times 9810)\times 0.3\times 3.6=\underline{8476\text{ N}}$.

$\Sigma F_y = 0 = W + (F_V)_o - (F_V)_w$

$$0 = S\times 9810\,\pi\times .6^2\times 6+\left(.36-\frac{.36\pi}{4}\right)\times 6\times(.8\times 9810)-9810\times\pi\times .18\times 6$$
$$-9810\times .8\times 2\times .6^2\times 6. \qquad \therefore S=\underline{0.955}.$$

b) $P=(F_H)_o=(0.8\times 62.4)\times 1\times 20\times 2=\underline{1996\text{ lb}}$.

$\Sigma F_y = 0 = W + (F_V)_o - (F_V)_w$

$$0 = S\times 62.4\times\pi\times 2^2\times 20+\left(4-\frac{4\pi}{4}\right)\times 20\times .8\times 62.4-62.4\times\pi\times 2\times 20$$
$$-62.4\times .8\times 2\times 2^2\times 20. \qquad \therefore S=\underline{0.955}.$$

2.63 The pressure in the dome is

a) $p = 60\,000 - 9810\times 3 - 0.8\times 9810\times 2 = 14\,870$ Pa or 14.87 kPa.

The force is $F = pA_{projected} = \pi\times 3^2\times 14.87 = \underline{420.4\text{ kN}}$.

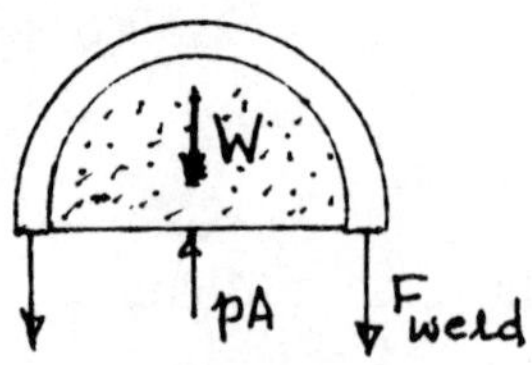

b) From a free-body diagram of the dome filled with oil:

$F_{weld} + W = pA$

Using the pressure from part (a):

$$F_{weld} = 14\,870\times\pi\times 3^2-(.8\times 9810)\times\frac{1}{2}\left(\frac{4}{3}\pi\times 3^3\right)=-23\,400\text{ N}$$

or $\underline{-23.4\text{ kN}}$.

2.64 A free-body diagram of the gate and water is shown.

$\frac{H}{3}F + d_w W = H\times P$.

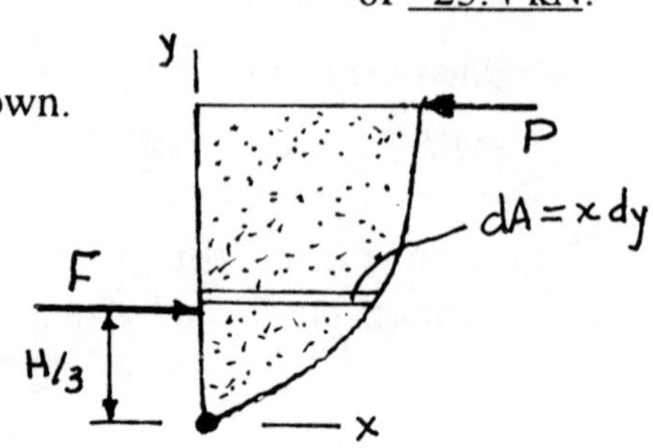

a) $H = 2$ m. $F = 9810 \times 1 \times 4 = 39\,240$ N.

$$W = 9810\int_0^2 2x\,dy = 9810\int_0^2 2\frac{y^{1/2}}{\sqrt{2}}dy = \frac{2\times 9810}{\sqrt{2}}\frac{2^{3/2}}{3/2} = 26\,160 \text{ N.}$$

$$d_w = \bar{x} = \frac{\int \frac{x}{2}x\,dy}{\int x\,dy} = \frac{\frac{1}{2}\int_0^1 4x^3dx}{\int_0^1 4x^2dx} = \frac{1}{2}\left(\frac{1/4}{1/3}\right) = 0.375 \text{ m.}$$

$$\therefore P = \frac{1}{3}\times 39\,240 + \frac{0.375}{2}\times 26\,160 = 17\,980 \text{ N} \quad \text{or} \quad \underline{17.98 \text{ kN}}.$$

b) $H = 8'$. $F = 62.4 \times 4 \times 32 = 7{,}987$ lb.

$$W = 62.4\int 4x\,dy = 62.4\times 4\int_0^2 4x^2dx = 62.4\times 16\times 2^3/3 = 2{,}662 \text{ lb.}$$

$$d_w = \bar{x} = \frac{\frac{1}{2}\int_0^2 4x^3dx}{\int_0^2 4x^2dx} = \frac{1}{2}\left(\frac{16/4}{8/3}\right) = 0.75'.$$

$\therefore P = \underline{2912 \text{ lb}}$.

2.65 W = weight of displaced water. a) $20\,000 + 250\,000 = 9810 \times 3\left[6\times d + \frac{d^2}{2}\right]$

$\therefore d^2 + 12d - 18.35 = 0$. $\therefore d = \underline{1.372 \text{ m}}$.

b) $270\,000 = 1.03\times 9810\times 3\,[6d + d^2/2]$

$d^2 + 12d - 17.81 = 0$. $\therefore d = \underline{1.336 \text{ m}}$.

2.66 $25 + F_B = 100$. $\therefore F_B = 75 = 9810\,V$. $\therefore V = 7.645 \times 10^{-3}$ m^3 or $\underline{7645 \text{ cm}^3}$

$\gamma \times 7.645 \times 10^{-3} = 100$.

$\therefore \gamma = \underline{13\,080 \text{ N/m}^3}$.

2.67 $1400 \times 9.81 \times 60 = 8 \times 100\,\Delta d \times 9810$. $\therefore \Delta d = 0.105$ m or $\underline{10.5 \text{ cm}}$.

2.67E $3000 \times 60 = 25 \times 300\,\Delta d \times 62.4$. $\therefore \Delta d = 0.3846'$ or $4.\underline{62''}$.

2.68 $100\,000 \times 9.81 + 6\,000\,000 = (12 \times 30 + 8h \times 30)\,9810$

$\therefore h = 1.465$ m. $\therefore$distance from top $= 2 - 1.465 = \underline{0.535 \text{ m}}$

2.69 $T + F_B = W$. (See Fig. 2.11 c.)

$T = 40\,000 - 1.59 \times 9810 \times 2 = 8804$ N or $\underline{8.804 \text{ kN}}$.

2.70 The forces acting on the balloon are its weight W, the buoyant force F_B, and the weight of the air in the balloon F_a. Sum forces:

$$F_B = W + F_a$$

$$\frac{4}{3}\pi R^3 \rho g = 1000 + \frac{4}{3}\pi R^3 \rho_a g$$

$$\frac{4}{3}\pi \times 5^3 \frac{100 \times 9.81}{.287 \times 293} = 1000 + \frac{4}{3}\pi \times 5^3 \frac{100 \times 9.81}{.287 T_a}. \qquad \therefore T_a = 350.4 \text{ K or } \underline{77.4°\text{C}}$$

2.71 The forces acting on the blimp are the payload F_p, the weight of the blimp W, the buoyant force F_B, and the weight of the helium F_h:

$$F_B = Fp + W + F_h$$

$$1500\pi \times 150^2 \times \frac{100 \times 9.81}{.287 \times 288} = F_p + 0.1\ F_p + 1500\ \pi \times 150^2 \times \frac{100 \times 9.81}{2.077 \times 288}$$

$$\therefore F_p = 9.86 \times 10^8 \text{ N}$$

$$N_{people} = \frac{9.86 \times 10^8}{800} = \underline{1.23 \times 10^6}$$

Of course equipment and other niceties such as gyms, pools, restaurants, etc., would add significant weight.

2.72 Neglect the bouyant force of air. A force balance yields

$$F_B = W + F$$
$$= 50 + 10 = 60$$
$$= 9800\ V. \qquad \therefore V = .006122 \text{ m}^3.$$

Density: $\rho g V = W$.

$$\rho \times 9.81 \times .006122 = 50. \qquad \therefore \rho = \underline{832.5 \text{ kg/m}^3}$$

Specific wt: $\gamma = \rho g = 832.5 \times 9.81 = \underline{8167 \text{ N/m}^3}$

Specific gravity: $S = \rho/\rho_{water} = 832.5/1000 = \underline{.8325}$

2.73 From a force balance $F_B = W + pA$.

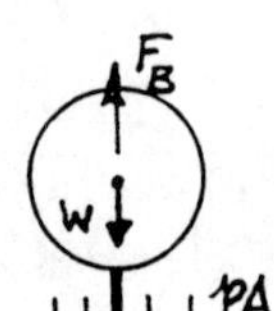

a) The buoyant force is found as follows ($h > 5.3$):

$$\cos\theta = \frac{h-5-R}{R}. \qquad \theta = \cos^{-1}\frac{h-5-R}{R}.$$

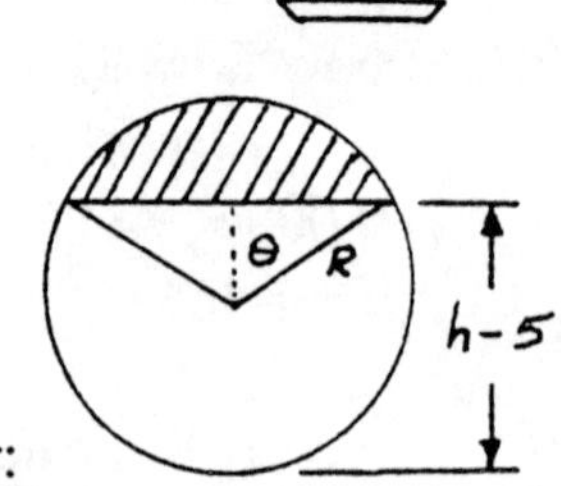

$$\text{Cross-hatched area} = \frac{2\theta}{2\pi}\pi R^2 - (h-R-5)R\sin\theta$$

$$\therefore F_B = \{\pi R^2 - \theta R^2 + (h-R-5)\ R \sin\theta\}\ 4 \times 9810.$$

$$F_B = 6000 + 9810\ h\ \pi \times .1^2.$$

The h that makes the 2 F_B's equal is found by trial-and-error:

$h = 5.5$: $7695 \stackrel{?}{=} 9880$ $\qquad h = 5.38$: $7658 \stackrel{?}{=} 7408$

$h = 5.4$: $7664 \stackrel{?}{=} 7857$ $\qquad \therefore h = \underline{5.39 \text{ m}}$

b) Modify the above expressions for the two F_B's ($h < 5.4$):

$F_B = \{\theta R^2 - (R-h+5)R \sin\theta\}\ 4 \times 9810.$

$F_B = 6000 + 9810\, h\, \pi \times .1^2.$ $\qquad \cos\theta = \dfrac{R-h+5}{R}$

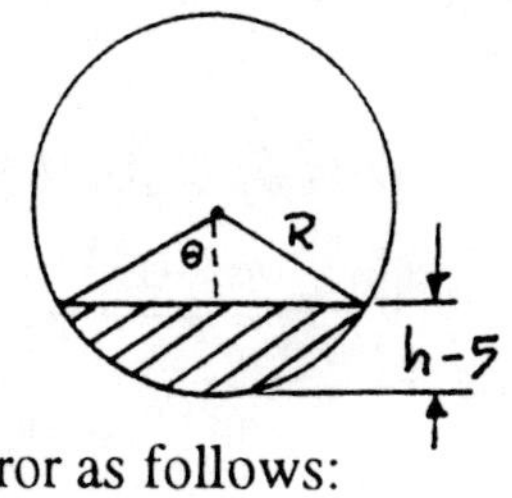

The h that makes the 2 F_B's equal is found by trial-and-error as follows:

$h = 5.2$: $\quad 7603 \stackrel{?}{=} 3856$ $\qquad h = 5.34$: $\quad 7646 \stackrel{?}{=} 7986$

$h = 5.3$: $\quad 7633 \stackrel{?}{=} 6756$ $\qquad \therefore h = \underline{5.33\text{ m}}$

c) Since we anticipate $h < 5.5$, we use the F_B's of part (b):

$h = 5.2$: $\quad 7603 \stackrel{?}{=} 4388$ $\qquad h = 5.29$: $\quad 7630 \stackrel{?}{=} 7418$

$h = 5.3$: $\quad 7633 \stackrel{?}{=} 7776$ $\qquad \therefore h = \underline{5.296\text{ m}}$

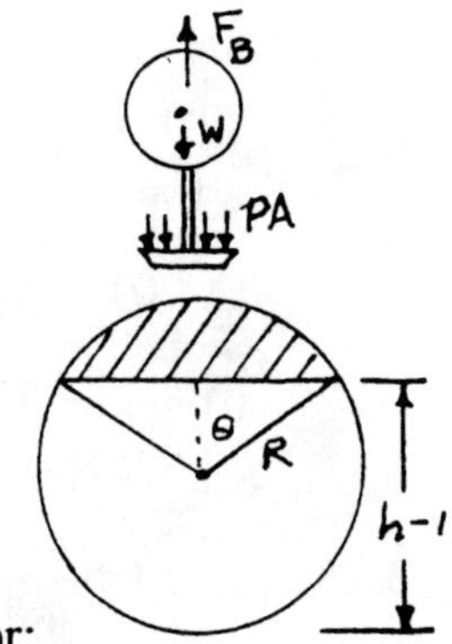

2.73E From a force balance $F_B = W + pA$.

a) The buoyant force is found as follows ($h > 16'$):

$\cos\theta = \dfrac{h-15-R}{R}$, $\qquad$ Area $= \theta R^2 - (h-15-R)\, R \sin\theta$

$\therefore F_B = 10 \times 62.4\ \{\pi R^2 - \theta R^2 + (h-15-R)\, R \sin\theta\}.$

$F_B = 1500 + \gamma h A.$

The h that makes the above 2 F_B's equal is found by trial-and-error:

$h = 16.5$: $\quad 1859 \stackrel{?}{=} 1577$ $\qquad h = 16.8$: $\quad 1866 \stackrel{?}{=} 1858$

$h = 17.0$: $\quad 1870 \stackrel{?}{=} 1960$ $\qquad \therefore h = \underline{16.82\text{ ft.}}$

b) Assume $h > 16\frac{1}{3}$ ft. and use the above equations with $R = 1.333'$:

$h = 16.4$: $\quad 1857 \stackrel{?}{=} 1853$ $\qquad \therefore h = \underline{16.4\text{ ft.}}$

c) Assume $h < 16\frac{2}{3}$ ft. With $R = 1.667'$

$F_B = 10 \times 62.4\ \{\theta R^2 - (R-h+15)\, R \sin\theta\}.$

$F_B = 1500 + \gamma h A.$ $\qquad \cos\theta = \dfrac{R-h+15}{R}$

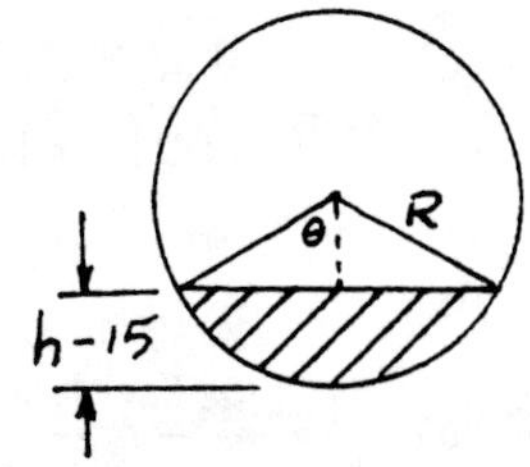

Trial-and-error for h:

$h = 16$: $\quad 1849 \stackrel{?}{=} 1374$ $\qquad h = 16.2$: $\quad 1853 \stackrel{?}{=} 1765$

$h = 16.4$: $\quad 1857 \stackrel{?}{=} 2170$ $\qquad \therefore h = \underline{16.25\text{ ft.}}$

2.74 a) $W = F_B$. $\left[0.01 + 13.6\times1000\times h\pi\times.015^2/4\right]\times9.81 = 9810\,\forall$.

$\forall = \frac{\pi\times.015^2}{4}\times.15 + \frac{\pi\times.005^2}{4}\times.06 = 2.769\times10^{-5}\text{ m}^3$. $\therefore h = 7.361\times10^{-3}$ m

$\therefore m_{Hg} = \underline{0.01769\text{ kg}}$. or 7.361 mm

b) $(.01 + .01769)\,9.81 = 9810\left[\frac{\pi\times.015^2}{4}\times.15 + \frac{\pi\times.005^2}{4}\times.12\right]S_x$.

$\therefore S_x = \underline{0.959}$.

c) $(.01 + .01769)\,9.81 = 9810\,\frac{\pi\times.015^2}{4}\times.15\;S_x$. $\therefore S_x = \underline{1.045}$.

2.75 $(.01 + m_{Hg})9.81 = 9810\left[\frac{\pi\times.015^2}{4}\times.15 + \frac{\pi\times.005^2}{4}\times.12\right]$. $\therefore m_{Hg} = 0.01886$.

a) $(.01 + .01886)\,9.81 = 9810\,\frac{\pi\times.015^2}{4}\times.15\;S_x$. $\therefore S_x = \underline{1.089}$.

b) $m_{Hg} = \underline{0.01886\text{ kg}}$.

2.76 a) $I_o = \frac{\pi d^4}{64} = \frac{\pi\times.25^4}{64} = 1.917\times10^{-4}\text{ m}^4$

$\forall = \frac{W}{r_{H_2O}} = \frac{.8\times9810\pi\times.25^2\times.3}{9810\times4} = 0.01178\text{ m}^3$. $\therefore$depth $= \frac{.01178}{\pi\times.25^2/4} = .24$ m

$\therefore \overline{GM} = 1.917\times10^{-4}/.01178 - (.15-.12) = -.0137$ m. $\therefore$ <u>It will not float with ends horizontal</u>.

b) $I_o = 1.917\times10^{-4}\text{ m}^4$. $\forall = 0.00982$. depth = 0.2 m.

$\overline{GM} = 1.917\times10^{-4}/.00982 - (.125-.1) = -0.00548$ m $\therefore$ <u>It will not float with ends horizontal</u>.

c) $I_o = 1.917\times10^{-4}\text{ m}^4$. $\forall = 0.00785$. depth = 0.16 m.

$\overline{GM} = 1.917\times10^{-4}/.00785 - (.1-.08) = 0.00442$ m $\therefore$ <u>It will float with ends horizontal</u>.

2.76E a) $I_o = \frac{\pi d^4}{64} = \frac{\pi\times(10/12)^4}{64} = 0.02367\text{ ft}^4$.

$\forall = \frac{W}{r_{H_2O}} = \frac{.8\times62.4\times\pi\times(5/12)^2\times12/12}{62.4} = 0.4363$. depth $= \frac{.4363}{\pi(5/12)^2} = 0.8'$

$\therefore \overline{GM} = .02367/.4363 - (.5-.4) = -.0457'$. $\therefore$ <u>It will not float with ends horizontal</u>.

b) $I_o = .02367 \text{ ft}^4$, $V = .3636 \text{ ft}^3$, depth = .6667'

$\overline{GM} = .02367/.3636 - (5-4)/12 = -.01823'$. $\therefore$ It will not float as given.

c) $V = .2909$, depth = 6.4", $= \overline{GM}\ \dfrac{.02367}{.2909} - \dfrac{4-3.2}{12} = .0147.$ $\therefore$ It will float.

2.77 With ends horizontal $I_o = \pi d^4/64$. The displaced volume is $V = \gamma_x \pi d^2 h / 4 \times 9800 = 8.014 \times 10^{-5} \gamma_x d^3$ since $h = d$. The depth the cylinder will sink is

$$\text{depth} = \frac{V}{A} = 8.014 \times 10^{-5} \gamma_x d^3 / \pi d^2 / 4 = 10.20 \times 10^{-5} \gamma_x d$$

The distance $\overline{CG}$ is

$$\overline{CG} = \frac{h}{2} - 10.2 \times 10^{-5} \gamma_x d / 2$$

Then $$\overline{GM} = \frac{\pi d^4/64}{8.014 \times 10^{-5} \gamma_x d^3} - \frac{d}{2} + 10.2 \times 10^{-5} \gamma_x d/2 > 0.$$

This gives (divide by d and multiply by γ_x):

$$612.5 - .5\,\gamma_x + 5.1 \times 10^{-5}\,\gamma_x^2 > 0.$$

Consequently,

$\gamma_x > 8369 \text{ N/m}^3$ or $\gamma_x < 1435 \text{ N/m}^3$

2.78 $I_o = \dfrac{d^4}{12}$. $V = \dfrac{W}{r_{H_2O}} = S\gamma_{H_2O} d^3 / \gamma_{H_2O} = Sd^3$. $\therefore h = Sd$.

$$\overline{GM} = \frac{d^4/12}{Sd^3} - (d/2 - Sd/2) = d\left(\frac{1}{12S} - \frac{1}{2} + \frac{S}{2}\right).$$

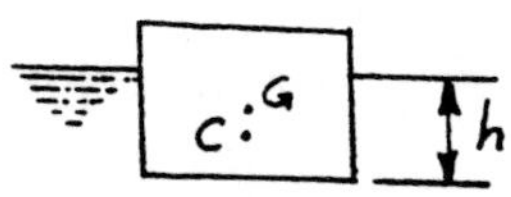

If $\overline{GM} = 0$ the cube is neutral. Then $6S^2 - 6S + 1 = 0$.

$$\therefore S = \frac{6 \pm \sqrt{36-24}}{12} = 0.7887, 0.2113.$$

The cube is unstable if $0.2113 < S < 0.7887$.

Note: Try $S = .8$ and $S = .1$ to see if $\overline{GM} > 0$. This indicates stability.

2.79 As shown, $\bar{y} = \dfrac{16 \times 9 + 16 \times 4}{16 + 16} = 6.5$ cm above the bottom edge.

$$G = \frac{4\gamma \times 9.5 + 16\gamma \times 8.5 + 16 S_A \gamma \times 4}{.5\gamma \times 8 + 2\gamma \times 8 + S_A \gamma \times 16} = 6.5 \text{ cm}. \quad \therefore 130 + 104\, S_A = 174 + 64\, S_A.$$

$\therefore$ $S_A = 1.1$.

2.80 a) $\bar{y} = \dfrac{16\times4+8\times1+8\times7}{16+8+8} = 4.$ $\quad \bar{x} = \dfrac{16\times1+8\times4+8\times4}{16+8+8} = 2.5.$

For G: $y = \dfrac{1.2\times16\times4+.5\times8\times1+1.5\times8\times7}{1.2\times16+.5\times8+1.5\times8} = 4.682.$

$$x = \frac{1.2\times16+.5\times8\times4+1.5\times8\times4}{1.2\times16+.5\times8+1.5\times8} = 2.364.$$

G must be directly under C:

$\tan\theta = \dfrac{.136}{.682}.$ $\quad \therefore \underline{\theta = 11.3°}.$

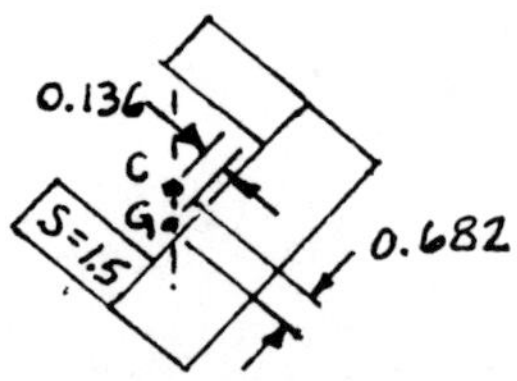

b) $\bar{y} = \dfrac{4\times2+2\times\frac{1}{2}+2\times3.5}{4+2+2} = 2.$ $\quad \bar{x} = \dfrac{4\times\frac{1}{2}+2\times2+2\times2}{4+2+2} = 1.25$

For G: $y = \dfrac{1.2\times4\times2+.5\times1+1.5\times7}{1.2\times4+.5\times2+1.5\times2} = 2.34.$ $x = \dfrac{1.2\times2+.5\times4+1.5\times4}{1.2\times4+.5\times2+1.5\times2} = 1.182$

$\Delta y = .34,$ $\quad \Delta x = .068.$ $\quad \tan\theta = \dfrac{.068}{.34}.$ $\quad \therefore \underline{\theta = 11.3°}.$

2.81 The centroid C is 1.5 m below the water surface. $\quad \therefore \overline{CG} = 1.5$ m.

Using Eq. 2.4.47: $\overline{GM} = \dfrac{\ell\times8^3/12}{\ell\times8\times3} - 1.5 = 1.777 - 1.5 = 0.277 > 0.$

$\therefore$ The barge is stable.

2.82 $\bar{y} = \dfrac{18.0\times3.414+16.97\times1}{18.0+16.97} = 2.24$ m. $\quad \therefore \overline{CG} = 2.24 - 1.5 = 0.74$ m.

Using Eq. 2.4.47: $\overline{GM} = \dfrac{\ell\times8.485^3/12}{34.97\ell} - .74 = 1.46 - .74 = 0.72.$ $\quad \therefore$ Stable.

2.83 a) $\tan\alpha = \dfrac{20}{9.81} = \dfrac{H}{4}.$ $\therefore H = 8.155$ m. $\quad p_{max} = 9810\,(8.155 + 2) = \underline{99\ 620\ \text{Pa}}$

b) $p_{max} = \rho(g + a_z)\,h = 1000\,(9.81 + 20) \times 2 = \underline{59\ 620\ \text{Pa}}$

c) $p_{max} = 1.94 \times 60\,(-12) - 1.94\,(32.2 + 60)\,(-6) = 2470$ psf or $\underline{17.15\ \text{psi}}$

d) $p_{max} = 1.94\,(32.2 + 60)\,(-6) = 1073$ psf or $\underline{7.45\ \text{psi}}$

2.84 The air volume is the same before and after.

$\therefore\ .5 \times 8 = hb/2.$ $\quad \tan\alpha = \dfrac{10}{9.81} = \dfrac{h}{b}.$

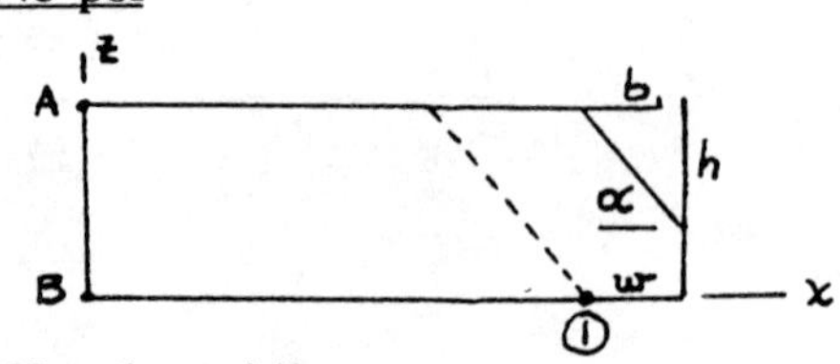

$4 = \dfrac{h}{2}\dfrac{9.81}{10}h.$ $\quad \therefore h = 2.856.$ $\quad \therefore$ Use dotted line.

$2.5w + \dfrac{1}{2}\times2.5\times2.452 = 4.$ $\quad \therefore w = 0.374$ m.

a) $p_A = -1000 \times 10\ (0 - 7.626) - 1000 \times 9.81 \times 2.5 = 51\ 740$ Pa or $\underline{51.74 \text{ kPa}}$
b) $p_B = -1000 \times 10\ (0 - 7.626) = 76\ 260$ Pa or $\underline{76.26 \text{ kPa}}$
c) $p_C = \underline{0}$. Air fills the space to the dotted line.

2.85 Use Eq. 2.5.2: Assume an air-water surface as shown.

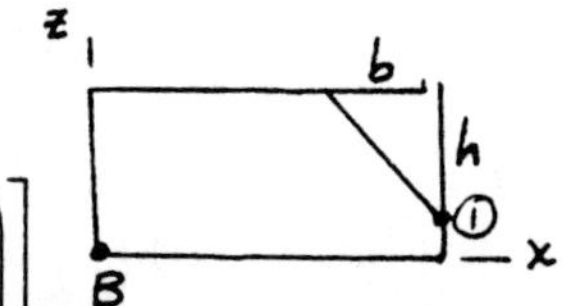

a) $$60\ 000 = -1000\ a_x\ (0\text{–}8) - 1000 \times 9.81 \left[0 - \left(2.5 - \sqrt{\frac{8a_x}{9.81}}\right)\right]$$

$$4 = \frac{h^2 \times 9.81}{2a_x} \qquad 60 = 8\ a_x + 24.52 - 9.81\sqrt{\frac{8a_x}{9.81}}. \qquad a_x - 4.435 = 1.1074\sqrt{a_x}.$$

$$a_x^2 - 10.1\ a_x + 19.67 = 0 \qquad \therefore a_x = 2.64,\ \underline{7.46 \text{ m/s}^2}$$

b) $$60\ 000 = -1000\ a_x\ (-8) - 1000\ (9.81 + 10)\left(-2.5 + \sqrt{\frac{8a_x}{9.81}}\right).$$

$$60 = 8\ a_x + 49.52 - 19.81\sqrt{\frac{8a_x}{19.81}}. \qquad a_x - 1.31 = 1.574\sqrt{a_x}.$$

$$a_x^2 - 5.1\ a_x + 1.44 = 0 \qquad \therefore a_x = 0.25,\ \underline{4.8 \text{ m/s}^2}$$

c) $$60\ 000 = -1000\ a_x\ (-8) - 1000\ (9.81 + 5)\ (-2.5 + \sqrt{\frac{8a_x}{14.81}}).$$

$$60 = 8\ a_x + 37.0 - 14.81\sqrt{\frac{8a_x}{14.81}}. \qquad a_x - 2.875 = 1.361\sqrt{a_x}.$$

$$a_x^2 - 7.6\ a_x + 8.266 = 0 \qquad \therefore a_x = 1.32,\ \underline{6.28 \text{ m/s}^2}$$

2.86 a) $a_x = 20 \times .866 = 17.32$ m/s^2, $a_z = 10$ m/s^2. Use Eq. 2.5.2 with the peep hole as ①. The x-axis is horizontal passing thru A. We have
$p_A = -1000 \times 17.32\ (0 - 1.232) - 1000\ (9.81 + 10)\ (0 - 1.866) = \underline{58\ 290 \text{ Pa}}$
b) $p_A = -1000 \times 8.66\ (0 - 1.848) - 1000\ (9.81 + 5)\ (0 - 2.799) = \underline{57\ 460 \text{ Pa}}$
c) The peep hole is located at (3.696, 5.598). Use Eq. 2.5.2:
$p_A = -1.94 \times 51.96\ (0 - 3.696) - 1.94\ (32.2 + 30)\ (0 - 5.598) = \underline{1048 \text{ psf}}$
d) The peep hole is located at (4.928, 7.464). Use Eq. 2.5.2:
$p_A = -1.94 \times 25.98\ (-4.928) - 1.94\ (32.2 + 15)\ (-7.464) = \underline{932 \text{ psf}}$

2.87 a) The pressure on the end AB (z is zero at B) is, using Eq. 2.5.2,
$p(z) = -1000 \times 10\ (-7.626) - 1000 \times 9.81(z) = 76\ 260 - 9810\ z$

$$\therefore F_{AB} = \int_0^{2.5} (76\ 260 - 9810z)4dz = 640\ 000 \text{ N} \quad \text{or} \quad \underline{640 \text{ kN}}$$

b) The pressure on the bottom BC is
$p(x) = -1000 \times 10\ (x - 7.626) = 76\ 260 - 10\ 000\ x.$

$$\therefore F_{BC} = \int_0^{7.626} (76\ 260 - 10\ 000x)4dx = 1.163 \times 10^6 \text{ N} \quad \text{or} \quad \underline{1163 \text{ kN}}$$

c) On the top $p(x) = -1000 \times 10\ (x - 5.174)$ where ① is on the top surface.

$$\therefore F_{top} = \int_0^{5.174} (51\ 740 - 10\ 000x)4dx = 5.35 \times 10^5 \text{ N or } \underline{535 \text{ kN}}$$

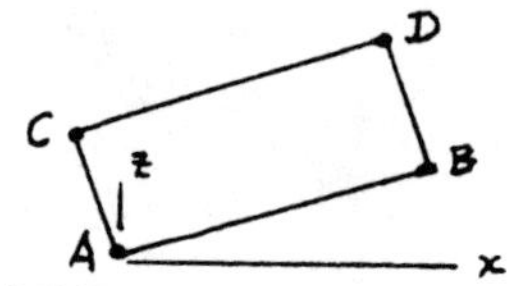

2.88 a) The pressure at A is 58.29 kPa. At B it is

$p_B = -1000 \times 17.32\ (1.732\text{–}1.232) - 1000\ (19.81)\ (1\text{–}1.866) = 8495$ Pa.

Since the pressure varies linearly over AB, we can use an average pressure times the area:

$$F_{AB} = \frac{58\ 290 + 8495}{2} \times 1.5 \times 2 = 100\ 200 \text{ N or } \underline{100.2 \text{ kN}}$$

b) $p_D = 0$. $p_C = -1000 \times 17.32\ (-.5\text{–}1.232) - 1000 \times 19.81\ (.866\text{–}1.866) = 49\ 810$ Pa.

$$F_{CD} = \frac{1}{2} \times 49\ 810 \times 1.5 \times 2 = 74\ 720 \text{ N or } \underline{74.72 \text{ kN}}.$$

c) $p_A = 58\ 290$ Pa. $p_C = 49\ 810$ Pa. $\therefore F_{AC} = \dfrac{58.29 + 49.81}{2} \times 1.5 = \underline{81.08 \text{ kN}}$.

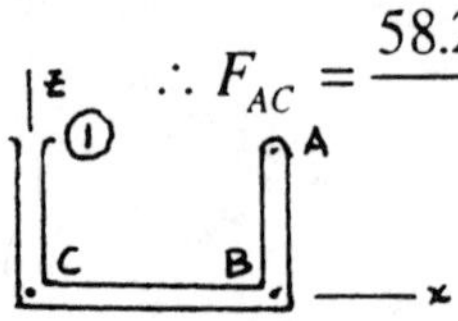

2.89 Use Eq. 2.5.2 with ① at the open end.

a) $p_A = \underline{0}$ since $z_2 = z_1$. $p_B = 1000 \times 19.81 \times 0.6 = \underline{11\ 890 \text{ Pa.}}$ $p_C = \underline{11\ 890 \text{ Pa}}$.

b) $p_A = -1000 \times 10\ (.9\text{–}0) = \underline{-9000 \text{ Pa}}$. $p_B = -1000 \times 10\ (.9) - 1000 \times 9.81(-.6) = \underline{-3114 \text{ Pa}}$

$p_C = -1000 \times 9.81 \times (-.6) = \underline{5886 \text{ Pa}}$.

c) $p_A = -1000 \times 20\ (.9) = \underline{-18\ 000 \text{ Pa}}$. $p_B = -1000 \times 20 \times .9 - 1000 \times 19.81(-.6) = \underline{-6110 \text{ Pa}}$

$p_C = \underline{11\ 890 \text{ Pa}}$

d) $p_A = \underline{0}$. $p_B = 1.94 \times (32.2\text{-}60)\left(\dfrac{25}{12}\right) = \underline{-112 \text{ psf}}$. $p_C = \underline{-112 \text{ psf}}$.

e) $p_A = 1.94 \times 60\left(-\dfrac{37.5}{12}\right) = \underline{-364 \text{ psf}}$.

$$p_B = 1.94 \times 60\left(-\frac{37.5}{12}\right) - 1.94 \times 32.2\left(-\frac{25}{12}\right) = \underline{-234 \text{ psf}}.$$

$$p_C = -1.94 \times 32.2\left(-\frac{25}{12}\right) = \underline{130 \text{ psf}}.$$

f) $p_A = 1.94 \times 30\left(\dfrac{37.5}{12}\right) = \underline{182 \text{ psf}}$.

$$p_B = -1.94(-30)\left(\frac{37.5}{12}\right) - 1.94 \times 62.2\left(-\frac{25}{12}\right) = 433 \underline{\text{ psf}}.$$

$$p_C = -1.94 \times 62.2 \times \left(-\frac{25}{12}\right) = \underline{251 \text{ psf}}.$$

2.90 Use Eq. 2.6.4 with ① at the open end.

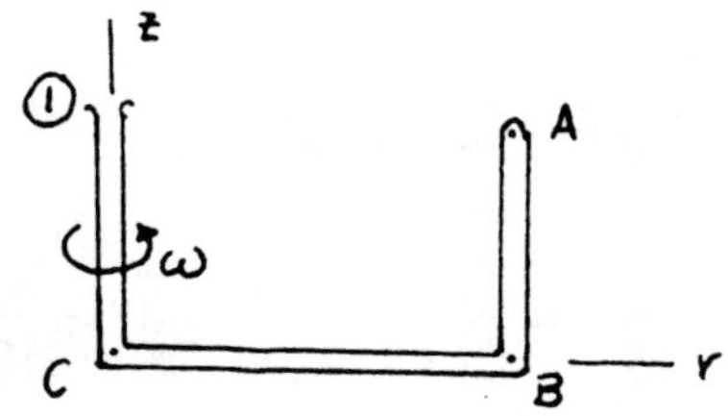

$\omega = \frac{50 \times 2\pi}{60} = 5.236$ rad/s.

a) $p_A = \frac{1000 \times 5.236^2}{2} \times (.6 \times 1.5)^2 = \underline{11\ 100 \text{ Pa}}$.

$p_B = \frac{1}{2} \times 1000 \times 5.236^2 \times .9^2 + 9810 \times .6 = \underline{16\ 990 \text{ Pa}}$. $p_C = 9810 \times .6 = \underline{5886 \text{ Pa}}$.

b) $p_A = \frac{1}{2} \times 1000 \times 5.236^2 \times .6^2 = \underline{4935 \text{ Pa.}}$

$p_B = \frac{1}{2} \times 1000 \times 5.236^2 \times .6^2 + 9810 \times .4 = \underline{8859 \text{ Pa}}$. $p_C = 9810 \times .4 = \underline{3924 \text{ Pa}}$.

c) $p_A = \frac{1}{2} \times 1.94 \times 5.236^2 \times \left(\frac{37.5}{12}\right)^2 = \underline{259.7 \text{ psf}}$.

$p_B = \frac{1}{2} \times 1.94 \times 5.236^2 \times \left(\frac{37.5}{12}\right)^2 + 62.4 \times \frac{25}{12} = \underline{389.7 \text{ psf}}$. $p_C = 62.4 \times \frac{25}{12} = \underline{130 \text{ psf}}$.

d) $p_A = \frac{1}{2} \times 1.94 \times 5.236^2 \times \left(\frac{22.5}{12}\right)^2 = \underline{93.5 \text{ psf}}$.

$p_B = \frac{1}{2} \times 1.94 \times 5.236^2 \times \left(\frac{22.5}{12}\right)^2 + 62.4 \times \frac{15}{12} = \underline{171.5 \text{ psf}}$. $p_C = 62.4 \times \frac{15}{12} = \underline{78 \text{ psf}}$.

2.91 Use Eq. 2.6.4 with ① at the open end.

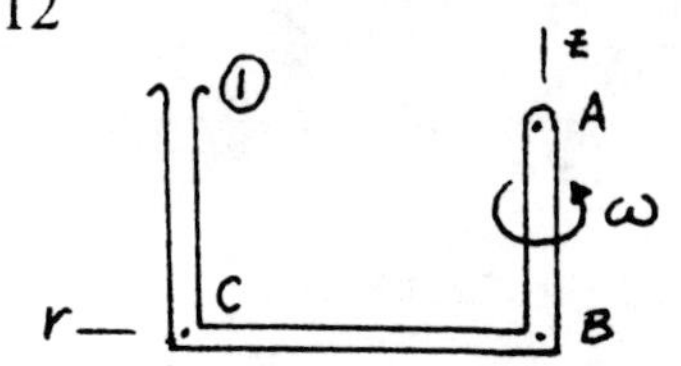

a) $p_A = \frac{1}{2} \times 1000 \times 10^2\ (0 - .9^2) = \underline{-40\ 500 \text{ Pa}}$.

$p_B = -40\ 500 + 9810 \times .6 = \underline{-34\ 600 \text{ Pa}}$. $p_C = 9810 \times .6 = \underline{5886 \text{ Pa}}$.

b) $p_A = \frac{1}{2} \times 1000 \times 10^2\ (0 - .6^2) = \underline{-18\ 000 \text{ Pa.}}$

$p_B = -18\ 000 + 9810 \times .4 = \underline{-14\ 080 \text{ Pa}}$. $p_C = 9810 \times .4 = \underline{3924 \text{ Pa}}$.

c) $p_A = \frac{1}{2} \times 1.94 \times 10^2 \left(0 - \frac{37.5^2}{144}\right) = \underline{-947 \text{ psf}}$.

$p_B = -947 + 62.4 \times \frac{25}{12} = \underline{-817 \text{ psf}}$. $p_C = 62.4 \times \frac{25}{12} = \underline{130 \text{ psf.}}$

d) $p_A = \frac{1}{2} \times 1.94 \times 10^2 \left(-\frac{22.5^2}{12^2}\right) = \underline{-341 \text{ psf}}$.

$p_B = -341 + 62.4 \times \frac{15}{12} = \underline{-263 \text{ psf}}$. $p_C = 62.4 \times \frac{15}{12} = \underline{78 \text{ psf.}}$

2.92 Use Eq. 2.6.4 with ① at the open end and ② at the origin. Given: $p_2 = 0$.

a) $0 = \frac{1}{2} \times 1000\ \omega^2\ (0 - .45^2) - 9810\ (0 - .6)$.

$\therefore \omega = \underline{7.62 \text{ rad/s}}$.

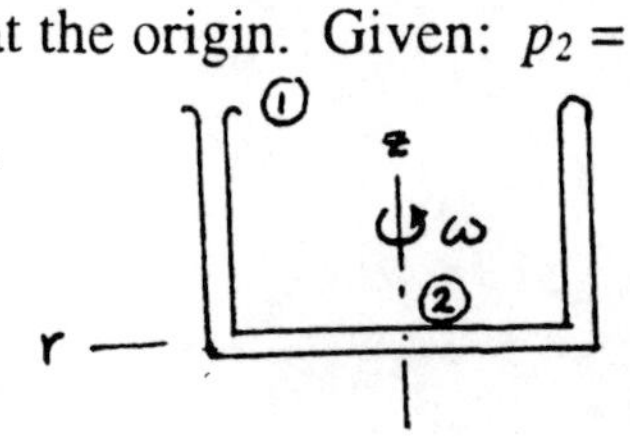

b) $0 = \frac{1}{2} \times 1000\, \omega^2 (0 - .3^2) - 9810\,(0 - .4)$. $\quad \therefore \omega = \underline{9.34 \text{ rad/s}}$.

c) $0 = \frac{1}{2} \times 1.94\, \omega^2 \left(0 - \frac{18.75^2}{12^2}\right) - 62.4\left(-\frac{25}{12}\right)$. $\quad \therefore \omega = \underline{7.41 \text{ rad/s}}$.

d) $0 = \frac{1}{2} \times 1.94\, \omega^2 \left(-\frac{11.25^2}{12^2}\right) - 62.4\left(-\frac{15}{12}\right)$. $\quad \therefore \omega = \underline{9.57 \text{ rad/s}}$.

2.93 The air volume before and after is equal.

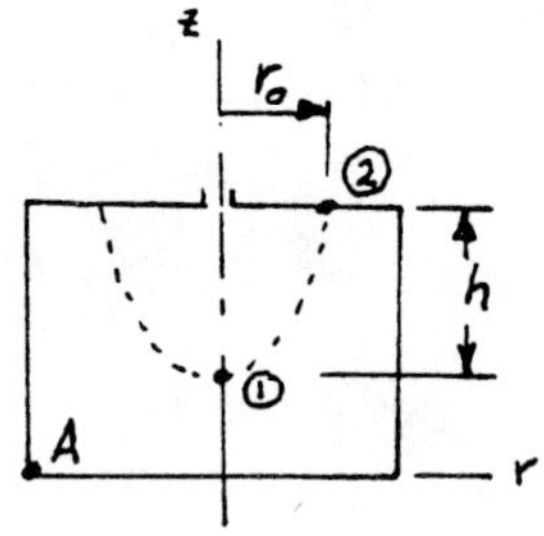

$\therefore \frac{1}{2}\pi r_0^2 h = \pi \times .6^2 \times .2$. $\quad \therefore r_0^2 h = 0.144$.

a) Using Eq. 2.6.5: $r_0^2 \times 5^2 / 2 = 9.81\, h$ $\quad \therefore h = 0.428$ m

$\therefore p_A = \frac{1}{2} \times 1000 \times 5^2 \times .6^2 - 9810\,(-.372) = \underline{8149 \text{ Pa}}$.

b) $r_0^2 \times 7^2 / 2 = 9.81\, h$. $\therefore h = 0.6$ m. $\therefore p_A = \frac{1000}{2} \times 7^2 \times .6^2 + 9810 \times .2 = \underline{10\,780 \text{ Pa}}$.

c) For $\omega = 10$, part of the bottom is bared.

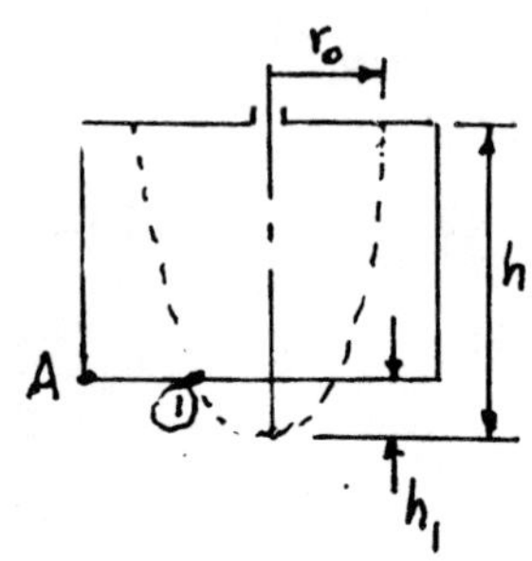

$\pi \times .6^2 \times .2 = \frac{1}{2}\pi r_0^2 h - \frac{1}{2}\pi r_1^2 h_1$.

Using Eq. 2.6.5:

$\frac{\omega^2 r_0^2}{2g} = h, \quad \frac{\omega^2 r_1^2}{2g} = h_1$.

$\therefore 0.144 = \frac{2g}{\omega^2} h^2 - \frac{2g}{\omega^2} h_1^2$ or $\quad h^2 - h_1^2 = \frac{0.144 \times 10^2}{2 \times 9.81}$.

Also, $h - h_1 = 0.8$. $\quad 1.6h - .64 = .7339$. $\quad \therefore h = 0.859$ m, $r_1 = 0.108$ m.

$\therefore p_A = \frac{1}{2} \times 1000 \times 10^2\,(.6^2 - .108^2) = \underline{17\,400 \text{ Pa}}$.

d) Following part (c): $h^2 - h_1^2 = \frac{0.144 \times 20^2}{2 \times 9.81}$. $\quad 1.6h - .64 = 2.936$. $\therefore h = 2.235$ m.

$\therefore p_A = \frac{1}{2} \times 1000 \times 20^2\,(.6^2 - .265^2) = \underline{57\,900 \text{ Pa}}$ $\quad r_1 = .265$ m

2.94 The answers to Problem 2.63 are increased by 25 000 Pa.

a) 33 150 Pa $\quad$ b) 35 780 Pa $\quad$ c) 42 400 Pa $\quad$ d) 82 900 Pa

2.95 $p(r) = \frac{1}{2}\rho\omega^2 r^2 - \rho g[0 - (.8 - h)].$

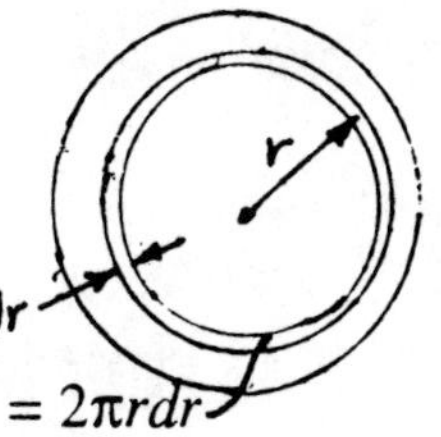

$p(r) = 500\omega^2 r^2 + 9810(.8 - h)$ if $h < .8$.

$p(r) = 500\omega^2 (r^2 - r_1^2)$ if $h > .8$.

a) $F = \int p2\pi r dr = 2\pi \int_0^{.6} (12\ 500r^3 + 3650r)dr = \underline{6670\text{ N}}$. (We used $h = .428$ m)

b) $F = \int p2\pi r dr = 2\pi \int_0^{.6} (24\ 500r^3 + 1962r)dr = \underline{7210\text{ N}}$. (We used $h = .6$ m)

c) $F = \int p2\pi r dr = 2\pi \int_{-.108}^{.6} (50\ 000(r^3 - .108^2 r)dr = \underline{9520\text{ N}}$. (We used $r_1 = .108$ m)

d) $F = \int p2\pi r dr = 2\pi \int_{.265}^{.6} (200\ 000(r^3 - .265^2 r)dr = \underline{26\ 400\text{ N}}$. (We used $r_1 = .265$ m)

THREE

Introduction to Fluids in Motion

3.1 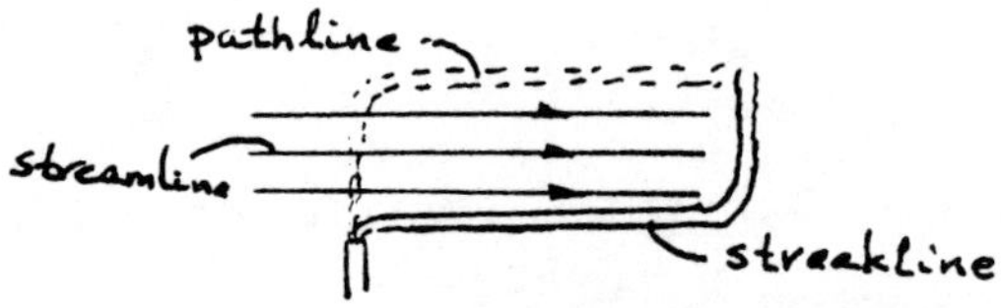

3.2 Pathlines: Release several at an instant in time and take a time exposure of the subsequent motions of the bulbs.

Sreakline: Continue to release the devises at a given location and after the last one is released, take a snapshot of the "line" of bulbs. Repeat this for several different release locations for additional streaklines.

3.3 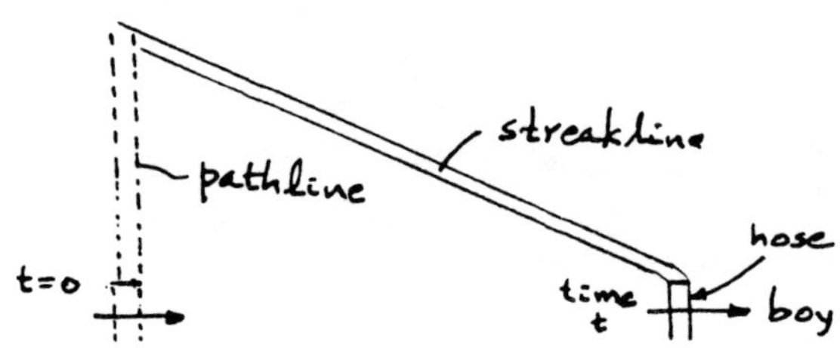

3.4 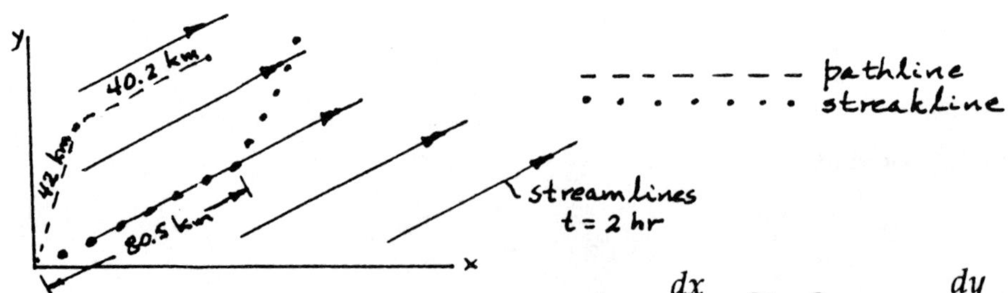

3.5 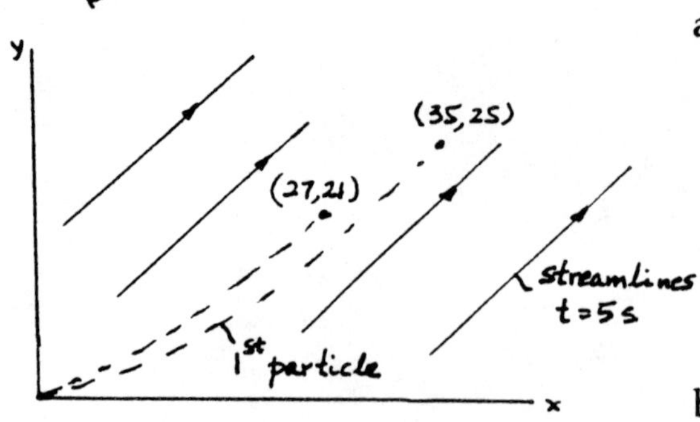

a) $u = \dfrac{dx}{dt} = 2t + 2 \qquad v = \dfrac{dy}{dt} = 2t$

$x = t^2 + 2t + \cancel{c_1}^{\,0} \qquad y = t^2 + \cancel{c_2}^{\,0}$

$= y + 2\sqrt{y}$

$\therefore x^2 - 2xy + y^2 = 4y$

$\therefore$ parabola.

b) $x = t^2 + 2t + c_1. \quad \therefore c_1 = -8$, and $c_2 = -4$.

$= y + 4 + 2(\pm\sqrt{y+4}) - 8$

$\therefore x^2 - 2xy + y^2 + 8x - 12y = 0.$

$\therefore$ parabola.

3.6 $\left.\begin{array}{l}\vec{V} = u\hat{i} + v\hat{j} + w\hat{k} \\ d\vec{r} = dx\hat{i} + dy\hat{j} + dz\hat{k}\end{array}\right\}$ $(\vec{V}\times d\vec{r})_z = \underline{udy - vdx}$ using $\hat{i}\times\hat{j} = \hat{k},\ \hat{j}\times\hat{i} = -\hat{k}$.

3.7 Lagrangian: Several college students would be hired to ride bikes around the various roads, making notes of quantities of interest.

Eulerian: Several college students would be positioned at each intersection and quantities would be recorded as a function of time.

3.8 a) At $t = 2$ and (0,0,0) $V = \sqrt{2^2} = \underline{2\text{ m/s}}$.

At $t = 2$ and (1,–2,0) $V = \sqrt{3^2 + 2^2} = \underline{3.606\text{ m/s}}$.

b) At $t = 2$ and (0,0,0) $V = \underline{0}$.

At $t = 2$ and (1,–2,0) $V = \sqrt{(-2)^2 + (-8)^2} = \underline{8.246\text{ m/s}}$.

c) At $t = 2$ and (0,0,0) $V = \sqrt{(-4)^2} = \underline{4\text{ m/s}}$.

At $t = 2$ and (1,–2,0) $V = \sqrt{2^2 + (-4)^2 + (-4)^2} = \underline{6\text{ m/s}}$.

3.9 a) $\cos\alpha = \vec{V}\cdot\hat{i}/V = (1+2)/\sqrt{3^2+2^2} = 0.832.$ $\quad\therefore \alpha = \underline{33.69^\circ}$

$\vec{V}\cdot\hat{n} = 0.\quad (3\hat{i} + 2\hat{j})\cdot(n_x\hat{i} + n_y\hat{j}) = 0.\quad \left.\begin{array}{r}3n_x + 2n_y = 0 \\ n_x^2 + n_y^2 = 1\end{array}\right\}\ \therefore\ \begin{array}{l} n_y = -\frac{3}{2}n_x \\ n_x^2 + \frac{9}{4}n_x^2 = 1\end{array}$

$\therefore n_x = \frac{2}{\sqrt{13}},\ n_y = -\frac{3}{\sqrt{13}}$ or $\underline{\hat{n} = \frac{1}{\sqrt{13}}(2\hat{i} - 3\hat{j})}$.

b) $\cos\alpha = \vec{V}\cdot\hat{i}/V = -2/\sqrt{(-2)^2 + (-8)^2} = -0.2425.$ $\quad\therefore \alpha = \underline{104^\circ}$

$\vec{V}\cdot\hat{n} = 0.\quad (-2\hat{i} - 8\hat{j})\cdot(n_x\hat{i} + n_y\hat{j}) = 0.\quad \left.\begin{array}{r}-2n_x - 8n_y = 0 \\ n_x^2 + n_y^2 = 1\end{array}\right\}\ \therefore\ \begin{array}{l} n_x = -4n_y \\ 16n_y^2 + n_y^2 = 1\end{array}$

$\therefore n_y = \frac{1}{\sqrt{17}},\ n_x = -\frac{4}{\sqrt{17}}$ or $\underline{\hat{n} = \frac{1}{\sqrt{17}}(-4\hat{i} + \hat{j})}$.

c) $\cos\alpha = \vec{V}\cdot\hat{i}/V = 5/\sqrt{5^2 + (-8)^2} = 0.6202.$ $\quad\therefore \alpha = \underline{-51.67^\circ}$

$\vec{V}\cdot\hat{n} = 0.\quad (5\hat{i} - 8\hat{j})\cdot(n_x\hat{i} + n_y\hat{j}) = 0.\quad \left.\begin{array}{r}5n_x - 8n_y = 0 \\ n_x^2 + n_y^2 = 1\end{array}\right\}\ \therefore\ \begin{array}{l} n_x = \frac{8}{5}n_y \\ \frac{64}{25}n_y^2 + n_y^2 = 1\end{array}$

$\therefore n_y = \frac{5}{\sqrt{89}},\ n_x = \frac{8}{\sqrt{89}}$ or $\underline{\hat{n} = \frac{1}{\sqrt{89}}(8\hat{i} + 5\hat{j})}$.

3.10 a) $\vec{V} \times d\vec{r} = 0.$ $\left[(x+2)\hat{i} + xt\hat{j}\right] \times (dx\hat{i} + dy\hat{j}) = 0.$

$\therefore (x+2)dy - xtdx = 0$ or $t\dfrac{xdx}{x+2} = dy.$

Integrate: $t\int \dfrac{xdx}{x+2} = \int dy.$ $t\left[x - 2\ell n|x+2|\right] = y + C.$

$2(1 - 2\ell n3) = -2 + C.$ $\therefore C = 0.8028.$

$\underline{t\left[x - 2\ell n|x+2|\right] = y + 0.8028}$

b) $\vec{V} \times d\vec{r} = 0.$ $\left[xy\hat{i} - 2y^2\hat{j}\right] \times (dx\hat{i} + dy\hat{j}) = 0.$

$\therefore xydy + 2y^2dx = 0$ or $\dfrac{2dx}{x} = -\dfrac{dy}{y}.$

Integrate: $2\ell nx = -\ell n(y/C).$ $2\ell n(1) = -\ell n(-2/C).$

$\therefore C = -2.$ $\ell nx^2 = -\ell n(y/-2).$ $\therefore \underline{x^2y = -2.}$

c) $\vec{V} \times d\vec{r} = 0.$ $\left[(x^2+4)\hat{i} - y^2t\hat{j}\right] \times (dx\hat{i} + dy\hat{j}) = 0.$

$(x^2+4)dy + y^2tdx = 0$ or $\dfrac{tdx}{x^2+4} = -\dfrac{dy}{y^2}.$

Integrate: $\dfrac{t}{2}\left(\tan^{-1}\dfrac{x}{2} + C\right) = \dfrac{1}{y}.$ $\dfrac{2}{2}\left(\tan^{-1}\dfrac{1}{2} + C\right) = -\dfrac{1}{2}.$

$\therefore C = -0.9636.$ $\underline{yt\left(\tan^{-1}\dfrac{x}{2} - 0.9636\right) = 2}$

3.11 a) $\dfrac{D\vec{V}}{Dt} = u\dfrac{\partial\vec{V}}{\partial x} + v\dfrac{\partial\vec{V}}{\partial y} + w\dfrac{\partial\vec{V}}{\partial z} + \dfrac{\partial\vec{V}}{\partial t} = \underline{0}.$

b) $u\dfrac{\partial\vec{V}}{\partial x} + v\dfrac{\partial\vec{V}}{\partial y} + w\dfrac{\partial\vec{V}}{\partial z} + \dfrac{\partial\vec{V}}{\partial t} = 2x(2\hat{i}) + 2y(2\hat{j}) = 4x\hat{i} + 4y\hat{j} = \underline{8\hat{i} - 4\hat{j}}$

c) $u\dfrac{\partial\vec{V}}{\partial x} + v\dfrac{\partial\vec{V}}{\partial y} + w\dfrac{\partial\vec{V}}{\partial z} + \dfrac{\partial\vec{V}}{\partial t} = x^2t(2xt\hat{i} + 2yt\hat{j}) + 2xyt(2xt\hat{j} + 2zt\hat{k}) + x^2\hat{i} + 2xy\hat{j}$

$+2yz\hat{k} = \underline{68\hat{i} - 100\hat{j} - 54\hat{k}}$

d) $u\dfrac{\partial\vec{V}}{\partial x} + v\dfrac{\partial\vec{V}}{\partial y} + w\dfrac{\partial\vec{V}}{\partial z} + \dfrac{\partial\vec{V}}{\partial t} = x(\hat{i} - 2yz\hat{j}) - 2xyz(-2xz\hat{j}) + tz(-2xy\hat{j} + t\hat{k}) + z\hat{k}$

$= x\hat{i} - (2yz - 4x^2yz^2 + 2xyzt)\hat{j} + (zt^2 + z)\hat{k}$

$= \underline{2\hat{i} - 114\hat{j} + 15\hat{k}}$

3.12 $\vec{\Omega} = \frac{1}{2}\left(\frac{\partial w}{\partial y} - \frac{\partial v}{\partial z}\right)\hat{i} + \frac{1}{2}\left(\frac{\partial u}{\partial z} - \frac{\partial w}{\partial x}\right)\hat{j} + \frac{1}{2}\left(\frac{\partial v}{\partial x} - \frac{\partial u}{\partial y}\right)\hat{k}.$

a) $\vec{\Omega} = -\frac{1}{2}\frac{\partial u}{\partial y}\hat{k} = 20y\hat{k} = \underline{-20\hat{k}}$

b) $\vec{\Omega} = \frac{1}{2}(0-0)\hat{i} + \frac{1}{2}(0-0)\hat{j} + \frac{1}{2}(0-0)\hat{k} = \underline{0}$

c) $\vec{\Omega} = \frac{1}{2}(2zt-0)\hat{i} + \frac{1}{2}(0-0)\hat{j} + \frac{1}{2}(2yt-0)\hat{k} = \underline{6\hat{i} - 2\hat{k}}$

d) $\vec{\Omega} = \frac{1}{2}(0+2xy)\hat{i} + \frac{1}{2}(0-0)\hat{j} + \frac{1}{2}(-2yz-0)\hat{k} = \underline{-2\hat{i} + 3\hat{k}}$

3.13 The vorticity $\vec{\omega} = 2\vec{\Omega}$. Using the results of Problem 3.7:

a) $\vec{\omega} = \underline{-40\hat{i}}$ b) $\vec{\omega} = \underline{0}$ c) $\vec{\omega} = \underline{12\hat{i} - 4\hat{k}}$ d) $\vec{\omega} = \underline{-4\hat{i} + 6\hat{k}}$

3.14 a) $\varepsilon_{xx} = \frac{\partial u}{\partial x} = 0, \quad \varepsilon_{yy} = \frac{\partial v}{\partial y} = 0, \quad \varepsilon_{zz} = \frac{\partial w}{\partial z} = 0.$

$$\varepsilon_{xy} = \frac{1}{2}\left(\frac{\partial u}{\partial y} + \frac{\partial v}{\partial x}\right) = -20y = 20, \quad \varepsilon_{xz} = \frac{1}{2}\left(\frac{\partial u}{\partial z} + \frac{\partial w}{\partial x}\right) = 0,$$

$$\varepsilon_{yz} = \frac{1}{2}\left(\frac{\partial v}{\partial z} + \frac{\partial w}{\partial y}\right) = 0. \qquad \therefore \text{rate-of strain} = \begin{bmatrix} 0 & 20 & 0 \\ 20 & 0 & 0 \\ 0 & 0 & 0 \end{bmatrix}$$

b) $\varepsilon_{xx} = 2, \ \varepsilon_{yy} = 2, \ \varepsilon_{zz} = 0.$
$\varepsilon_{xy} = 0, \ \varepsilon_{xz} = 0, \ \varepsilon_{yz} = 0.$

$$\text{rate-of strain} = \begin{bmatrix} 2 & 0 & 0 \\ 0 & 2 & 0 \\ 0 & 0 & 0 \end{bmatrix}$$

c) $\varepsilon_{xx} = 2xt = 8, \ \varepsilon_{yy} = 2xt = 8, \ \varepsilon_{zz} = 2yt = -4.$

$$\varepsilon_{xy} = \frac{1}{2}(2yt) = -2, \ \varepsilon_{xz} = \frac{1}{2}(0) = 0, \ \varepsilon_{yz} = \frac{1}{2}(2zt) = 6.$$

$$\text{rate-of strain} = \begin{bmatrix} 8 & -2 & 0 \\ -2 & 8 & 6 \\ 0 & 6 & -4 \end{bmatrix}$$

d) $\varepsilon_{xx} = 1,\ \varepsilon_{yy} = -2xz = -12,\ \varepsilon_{zz} = t = 2.$

$$\varepsilon_{xy} = \frac{1}{2}(-2yz) = 3,\ \ \varepsilon_{xz} = \frac{1}{2}(0) = 0,\ \ \varepsilon_{yz} = \frac{1}{2}(-2xy) = 2.$$

$$\text{rate-of strain} = \begin{bmatrix} 1 & 3 & 0 \\ 3 & -12 & 2 \\ 0 & 2 & 2 \end{bmatrix}$$

3.15 a) $a_r = \left(10 - \frac{40}{r^2}\right)\cos\theta\left(\frac{80}{r^3}\right)\cos\theta - \left(10 + \frac{40}{r^2}\right)\frac{\sin\theta}{r}\left(1 - \frac{40}{r^2}\right)(-\sin\theta)$

$$-\frac{1}{r}\left(10 + \frac{40}{r^2}\right)^2 \sin^2\theta = (10 - 2.5)(-1)1.25(-1) = \underline{9.375 \text{ m/s}^2}.$$

$$a_\theta = \left(10 - \frac{40}{r^2}\right)\cos\theta\left(\frac{80}{r^3}\right)\sin\theta + \left(10 + \frac{40}{r^2}\right)\frac{\sin\theta}{r}\left(10 + \frac{40}{r^2}\right)\cos\theta$$

$$-\frac{1}{r}\left(100 - \frac{1600}{r^4}\right)\sin\theta\cos\theta = \underline{0} \quad \text{since } \sin 180° = 0.$$

$a_\phi = \underline{0}.$

b) $\omega_r = 0,\ \omega_\theta = 0,\ \omega_z = \frac{1}{r}\left(-10 + \frac{40}{r^2}\right)\sin\theta - \frac{1}{r}\left(10 - \frac{40}{r^2}\right)(-\sin\theta) = 0.$

At (4, 180°) $\quad \vec{\omega} = \underline{0} \quad$ since $\vec{\omega} = 0$ everywhere.

3.16 a) $a_r = \left(10 - \frac{80}{r^3}\right)\cos\theta\left(\frac{240}{r^4}\right)\cos\theta - \left(10 + \frac{80}{r^3}\right)\frac{\sin\theta}{r}(-\sin\theta)\left(10 - \frac{80}{r^3}\right)$

$$-\left(10 + \frac{80}{r^3}\right)^2 \frac{\sin^2\theta}{r} = 8.75(-1)(.9375)(-1) = \underline{8.203 \text{ m/s}^2}$$

$a_\theta = \underline{0}$ since sin 180° = 0. $\quad a_\phi = \underline{0}$ since $v_\phi = 0$.

b) $\omega_r = \underline{0},\ \omega_\theta = \underline{0},\ \omega_\phi = \underline{0},$ since sin 180° = 0.

3.17 $\vec{a} = \frac{\partial \vec{V}}{\partial t} + u\frac{\partial \vec{V}}{\partial x} + v\frac{\partial \vec{V}}{\partial y} + w\frac{\partial \vec{V}}{\partial z} = \underline{\frac{\partial u}{\partial t}\hat{i}}.$ (terms $u\frac{\partial \vec{V}}{\partial x}$, v, w struck out and marked 0)

For steady flow $\partial u / \partial t = 0$ so that $\underline{\vec{a} = 0}.$

3.18 Assume $u(r,x)$ and $v(r,x)$ are not zero. Then, replacing z with x in Eqs. 3.2.13 and recognizing that $v_\theta = 0$ and $\partial / \partial\theta = 0$:

$$a_r = v\frac{\partial v}{\partial r} + u\frac{\partial v}{\partial x}$$

$$a_x = v\frac{\partial u}{\partial r} + u\frac{\partial u}{\partial x}$$

3.19 a) $u = 2(1-0)(1-e^{-t/10}) = \underline{2\text{ m/s}}$ at $t = \infty$.

$$a_x = \frac{\partial u}{\partial t} = 2(1-0)\left(\frac{1}{10}\right)\left(e^{-t/10}\right) = \underline{0.2\text{ m/s}^2} \text{ at } t = 0.$$

b) $u = 2(1-0.5^2)(1-e^{-t/10}) = \underline{1.875\text{ m/s}}$ at $t = \infty$.

$$a_x = 2(1-0.5^2/2^2)\left(\frac{1}{10}e^{-t/10}\right) = \underline{0.0125\text{ m/s}^2} \text{ at } t = 0.$$

c) $u = 2(1-2^2/2^2)(1-e^{-t/10}) = \underline{0}$ for all t.

$$a_x = 2(1-2^2/2^2)\left(\frac{1}{10}e^{-t/10}\right) = \underline{0} \text{ for all } t.$$

3.20 $$\frac{DT}{Dt} = u\frac{\partial T}{\partial x} + v\frac{\partial T}{\partial y} + w\frac{\partial T}{\partial z} + \frac{\partial T}{\partial t} = 20(1-y^2)\left(-\frac{\pi}{100}\right)\sin\frac{\pi t}{100} = -\frac{\pi}{5}\times 0.5878$$

$$= \underline{-0.3693\ ^\circ\text{C/s}}.$$

3.21 $$\frac{D\rho}{Dt} = u\frac{\partial \rho}{\partial x} + v\frac{\partial \rho}{\partial y} + w\frac{\partial \rho}{\partial z} + \frac{\partial \rho}{\partial t} = 10(-1.23\times 10^{-4}e^{-3000\times 10^{-4}}) = \underline{-9.11\times 10^{-4}\frac{\text{kg}}{\text{m}^3\cdot\text{s}}}.$$

3.22 $$\frac{D\rho}{Dt} = u\frac{\partial \rho}{\partial x} + v\frac{\partial \rho}{\partial y} + w\frac{\partial \rho}{\partial z} + \frac{\partial \rho}{\partial t} = 10\left(-\frac{1000}{4}\right) = \underline{-2500\frac{\text{kg}}{\text{m}^3\cdot\text{s}}}.$$

3.23 $$\frac{D\rho}{Dt} = u\frac{\partial \rho}{\partial x} = 4\times(.01) = \underline{0.04\text{ kg/m}^3\text{·s}}$$

3.24 $\dfrac{D}{Dt} = \vec{V}\cdot\vec{\nabla} + \dfrac{\partial}{\partial t}$ observing that the dot product of two vectors $\vec{A} = A_x\hat{i} + A_y\hat{j} + A_z\hat{k}$ and $\vec{B} = B_x\hat{i} + B_y\hat{j} + B_z\hat{k}$ is $\vec{A}\cdot\vec{B} = A_xB_x + A_yB_y + A_zB_z$.

3.25 $$\left.\begin{aligned} a_x &= \frac{\partial u}{\partial t} + \vec{V}\cdot\vec{\nabla}u \\ a_y &= \frac{\partial v}{\partial t} + \vec{V}\cdot\vec{\nabla}v \\ a_z &= \frac{\partial w}{\partial t} + \vec{V}\cdot\vec{\nabla}w \end{aligned}\right\} \quad \therefore \vec{a} = \frac{\partial \vec{V}}{\partial t} + (\vec{V}\cdot\vec{\nabla})\vec{V}$$

3.26 Using Eq. 3.2.15:

a) $$\vec{A} = \vec{a} + \frac{d^2\vec{s}}{dt^2} + 2\vec{\Omega}\times\vec{V} + \vec{\Omega}\times(\vec{\Omega}\times\vec{r}) + \frac{d\vec{\Omega}}{dt}\times\vec{r}$$

$$= 2(20\hat{k}\times 4\hat{i}) + 20\hat{k}\times(20\hat{k}\times 1.5\hat{i}) = \underline{160\hat{j} - 600\hat{i}\text{ m}^2/\text{s}}$$

b) $$\vec{A} = 2\vec{\Omega}\times\vec{V} + \vec{\Omega}\times(\vec{\Omega}\times\vec{r}) = 2(20\hat{k}\times -20\cos 30^\circ\hat{j}) + 20\hat{k}\times(20\hat{k}\times 3\hat{i}) = \underline{-507\hat{i}}$$

3.27 $\vec{\Omega} = \dfrac{2\pi}{24 \times 60 \times 60}\hat{k} = 7.272 \times 10^{-5}\hat{k}$ rad/s.

$\vec{v} = 5(-.707\hat{i} - .707\hat{k}) = -3.535\hat{i} - 3.535\hat{k}$ m/s.

$\vec{A} = 2\vec{\Omega} \times \vec{V} + \vec{\Omega} \times (\vec{\Omega} \times \vec{r})$

$= 2 \times 7.272 \times 10^{-5}\hat{k} \times (-3.535\hat{i} - 3.535\hat{k}) + 7.272 \times 10^{-5}\hat{k} \times$

$[7.272 \times 10^{-5}\hat{k} \times 6 \times 10^{6}(-.707\hat{i} + .707\hat{k})] = \underline{-51.4 \times 10^{-5}\hat{j} + 0.0224\hat{i} \text{ m/s}^2}.$

Note: We have neglected the acceleration of the earth relative to the sun since it is quite small (it is $d^2\bar{s}/dt^2$). The component $(-51.4 \times 10^{-5}\hat{j})$ is the Coriolis acceleration and causes air motions to move c.w. or c.c.w. in the two hemispheres.

3.28 a) two-dimensional (r,z)
b) two-dimensional (x,y)
c) two-dimensional (r,z)
d) two-dimensional (r,z)
e) three-dimensional (x,y,z)
f) three-dimensional (x,y,z)
g) two-dimensional (r,z)
h) one-dimensional (r)

3.29 Steady: a, c, e, f, h
Unsteady: b, d, g

3.30 b. It is an unsteady plane flow.

3.31 a) 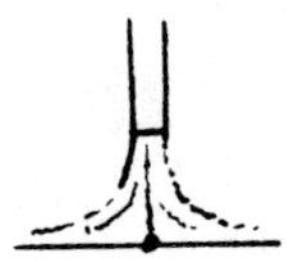 d)  e) 

3.32 f, h

3.33 a) inviscid.
b) inviscid.
c) inviscid.
d) viscous inside the boundary layer.
e) viscous inside the boundary layers and separated regions.
f) viscous.
g) viscous.
h) viscous.

3.34 d and e. Each flow possesses a stagnation point.

3.35 Re = $VL/\nu = 2 \times .015/.77 \times 10^{-6} = 39\,000$. $\therefore$ Turbulent.

3.36 $\text{Re} = \dfrac{VL}{\nu} = .2 \times .8/1.4 \times 10^{-5} = 11\,400$. $\therefore$ Turbulent.

3.37 $\text{Re} = \dfrac{VL}{\nu} = \dfrac{4 \times .06}{1.7 \times 10^{-5}} = 14\,100$. $\therefore$ Turbulent.

Note: We used the smallest dimension to be safe!

3.38 a) $\text{Re} = \dfrac{VD}{\nu} = \dfrac{1.2 \times 0.01}{1.51 \times 10^{-5}} = 795$. Always laminar.

b) $\text{Re} = \dfrac{VD}{\nu} = \dfrac{1.2 \times 1}{1.51 \times 10^{-5}} = 79\,500$. May not be laminar.

3.39 $\text{Re} = 3 \times 10^5 = Vx_T/\nu$. $\nu = \mu/\rho$ where $\mu = \mu(T)$.

a) $T = 223$ K or -50°C. $\therefore \mu = 1.5 \times 10^{-5}\ \text{N}\cdot\text{s/m}^2$.

$$\therefore \nu = \frac{1.5 \times 10^{-5}}{.3376 \times 1.23} = 2.5 \times 10^{-5}\ \text{m}^2/\text{s}.$$

$$3 \times 10^5 = \frac{900 \times 1000 x_T}{3600 \times 2.5 \times 10^{-5}}. \quad \therefore x_T = 0.03 \text{ m} \quad \text{or} \quad \underline{3 \text{ cm}}$$

b) $T = -48$°F. $\therefore \mu = 3.3 \times 10^{-7}$ lb-sec/ft². $\nu = \dfrac{3.3 \times 10^{-7}}{.00089} = 3.7 \times 10^{-4}$ ft²/sec.

$$3 \times 10^5 = \frac{600 \times 5280 x_T}{3600 \times 3.7 \times 10^{-4}}. \quad \therefore x_T = 0.13' \quad \text{or} \quad \underline{1.5''}$$

3.40 Assume the flow is parallel to the leaf. Then $3 \times 10^5 = Vx_T/\nu$.

$\therefore x_T = 3 \times 10^5 \nu / V = 3.5 \times 10^5 \times 1.4 \times 10^{-4} / 6 = 8.17$ m.

The flow is expected to be laminar.

3.41 a) $M = \dfrac{V}{c} = \dfrac{100}{\sqrt{1.4 \times 287 \times 236}} = 0.325$. For accurate calculations the flow is compressible. Assume incompressible flow if an error of 4%, or so, is acceptable.

b) $M = \dfrac{V}{c} = \dfrac{80}{\sqrt{1.4 \times 287 \times 288}} = 0.235$. $\therefore$ Assume incompressible.

c) $M = \dfrac{V}{c} = \dfrac{100}{\sqrt{1.4 \times 287 \times 373}} = 0.258$. $\therefore$ Assume incompressible.

3.42 $\frac{D\rho}{Dt} = u\frac{\partial \rho}{\partial x} + v\frac{\partial \rho}{\partial y} + w\frac{\partial \rho}{\partial z} + \frac{\partial \rho}{\partial t} = 0.$ For a steady, plane flow $\partial \rho / \partial t = 0$ and $w = 0$. Then

$$u\frac{\partial \rho}{\partial x} + v\frac{\partial \rho}{\partial y} = 0.$$

3.43 $\frac{D\rho}{Dt} = u\frac{\partial \rho}{\partial x} + v\frac{\partial \rho}{\partial y} + w\frac{\partial \rho}{\partial z} + \frac{\partial \rho}{\partial t} = 0.$ (each term struck through, $\to 0$) $\quad \therefore$ incompressible.

3.44 $\frac{V^2}{2} = \frac{p}{\rho}.$ Use $\rho = 1.12$ kg/m³.

a) $v = \sqrt{2p/\rho} = \sqrt{2 \times 2000 / 1.12} = \underline{59.8 \text{ m/s}}.$

b) $v = \sqrt{2p/\rho} = \sqrt{2 \times 6000 / 1.12} = \underline{104 \text{ m/s}}.$

c) $v = \sqrt{2p/\rho} = \sqrt{2 \times 600 / 1.12} = \underline{32.7 \text{ m/s}}.$

3.44E $\frac{V^2}{2} = \frac{p}{\rho}.$ Use $\rho = 0.0021$ slug/ft³.

a) $v = \sqrt{2p/\rho} = \sqrt{2 \times .3 \times 144/.0021} = \underline{203 \text{ ft/sec}}.$

b) $v = \sqrt{2p/\rho} = \sqrt{2 \times .9 \times 144/.0021} = \underline{351 \text{ ft/sec}}.$

c) $v = \sqrt{2p/\rho} = \sqrt{2 \times .09 \times 144/.0021} = \underline{111 \text{ ft/sec}}.$

3.45 $p = \rho\frac{V^2}{2} = 1.23\left(\frac{120 \times 1000}{3600}\right)^2 / 2 = 683 \text{ Pa}.$

$\therefore F = pA = 683\,\pi \times .075^2 = \underline{12.1 \text{ N}}.$

3.46 $\frac{V^2}{2} + \frac{p}{\rho} = 0.$ $\quad \therefore V = \sqrt{\frac{-2p}{\rho}} = \sqrt{\frac{2 \times 2000}{1.23}} = \underline{57.0 \text{ m/s}}$

3.47 a) $\frac{V^2}{2} + \frac{p}{\rho} = \frac{V_0^2}{2} + \frac{p_o}{\rho}.$ ($V_0 \to 0$) $\quad \frac{(-10x)^2}{2} + \frac{p}{\rho} = \frac{p_o}{\rho}.$ $\quad \therefore p = \underline{p_o - 50x^2\rho}$

b) $\frac{V^2}{2} + \frac{p}{\rho} = \frac{V_0^2}{2} + \frac{p_o}{\rho}.$ ($V_0 \to 0$) $\quad \frac{(10y)^2}{2} + \frac{p}{\rho} = \frac{p_o}{\rho}.$ $\quad \therefore p = \underline{p_o - 50y^2\rho}$

3.48 $\frac{V^2}{2}+\frac{p}{\rho}=\frac{U_\infty^2}{2}+\frac{p_\infty}{\rho}$. ($p_\infty$ struck out, $= 0$)

a) $v_\theta=0$ and $\theta=180°$: $v_r=U_\infty(1-r_c^2/r^2)(-1)$.

$$\therefore p=\frac{\rho}{2}\left(U_\infty^2-v_r^2\right)=\underline{\frac{\rho}{2}U_\infty^2\left[2\frac{r_c^2}{r^2}-\left(\frac{r_c}{r}\right)^4\right]}.$$

b) Let $r=r_c$: $\underline{p_T=\frac{\rho}{2}U_\infty^2}$

c) $v_r=0$ and $r=r_c$: $v_\theta=-U_\infty 2\sin\theta$. $\therefore p=\frac{\rho}{2}\left(U_\infty^2-v_\theta^2\right)=\underline{\frac{\rho}{2}U_\infty^2[1-4\sin^2\theta]}$

d) Let $\theta=90°$: $\underline{p_{90}=-\frac{3}{2}\rho U_\infty^2}$

3.49 $\frac{V^2}{2}+\frac{p}{\rho}=\frac{U_\infty^2}{2}+\frac{p_\infty}{\rho}$. ($p_\infty$ struck out, $= 0$)

a) $v_\theta=0$ and $\theta=180°$: $p=\frac{\rho}{2}\left(U_\infty^2-v_r^2\right)=\underline{\frac{\rho}{2}U_\infty^2\left[2\left(\frac{r_c}{r}\right)^3-\left(\frac{r_c}{r}\right)^6\right]}.$

b) Let $r=r_c$: $\underline{p_T=\frac{1}{2}\rho U_\infty^2}$.

c) $v_r=0$ and $r=r_c$: $p=\frac{\rho}{2}\left(U_\infty^2-v_\theta^2\right)=\underline{\frac{\rho}{2}U_\infty^2[1-4\sin^2\theta]}$

d) Let $\theta=90°$: $\underline{p_{90}=-\frac{3}{2}\rho U_\infty^2}$

3.50 $\frac{V^2}{2}+\frac{p}{\rho}=\frac{U_\infty^2}{2}+\frac{p_\infty}{\rho}$. ($p_\infty$ struck out, $= 0$)

a) $p=\frac{\rho}{2}\left(U_\infty^2-u^2\right)=\frac{\rho}{2}\left[10^2-\left(10+\frac{20\pi}{2\pi x}\right)^2\right]=50\rho\left[1-\left(1+\frac{1}{x}\right)^2\right]$

$$=\underline{-50\rho\left(\frac{2}{x}+\frac{1}{x^2}\right)}$$

b) $u=0$ when $x=-1$. $p_{-1}=-50\rho(-2+1)=\underline{50\rho}$

c) $p=\frac{\rho}{2}\left(U_\infty^2-u^2\right)=\frac{\rho}{2}\left[30^2-\left(30+\frac{60\pi}{2\pi x}\right)^2\right]=450\rho\left[1-\left(1+\frac{1}{x}\right)^2\right]$

$$=\underline{-450\rho\left(\frac{2}{x}+\frac{1}{x^2}\right)}$$

d) $u=0$ when $x=-1$. $p_{-1}=-450\rho(-2+1)=\underline{450\rho}$

3.51 $\frac{V_1^2}{2}+\frac{p_1}{\rho}=\frac{V_2^2}{2}+\frac{p_2}{\rho}$. $V_1 = 0$ and $p_1 - p_2 = 20$ kPa.

$$V_2^2=\frac{2}{\rho}(p_1-p_2)=\frac{2}{1000}(20\ 000)=40. \quad \therefore V_2=\underline{6.32 \text{ m/s}}$$

3.52 Assume the velocity in the plenum is zero. Then

$$\frac{V_1^2}{2}+\frac{p_1}{\rho}=\frac{V_2^2}{2}+\frac{p_2}{\rho} \text{ or } V_2^2=\frac{2}{1.13}(60-10.2). \quad \therefore V_2=\underline{9.39 \text{ m/s}}$$

We found $\rho = 1.13 \text{ kg/m}^3$ in Table B.2.

3.53 Bernoulli from the stream to the pitot probe:

$$p_T=\rho\frac{V^2}{2}+p.$$

Manometer: $p_T+\gamma H-\gamma_{Hg}H-\gamma h=p-\gamma h.$

Then, $\rho\frac{V^2}{2}+p+\gamma H-\gamma_{Hg}H=p$

$$V^2=\frac{\gamma_{Hg}-\gamma}{\rho}(2H)$$

a) $V^2=\frac{(13.6-1)9800}{1000}(2\times 0.04). \quad \therefore V=\underline{3.14 \text{ m/s}}$

b) $V^2=\frac{(13.6-1)9800}{1000}(2\times 0.1). \quad \therefore V=\underline{4.97 \text{ m/s}}$

c) $V^2=\frac{(13.6-1)62.4}{1.94}(2\times 2/12). \quad \therefore V=\underline{11.62 \text{ fps}}$

d) $V^2=\frac{(13.6-1)62.4}{1.94}(2\times 4/12). \quad \therefore V=\underline{16.44 \text{ fps}}$

3.54 The pressure at 90° from Problem 3.48 is $p_{90}=-\frac{3}{2}\rho U_\infty^2$. The pressure at the stagnation point is $p_T=\frac{1}{2}\rho U_\infty^2$. The manometer provides:

$$p_T-\gamma H=p_{90}$$

$$\frac{1}{2}\times 1.204U_\infty^2-9800\times 0.04=-\frac{3}{2}\times 1.204U_\infty^2. \quad \therefore U_\infty=\underline{12.76 \text{ m/s}}$$

3.55 The pressure at 90° from Problem 3.49 is $p_{90} = -\frac{3}{2}\rho U_\infty^2$. The pressure at the stagnation point is $p_T = \frac{1}{2}\rho U_\infty^2$. The manometer provides:

$$p_T - \gamma H = p_{90}$$

$$\frac{1}{2} \times 1.204 U_\infty^2 - 9800 \times 0.04 = -\frac{3}{2} \times 1.204 U_\infty^2. \quad \therefore U_\infty = \underline{12.76 \text{ m/s}}$$

3.56 Assume an incompressible flow with point 1 outside in the room where $p_1 = 0$ and $v_1 = 0$. The Bernoulli's equation gives, with $p_2 = \gamma_w h_2$,

$$\frac{V_1^2}{2} + \frac{p_1}{\rho} = \frac{V_2^2}{2} + \frac{p_2}{\rho}.$$

a) $0 = \frac{V_2^2}{2} + \frac{-9800 \times 0.02}{1.204}. \quad \therefore V_2 = \underline{18.04 \text{ m/s}}$

b) $0 = \frac{V_2^2}{2} + \frac{-9800 \times 0.08}{1.204}. \quad \therefore V_2 = \underline{36.1 \text{ m/s}}$

c) $0 = \frac{V_2^2}{2} + \frac{-62.4 \times 1/12}{0.00233}. \quad \therefore V_2 = \underline{66.8 \text{ fps}}$

d) $0 = \frac{V_2^2}{2} + \frac{-62.4 \times 4/12}{0.00233}. \quad \therefore V_2 = \underline{133.6 \text{ fps}}$

3.57 Assume incompressible flow ($V < 100$ m/s) with point 1 outside the wind tunnel where $p_1 = 0$ and $V_1 = 0$. Bernoulli's equation gives

$$0 = \frac{V_2^2}{2} + \frac{p_2}{\rho}. \quad \therefore p_2 = -\frac{1}{2}\rho V_2^2$$

a) $\rho = \frac{p}{RT} = \frac{90}{0.287 \times 253} = 1.239 \text{ kg/m}^3. \quad \therefore p_2 = -\frac{1}{2} \times 1.239 \times 100^2 = \underline{-6195 \text{ Pa}}$

b) $\rho = \frac{p}{RT} = \frac{95}{0.287 \times 273} = 1.212 \text{ kg/m}^3. \quad \therefore p_2 = -\frac{1}{2} \times 1.212 \times 100^2 = \underline{-6060 \text{ Pa}}$

c) $\rho = \frac{p}{RT} = \frac{92}{0.287 \times 293} = 1.094 \text{ kg/m}^3. \quad \therefore p_2 = -\frac{1}{2} \times 1.094 \times 100^2 = \underline{-5470 \text{ Pa}}$

d) $\rho = \frac{p}{RT} = \frac{100}{0.287 \times 313} = 1.113 \text{ kg/m}^3. \quad \therefore p_2 = -\frac{1}{2} \times 1.113 \times 100^2 = \underline{-5566 \text{ Pa}}$

3.58 a) $p_A = \gamma h = 9800 \times 4 = 39\ 200$ Pa, $V_A = 0$. Using $h_A = h_2$,

$$\frac{V_A^2}{2g} + \frac{p_A}{\gamma} + h_A = \frac{V_2^2}{2g} + \frac{p_2}{\gamma} + h_2. \quad p_2 = p_A - \frac{V_2^2}{2g}\gamma$$

$$= 39\ 200 - \frac{14^2}{2 \times 9.81} \times 9800 = \underline{-58\ 700 \text{ Pa}}$$

b) $p_B = 0$ and $V_B = 0$. Bernoulli's equation gives, with the datum through the pipe,

$$\frac{V_B^2}{2g} + \frac{p_B}{\gamma} + h_B = \frac{V_2^2}{2g} + \frac{p_2}{\gamma} + h_2. \quad p_2 = \left(4 - \frac{14^2}{2 \times 9.81}\right) 9800 = \underline{-58\ 700 \text{ Pa}}$$

3.59 Bernoulli: $$\frac{V_2^2}{2g} + \frac{p_2}{\gamma} = \frac{V_1^2}{2g} + \frac{p_1}{\gamma}$$

Manometer: $$p_1 + \gamma z + \gamma_{Hg} H - \gamma H - \gamma z = \frac{V_2^2}{2g}\gamma + p_2$$

Substitute Bernoulli's into the manometer eq.:

$$p_1 + (\gamma_{Hg} - \gamma)H = \frac{V_1^2}{2g}\gamma + p_1.$$

a) Use $H = 0.01$ m: $\frac{V_1^2 \times 9800}{2 \times 9.81} = (13.6 - 1)9800 \times 0.01 \quad \therefore V_1 = \underline{1.572 \text{ m/s}}$

Substitute into Bernoulli:

$$p_1 = \frac{V_2^2 - V_1^2}{2g}\gamma = \frac{20^2 - 1.572^2}{2 \times 9.81} \times 9800 = \underline{198\ 600 \text{ Pa}}$$

b) Use $H = 0.05$ m: $\frac{V_1^2 \times 9800}{2 \times 9.81} = (13.6 - 1)9800 \times 0.05 \quad \therefore V_1 = \underline{3.516 \text{ m/s}}$

Substitute into Bernoulli:

$$p_1 = \frac{V_2^2 - V_1^2}{2g}\gamma = \frac{20^2 - 3.516^2}{2 \times 9.81} \times 9800 = \underline{193\ 600 \text{ Pa}}$$

c) Use $H = 0.1$ m: $\frac{V_1^2 \times 9800}{2 \times 9.81} = (13.6 - 1)9800 \times 0.1 \quad \therefore V_1 = \underline{4.972 \text{ m/s}}$

Substitute into Bernoulli:

$$p_1 = \frac{V_2^2 - V_1^2}{2g}\gamma = \frac{20^2 - 4.972^2}{2 \times 9.81} \times 9800 = \underline{187\ 400 \text{ Pa}}$$

3.60 Bernoulli: $\frac{V_1^2}{2}+\frac{p_1}{\rho}=\frac{V_2^2}{2}+\frac{p_2}{\rho}$. $\therefore V_2=\sqrt{2p_1/\rho}$

across nozzle

Bernoulli : $\frac{V_1^2}{2g}+\frac{p_1}{\gamma}+h_1=\frac{V_2^2}{2g}+\frac{p_2}{\gamma}+h_2$. $\therefore h_2=p_1/\gamma$.

to max. height

a) $V_2=\sqrt{2p_1/\rho}=\sqrt{2\times 700\,000/1000}=\underline{37.42\text{ m/s}}$
$h_2=p_1/\gamma=700\,000/9800=\underline{71.4\text{ m}}$

b) $V_2=\sqrt{2p_1/\rho}=\sqrt{2\times 1\,400\,000/1000}=\underline{52.92\text{ m/s}}$
$h_2=p_1/\gamma=1\,400\,000/9800=\underline{142.9\text{ m}}$

c) $V_2=\sqrt{2p_1/\rho}=\sqrt{2\times 100\times 144/1.94}=\underline{121.8\text{ fps}}$
$h_2=p_1/\gamma=100\times 144/62.4=\underline{231\text{ ft}}$

d) $V_2=\sqrt{2p_1/\rho}=\sqrt{2\times 200\times 144/1.94}=\underline{172.3\text{ fps}}$
$h_2=p_1/\gamma=200\times 144/62.4=\underline{462\text{ ft}}$

3.61 a) Apply Bernoulli's eq. from the surface to a point on top of the downstream flow:

$$\frac{V_1^2}{2g}+\frac{p_1}{\gamma}+h_1=\frac{V_2^2}{2g}+\frac{p_2}{\gamma}+h_2. \quad \therefore \underline{V_2=\sqrt{2g(H-h)}}$$

b) Apply Bernoulli's eq. from a point near the bottom upstream to a point on the bottom of the downstream flow:

$$\frac{V_1^2}{2g}+\frac{p_1}{\gamma}+h_1=\frac{V_2^2}{2g}+\frac{p_2}{\gamma}+h_2.$$

Using $p_1=\gamma H$, $p_2=\gamma h$ and $h_1=h_2$, $\underline{V_2=\sqrt{2g(H-h)}}$

3.62 $\frac{V_1^2}{2}+\frac{p_1}{\rho}=\frac{V_2^2}{2}+\frac{p_2}{\rho}$. p_2 = -100 000 Pa, the lowest possible pressure.

a) $\frac{600\,000}{1000}=\frac{V_2^2}{2}-\frac{100\,000}{1000}$. $\therefore V_2=\underline{37.4\text{ m/s}}$.

b) $\frac{300\,000}{1000}=\frac{V_2^2}{2}-\frac{100\,000}{1000}$. $\therefore V_2=\underline{28.3\text{ m/s}}$.

c) $\frac{80 \times 144}{1.94} = \frac{V_2^2}{2} - \frac{14.7 \times 144}{1.94}$. $\therefore V_2 = \underline{118.6 \text{ ft/sec}}$.

d) $\frac{40 \times 144}{1.94} = \frac{V_2^2}{2} - \frac{14.7 \times 144}{1.94}$. $\therefore V_2 = \underline{90.1 \text{ ft/sec}}$.

3.63 A water system must never have a negative pressure, since a leak could ingest impurities. $\therefore$ The least pressure is zero gage.

$$\frac{V_1^2}{2} + \frac{p_1}{\rho} + gz_1 = \frac{V_2^2}{2} + \frac{p_2}{\rho} + gz_2. \quad V_1 = V_2. \quad \text{Let } z_1 = 0, \text{and } p_2 = 0.$$

$$\frac{500\,000}{1000} = 9.81\, z_2. \quad \therefore z_2 = \underline{51.0 \text{ m}}.$$

3.64 $\frac{V_1^2}{2} + \frac{p_1}{\rho} = \frac{V_2^2}{2} + \frac{p_2}{\rho}$. $p_1 = \frac{\rho}{2}\left(V_2^2 - V_1^2\right) = \frac{1.23}{2}\left(2^2 - 10^2\right) = \underline{-59.0 \text{ Pa}}$

3.65 $\frac{V_1^2}{2} + \frac{p_1}{\rho} = \frac{V_2^2}{2} + \frac{p_2}{\rho}$. $p_1 = \frac{\rho}{2}\left(V_2^2 - V_1^2\right) = \frac{1.23}{2}\left(2^2 - 8^2\right) = \underline{-36.9 \text{ Pa}}$

3.66 Apply Bernoulli's equation between the exit (point 2) where the radius is R and a point 1 in between the exit and the center of the tube at a radius less than R:

$$\frac{V_1^2}{2} + \frac{p_1}{\rho} = \frac{V_2^2}{2} + \frac{p_2}{\rho}^{\,0}. \quad \therefore p_1 = \rho\frac{V_2^2 - V_1^2}{2}.$$

Since $V_2 < V_1$, we see that p_1 is negative (a vacuum) so that the envelope would tend to rise due to the negative pressure across most of its area (except for a small area near the end of the tube).

3.67 $\text{Re} = \frac{VD}{\nu}$. For air $\nu \cong 1.5 \times 10^{-5}$. Use reasonable values from your experience!

a) $\text{Re} = \frac{20 \times 0.03}{1.5 \times 10^{-5}} = 4 \times 10^4$. $\therefore$ Separate

b) $\text{Re} = \frac{20 \times 0.005}{1.5 \times 10^{-5}} = 6700$. $\therefore$ Separate

c) $\text{Re} = \frac{20 \times 2}{1.5 \times 10^{-5}} = 2.7 \times 10^6$. $\therefore$ Separate

d) $Re = \dfrac{5 \times 0.002}{1.5 \times 10^{-5}} = 670.$ $\therefore$ Separate

e) $Re = \dfrac{20 \times 2}{1.5 \times 10^{-5}} = 2.7 \times 10^{6}.$ $\therefore$ Separate

f) $Re = \dfrac{100 \times 3}{1.5 \times 10^{-5}} = 2 \times 10^{7}.$ $\therefore$ It will tend to separate, except streamlining the components eliminates separation.

3.68

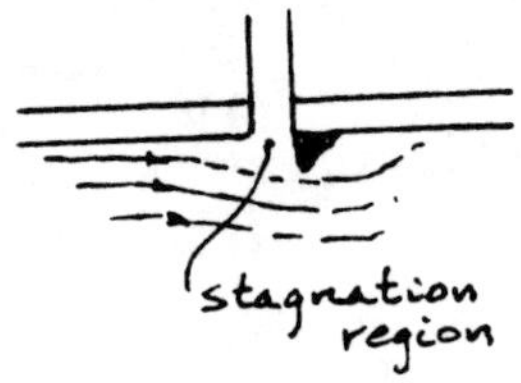

A burr downstream of the opening will create a region that acts similar to a stagnation region thereby creating a high pressure since the velocity will be relatively low in that region.

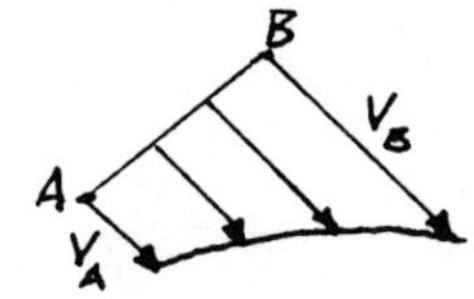

3.69 $\Delta p = \rho \dfrac{V^2}{R} \Delta n = 1000 \dfrac{10^2}{0.05} \times 0.02 = \underline{40\,000 \text{ Pa}}$

Along AB, we expect $V_A > 10$ m/s and $V_B < 10$ m/s.

3.70 The higher pressure at B will force the fluid toward the lower pressure at A, especially in the wall region of slow moving fluid, thereby causing a secondary flow normal to the pipe's axis. This results in a relatively high loss for an elbow.

3.71 Refer to Bernoulli's equation: $\dfrac{V_1^2}{2} + \dfrac{p_1}{\rho} = \dfrac{V_2^2}{2} + \dfrac{p_2}{\rho}$

$p_A > p_B$ since $V_A < V_B$

$p_C < p_D$ since $V_C > V_D$

$p_B > p_D$ since $V_D > V_B$

FOUR

The Integral Forms of the Fundamental Laws

4.1 a) No net force may act on the system: $\underline{\Sigma \vec{F} = 0}$.

b) The energy transfered to or from the system must be zero: $\underline{Q - W = 0}$.

c) If $\underline{\vec{V}}$ <u>is the same for all volume elements</u> then $\Sigma \vec{F} = \frac{D}{Dt} \vec{V} \int dm$, or

$$\Sigma \vec{F} = \frac{D}{Dt}(m\vec{V}). \text{ Since mass is constant for a system } \Sigma \vec{F} = m\frac{D\vec{V}}{Dt}.$$

Since $\frac{D\vec{V}}{Dt} = \vec{a}, \quad \Sigma \vec{F} = m\vec{a}.$

4.2 Extensive properties: Mass, m; Momentum, $m\vec{V}$; kinetic energy, $\frac{1}{2}mV^2$; potential energy, mgh; enthalpy, H.

Associated Intensive properties (divide by the mass): unity, 1; velocity, $\vec{V}$; $V^2/2$; gh; $H/m = h$ (specific enthalpy).

Intensive properties: Temperature, T; time, t; pressure, p; density, ρ; viscosity, μ.

4.3

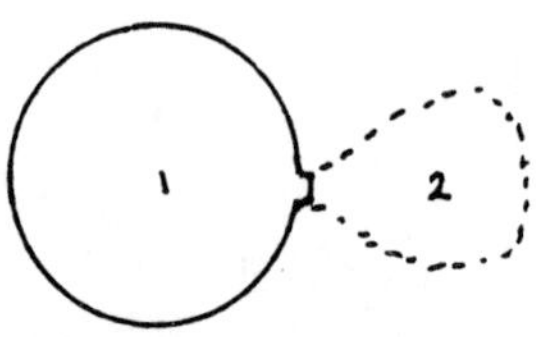

System $(t) = \forall_1$ C.V. $(t) = \forall_1$

System $(t + \Delta t) = \forall_1 + \forall_2$ C.V. $(t + \Delta t) = \forall_1$

4.4

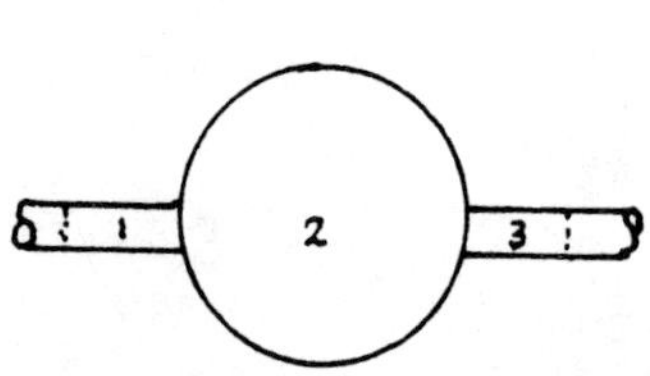

System $(t) = \forall_1 + \forall_2$ C.V. $(t) = \forall_1 + \forall_2$

System $(t + \Delta t) = \forall_2 + \forall_3$ C.V. $(t + \Delta t) = \forall_1 + \forall_2$

4.5 a) The energy equation (the 1st law of Thermo).

b) The conservation of mass.

c) Newton's 2nd law.

d) The energy equation.

e) The energy equation.

4.6

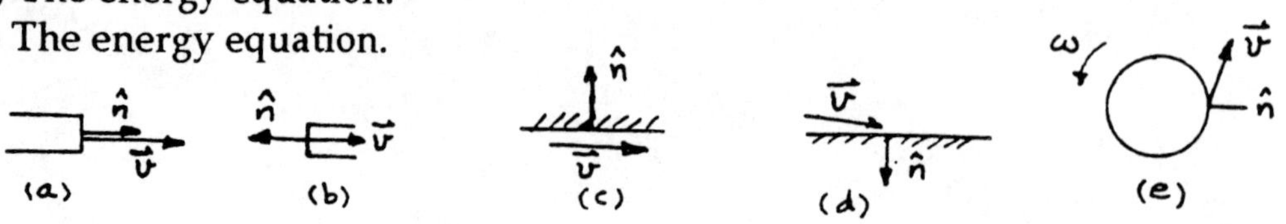

4.7 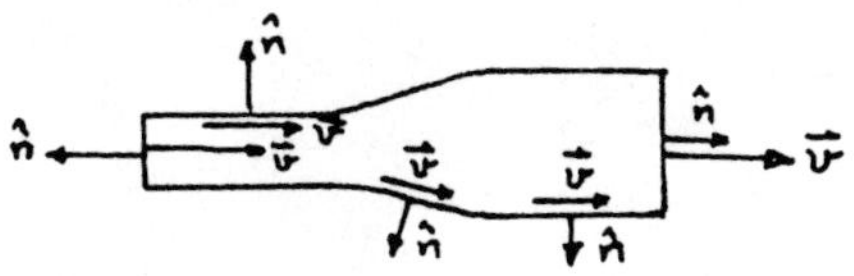

4.8 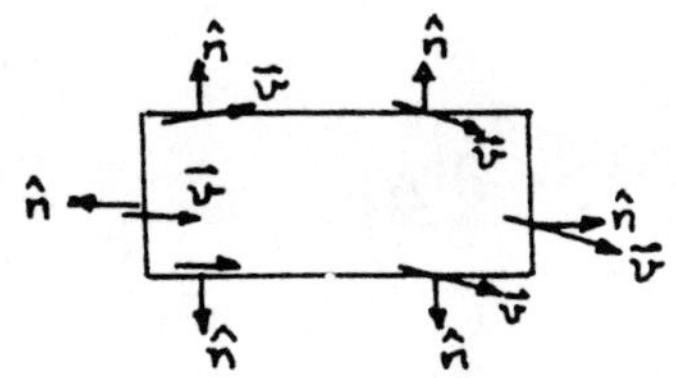

4.9 $\hat{n}_1 = -\frac{1}{\sqrt{2}}\hat{i} - \frac{1}{\sqrt{2}}\hat{j} = \underline{-0.707(\hat{i}+\hat{j})}$

$\hat{n}_2 = \underline{0.866\hat{i} - 0.5\hat{j}}$

$\hat{n}_3 = \underline{-\hat{j}}$

$V_{1n} = \vec{V}_1 \cdot \hat{n}_1 = 10\hat{i} \cdot [-0.707(\hat{i}+\hat{j})] = \underline{-7.07 \text{ fps}}$

$V_{2n} = \vec{V}_2 \cdot \hat{n}_2 = 10\hat{i} \cdot (0.866\hat{i} - 0.5\hat{j}) = \underline{8.66 \text{ fps}}$

$V_{3n} = \vec{V}_3 \cdot \hat{n}_2 = 10\hat{i} \cdot (-\hat{j}) = \underline{0}$

4.10 flux = $\eta\rho\hat{n} \cdot \vec{V}A$

$\text{flux}_1 = \eta\rho[-0.707(\hat{i}+\hat{j})] \cdot 10\hat{i}A / 0.707 = \underline{-10\eta\rho A}$

$\text{flux}_2 = \eta\rho(0.866\hat{i} - 0.5\hat{j}) \cdot 10\hat{i}A / 0.866 = \underline{10\eta\rho A}$

$\text{flux}_3 = \eta\rho(-\hat{j}) \cdot 10\hat{i}A_3 = \underline{0}$

4.11 $(\vec{B} \cdot \hat{n})A = 15(0.5\hat{i} + 0.866\hat{j}) \cdot \hat{j}(10 \times 12)$

$= 15 \times 0.866 \times 120 = 1559 \text{ cm}^3$

Volume = $15 \sin 60° \times 10 \times 12 = 1559 \text{ cm}^3$

4.12 The control volume must be independent of time. Since all space coordinates are integrated out on the left only time remains; thus, we use an ordinary derivative to differentiate a function of time. But, on the right, we note that ρ and η may be functions of (x,y,z,t); hence, the partial derivative is used.

4.13 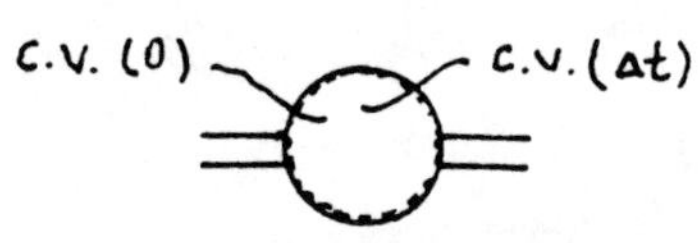

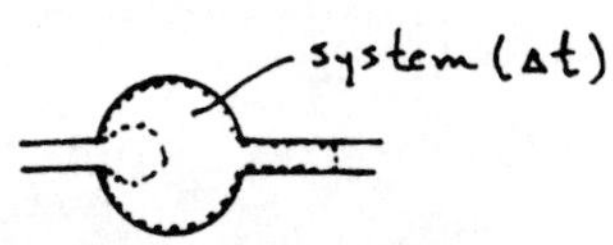

water is moving through the pump.

4.14 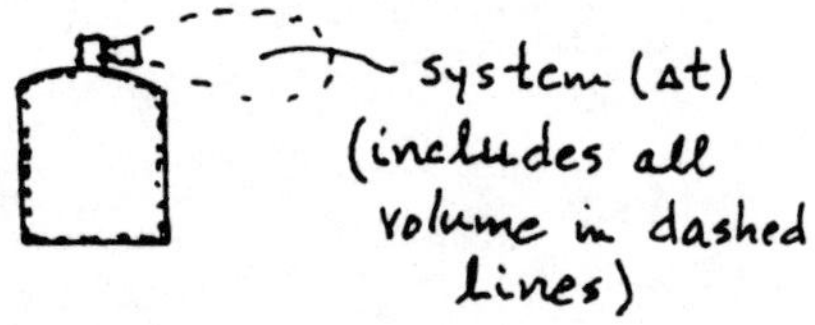

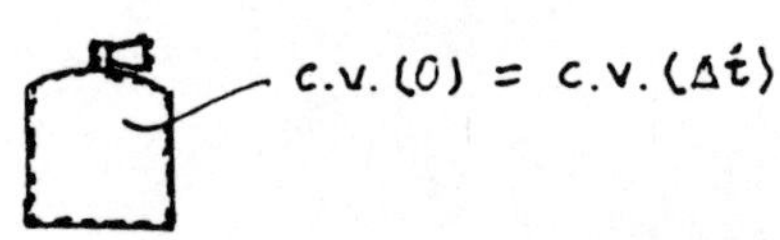

4.15 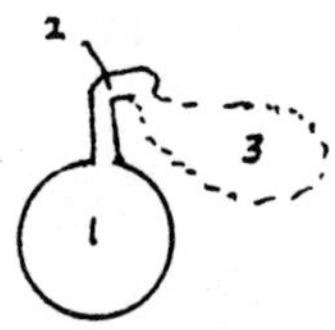

system $(\Delta t) = \forall_1 + \forall_2 + \forall_3$

c.v. $(\Delta t) = \forall_1 + \forall_2$

4.16 If fluid crosses the control surface only on areas A_1 and A_2,

$$\int_{c.s.} \rho \hat{n} \cdot \vec{V} dA = \int_{A_1} \rho \hat{n} \cdot \vec{V} dA + \int_{A_2} \rho \hat{n} \cdot \vec{V} dA = 0$$

For uniform flow all quantities are constant over each area:

$$\rho_1 \hat{n}_1 \cdot \vec{V}_1 \int_{A_1} dA + \rho_2 \hat{n}_2 \cdot \vec{V}_2 \int_{A_2} dA = 0$$

Let A_1 be the inlet so $\hat{n}_1 \cdot \vec{V}_1 = -V_1$ and A_2 be the outlet so $\hat{n}_2 \cdot \vec{V}_2 = V_2$. Then

$$-\rho_1 V_1 A_1 + \rho_2 V_2 A_2 = 0$$

or

$$\rho_2 A_2 V_2 = \rho_1 A_1 V_1$$

4.17 Use Eq. 4.3.2 with m_V representing the mass in the volume:

$$0 = \frac{dm_V}{dt} + \int_{c.s.} \rho \hat{n} \cdot \vec{V} dA$$

$$= \frac{dm_V}{dt} + \rho A_2 V_2 - \rho A_1 V_1$$

$$= \frac{dm_V}{dt} + \rho Q - \dot{m}.$$

Finally,

$$\underline{\frac{dm_V}{dt} = \dot{m} - \rho Q.}$$

4.18 Use Eq. 4.3.2 with m_S representing the mass in the sponge:

$$0 = \frac{dm_S}{dt} + \int \rho \hat{n} \cdot \vec{V} dA$$

$$= \frac{dm_S}{dt} + \rho A_2 V_2 + \rho A_3 V_3 - \rho A_1 V_1$$

$$= \frac{dm_S}{dt} + \dot{m}_2 + \rho A_3 V_3 - \rho Q_1.$$

Finally,

$$\underline{\frac{dm_S}{dt} = \rho Q_1 - \dot{m}_2 - \rho A_3 V_3.}$$

4.19 $A_1V_1 = A_2V_2$. $\pi \times .03^2 \times 20 = \pi \times .06^2 \times V_2$. $\therefore V_2 = \underline{5 \text{ m/s}}$.

$\dot{m} = \rho AV = 1000\pi \times .03^2 \times 20 = \underline{56.55 \text{ kg/s}}$. $Q = AV = \pi \times .03^2 \times 20 = \underline{0.05655 \text{ m}^3/\text{s}}$.

4.19E $A_1V_1 = A_2V_2$. $\pi \times \dfrac{1.25^2}{144} \times 60 = \pi \times \dfrac{2.5^2}{144} V_2$. $\therefore V_2 = \underline{15 \text{ ft/sec}}$.

$$\dot{m} = \rho AV = 1.94\pi \frac{1.25^2}{144} \times 60 = \underline{3.968 \text{ slug/sec}}. \quad Q = AV = \pi \frac{1.25^2}{144} \times 60 = \underline{2.045 \text{ ft}^3/\text{sec}}.$$

4.20 $A_1V_1 = A_2V_2$. $\pi \times .025^2 \times 10 = (2\pi \times .6 \times .003)V_2$. $\therefore V_2 = \underline{1.736 \text{ m/s}}$.

$\dot{m} = \rho AV = 1000\pi \times .025^2 \times 10 = \underline{19.63 \text{ kg/s}}$. $Q = AV = \pi \times .025^2 \times 10 = \underline{0.01963 \text{ m}^3/\text{s}}$.

4.21 $\dot{m}_{in} = \rho A_1V_1 + \rho A_2V_2$. $200 = 1000\,\pi \times .025^2 \times 25 + 1000\,Q_2$. $\therefore Q_2 = \underline{0.1509 \text{ m}^3/\text{s}}$.

4.22 $\rho_1 = \dfrac{p_1}{RT_1} = \dfrac{200}{0.287 \times 293} = 2.378 \text{ kg/m}^3$. $\rho_2 = \dfrac{50}{287 \times 353} = .4935 \text{ kg/m}^3$.

$\dot{m} = \rho AV$. $\therefore V_1 = \dfrac{\dot{m}}{\rho_1 A_1} = \dfrac{2}{2.378\pi \times .05^2} = \underline{107.1 \text{ m/s}}$.

$\dot{m}_2 = \dot{m}_1 = 2 = .4935 \times .05 \times .07V_2$. $\therefore V_2 = \underline{1158 \text{ m/s}}$.

4.22E $\rho_1 = \dfrac{p_1}{RT_1} = \dfrac{40 \times 144}{1716 \times 520} = .006455 \text{ slug/ft}^3$. $\rho_2 = \dfrac{7 \times 144}{1716 \times 610} = .000963 \text{ slug/ft}^3$.

$\dot{m} = \rho AV$. $\therefore V_1 = \dfrac{\dot{m}}{\rho_1 A_1} = \dfrac{.2}{(\pi \times 2^2/144).006455}$. $\therefore V_2 = \underline{355 \text{ fps}}$.

$\dot{m}_2 = 0.2 = .000963 \times (2 \times 3/144)V_2$. $\therefore V_2 = \underline{4984 \text{ fps}}$.

4.23 $\rho_1 A_1 V_1 = \rho_2 A_2 V_2$. $\rho_1 = \dfrac{p_1}{RT} = \dfrac{500}{.287 \times 393} = 4.433 \dfrac{\text{kg}}{\text{m}^3}$. $\rho_2 = \dfrac{1246}{.287 \times 522} = 8.317 \dfrac{\text{kg}}{\text{m}^3}$

$4.433\,\pi \times .05^2 \times 600 = 8.317\,\pi \times .05^2\,V_2$. $\therefore V_2 = \underline{319.8 \text{ m/s}}$.

$\dot{m} = \rho_1 A_1 V_1 = \underline{20.89 \text{ kg/s}}$. $Q_1 = A_1V_1 = \underline{4.712 \text{ m}^3/\text{s}}$. $Q_2 = \underline{2.512 \text{ m}^3/\text{s}}$.

4.24 $\rho_1 A_1 V_1 = \rho_2 A_2 V_2$

$$\frac{p_1}{RT_1} A_1 V_1 = \frac{p_2}{RT_2} A_2 V_2$$

$$\frac{200}{293} \pi \times 0.05^2 \times 40 = \frac{120}{T_2} \pi \times 0.03^2 \times 120.$$

$\therefore T_2 = 189.9 \text{ K}$ or $\underline{-83^\circ\text{C}}$.

4.25 a) $A_1V_1 = A_2V_2$. $(2 \times 1.5 + 1.5 \times 1.5)\,3 = \pi \dfrac{d_2^2}{4} \times 2.$ $\therefore d_2 = \underline{3.167 \text{ m}}$

b) $(2 \times 1.5 + 1.5 \times 1.5)\,3 = \pi \dfrac{d_2^2}{4} \times \dfrac{2}{2}.$ $\therefore d_2 = \underline{4.478 \text{ m}}$

c) $(2 \times 1.5 + 1.5 \times 1.5)\,3 = \left(\dfrac{1}{3}\pi R^2 - \dfrac{R}{2} \times .866R\right) \times 2.$

$\therefore R = 3.581 \text{ m}.$ $\therefore d_2 = \underline{7.162 \text{ m}}$

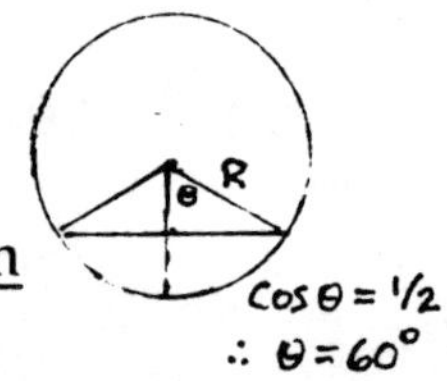

4.26 a) $v = 10\left(1 - \dfrac{r}{r_0}\right).$ $\pi r_0^2 V = \int_0^{r_0} v\,dA = \int_0^{r_0} 10\left(1 - \dfrac{r}{r_0}\right) 2\pi r\,dr = 20\pi \int_0^{r_0} \left(r - \dfrac{r^2}{r_0}\right) dr.$

$$\therefore V = \frac{20}{r_0^2}\left(\frac{r_0^2}{2} - \frac{r_0^2}{3}\right) = \frac{10}{3} = \underline{3.333 \text{ m/s}}.$$

$\dot{m} = \rho A V = 1000 \times \pi \times .04^2 \times 3.33 = \underline{16.75 \text{kg/s}}.$ $Q = AV = \underline{0.01675 \text{ m}^3/\text{s}}.$

b) $v = 10\left(1 - \dfrac{r^2}{r_0^2}\right).$ $\pi r_0^2 V = \int_0^{r_0} 10\left(1 - \dfrac{r^2}{r_0^2}\right) 2\pi r\,dr = 20\pi \left(\dfrac{r_0^2}{2} - \dfrac{r_0^2}{4}\right).$ $\therefore V = \underline{5 \text{ m/s}}$

$\dot{m} = \rho A V = 1000 \times \pi \times .04^2 \times 5 = \underline{25.13 \text{ kg/s}}.$ $Q = AV = \underline{0.02513 \text{ m}^3/\text{s}}.$

c) $v = 20\left(1 - \dfrac{r}{r_0}\right).$ $\pi r_0^2 V = \int_{r_0/2}^{r_0} 20\left(1 - \dfrac{r}{r_0}\right) 2\pi r\,dr + 10\pi r_0^2/4.$ $\therefore V = \underline{5.833 \text{ m/s}}$

$\dot{m} = \rho A V = 1000 \times \pi \times .04^2 \times 5.833 = \underline{29.32 \text{ kg/s}}.$ $Q = \underline{0.02932 \text{ m}^3/\text{s}}.$

4.27 a) Since the area is rectangular $V = \underline{5 \text{ m/s}}.$

$\dot{m} = \rho A V = 1000 \times .08 \times .8 \times 5 = \underline{320 \text{ kg/s}}.$ $Q = \dfrac{\dot{m}}{\rho} = \underline{0.32 \text{ m}^3/\text{s}}.$

b) $v = 40\left(\dfrac{y}{h} - \dfrac{y^2}{h^2}\right)$ with $y = 0$ at the lower wall.

$\therefore Vhw = \int_0^h 40\left(\dfrac{y}{h} - \dfrac{y^2}{h^2}\right) w\,dy = 40 \times \dfrac{h}{6} w.$ $\therefore V = \underline{6.667 \text{ m/s}}.$

$\dot{m} = \rho A V = 1000 \times .08 \times .8 \times 6.667 = \underline{426.7 \text{ kg/s}}.$ $Q = \underline{0.4267 \text{ m}^3/\text{s}}.$

c) $V \times .08 = 10 \times .04 + 5 \times .02 + 5 \times .02.$ $\therefore V = \underline{7.5 \text{ m/s}}.$

$\dot{m} = \rho A V = 1000 \times .08 \times .8 \times 7.5 = \underline{480 \text{ kg/s}}.$ $Q = \dfrac{\dot{m}}{\rho} = \underline{0.48 \text{ m}^3/\text{s}}.$

4.28 a) $A_1V_1 = \int v_2\,dA.$ $\pi \times .01^2 \times 2 = \int_0^{r_0} v_{\text{max}}\left(1 - \dfrac{r^2}{r_0^2}\right) 2\pi r\,dr = 2\pi v_{\text{max}} \dfrac{r_0^2}{4}.$

With $r_0 = .01$, $v_{\text{max}} = 4 \text{ m/s}.$ $\therefore v(r) = \underline{4\left(1 - \dfrac{r^2}{0.01^2}\right) \text{ m/s}}.$

b) $A_1V_1 = \int v_2 dA.$ $\quad .02w \times 2 = \int_{-h}^{h} v_{\max}\left(1-\frac{y^2}{h^2}\right)w dy = v_{\max} w \frac{4h}{3}.$

With $h = .01$, $v_{max} = 3$ m/s. $\quad \therefore v(y) = \underline{3\left(1-\frac{y^2}{0.01^2}\right) \text{m/s}}.$

4.28E a) $A_1V_1 = \int v_2 dA.$ $\quad \pi \times \left(\frac{1}{24}\right)^2 \times 6 = \int_0^{r_0} v_{\max}\left(1-\frac{r^2}{r_0^2}\right) 2\pi r dr = 2\pi v_{\max}\frac{r_0^2}{4}.$

With $r_0 = \frac{1}{24}$, $\quad v_{\max} = 12$ fps. $\quad \therefore v(r) = \underline{12(1-576r^2) \text{ fps}}.$

b) $A_1V_1 = \int v_2 dA.$ $\quad \frac{1}{12} \times w \times 6 = \int_{-h}^{h} v_{\max}\left(1-\frac{y^2}{h^2}\right) w dy = v_{\max} w \frac{4h}{3}.$

With $h = \frac{1}{24}$, $\quad v_{\max} = 9$ fps. $\quad \therefore v(y) = \underline{9(1-576y^2) \text{ fps}}.$

4.29 If $dm/dt = 0$, then $\rho_1 A_1 V_1 = \rho_2 A_2 V_2 + \rho_3 A_3 V_3$. In terms of $\dot{m}_2$ and Q_3 this becomes, letting $\rho_1 = \rho_2 = \rho_3$,

$$1000 \times \pi \times 0.02^2 \times 12 = \dot{m}_2 + 1000 \times 0.01. \qquad \therefore \dot{m}_2 = \underline{5.08 \text{ kg/s}}.$$

4.30 $\int_0^{r_1} v_1 dA = A_2 V_2.$ $\quad \int_0^{r_1} v_{\max}\left(1-\frac{r^2}{r_1^2}\right) 2\pi r dr = \pi \times .0025^2 \times 2.$

$\therefore 2\pi v_{\max}\frac{.005^2}{4} = \pi \times .0025^2 \times 2.$ $\quad \therefore v_{\max} = 1$ m/s. $\quad \therefore v(r) = \underline{\left(1-\frac{r^2}{.005^2}\right) \text{m/s}}.$

4.31 $\dot{m}_{in} = \dot{m}_{out} + \dot{m}.$ $\quad \rho \times .2 \times 2 \times 10 = \left[\rho \int_0^{.1} 10(20y - 100y^2) 2dy + \rho \times .1 \times 2 \times 10\right] + \dot{m}.$

Note: We see that at $y = 0.1$ m the velocity $u(.1) = 10$ m/s. Thus we integrate to $y = 0.1$, and between $y = 0.1$ and 0.2 the velocity $u = 10$.

$4\rho = \left[\frac{4}{3}\rho + 2\rho\right] + \dot{m}.$ $\quad \therefore \dot{m} = 0.6667\rho = \underline{0.82 \text{ kg/s}}.$

4.32 $V_1 h_1 = \int_0^h u(y) dy.$ $\quad 10 \times .05 = \int_0^h 10(20y - 100y^2) dy$

$$= 10\left(10h^2 - \frac{100}{3}h^3\right).$$

$\therefore 666.7\, h^3 - 200\, h^2 = -1$. This can be solved by trial-and-error:

$h = .06$: $\quad -.576 \stackrel{?}{=} -1.$ $\qquad h = .07$: $\quad -.751 \stackrel{?}{=} -1.$

$h = .08$: $\quad -.939 \stackrel{?}{=} -1.$ $\qquad h = .083$: $\quad -.997 \stackrel{?}{=} -1.$

$h = .084$: $\quad -1.016 \stackrel{?}{=} -1.$ $\qquad \therefore h = 0.0832$: $\quad$ or $\quad$ 8.32 cm.

Note: Fluid does not cross a streamline so all the flow that enters on the left leaves on the right. The streamline simply moves further from the wall.

4.33 $\dot{m} = \int \rho V dA = \int_0^{.1} 1200\left(1-\frac{5}{3}y\right)0.5(20y-100y^2)1.5dy$

$$= 18\,000\int_0^{.1}\left(y-\frac{5}{3}y^2-5y^2+\frac{25}{3}y^3\right)dy = \underline{53.75 \text{ kg/s}}.$$

$\overline{V} = \frac{2}{3}u_{max} = \frac{2}{3}\times.5 = \frac{1}{3}\text{m/s}.$ (See Prob. 4.28 b).

$\overline{\rho} = \frac{1200+1000}{2} = 1100 \text{ kg/m}^3.$ $\quad \therefore \overline{\rho}\overline{V}A = 1100\times\frac{1}{3}\times(.1\times1.5) = \underline{55 \text{ kg/s}}.$

Thus we see that $\overline{\rho}\overline{V}A \neq \dot{m}$, if ρ and V depend on y.

4.33E $\dot{m} = \int \rho V dA = \int_0^{1/3} 2.2(1-.3545y)(6y-9y^2)2\times5dy$

$$= 22\int_0^{1/3}\left(6y-2.127y^2-9y^2+3.19y^3\right)dy = \underline{4.528 \text{ slug/sec}}.$$

$\overline{V} = \frac{2}{3}u_{max} = \frac{2}{3}\times2 = \frac{4}{3}\text{fps}.$ (See Prob. 4.28 b).

$\overline{\rho} = \frac{2.2+1.94}{2} = 2.07 \text{ slug/ft}^3.$ $\quad \therefore \overline{\rho}\overline{V}A = 2.07\times\frac{4}{3}\times\left(5\times\frac{1}{3}\right) = \underline{4.6 \text{ slug/sec}}.$

Thus, $\overline{\rho}\overline{V}A \neq \dot{m}$ since $\rho = \rho(y)$ and $V = V(y)$.

4.34 $A_1V_1 = A_2V_2.$ $\qquad \pi\times.01^2\times8 = (2\pi\times.2\times.04)V_2\cos30°.$

$$\therefore V_2 = \underline{0.05774 \text{ m/s}}.$$

4.35 $2000\times\frac{4}{3}\pi\times.0015^3\frac{\text{m}^3\text{ of H}_2\text{O}}{\text{m}^3\text{ of air}}\times9000\times5\frac{\text{m}^3\text{ of air}}{\text{s}} = 1.5\times(1.5h).$ $\quad \therefore h = \underline{0.565 \text{ m}}.$

4.36 Use Eq. 4.3.3: $0 = \int \frac{\partial \rho}{\partial t} d\Psi + \rho_1 \vec{V}_1 \cdot \hat{n} A_1.$ $\vec{V}_1 \cdot \hat{n}_1 = -V_1.$

$\therefore \rho_1 A_1 V_1 = \frac{\partial \rho}{\partial t} \Psi_{tire}.$ $\left(\frac{250+100}{.287 \times 293}\right) \pi \times .003^2 \times 60 = \frac{\partial \rho}{\partial t} \times 0.5.$

$\therefore \frac{\partial \rho}{\partial t} = \underline{0.0141 \frac{\text{kg}}{\text{m}^3 \cdot \text{s}}}.$

4.36E Use Eq. 4.3.3: $0 = \int \frac{\partial \rho}{\partial t} d\Psi + \rho_1 \vec{V}_1 \cdot \hat{n}_1 A_1.$ $\vec{V}_1 \cdot \hat{n}_1 = -V_1.$

$\therefore \rho_1 A_1 V_1 = \frac{\partial \rho}{\partial t} \Psi_{tire}.$ $\frac{(37+14.7)144}{1716 \times 520} \times \pi \times \left(\frac{1}{96}\right)^2 \times 180 = \frac{\partial \rho}{\partial t} \times 17.$

$\therefore \frac{\partial \rho}{\partial t} = \underline{3.01 \times 10^{-5} \frac{\text{slug}}{\text{ft}^3 - \text{sec}}}.$

4.37 $\dot{m}_{in} = \dot{m}_2 + \dot{m}_3.$ $\bar{V}_1 = 20$ m/s (see Prob. 4.28a).

$20 \times 1000\pi \times .02^2 = 10 + 1000\pi \times .02^2 \bar{V}_3.$ $\therefore \bar{V}_3 = \underline{12.04 \text{ m/s}}.$

4.38 $0 = \frac{d}{dt} m_{c.v.} + \dot{m}_{net} = \frac{d}{dt} m_{c.v.} + \dot{m}_2 + \dot{m}_3 - \dot{m}_1$

$\therefore \frac{d}{dt} m_{c.v.} = \dot{m}_1 - \dot{m}_2 - \dot{m}_3 = 1000 \times \pi \times .02^2 \times 20 - 10 - 1000\pi \times .02^2 \times 10$

$= \underline{2.57 \text{ kg/s}}.$

4.39 The control surface is close to the interface at the instant shown.
$\therefore V_i$ = interface velocity.

$\rho_e A_e V_e = \rho_i A_i V_i.$ $1.5 \times \pi \times .15^2 \times 300 = \frac{8000}{.287 \times 673} \pi \times 1^2 V_i.$

$\therefore V_i = \underline{0.244 \text{ m/s}}.$

4.40 Assume an incompressible flow:

$4Q_1 = A_2 V_2$

$4 \times 50 / 60 = (0.6 \times 1.2) V_2.$ $\therefore V_2 = \underline{4.63 \text{ m/s}}.$

4.40E Assume an incompressible flow:

$4Q_1 = A_2 V_2$

$4 \times 1500 / 60 = (2 \times 4) V_2.$ $\therefore V_2 = \underline{12.5 \text{ fps}}.$

4.41 For an incompressible flow (low speed air flow)

$$\int_{A_1} u dA = A_2 V_2$$

$$\int_0^{0.2} 20y^{1/5} \times 0.8dy = \pi \times 0.15^2 V_2.$$

$$20 \times 0.8 \frac{5}{6} 0.2^{6/5} = \pi \times 0.15^2 V_2. \qquad \therefore V_2 = \underline{27.3 \text{ m/s}}.$$

4.42 $A_1 V_1 + \int v_2 dA = A_e V_e$

$$\pi(0.1^2 - 0.025^2) \times 4 + \int_0^{0.025} 200\left(1 - \frac{r^2}{0.025^2}\right) 2\pi r dr = \pi \times 0.1^2 V_e$$

$$0.1178 + 0.1963 = 0.0314 V_e. \qquad \therefore V_e = \underline{10.0 \text{ m/s}}.$$

4.43 Draw a control volume around the entire set-up:

$$0 = \frac{dm_{tissue}}{dt} + \rho V_2 A_2 - \rho V_1 A_1$$

$$= \dot{m}_{tissue} + \rho\pi\left(\frac{d_2^2 - d^2}{4}\right)\dot{h}_2 - \rho\pi(h_1 \tan\phi)^2 \dot{h}_1$$

or

$$\underline{\dot{m}_{tissue} = \rho\pi\left[\frac{d^2 - d_2^2}{4}\dot{h}_2 + h_1^2 \dot{h}_1 \tan^2\phi\right]}.$$

4.44 The width w of the channel is constant throughout the flow. Then

$$0 = \frac{dm}{dt} + \rho A_2 V_2 - \rho A_1 V_1$$

$$0 = \frac{d}{dt}(\rho w h L) + \rho A_2 V_2 - \rho A_1 V_1$$

$$0 = \rho \frac{dh}{dt} w \times 100 + \rho 0.2 w \times 8 - \rho 4 w \times 0.2. \qquad \therefore \dot{h} = \underline{0.008 \text{ m/s}}.$$

4.45 $0 = \frac{dm}{dt} + \rho A_2 V_2 - \rho A_1 V_1$

$$= \dot{m} + 1000(\pi \times 0.003^2 \times 0.02 - 10 \times 10^{-6} / 60). \qquad \therefore \dot{m} = \underline{3.99 \times 10^{-4} \text{ kg/s}}.$$

4.46 $\rho_1 A_1 V_1 = \rho_2 A_2 V_2$

$$\dot{m}_1 = \rho_2 A_2 V_2$$

$$400 e^{-10/100} \times 10^{-6} \times 900 = 0.2 \times \pi \times 0.05^2 V_e. \qquad \therefore V_e = \underline{207 \text{ m/s}}.$$

4.47 $0 = \frac{dm}{dt} + \rho_3 Q_3 - \rho_1 A_1 V_1 - \dot{m}_2$ where $m = \rho A h$.

a) $0 = 1000\pi \times 0.6^2 \dot{h} + 1000 \times 0.6/60 - 1000\pi \times 0.02^2 \times 10 - 10.$

$\therefore \dot{h} = 0.0111 \text{ m/s}$ or $\underline{11.1 \text{ mm/s}}$

b) $0 = 1000\pi \times 0.6^2 \dot{h} + 1000 \times 0.01 - 0 - 20.$

$\therefore \dot{h} = 0.00884 \text{ m/s}$ or $\underline{8.84 \text{ mm/s}}.$

c) $0 = 1000\pi \times 0.6^2 \dot{h} + 1000 \times 1.0/60 - 1000\pi \times 0.02^2 \times 5 - 10.$

$\therefore \dot{h} = -0.000339 \text{ m/s}$ or $\underline{-0.339 \text{ mm/s}}.$

4.48 $A_1 V_1 = A_2 V_2$ where A_2 is an area just under the top surface.

a) $\pi \times 0.02^2 \times 10e^{-t/10} = \pi \times (h \tan 60^\circ)^2 \frac{dh}{dt}$

$\therefore h^2 dh = 0.001333 e^{-t/10} dt.$ $\qquad \therefore h^3 = -0.04 e^{-t/10} + 0.04.$

Finally,

$h(t) = \underline{0.342(1 - e^{-t/10})^{1/3}}.$

b) $0.04 \times 10 \times 10e^{-t/10} = (h \tan 60^\circ) \times 10\dot{h}$

$\therefore h dh = 0.2309 e^{-t/10} dt.$ $\qquad \therefore h^2 = -4.62 e^{-t/10} + 4.62.$

Finally,

$h(t) = \underline{4.62(1 - e^{-t/10})^{1/2}}.$

4.49 $\dot{W} = T\omega + pAV + \mu \frac{du}{dy} A_{belt}$

$= 20 \times 500 \times 2\pi/60 + 400 \times 0.4 \times 0.5 \times 10 + 1.81 \times 10^{-5} \times 100 \times 0.5 \times 0.8$

$= 1047 + 800 + 0.000724 = \underline{1847 \text{ W}}$

4.50 If the temperature is essentially constant, the internal energy of the c.v. does not change and the flux of internal energy into the pipe is the same as that leaving the pipe. Hence, the two integral terms are zero. The losses are equal to the heat transfer exiting the pipe.

4.51 80% of the power is used to increase the pressure while 20% increases the internal energy ($\dot{Q} = 0$ because of the insulation). Hence,

$\dot{m}\Delta\tilde{u} = 0.2\dot{W}$

$1000 \times 0.02 \times 4.18\Delta T = 0.2 \times 500.$ $\qquad \therefore \Delta T = \underline{0.836 \text{ °C}}.$

4.52 $\dot{W}_P = \frac{Q\gamma H_P}{\eta_P}.$ $\qquad 5 \times 746 = \frac{Q \times 9800 \times 20}{0.87}.$ $\qquad \therefore Q = \underline{0.01656 \text{ m}^3/\text{s}}.$

4.53 $-\dfrac{\dot{W}_T}{\dot{m}g} = -40 \times 0.89.$

a) $\dot{W}_T = 40 \times 0.89 \times 200 \times 9.81 = \underline{69\ 850\ \text{W}}$

b) $\dot{W}_T = 40 \times 0.89 \times (90\ 000 / 60) \times 9.81 = \underline{523\ 900\ \text{W}}$

c) $\dot{W}_T = 40 \times 0.89 \times (8 \times 10^6 / 3600) \times 9.81 = \underline{776\ 100\ \text{W}}$

4.54 $-\dfrac{\dot{W}_T}{\rho A V g} = \eta_T \Delta z. \quad \dfrac{10\ 000\ 000}{1000 \times 3 \times 60 \times V \times 9.81} = 0.89 \times 50. \quad \therefore V = \underline{1.273\ \text{m/s}}.$

4.55 $\dfrac{V_1^2}{2g} + \dfrac{p_1}{\gamma} + z_1 = \dfrac{V_2^2}{2g} + \dfrac{p_2}{\gamma} + z_2.$

$\dfrac{4^2}{2 \times 9.81} + 2 = \dfrac{16}{2 \times 9.81 h_2^2} + h_2.$

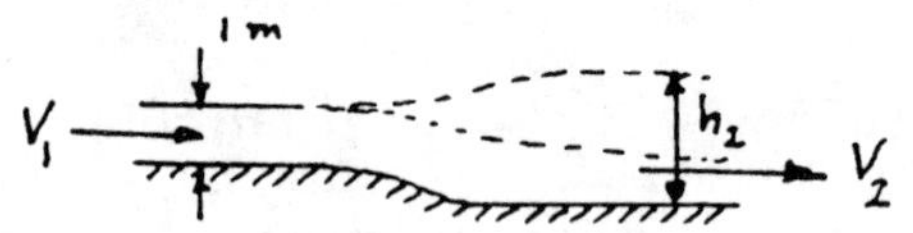

Continuity: $1 \times 4 = V_2 h_2.$

$2.815 = 0.815 / h_2^2 + h_2.$

This cubic can be solved by trial-and-error.

$h_2 = 2.6$: $2.82 \overset{?}{=} 2.72$

$h_2 = 2.7$: $2.82 \overset{?}{=} 2.81$

$h_2 = .6$: $2.82 \overset{?}{=} 2.86 \Rightarrow \therefore h_2 = \underline{0.61\ \text{m}}$ $\qquad \therefore h_2 = \underline{2.71\ \text{m}}$

4.55E $\dfrac{V_1^2}{2g} + \dfrac{p_1}{\gamma} + z_1 = \dfrac{V_2^2}{2g} + \dfrac{p_2}{\gamma} + z_2.$

$\dfrac{12^2}{2 \times 32.2} + 6 = \dfrac{36^2}{64.4 h_2^2} + h_2.$

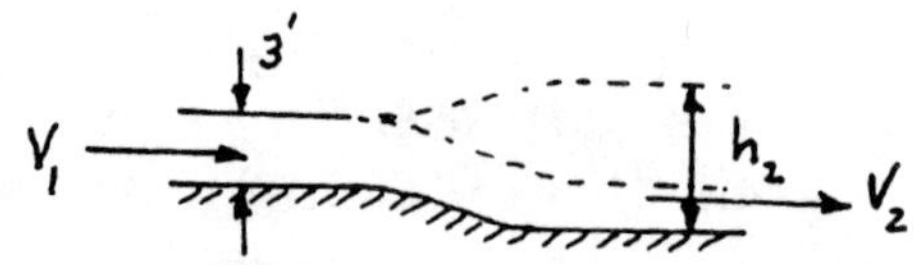

Continuity: $3 \times 12 = h_2 V_2.$

$8.236 = \dfrac{20.1}{h_2^2} + h_2.$

This can be solved by trial-and-error.

$h_2 = 8'$: $8.24 \overset{?}{=} 8.31$ $\qquad h_2 = 7.9'$: $8.24 \overset{?}{=} 8.22.$ $\qquad \because h_2 = \underline{7.93'}.$

$h_2 = 1.8'$: $8.24 \overset{?}{=} 8.00$ $\qquad h_2 = 1.75'$: $8.24 \overset{?}{=} 8.31.$ $\qquad \because h_2 = \underline{1.76'}.$

4.56 $\frac{V_1^2}{2g} + z_1 = \frac{V_2^2}{2g} + z_2 + h_L$. $\quad \therefore \frac{4^2}{2 \times 9.81} + 2 = \frac{16}{2 \times 9.81 h_2^2} + h_2 + 0.2$.

$\therefore 2.615 = 0.815 / h_2^2 + h_2$.

Trial-and-error provides the following:

$h_2 = 2.5$: $2.615 \overset{?}{=} 2.63$ $\quad h_2 = 2.45$: $2.615 \overset{?}{=} 2.59$. $\quad \therefore h_2 = \underline{2.47 \text{ m}}$

$h_2 = 0.65$: $2.615 \overset{?}{=} 2.58$ $\quad h_2 = 0.64$: $2.615 \overset{?}{=} 2.63$. $\quad \therefore h_2 = \underline{0.646 \text{ m}}$

4.56E $\frac{V_1^2}{2g} + z_1 = \frac{V_2^2}{2g} + z_2 + h_L$. $\quad \therefore \frac{12^2}{64.4} + 6 = \frac{36^2}{64.4 h_2^2} + h_2 + 0.6$.

$$7.636 = \frac{20.1}{h_2^2} + h_2.$$

Trial-and-error provides the following:

$h_2 = 7.2$: $7.64 \overset{?}{=} 7.59$ $\quad h_2 = 7.3$: $7.64 \overset{?}{=} 7.68$. $\quad \therefore h_2 = \underline{7.25'}$

$h_2 = 1.86$: $7.64 \overset{?}{=} 7.67$ $\quad h_2 = 1.85$: $7.64 \overset{?}{=} 7.72$. $\quad \therefore h_2 = \underline{1.862'}$

4.57 Manometer: Position the datum at the top of the right mercury level.

$$9810 \times .4 + 9810 z_2 + p_2 + \frac{V_2^2}{2} \times 1000 = (9810 \times 13.6) \times .4 + 9810 \times 2 + p_1$$

Divide by $\gamma = 9810$: $\quad .4 + z_2 + \frac{p_2}{\gamma} + \frac{V_2^2}{2g} = 13.6 \times .4 + 2 + \frac{p_1}{\gamma}$. ①

Energy: $\quad \frac{V_1^2}{2g} + \frac{p_1}{\gamma} + z_1 = \frac{V_2^2}{2g} + \frac{p_2}{\gamma} + z_2$. ②

Subtract ② from ①: With $z_1 = 2$ m, $\quad \frac{V_1^2}{2g} = 12.6 \times .4$. $\quad \therefore V_1 = \underline{9.86 \text{ m/s}}$

4.58 The manometer equation (see Prob. 4.57) is

$$0.4 + z_2 + \frac{p_2}{\gamma} + \frac{V_2^2}{2g} = 13.6 \times .4 + 2 + \frac{p_1}{\gamma}. \quad ①$$

Energy: $\quad \frac{V_1^2}{2g} + \frac{p_1}{\gamma} + z_1 = \frac{V_2^2}{2g} + \frac{p_2}{\gamma} + z_2 + 0.05 \frac{V_2^2}{2g}$. ②

Subtract ② from ①: With $z_1 = 2$ m, and with $V_2 = 4V_1$ (continuity)

$\frac{1.8 V_1^2}{2g} = 12.6 \times 0.4$. $\quad \therefore V_1 = \underline{7.41 \text{ m/s}}$.

4.59 Continuity: $\pi \times .025^2 V_1 = \pi \times .04^2 V_2$. $\therefore V_1 = 2.56\ V_2$.

$Q = 600 \times 10^{-3}/60 = \pi \times .025^2\ V_1$. $\therefore V_1 = 5.09$ m/s, & $V_2 = 1.99$ m/s.

Energy: $$\frac{V_1^2}{2g} + \frac{p_1}{\gamma} = \frac{V_2^2}{2g} + \frac{p_2}{\gamma} + 0.37\frac{V_1^2}{2g}.$$

$$\therefore p_2 = 400 + 9810\left[\frac{5.09^2}{19.62} - \frac{1.99^2}{19.62} - .37\frac{5.09^2}{19.62}\right] / 1000 = \underline{406.2 \text{ kPa}}$$

Note: The factor of 1000 is to convert Pa to kPa.

4.59E $Q = 120 \times 0.002228 = \pi \times \left(\frac{1}{12}\right)^2 V_1$. $\therefore V_1 = 12.25$ fps.

Continuity: $\pi \times \left(\frac{1}{12}\right)^2 V_1 = \pi \times \left(\frac{1.5}{12}\right)^2 V_2$. $\therefore V_2 = 5.44$ fps.

Energy: $$\frac{V_1^2}{2g} + \frac{p_1}{\gamma} = \frac{V_2^2}{2g} + \frac{p_2}{\gamma} + 0.37\frac{V_1^2}{2g}.$$

$$\therefore p_2 = 60 \times 144 + 62.4\left[0.63\frac{12.25^2}{64.4} - \frac{5.44^2}{64.4}\right] = 8702.9 \text{ psf or } \underline{60.44 \text{ psi}}$$

4.60 $Q = 600 \times 10^{-3}/60 = \pi \times .02^2\ V_1$. $\therefore V_1 = 7.958$ m/s.

$$V_2 = \frac{A_1 V_1}{A_2} = \frac{.04^2 \times 7.958}{.06^2} = 3.537 \text{ m/s}.$$

Energy: $$\frac{V_1^2}{2g} + \frac{p_1}{\gamma} = \frac{V_2^2}{2g} + \frac{p_2}{\gamma} + h_L.$$

$$\therefore h_L = \frac{7.958^2 - 3.537^2}{2 \times 9.81} + \frac{690\,000 - 700\,000}{9810} = \underline{1.571 \text{ m}}$$

4.61 $V_1 = Q / A_1 = \frac{0.08}{\pi \times .03^2} = 28.29$ m/s. $\therefore V_2 = 9V_1 = 254.6$ m/s.

Energy: $$\frac{V_1^2}{2g} + \frac{p_1}{\gamma} = \frac{V_2^2}{2g} + \frac{p_2}{\gamma} + .2\frac{V_1^2}{2g}.$$

$$\therefore p_1 = 9810\left[\frac{254.6^2}{2 \times 9.81} - 0.8\frac{28.29^2}{2 \times 9.81}\right] = \underline{32.1 \times 10^6 \text{ Pa}}.$$

4.62 a) Across the nozzle: $\pi \times .07^2 V_1 = \pi \times .025^2 V_2$. $\therefore V_2 = 7.84\ V_1$.

Energy: $$\frac{V_1^2}{2g} + \frac{p_1}{\gamma} = \frac{V_2^2}{2g} + \frac{p_2}{\gamma}. \qquad \therefore p_1 = 9810\frac{7.84^2 - 1}{2 \times 9.81} V_1^2.$$

For the contraction: $\pi\times.07^2 V_1 = \pi\times.05^2 V_3$. $\therefore V_3 = 1.96\ V_1$.

Energy: $\frac{V_1^2}{2g}+\frac{p_1}{\gamma}=\frac{V_3^2}{2g}+\frac{p_3}{\gamma}$.

Manometer: $\gamma\times.15+p_1 = 13.6\gamma\times.15+p_3$. $\therefore \frac{p_1}{\gamma}=12.6\times.15+\frac{p_3}{\gamma}$.

Subtract the above 2 eqns: $\frac{V_1^2}{2g}+12.6\times.15=\frac{V_3^2}{2g}=1.96^2\frac{V_1^2}{2g}$.

$\therefore (1.96^2-1)V_1^2 = 12.6\times.15\times 2g$. $\therefore V_1 = 3.612$ m/s. $\therefore p_1 = \underline{394\ 400\ \text{Pa}}$.

From the reservoir surface to section ①:

$$\frac{V_0^2}{2g}+\frac{p_0}{\gamma}+z_0=\frac{V_1^2}{2g}+\frac{p_1}{\gamma}+z_1$$

$$H=\frac{3.612^2}{19.62}+\frac{394\ 400}{9810}=\underline{40.0\ \text{m}}.$$

b) Manometer: $\gamma\times.2+p_1 = 13.6\gamma\times.2+p_3$. $\therefore \frac{p_1}{\gamma}=12.6\times.2+\frac{p_3}{\gamma}$.

Energy: $\frac{V_1^2}{2g}+\frac{p_1}{\gamma}=\frac{V_3^2}{2g}+\frac{p_3}{\gamma}$. Also, $V_3 = 1.96\ V_1$.

$\therefore \frac{V_1^2}{2g}+12.6\times.2=\frac{1.96^2 V_1^2}{2g}$. $\therefore V_1 = 4.171$ m/s.

The nozzle is the same as in part (a): $\therefore p_1 = \underline{534\ 700\ \text{Pa}}$.

From the reservoir surface to the nozzle exit:

$$\frac{V_0^2}{2g}+\frac{p_0}{\gamma}+z_0=\frac{V_2^2}{2g}+\frac{p_2}{\gamma}+z_2. \qquad \therefore H=\frac{V_2^2}{2g}=\frac{32.7^2}{2\times 9.81}=\underline{54.5\ \text{m}}.$$

4.63 a) Energy: $\frac{V_0^2}{2g}+\frac{p_0}{\gamma}+z_0=\frac{V_2^2}{2g}+\frac{p_2}{\gamma}+z_2$. $\therefore V_2=\sqrt{2gz_0}=\sqrt{2\times 9.81\times 2.4}=6.862$ m/s.

$Q = AV = .8\times 1\times 6.862 = \underline{5.49\ \text{m}^3/\text{s}}$.

For the second geometry the pressure on the surface is zero but it increases with depth. The elevation of the surface is 0.8 m.

$\therefore z_0=\frac{V_2^2}{2g}+h$. $\therefore V_2=\sqrt{2g(z_0-h)}=\sqrt{2\times 9.81\times 2}=6.264$ m/s.

$\therefore Q = .8\times 6.264 = \underline{5.01\ \text{m}^3/\text{s}}$.

Note: z_0 is measured from the channel bottom in the 2nd geometry. $\therefore z_0 = H + h$.

b) $\frac{V_0^2}{2g}+\frac{p_0}{\gamma}+z_0=\frac{V_2^2}{2g}+\frac{p_2}{\gamma}+z_2$. $\therefore V_2=\sqrt{2gz_0}=\sqrt{2\times 32.2\times\left(6+\frac{2}{2}\right)}=21.23$ fps.

$\therefore Q=AV=(2\times 1)\times 21.23=$ 42.5 cfs.

For the second geometry, the bottom is used as the datum:

$\therefore z_0=\frac{V_2^2}{2g}+0+h$. $\therefore \frac{V_2^2}{2g}=(H+h)-h$.

$\therefore V_2=\sqrt{2gH}=\sqrt{2\times 32.2\times 6}=19.66$ fps. $\therefore Q=$ 39.3 cfs.

4.64 From the reservoir surface to the exit:

$$\frac{V_0^2}{2g}+\frac{p_0}{\gamma}+z_0=\frac{V_2^2}{2g}+\frac{p_2}{\gamma}+z_2+K\frac{V_1^2}{2g}.$$

Continuity: $V_1=V_2\frac{.03^2}{.08^2}=.1406\ V_2$.

$$10=\frac{V_2^2}{2g}+5\times\frac{.1406^2V_2^2}{2g}$$

$\therefore V_2=13.36$ m/s. $\therefore Q=13.36\times\pi\times .015^2=$ 0.00944 m^3/s.

The velocity in the pipe is $V_1=1.878$ m/s.

Energy $0\to A$: $10=\frac{1.878^2}{2\times 9.81}+\frac{p_A}{9810}+.8\frac{1.878^2}{2\times 9.81}+3$. $\therefore p_A=$ 65 500 Pa.

Energy $0\to B$: $10=\frac{1.878^2}{2\times 9.81}+\frac{p_B}{9810}+2.0\frac{1.878^2}{2\times 9.81}+10$. $\therefore p_B=$ -5290 Pa.

Energy $0\to C$: $10=\frac{1.878^2}{2\times 9.81}+\frac{p_C}{9810}+12+2.8\frac{1.878^2}{2\times 9.81}$. $\therefore p_C=$ - 26 300 Pa.

Energy $0\to D$: $10=\frac{1.878^2}{2\times 9.81}+\frac{p_D}{9810}+0+5\frac{1.878^2}{2\times 9.81}$. $\therefore p_D=$ 87 500 Pa.

4.65 $\frac{V_0^2}{2g}+\frac{p_0}{\gamma}+z_0=\frac{V_2^2}{2g}+\frac{p_2}{\gamma}+z_2$. $\frac{80\,000}{9810}+4=\frac{V_2^2}{2\times 9.81}$.

$\therefore V_2=19.04$ m/s.

a) $Q=A_2V_2=\pi\times .025^2\times 19.04=$ 0.0374 m^3/s.

b) $Q=A_2V_2=\pi\times .09^2\times 19.04=$ 0.485 m^3/s.

c) $Q=A_2V_2=\pi\times .05^2\times 19.04=$ 0.1495 m^3/s.

4.66 a) $\frac{p_0}{\gamma}+z_0=\frac{V_2^2}{2g}+1.54\frac{V_1^2}{2g}$. $\frac{80\,000}{9810}+4=\frac{16V_1^2}{2g}+1.54\frac{V_1^2}{2g}$. $\therefore V_1=3.687$ m/s.

$Q=A_1V_1=\pi\times .05^2\times 3.687=$ 0.0290 m^3/s.

b) $A_1V_1 = A_2V_2$. $V_1 = \frac{.09^2}{.05^2}V_2 = 3.24V_2$.

$$\frac{80\,000}{9810} + 4 = \frac{V_2^2}{2g} + 2.3\frac{3.24^2 V_2^2}{2g}.$$ $\therefore V_2 = 3.08$ m/s.

$\therefore Q = A_2V_2 = \underline{0.0784 \text{ m}^3/\text{s}}$.

c) $$\frac{80\,000}{9810} + 4 = \frac{V_2^2}{2g} + 1.5\frac{V_2^2}{2g}.$$ $\therefore V_2 = 9.77$ m/s.

$\therefore Q = A_2V_2 = \underline{0.0767 \text{ m}^3/\text{s}}$.

4.67 Manometer: $\gamma H + \gamma z + p_1 = 13.6\gamma H + \gamma z + p_2$. $\therefore \frac{p_1}{\gamma} = 12.6H + \frac{p_2}{\gamma}$.

Energy: $$\frac{p_1}{\gamma} + \frac{V_1^2}{2g} = \frac{p_2}{\gamma} + \frac{V_2^2}{2g}.$$

Combine energy and manometer: $12.6H = \frac{V_2^2 - V_1^2}{2g}$.

Continuity: $V_2 = \frac{d_1^2}{d_2^2}V_1$. $\therefore V_1^2 = 12.6H \times 2g \Big/ \left(\frac{d_1^4}{d_2^4} - 1\right)$.

$$\therefore Q = V_1\pi\frac{d_1^2}{4} = \frac{\pi}{4}\left(\frac{12.6H \times 2g}{d_1^4/d_2^4 - 1}\right)^{1/2} d_1^2 = \underline{12.35 d_1^2 d_2^2 \left(\frac{H}{d_1^4 - d_2^4}\right)^{1/2}}$$

4.68 Use the result of Problem 4.67:

a) $Q = 12.35 \times .16^2 \times .08^2 \left(\frac{.2}{.16^4 - .08^4}\right)^{1/2} = \underline{0.0365 \text{ m}^3/\text{s}}$.

b) $Q = 12.35 \times .24^2 \times .08^2 \left(\frac{.4}{.24^4 - .08^4}\right)^{1/2} = \underline{0.0503 \text{ m}^3/\text{s}}$.

c) Using English units with $g = 32.2$: $Q = 22.37 d_1^2 d_2^2 \left(\frac{H}{d_1^4 - d_2^4}\right)^{1/2}$.

$$Q = 22.37 \times \left(\frac{1}{2}\right)^2 \left(\frac{1}{4}\right)^2 \left(\frac{10/12}{.5^4 - .25^4}\right)^{1/2} = \underline{1.318 \text{ cfs}}.$$

d) $Q = 22.37 \times 1^2 \times \left(\frac{1}{3}\right)^2 \left(\frac{15/12}{1^4 - .3333^4}\right)^{1/2} = \underline{2.796 \text{ cfs}}$.

4.69 a) Energy from surface to outlet: $\frac{V_2^2}{2g} = H$. $\therefore V_2^2 = 2gH$.

Energy from constriction to outlet: $$\frac{p_1}{\gamma} + \frac{V_1^2}{2g} = \frac{p_2}{\gamma} + \frac{V_2^2}{2g}.$$

Continuity: $V_1 = 4V_2$. With $p_1 = p_v = 2450$ Pa and $p_2 = 100\ 000$ Pa,

$$\frac{2450}{9810} + \frac{16}{2 \times 9.81} \times 2gH = \frac{100\ 000}{9810} + \frac{1}{2 \times 9.81} \times 2gH. \qquad \therefore H = \underline{0.663 \text{ m}}.$$

b) With $p_1 = 0.34$ psia, $p_2 = 14.7$ psia,

$$\frac{.34 \times 144}{62.4} + \frac{16}{2g} 2gH = \frac{14.7 \times 144}{62.4} + \frac{1}{2g} 2gH. \qquad \therefore H = \underline{2.21 \text{ ft}}.$$

4.70 Continuity: $V_1 = 4V_2$. Energy — surface to exit: $\frac{V_2^2}{2g} = H$.

Energy-constriction to exit: $\frac{p_v}{\gamma} + \frac{V_1^2}{2g} = \frac{p_2}{\gamma} + \frac{V_2^2}{2g}$.

$$\therefore p_v = p_2 + \frac{V_2^2 - 16V_2^2}{2g}\gamma = p_2 - 15H\gamma = 100\ 000 - 15 \times .65 \times 9810 = 4350 \text{ Pa}.$$

From Table B.1, $T = \underline{30°C}$.

4.71 Energy — surface to surface: $z_0 = z_2 + h_L$. $\qquad \therefore 30 = 20 + 2\frac{V_2^2}{2g}$.

Continuity: $V_1 = 4V_2$. $\qquad \therefore V_1^2 = 160\,g$. $\qquad \therefore V_2^2 = 10\,g$.

Energy — surface to constriction: $30 = \frac{160g}{2g} + \frac{(-94\ 000)}{9810} + z_1$

$\therefore z_1 = -40.4$ m. $\qquad \therefore H = 40.4 + 20 = \underline{60.4 \text{ m}}$.

4.72 Continuity: $V_2 = \frac{10^2}{6^2} V_1 = 2.778\ V_1$.

Energy: $\frac{V_1^2}{2g} + \frac{p_1}{\gamma} = \frac{V_2^2}{2g} + \frac{p_2}{\gamma}$. $\qquad \frac{V_1^2}{2g} + \frac{200\ 000}{9810} = \frac{2.778^2 V_1^2}{2g} + \frac{2450}{9810}$.

$\therefore V_1 = 7.67$ m/s. $\qquad \therefore Q = \pi \times .05^2 \times 7.67 = \underline{0.0602 \text{ m}^3/\text{s}}$.

4.73 Velocity at exit = V_e. Velocity in constriction = V_1. Velocity in pipe = V_2.

Energy – surface to exit: $\frac{V_e^2}{2g} = H$. $\qquad \therefore V_e^2 = 2gH$.

Continuity across nozzle: $V_2 = \frac{D^2}{d^2} V_e$. Also, $V_1 = 4V_2$.

Energy – surface to constriction: $H = \frac{V_1^2}{2g} + \frac{p_v}{\gamma}$.

a) $5 = \frac{1}{2g}\left(16 \times \frac{D^4}{.2^4} \times 2g \times 5\right) + \frac{-97\,550}{9810}$. $\quad \therefore D = \underline{0.131 \text{ m}}$

b) $15 = \frac{1}{2g}\left(16 \frac{D^4}{(8/12)^4} \times 2g \times 15\right) + \frac{(.34 - 14.7)144}{62.4}$. $\quad \therefore D = 0.446'$ or $\underline{5.35''}$

4.74 Energy — surface to exit: $3 = \frac{V_2^2}{2g} + 4\frac{V_2^2}{2g}$. $\quad \therefore V_2^2 = 11.77$.

Energy — surface to "A": $3 = \frac{11.77}{2 \times 9.81} + \frac{1176 - 100\,000}{9810} + (H + 3) + 1.5\frac{11.77}{2 \times 9.81}$.

$\therefore H = \underline{8.57 \text{ m}}$.

4.75 $\dot{m} = \rho A V = 1000 \times \pi \times .02^2 \times 40 = 50.27 \text{ kg/s}$.

$\dot{W}_P = 50.27 \times 9.81\left[\frac{10^2 - 40^2}{2 \times 9.81} + \frac{800\,000}{9810}\right] / 0.85 = \underline{2960 \text{ W}}$.

4.75E $\dot{m} = \rho A V = 1.94 \times \pi \times \left(\frac{1}{12}\right)^2 \times 120 = 5.079 \text{ slug/sec}$.

$\dot{W}_P = 5.079 \times 32.2\left[\frac{30^2 - 120^2}{2 \times 32.2} + \frac{120 \times 144}{62.4}\right] / 0.85 = 12{,}950 \frac{\text{ft-lb}}{\text{sec}}$ or $\underline{23.5 \text{ Hp}}$.

4.76 $\dot{m} = \rho A V = 1000 \times \pi \times .02^2 \times 40 = 50.27 \text{ kg/s}$.

$20\,000 = 50.27 \times 9.81\left[\frac{10^2 - 40^2}{2 \times 9.81} + \frac{\Delta p}{9810}\right] / 0.82$. $\quad \therefore \Delta p = \underline{1.088 \times 10^6 \text{ Pa}}$.

4.77 $-\dot{W}_T = 2 \times 1000 \times 9.81\left[\frac{0 - 10.2^2}{2 \times 9.81} + \frac{-600\,000}{9810}\right] \times 0.87$. $\quad \therefore \dot{W}_T = \underline{1.304 \times 10^6 \text{ W}}$.

We used $V_2 = Q / A_2 = \frac{2}{\pi \times .25^2} = 10.2 \text{ m/s}$.

4.78 $-10\,000\,000 = 15 \times 1000 \times 9.81\left[\frac{3.06^2 - 4.77^2}{2 \times 9.81} + \frac{120\,000 - 900\,000}{9810}\right]\eta_T$.

$V_1 = \frac{15}{\pi \times 1^2} = 4.77 \frac{\text{m}}{\text{s}}$. $\quad V_2 = \frac{15}{\pi \times 1.25^2} = 3.06 \frac{\text{m}}{\text{s}}$. $\quad \therefore \eta_T = \underline{0.847}$

4.78E $V_1 = \dfrac{450}{\pi \times 3^2} = 15.9$ fps. $\quad V_2 = \dfrac{450}{\pi \times 3.75^2} = 10.19$ fps.

$$-10{,}000\left(\frac{1}{.746}\right) \times 550 = 450 \times 1.94 \times 32.2\left[\frac{10.19^2 - 15.9^2}{2 \times 32.2} + \frac{(18-140)144}{62.4}\right]\eta_T.$$

$$\therefore\ \eta_T = \underline{0.924}$$

4.79 a) $\dot{Q} - \dot{W}_S = \dot{m}g\left[\dfrac{V_2^2 - V_1^2}{2g} + \dfrac{p_2}{\gamma_2} - \dfrac{p_1}{\gamma_1} + z_2 - z_1 + \dfrac{c_v}{g}(T_2 - T_1)\right].$

The above is Eq. 4.4.17 with Eq. 4.4.18 and Eq. 1.7.16.

$\gamma_1 = \dfrac{p_1 g}{RT_1} = \dfrac{85 \times 9.81}{.287 \times 293} = 9.92\ \text{N/m}^3.\quad \gamma_2 = \dfrac{600 \times 9.81}{.287\ T_2} = \dfrac{20\ 500}{T_2}.$

$$\therefore -(-1\ 500\ 000) = 5 \times 9.81\left[\frac{200^2}{2 \times 9.81} + \frac{600\ 000 T_2}{20\ 500} - \frac{85\ 000}{9.92} + \frac{716.5}{9.81}(T_2 - 293)\right].$$

$\therefore T_2 = 572$ K or $\underline{299^\circ\text{C}}$.

Be careful of units! $p_2 = 600\ 000$ Pa, $c_v = 716.5 \dfrac{\text{J}}{\text{K} \cdot \text{kg}}$

b) -60 000 + 1 500 000 = same as above. $\quad \therefore T_2 = 560$ K or $\underline{287^\circ\text{C.}}$

4.80 $\gamma_1 = \dfrac{p_1 g}{RT} = \dfrac{100 \times 9.81}{.287 \times 288} = 11.87 \dfrac{\text{N}}{\text{m}^3}.\quad \gamma_2 = \dfrac{400 \times 9.81}{.287 \times 423} = 32.32 \dfrac{\text{N}}{\text{m}^3}.\quad c_v = 716.5\ \text{J/kg} \cdot \text{K}.$

Use Eq. 4.4.17 with 4.4.18 and 1.7.16. $\dot{m}g = \gamma_2 A_2 V_2 = 32.32\pi \times .01^2 \times 200$
$= 2.03$ N/s.

$$\dot{Q} + \dot{W}_c = \dot{m}g\left[\frac{V_2^2 - V_1^2}{2g} + \frac{p_2}{\gamma_2} - \frac{p_1}{\gamma_1} + \frac{c_v}{g}(T_2 - T_1) + z_2 - z_1\right].$$

$$-20\ 000 \times \frac{2.03}{9.81} + \dot{W}_c = 2.03\left[\frac{200^2}{2 \times 9.81} + \frac{400\ 000}{32.32} - \frac{100\ 000}{11.87} + \frac{716.5}{9.81}(150 - 15)\right].$$

$\therefore \dot{W}_c = \underline{36\ 300\ \text{W}}.$

4.80E $\gamma_1 = \dfrac{p_1 g}{RT_1} = \dfrac{14.7 \times 144 \times 32.2}{1716 \times 520} = 0.0764 \dfrac{\text{lb}}{\text{ft}^3}.\quad \gamma_2 = \dfrac{60 \times 144 \times 32.2}{1716 \times 760} = 0.213 \dfrac{\text{lb}}{\text{ft}^3}.$

$c_v = 4296 \dfrac{\text{ft - lb}}{\text{slug -}^\circ\text{R}}.\quad \dot{m}g = \rho AVg = \gamma AV = .213 \times \pi \times \left(\dfrac{1}{24}\right)^2 \times 600 = .697$ lb/sec.

Use Eq. 4.4.17 with Eqs. 4.4.18 and 1.7.16:

$$\dot{Q}+\dot{W}_c = \dot{m}g\left[\frac{V_2^2-V_1^2}{2g}+\frac{p_2}{\gamma_2}-\frac{p_1}{\gamma_1}+\frac{c_v}{g}(T_2-T_1)+z_2-z_1\right].$$

$$-10\times 778\times .697+\dot{W}_c = .697\left[\frac{600^2}{2\times 32.2}+\frac{60\times 144}{.213}-\frac{14.7\times 144}{.0764}+\frac{4296}{32.2}(300-60)\right]$$

$$\therefore \dot{W}_c = 40\ 600\frac{\text{ft - lb}}{\text{sec}} \quad \text{or} \quad \underline{73.8 \text{ Hp}}.$$

4.81 Energy — surface to exit: $-\dot{W}_T\eta_T = \dot{m}g\left[\frac{V_2^2}{2g}-20+4.5\frac{V_2^2}{2g}\right].$

$$V_2 = \frac{15}{\pi\times .6^2} = 13.26 \text{ m/s.} \qquad \dot{m}g = Q\gamma = 15\times 9810 = 147\ 150 \text{ N/s.}$$

$$-\dot{W}_T\times 0.8 = 147\ 150\left[\frac{13.26^2}{2\times 9.81}-20+4.5\frac{13.26^2}{2\times 9.81}\right]. \qquad \therefore \dot{W}_T = \underline{5390 \text{ kW}}.$$

4.82 Energy — surface to "C":

$$\dot{W}_P\times .8+\dot{m}g\times 10 = \left[\frac{10^2}{2\times 9.81}+\frac{200\ 000}{9810}+7.7\frac{10^2}{2\times 9.81}\right]770.5.$$

$(\dot{m}g = \rho AVg = 1000\times \pi\times .05^2\times 10\times 9.81 = 770.5 \text{ N/s.})$ $\therefore \dot{W}_P = \underline{52\ 700 \text{ W}}.$

Energy — surface to "A": $30 = \frac{10^2}{2\times 9.81}+\frac{p_A}{9810}+1.5\frac{10^2}{2\times 9.81}.$ $\therefore p_A = \underline{169\ 300 \text{ Pa}}.$

Energy — surface to "B": $\dot{W}_P\eta_P = \dot{m}g\left[\frac{V_B^2-V_O^2}{2g}+\frac{p_B-p_O}{\gamma}+z_B-z_O+K\frac{V_B^2}{2g}\right]$

$$52\ 700\times .8 = 770.5\left[\frac{10^2}{2\times 9.81}+\frac{p_B}{9810}-30+1.5\frac{10^2}{2\times 9.81}\right]. \qquad \therefore p_B = \underline{706\ 100 \text{ Pa}}.$$

4.83 Manometer: $9810\times .8+\gamma z_1+p_1 = (9810\times 13.6)\times .8+\gamma z_2+p_2+\frac{V_2^2}{2}\rho.$

$$\therefore .8+z_1+\frac{p_1}{\gamma} = 13.6\times .8+z_2+\frac{p_2}{\gamma}+\frac{V_2^2}{2g}. \qquad \left(V_1 = \frac{.6}{\pi\times .1^2} = 19.1\frac{\text{m}}{\text{s}}\right)$$

Energy: $\frac{V_1^2}{2g}+z_1+\frac{p_1}{\gamma} = H_T+z_2+\frac{p_2}{\gamma}+\frac{V_2^2}{2g}.$

$$\therefore H_T = 12.6\times .8+\frac{19.1^2}{2\times 9.81} = 28.67. \qquad \therefore \dot{W}_T = \gamma Q\eta_T H_T = \underline{152\ 600 \text{ W}}.$$

4.83E Manometer: $\gamma \times \frac{20}{12} + \gamma z_1 + p_1 = 13.6\gamma \times \frac{20}{12} + \gamma z_2 + p_2 + \frac{V_2^2}{2}\rho.$

$$\therefore \frac{20}{12} + z_1 + \frac{p_1}{\gamma} = 13.6 \times \frac{20}{12} + z_2 + \frac{p_2}{\gamma} + \frac{V_2^2}{2g}.$$

Energy: $\frac{V_1^2}{2g} + z_1 + \frac{p_1}{\gamma} = H_T + z_2 + \frac{p_2}{\gamma} + \frac{V_2^2}{2g}.$

$$\therefore \frac{20}{12} = 13.6 \times \frac{20}{12} + \frac{V_1^2}{2g} - H_T. \qquad V_1 = \frac{18}{\pi \times \left(\frac{1}{3}\right)^2} = 51.6 \text{ fps.}$$

$$\therefore H_T = 12.6 \times \frac{20}{12} + \frac{51.6^2}{2 \times 32.2} = 62.3'. \qquad \dot{W}_T = \gamma Q \eta_T H_T = 62.4 \times 18 \times .9 \times 62.3$$

$$= 62{,}980 \ \frac{\text{ft - lb}}{\text{sec}} \quad \text{or} \quad \underline{115 \text{ Hp.}}$$

4.84 Energy — across the nozzle: $\frac{p_1}{\gamma} + \frac{V_1^2}{2g} = \frac{p_2}{\gamma} + \frac{V_2^2}{2g}.$ $\quad V_2 = \frac{5^2}{2^2} V_1 = 6.25 V_1.$

$$\therefore \frac{400\ 000}{9810} + \frac{V_1^2}{2 \times 9.81} = \frac{6.5^2 V_1^2}{2 \times 9.81}. \qquad \therefore V_1 = 4.404 \text{ m/s}$$

and

$$V_A = 6.88 \text{ m/s.} \qquad V_2 = 27.53 \text{ m/s.}$$

Energy — surface to exit:

$$H_P + 15 = \frac{27.53^2}{2 \times 9.81} + 1.5 \frac{4.404^2}{2 \times 9.81} + 3.2 \frac{6.88^2}{2 \times 9.81}. \qquad \therefore H_P = 32.8 \text{ m.}$$

$\therefore \dot{W}_P = \gamma Q H_P / \eta_P = 9810 \times (\pi \times .01^2) \times 27.53 \times 36.0 / .85 = \underline{3730 \text{ W}}.$

Energy — surface to "A":

$$15 = \frac{6.88^2}{2 \times 9.81} + \frac{p_A}{9810} + 3.2 \frac{6.88^2}{2 \times 9.81}. \qquad \therefore p_A = \underline{47\ 800 \text{ Pa}}$$

Energy — surface to "B":

$$32.8 + 15 = \frac{4.404^2}{2 \times 9.81} + \frac{p_B}{9810} + 3.2 \frac{6.88^2}{2 \times 9.81}. \qquad \therefore p_B = \underline{383\ 000 \text{ Pa}}$$

4.85 $V = \frac{Q}{A} = \frac{0.1}{\pi \times .04^2} = 19.89 \text{ m/s.}$

Energy — surface to entrance: $\quad H_P = \frac{V_2^2}{2g} + \frac{p_2}{\gamma} + z_2 + K \frac{V_2^2}{2g}.$

$$\therefore H_P = \frac{19.89^2}{2 \times 9.81} + \frac{180\ 000}{9810} + 50 + 5.6 \frac{19.89^2}{2 \times 9.81} = 201.4 \text{ m.}$$

$\therefore \dot{W}_P = \gamma Q H_P / \eta_P = 9810 \times 0.1 \times 201.4 / 0.75 = \underline{263\ 000 \text{ W}}.$

4.86 Energy —surface to exit: $10 = \frac{V_2^2}{2g} + \frac{p_2}{\gamma} + z_2 + 2.2\frac{V_2^2}{2g}.$

$\therefore V_2 = 7.83 \text{ m/s}. \quad Q = 0.02 = 7.83 \times \pi d_2^2 / 4. \quad \therefore d_2 = \underline{0.0570 \text{ m}}.$

4.87 Depth on raised section = y_2. Continuity: $3 \times 3 = V_2 y_2$.

Energy (see Eq. 4.4.21): $\frac{3^2}{2g} + 3 = \frac{V_2^2}{2g} + (0.4 + y_2).$

$\therefore 3.059 = \frac{9^2}{2g\,y_2^2} + y_2, \quad \text{or} \quad y_2^3 - 3.059\,y_2^2 + 4.128 = 0$

Trial-and-error:

$$\left.\begin{array}{ll} y_2 = 2.0: & -.11 \stackrel{?}{=} 0. \\ y_2 = 1.8: & +.05 \stackrel{?}{=} 0. \end{array}\right\} \quad \therefore y_2 = \underline{1.85 \text{ m}}.$$

$$\left.\begin{array}{ll} y_2 = 2.1: & -.1 \stackrel{?}{=} 0. \\ y_2 = 2.3: & +.1 \stackrel{?}{=} 0. \end{array}\right\} \quad \therefore y_2 = \underline{2.22 \text{ m}}.$$

The depth that actually occurs depends on the downstream conditions. We cannot select a "correct" answer between the two.

4.88 Mass flux occurs as shown. The velocity of all fluid elements leaving the top and bottom is approximately 32 m/s. The distance where $u = 32$ m/s is $y = \pm 2$ m. To find $\dot{m}_3$ use continuity:

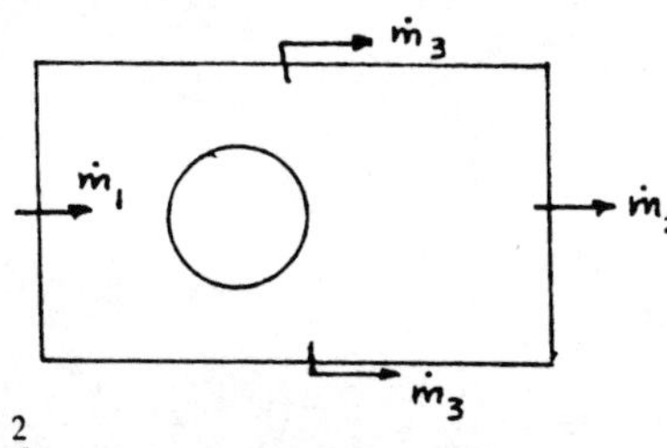

$\dot{m}_1 = \dot{m}_2 + 2\dot{m}_3. \qquad \rho 4 \times 10 \times 32 = \rho 2\int_0^2 (28 + y^2) 10 dy + 2\dot{m}_3.$

$\therefore \dot{m}_3 = 640\rho - 10\rho\left[28 \times 2 + \frac{8}{3}\right] = 53.3\rho.$

$\text{Rate of K.E. loss} = \dot{m}_1\frac{V_1^2}{2} - 2\dot{m}_3\frac{V_1^2}{2} - \rho 2\int_0^2 \frac{u^3}{2} 10 dy$

$= 1280\rho\frac{32^2}{2} - 53.3\rho 32^2 - 10\rho\int_0^2 (28 + y^2)^3 dy$

$= \rho[655\,360 - 54\,579 - 507\,320] = \underline{115\,000 \text{ W}}.$

4.89 The average velocity at section 2 is also 8 m/s. The kinetic-energy-correction factor for a parabola is 2 (see Example 4.9). The energy equation is:

$$\frac{V_1^2}{2g}+\frac{p_1}{\gamma}=\alpha_2\frac{\overline{V}_2^2}{2g}+\frac{p_2}{\gamma}+h_L.$$

$$\frac{8^2}{2\times 9.81}+\frac{150\ 000}{9810}=2\frac{8^2}{2\times 9.81}+\frac{110\ 000}{9810}+h_L.$$

$$\therefore h_L=\underline{0.815\text{ m}}.$$

4.90 $\overline{V}=\frac{1}{A}\int VdA=\frac{1}{2}\int_0^2(28+y^2)dy=\frac{1}{2}\left[28\times 2+\frac{2^3}{3}\right]=29.33\text{ m/s}$

$$\alpha=\frac{1}{A\overline{V}^3}\int V^3dA=\frac{1}{2\times 29.33^3}\int_0^2(28+y^2)^3dy$$

$$=\frac{1}{2\times 29.33^3}\left[28^3\times 2+3\times 28^2\times 2^3/3+3\times 28\times 2^5/5+2^7/7\right]=\underline{1.005}$$

4.91 a)

$$\overline{V}=\frac{1}{A}\int VdA=\frac{1}{\pi\times 0.01^2}\int_0^{0.01}10\left(1-\frac{r^2}{0.01^2}\right)2\pi rdr=\frac{20}{0.01^2}\left[\frac{0.01^2}{2}-\frac{0.01^4}{4\times 0.01^2}\right]=5\text{ m/s}$$

$$\alpha=\frac{1}{A\overline{V}^3}\int V^3dA=\frac{1}{\pi\times 0.01^2\times 5^3}\int_0^{0.01}10^3\left(1-\frac{r^2}{0.01^2}\right)^3 2\pi rdr$$

$$=\frac{2000}{0.01^2\times 5^3}\left(\frac{0.01^2}{2}-\frac{3\times 0.01^4}{4\times 0.01^2}+\frac{3\times 0.01^6}{6\times 0.01^4}-\frac{0.01^8}{8\times 0.01^6}\right)=\underline{2.00}$$

b)

$$\overline{V}=\frac{1}{A}\int VdA=\frac{1}{0.02w}\int_0^{0.02}10\left(1-\frac{y^2}{0.02^2}\right)wdy=\frac{10}{0.02}\left(0.02-\frac{0.02^3}{3\times 0.02^2}\right)=6.67\text{ m/s}$$

$$\alpha=\frac{1}{A\overline{V}^3}\int V^3dA=\frac{1}{0.02\times w\times 6.67^3}\int_0^{0.02}10^3\left(1-\frac{y^2}{0.02^2}\right)^3 wdy$$

$$=\frac{1000}{0.02\times 6.67^3}\left(0.02-\frac{3\times 0.02^3}{3\times 0.02^2}+\frac{3\times 0.02^5}{5\times 0.02^4}-\frac{0.02^7}{7\times 0.02^6}\right)=\underline{1.541}$$

4.92 $\overline{V}=\frac{1}{A}\int VdA=\frac{1}{\pi R^2}\int_0^R u_{max}\left(1-\frac{r}{R}\right)^{1/n}2\pi rdr=-2u_{max}\left(\frac{n}{2n+1}-\frac{n}{n+1}\right)$

$$K.E.=\rho\int V\left(\frac{V^2}{2}\right)dA=\frac{\rho}{2}\int_0^R u_{max}^3\left(1-\frac{r}{R}\right)^{3/n}2\pi rdr=\rho\pi u_{max}^3\left[-R^2\left(\frac{n}{3+2n}-\frac{n}{3+n}\right)\right]$$

a) $\overline{V} = -2u_{\max}\left(\frac{5}{11} - \frac{5}{6}\right) = 0.758\ u_{\max}$

$$K.E. = \rho\pi R^2 u_{\max}^3\left(\frac{5}{8} - \frac{5}{13}\right) = \underline{0.24\ \rho\pi R^2 u_{\max}^3}$$

$$\alpha = \frac{K.E.}{\frac{1}{2}\rho A\overline{V}^3} = \frac{0.24\rho\pi R^2 u_{\max}^3}{\frac{1}{2}\rho\pi R^2 \times 0.758^3 u_{\max}^3} = \underline{1.102}$$

b) $\overline{V} = -2u_{\max}\left(\frac{7}{15} - \frac{7}{8}\right) = 0.817\ u_{\max}$

$$K.E. = \rho\pi u_{\max}^3 R^2\left(\frac{7}{10} - \frac{7}{17}\right) = \underline{0.288\ \rho\pi R^2 u_{\max}^3}$$

$$\alpha = \frac{K.E.}{\rho A\overline{V}\left(\frac{\overline{V}^2}{2}\right)} = \frac{0.288\rho\pi R^2 u_{\max}^3}{\rho\pi R^2 \times 0.817 u_{\max}\left(\frac{0.817^2 u_{\max}^2}{2}\right)} = \underline{1.056}$$

c) $\overline{V} = -2u_{\max}\left(\frac{9}{19} - \frac{9}{10}\right) = 0.853\ u_{\max}$

$$K.E. = \rho\pi R^2 u_{\max}^3\left(\frac{9}{12} - \frac{9}{21}\right) = \underline{0.321\ \rho\pi R^2 u_{\max}^3}$$

$$\alpha = \frac{K.E.}{\frac{1}{2}\rho A\overline{V}^3} = \frac{0.321\rho\pi R^2 u_{\max}^3}{\frac{1}{2}\rho\pi R^2 \times 0.853^3 u_{\max}^3} = \underline{1.034}$$

4.93 Engine power = $F_D \times V_\infty + \dot{m}\left(\frac{V_2^2 - V_1^2}{2} + \tilde{u}_2 - \tilde{u}_1\right)$

$$\dot{m}_f g_f = F_D V_\infty + \dot{m}\left[\frac{V_2^2 - V_1^2}{2} + c_v(T_2 - T_1)\right]$$

4.94 $\dot{W}\eta = F_D \times V$

$$\frac{10^{-3}}{5}\left(\frac{\text{m}^3}{\text{km}}\right) \times 930\left(\frac{\text{kg}}{\text{m}^3}\right) \times q_f\left(\frac{\text{kJ}}{\text{kg}}\right) \times \frac{100}{3600}\left(\frac{\text{km}}{\text{s}}\right) \times 0.15 = \frac{1340}{1000} \times \frac{100\,000}{3600}$$

$$\therefore q_f = \underline{48\,030\ \text{kJ/kg}}$$

4.95 $0=\alpha_2\frac{V_2^2}{2g}+\frac{p_2}{\gamma}+z_2-\frac{V_1^2}{2g}-\frac{p_1}{\gamma}-z_1+32\frac{\nu LV}{gD^2}$

$$0=2\frac{V^2}{2\times 9.81}-0.35+32\times\frac{10^{-6}\times 180V}{9.81\times 0.02^2}.$$

$V^2+14.4V-3.434=0.\quad \therefore V=0.235\text{ m/s}$ and $Q=\underline{7.37\times 10^{-5}\text{ m}^3/\text{s}}$

4.96 Energy from surface to surface: $H_P=\frac{V_2^2}{2g}+\frac{p_2}{\gamma}+z_2-\frac{V_1^2}{2g}-\frac{p_1}{\gamma}-z_1+K\frac{V^2}{2g}.$

a) $H_P=40+5\frac{Q^2}{\pi\times 0.04^2\times 2\times 9.81}=40+50.7\,Q^2$

Try $Q=0.25$: $H_P=43.2$ (energy). $H_P=58$ (curve)

Try $Q=0.30$: $H_P=44.6$ (energy). $H_P=48$ (curve)

Solution: $Q=\underline{0.32\text{ m}^3/\text{s}}$.

b) $H_P=40+\frac{20\,Q^2}{\pi\times 0.04^2\times 2\times 9.81}=40+203\,Q^2$

Try $Q=0.25$: $H_P=52.7$ (energy). $H_P=58$ (curve)

Solution: $Q=\underline{0.27\text{ m}^3/\text{s}}$

Note: The curve does not allow for significant accuracy.

4.97 Continuity: $A_1V_1=A_2V_2+A_3V_3$

$$\pi\times 0.06^2\times 5=\pi\times 0.02^2\times 20+\pi\times 0.03^2V_3.\qquad \therefore V_3=11.11\text{ m/s}$$

Energy: energy in + pump energy = energy out

$$\dot m_1\left(\frac{V_1^2}{2}+\frac{p_1}{\rho}\right)+\dot W_P\times\eta_P=\dot m_2\left(\frac{V_2^2}{2}+\frac{p_2}{\rho}\right)+\dot m_3\left(\frac{V_3^2}{2}+\frac{p_3}{\rho}\right)$$

$$1000\pi\times 0.06^2\times 5\left(\frac{5^2}{2}+\frac{120\,000}{1000}\right)+0.85\dot W_P=1000\pi\times 0.02^2\times 20\left(\frac{20^2}{2}+\frac{300\,000}{1000}\right)$$
$$+1000\pi\times 0.03^2\times 11.11\left(\frac{11.11^2}{2}+\frac{500\,000}{1000}\right)$$

$$\therefore \dot W_P=\underline{26\,700\text{ W}}$$

4.98 $\frac{V_1^2}{2g}+\frac{p_1}{\gamma}=\frac{V_2^2}{2g}+\frac{p_2}{\gamma}.\qquad V_2=\frac{6^2}{3^2}V_1=4\,V_1.$

a) $\frac{V_1^2}{2\times 9.81}+\frac{200\,000}{9810}=\frac{16\,V_1^2}{2\times 9.81}.\qquad \therefore V_1=5.164\text{ m/s}.$

$p_1A_1 - F = \dot{m}(V_2 - V_1).$

$200\ 000\pi \times .03^2 - F = 1000\pi \times .03^2 \times 5.164(4 \times 5.164 - 5.164).$

$\therefore F = \underline{339 \text{ N}}.$

b) $\frac{V_1^2}{2 \times 9.81} + \frac{400\ 000}{9810} = \frac{16\ V_1^2}{2 \times 9.81}.$ $\therefore V_1 = 7.303 \text{ m / s}.$

$400\ 000\pi \times .03^2 - F = 1000\pi \times .03^2 \times 7.303(4 \times 7.303 - 7.303).$

$\therefore F = \underline{679 \text{ N}}.$

c) $\frac{V_1^2}{2 \times 9.81} + \frac{600\ 000}{9810} = \frac{16\ V_1^2}{2 \times 9.81}.$ $\therefore V_1 = 8.944 \text{ m / s}.$

$600\ 000\pi \times .03^2 - F = 1000\pi \times .03^2 \times 8.944(4 \times 8.944 - 8.944).$

$\therefore F = \underline{1018 \text{ N}}.$

4.98E $\frac{V_1^2}{2g} + \frac{p_1}{\gamma} = \frac{V_2^2}{2g} + \frac{p_2}{\gamma}.$ $V_2 = \frac{2.5^2}{1.2^2} V_1 = 4.34\ V_1.$

$\frac{V_1^2}{2g} + \frac{p_1}{\gamma} = \frac{18.84}{2g} V_1^2.$ $V_1^2 = \frac{2g}{\gamma} \times \frac{p_1}{17.84}.$

$\Sigma F_x = \dot{m}(V_{2x} - V_{1x}).$

$\therefore p_1A_1 - F = \rho A_1 V_1 (4.34 V_1 - V_1) = \rho A_1 V_1^2 \times 3.34.$

a) $V_1^2 = \frac{2 \times 32.2}{62.4} \frac{30 \times 144}{17.84} = 249.9.$

$\therefore F = p_1A_1 - 3.34 \rho A V_1^2 = 30\pi \times 1.25^2 - 3.34 \times 1.94 \times \frac{\pi \times 1.25^2}{144} \times 249.9 = \underline{92.1 \text{ lb.}}$

b) $V_1^2 = \frac{2 \times 32.2}{62.4} \frac{60 \times 144}{17.84} = 499.8.$

$\therefore F = p_1A_1 - 3.34 \rho A V_1^2 = 60\pi \times 1.25^2 - 3.34 \times 1.94 \times \frac{\pi \times 1.25^2}{144} \times 499.8 = \underline{184.1 \text{ lb.}}$

c) $V_1^2 = \frac{2 \times 32.2}{62.4} \frac{90 \times 144}{17.84} = 749.7.$

$\therefore F = p_1A_1 - 3.34 \rho A V_1^2 = 90\pi \times 1.25^2 - 3.34 \times 1.94 \times \frac{\pi \times 1.25^2}{144} \times 749.7 = \underline{276.2 \text{ lb.}}$

4.99 $\frac{V_1^2}{2g} + \frac{p_1}{\gamma} = \frac{V_2^2}{2g} + \frac{p_2}{\gamma}.$ $V_2 = \frac{9^2}{3^2} V_1 = 9V_1.$

$\frac{V_1^2}{2 \times 9.81} + \frac{2\ 000\ 000}{9810} = \frac{81\ V_1^2}{2 \times 9.81}.$ $\therefore V_1^2 = 50.$

$$p_1 A_1 - F = \dot{m}(V_2 - V_1) = \dot{m} 8 V_1$$

$$2\ 000\ 000\pi \times .045^2 - F = 1000\pi \times .045^2 \times 8 \times 50$$

$$\therefore F = \underline{10\ 180\ \text{N}}.$$

4.100 $\dfrac{V_1^2}{2g} + \dfrac{p_1}{\gamma} = \dfrac{V_2^2}{2g} + \dfrac{p_2}{\gamma}$. $\quad V_2 = \dfrac{A_1}{A_2} V_1$. $\quad \Sigma F_x = \dot{m}(V_{2x} - V_{1x})$.

a) $V_2 = \dfrac{10^2}{8^2} V_1 = 1.562\ V_1$. $\quad \dfrac{V_1^2}{2 \times 9.81} + \dfrac{400\ 000}{9810} = \dfrac{2.441\ V_1^2}{2 \times 9.81}$. $\quad \therefore V_1 = 23.56$ m / s.

$\therefore p_1 A_1 - F = \dot{m}(V_2 - V_1)$.

$400\ 000\pi \times .05^2 - F = 1000\pi \times .05^2 \times 23.56(.562 \times 23.56)$. $\quad \therefore F = \underline{692\ \text{N}}$.

b) $V_2 = \dfrac{10^2}{6^2} V_1 = 2.778\ V_1$. $\quad \dfrac{V_1^2}{2g} + \dfrac{400\ 000}{9810} = \dfrac{7.716\ V_1^2}{2g}$. $\quad \therefore V_1 = 10.91$ m / s.

$400\ 000\pi \times .05^2 - F = 1000\pi \times .05^2 \times 10.91(1.778 \times 10.91)$. $\quad \therefore F = \underline{1479\ \text{N}}$.

c) $V_2 = \dfrac{10^2}{4^2} V_1 = 6.25\ V_1$. $\quad \dfrac{V_1^2}{2g} + \dfrac{400\ 000}{9810} = \dfrac{39.06\ V_1^2}{2g}$. $\quad \therefore V_1 = 4.585$ m / s.

$400\ 000\pi \times .05^2 - F = 1000\pi \times .05^2 \times 4.585(5.25 \times 4.585)$. $\quad \therefore F = \underline{2275\ \text{N}}$.

d) $V_2 = \dfrac{10^2}{2^2} V_1 = 25\ V_1$. $\quad \dfrac{V_1^2}{2g} + \dfrac{400\ 000}{9810} = \dfrac{625\ V_1^2}{2g}$. $\quad \therefore V_1 = 1.132$ m / s.

$400\ 000\pi \times .05^2 - F = 1000\pi \times .05^2 \times 1.132(24 \times 1.132)$. $\quad \therefore F = \underline{2900\ \text{N}}$.

4.101 $V_2 = \dfrac{8^2}{4^2} \times 10 = 40$ m / s. $\quad \dfrac{V_1^2}{2g} + \dfrac{p_1}{\gamma} = \dfrac{V_2^2}{2g} + \dfrac{p_2}{\gamma}$.

$$\therefore p_1 = 9810\left[\frac{40^2}{2 \times 9.81} - \frac{10^2}{2 \times 9.81}\right] = 750\ 000\ \text{Pa}.$$

$p_1 A_1 - F = \dot{m}(V_{2x} - V_{1x})$. $\quad 750\ 000\pi \times .04^2 - F = 1000\pi \times .04^2 \times 10(-40 - 10)$.

$$\therefore F = \underline{6280\ \text{N}}.$$

4.101E $V_2 = 4V_1 = 120$ fps. $\quad p_1 = \gamma\left[\dfrac{V_2^2 - V_1^2}{2g}\right] = 62.4\left[\dfrac{120^2 - 30^2}{2 \times 32.2}\right] = 13{,}080$ psf.

$$F = p_1 A_1 - \dot{m}(V_{2x} - V_{1x}) = 13{,}080\,\frac{\pi}{4}\left(\frac{3}{12}\right)^2 - 1.94 \times \frac{\pi}{4}\left(\frac{3}{12}\right)^2 \times 30(-120 - 30) = \underline{1072\ \text{lb}}.$$

4.102 $V_2 = 4\ V_1$. $\quad \dfrac{V_1^2}{2g} + \dfrac{p_1}{\gamma} = \dfrac{V_2^2}{2g} + \dfrac{p_2}{\gamma}$. $\quad \therefore \dfrac{15\ V_1^2}{2g} = \dfrac{p_1}{\gamma}$.

a) $V_1^2 = \frac{2 \times 9.81}{15 \times 9810} \times 200\ 000 = 26.67. \quad \therefore V_1 = 5.16 \text{ m/s}, \quad V_2 = 20.7 \text{ m/s}.$

$p_1 A_1 - F_x = \dot{m}(V_{2x} - V_{1x}). \quad \therefore F_x = 200\ 000\pi \times .04^2 + 1000\pi \times .04^2 \times 5.16^2 = \underline{1139 \text{ N}}.$

$F_y = \dot{m}(V_{2y} - V_{1y}). \quad \therefore F_y = 1000\pi \times .04^2 \times 5.16(20.7) = \underline{537 \text{ N}}.$

b) $V_1^2 = \frac{2 \times 9.81}{15 \times 9810} \times 400\ 000 = 53.33. \quad \therefore V_1 = 7.30 \text{ m/s}, \quad V_2 = 29.2 \text{ m/s}.$

$p_1 A_1 - F_x = \dot{m}(V_{2x} - V_{1x}). \quad \therefore F_x = 400\ 000\pi \times .04^2 + 1000\pi \times .04^2 \times 7.3^2 = \underline{2280 \text{ N}}.$

$F_y = \dot{m}(V_{2y} - V_{1y}) = 1000\pi \times .04^2 \times 7.3 \times (29.2) = \underline{1071 \text{ N}}.$

c) $V_1^2 = \frac{2 \times 9.81}{15 \times 9810} \times 800\ 000 = 106.7. \quad \therefore V_1 = 10.33 \text{ m/s}, \quad V_2 = 41.3 \text{ m/s}.$

$F_x = p_1 A_1 + \dot{m} V_1^2 = 800\ 000\pi \times .04^2 + 1000\pi \times .04^2 \times 10.33^2 = \underline{4560 \text{ N}}.$

$F_y = \dot{m}(V_{2y}) = 1000\pi \times .04^2 \times 10.33(41.3) = \underline{2140 \text{ N}}.$

4.103 $V_2 = \frac{40^2}{10^2} V_1 = 80 \text{ m/s}. \qquad \frac{V_1^2}{2g} + \frac{p_1}{\gamma} = \frac{V_2^2}{2g} + \frac{p_2}{\gamma}$

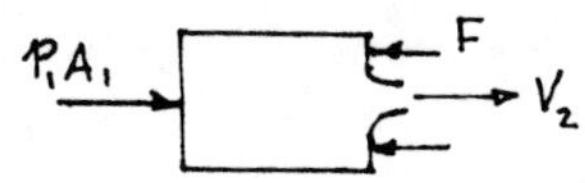

$\therefore p_1 = 9810\left[\frac{80^2}{2 \times 9.81} - \frac{5^2}{2 \times 9.81}\right] = 3.19 \times 10^6 \text{ Pa}.$

$p_1 A_1 - F = \dot{m}(V_{2x} - V_{1x}). \quad \therefore F = 3.19 \times 10^6 \pi \times .2^2 - 1000\pi \times .2^2 \times 5(80 - 5) = \underline{353\ 000 \text{ N}}.$

4.104 $A_1 V_1 = A_2 V_2. \qquad \pi \times .025^2 \times 4 = \pi(.025^2 - .02^2)V_2. \qquad \therefore V_2 = 11.11 \text{ m/s}.$

$\frac{p_1}{\gamma} + \frac{V_1^2}{2g} = \frac{p_2}{\gamma} + \frac{V_2^2}{2g}.$

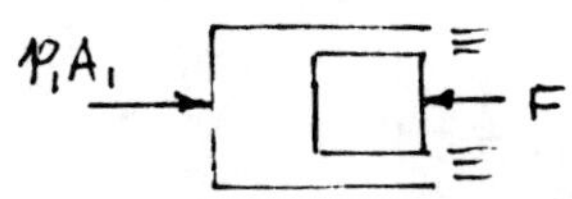

$p_1 = 9810\left[\frac{11.11^2 - 4^2}{2 \times 9.81}\right] = 53\ 700 \text{ Pa}.$

$p_1 A_1 - F = \dot{m}(V_2 - V_1). \quad \therefore F = 53\ 700\pi \times .025^2 - 1000\pi \times .025^2 \times 4(11.11 - 4) = \underline{49.6 \text{ N}}.$

4.105 Continuity: $.7\ V_1 = .1\ V_2. \qquad \therefore V_2 = 7\ V_1.$

Energy: $\frac{V_1^2}{2g} + \frac{p_1}{\gamma} + z_1 = \frac{V_2^2}{2g} + \frac{p_2}{\gamma} + z_2$

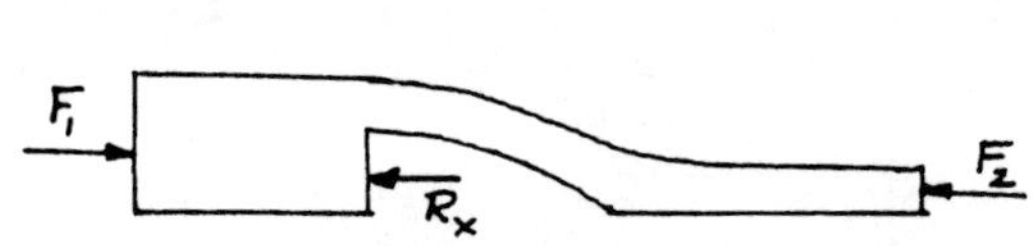

$\frac{V_1^2}{2 \times 9.81} + .7 = \frac{49 V_1^2}{2 \times 9.81} + .1. \qquad \therefore V_1 = 0.495, \quad V_2 = 3.467 \text{ m/s}.$

Momentum: $F_1 - F_2 - R_x = \dot{m}(V_2 - V_1)$

$$9810 \times .35(.7 \times 1.5) - 9810 \times .05(0.1 \times 1.5) - R_x = 1000 \times (.1 \times 1.5) \times 3.467(3.467 - .495)$$

$\therefore R_x = \underline{1986 \text{ N}}.$ $\therefore R_x$ acts to the left on the water, and to the right on the obstruction.

4.106 Continuity: $6\, V_1 = .2\, V_2$. $\therefore V_2 = 30\, V_1$.

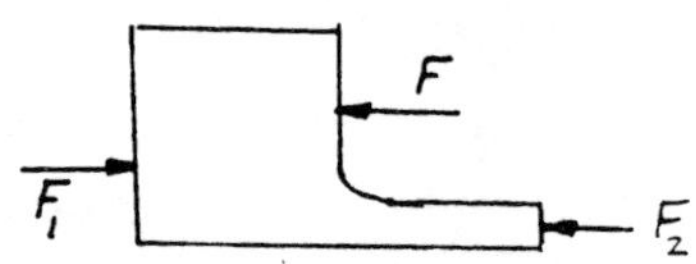

Energy: (along bottom streamline) $\frac{V_1^2}{2g} + \frac{p_1}{\gamma} + z_1 = \frac{V_2^2}{2g} + \frac{p_2}{\gamma} + z_2$

$$\frac{V_2^2/900}{2 \times 9.81} + 6 = \frac{V_2^2}{2 \times 9.81} + .2. \quad \therefore V_2 = 10.67, \quad V_1 = .36 \text{ m/s}.$$

Momentum: $F_1 - F_2 - F = \dot{m}(V_2 - V_1)$

$$9810 \times 3(6 \times 4) - 9810 \times .1(.2 \times 4) - F = 1000 \times (.2 \times 4) \times 10.67(10.67 - .36)$$

$\therefore F = \underline{618\ 000 \text{ N}}.$ (F acts to the right on the gate.)

4.107 a) $8 \times .6 = V_2 y_2$. $F_1 - F_2 = \dot{m}(V_2 - V_1)$.

$$\gamma \times .3 \times .6w - \gamma \frac{y_2}{2} y_2 w = \rho .6w \times 8\left(\frac{8 \times .6}{y_2} - 8\right).$$

$$\frac{\gamma}{2}(.36 - y_2^2) = 4.8\rho \times 8 \frac{.6 - y_2}{y_2}. \quad \therefore (.6 + y_2)y_2 = \frac{4.8 \times 8 \times 2}{9.81}.$$

$y_2^2 + .6y_2 - 7.829 = 0.$ $\therefore y_2 = \underline{2.51 \text{ m}}.$ (See Example 4.15.)

b) $$y_2 = \frac{1}{2}\left[-y_1 + \sqrt{y_1^2 + \frac{8}{g} y_1 V_1^2}\right] = \frac{1}{2}\left[-.4 + \sqrt{.4^2 + \frac{8}{9.81} \times .4 \times 12^2}\right] = \underline{3.23 \text{ m}}.$$

c) $$y_2 = \frac{1}{2}\left[-y_1 + \sqrt{y_1^2 + \frac{8}{g} y_1 V_1^2}\right] = \frac{1}{2}\left[-2 + \sqrt{2^2 + \frac{8}{32.2} \times 2 \times 20^2}\right] = \underline{6.12 \text{ ft}}.$$

d) $$y_2 = \frac{1}{2}\left[-y_1 + \sqrt{y_1^2 + \frac{8}{g} y_1 V_1^2}\right] = \frac{1}{2}\left[-3 + \sqrt{3^2 + \frac{8}{32.2} \times 3 \times 30^2}\right] = \underline{11.54 \text{ ft}}.$$

4.108 Continuity: $V_2 y_2 = V_1 y_1 = 4V_2 y_1$. $\therefore y_2 = 4y_1$.

Use the result of Example 4.15: $$y_2 = \frac{1}{2}\left[-y_1 + \left(y_1^2 + \frac{8}{g} y_1 V_1^2\right)^{1/2}\right]$$

a) $y_2 = 4 \times .8 = \underline{3.2 \text{ m}}.$

$$3.2 = \frac{1}{2}\left[-.8 + \left(.8^2 + \frac{8}{9.81} \times .8 \times V_1^2\right)^{1/2}\right]. \quad \therefore V_1 = \underline{8.86 \text{ m/s}}.$$

b) $y_2 = 4 \times 2 = \underline{8 \text{ ft}}.$

$$8 = \frac{1}{2}\left[-2 + \left(2^2 + \frac{8}{32.2} \times 2 \times V_1^2\right)^{1/2}\right]. \qquad \therefore V_1 = \underline{25.4 \text{ fps}}.$$

4.109 $V = \frac{9}{3 \times 3} = 1 \text{ m/s}. \qquad \frac{1^2}{2 \times 9.81} + 3 = \frac{V_1^2}{2 \times 9.81} + y_1. \quad V_1 y_1 = 1 \times 3.$

$$\therefore 3.05 = \frac{V_1^2}{19.62} + \frac{3}{V_1}. \quad \text{Trial-and-error:} \quad \left.\begin{matrix} V_1 = 7: & 3.05 \overset{?}{=} 2.93 \\ V_1 = 7.2: & 3.05 \overset{?}{=} 3.06 \end{matrix}\right\} \begin{matrix} V_1 = 7.19 \text{ m/s}. \\ y_1 = .417 \text{ m}. \end{matrix}$$

$$y_2 = \frac{1}{2}\left[-.417 + \left(.417^2 + \frac{8}{9.81} \times .417 \times 7.19^2\right)^{1/2}\right] = \underline{1.90 \text{ m}}.$$

$V_2 \times 1.9 = 7.19 \times .417. \qquad V_2 = \underline{1.58 \text{ m/s}}.$

4.110 Refer to Example 4.15: $\gamma \frac{y_1}{2} y_1 w - \gamma \times 1 \times 2w = \rho \times 2w \times 3\left(3 - \frac{6}{y_1}\right). \quad (V_1 y_1 = 3 \cdot 2).$

$$\therefore \frac{\gamma}{2}(y_1^2 - 4) = 18\rho\left(\frac{y_1 - 2}{y_1}\right). \quad \therefore (y_1 + 2)y_1 = \frac{36}{9.81} = 3.67. \quad \therefore y_1 = \underline{1.16 \text{ m}}, \ V_1 = \underline{5.17 \text{ m/s}}.$$

4.110E Refer to Example 4.15: $\gamma \frac{y_1}{2} y_1 w - \gamma \times 3 \times 6w = \rho \times 6w \times 10\left(10 - \frac{60}{y_1}\right). \quad (V_1 y_1 = 6 \cdot 10).$

$$\therefore \frac{\gamma}{2}(y_1^2 - 36) = 600\rho\left(\frac{y_1 - 6}{y_1}\right). \quad \therefore (y_1 + 6)y_1 = \frac{1200}{32.2} = 37.27. \quad \therefore y_1 = \underline{3.8 \text{ ft}}, \ V_1 = \underline{15.8 \text{ fps}}.$$

4.111 Continuity: $20 \times \pi \times .015^2 = V_2 \pi \times .03^2.$

$\therefore V_2 = 5 \text{ m/s}.$

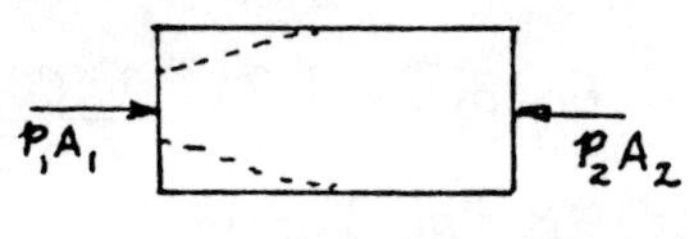

Momentum: $p_1 A_1 - p_2 A_2 = \dot{m}(V_2 - V_1).$

$$60\,000\pi \times .03^2 - p_2 \pi \times .03^2 = 1000\pi \times .015^2 \times 20(5 - 20). \quad \therefore p_2 = \underline{135 \text{ kPa}}.$$

4.112 $V_1 A_1 = 2 V_2 A_2. \qquad V_2 = 15\frac{\pi \times .05^2}{2\pi \times .025^2} = 30 \text{ m/s}.$

$$\frac{p_1}{\gamma} + \frac{V_1^2}{2g} = \frac{p_2}{\gamma} + \frac{V_2^2}{2g}. \qquad \therefore p_1 = 9810\frac{30^2 - 15^2}{2 \times 9.81} = 337\,500 \text{ Pa}.$$

$\Sigma F_x = \dot{m}(V_{2x} - V_{1x}). \qquad p_1 A_1 - F = \dot{m}(-V_1).$

$$\therefore F = p_1 A_1 + \dot{m} V_1 = 337\,500\pi \times .05^2 + 1000\pi \times .05^2 \times 15^2 = \underline{4420 \text{ N}}.$$

4.112E $V_1 A_1 = 2V_2 A_2$. $\quad V_2 = 45 \dfrac{\pi(2/12)^2}{2\pi(1/12)^2} = 90$ fps.

$$\frac{p_1}{\gamma} + \frac{V_1^2}{2g} = \frac{p_2}{\gamma} + \frac{V_2^2}{2g}. \qquad \therefore p_1 = 62.4\frac{90^2 - 45^2}{2 \times 32.2} = 5886 \text{ psf.}$$

$$\Sigma F_x = \dot{m}(V_{2x} - V_{1x}). \qquad p_1 A_1 - F = \dot{m}(-V_1).$$

$$\therefore F = p_1 A_1 + \dot{m} V_1 = 5886\pi\left(\frac{2}{12}\right)^2 + 1.94\pi\left(\frac{2}{12}\right)^2 \times 45^2 = \underline{857 \text{ lb}}.$$

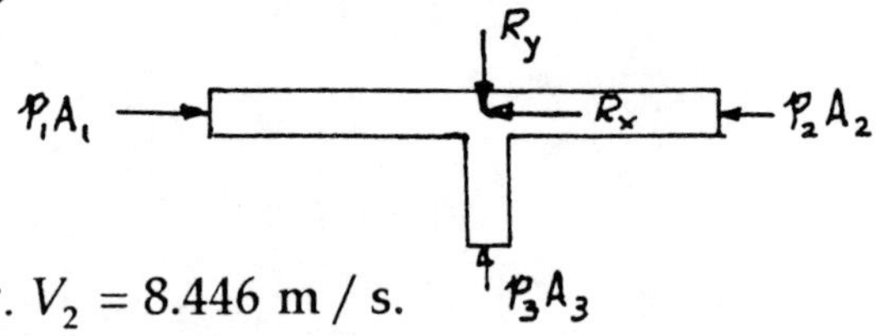

4.113 $\dot{m}_1 = 1000\pi \times .03^2 \times 12 = 33.93$ kg / s.

$\dot{m}_3 = 1000\pi \times .02^2 \times 8 = 10.05$ kg / s.

$\therefore \dot{m}_2 = \dot{m}_1 - \dot{m}_3 = 23.88 = 1000\pi \times .03^2 V_2$. $\quad \therefore V_2 = 8.446$ m / s.

Energy — ① → ②: $\dfrac{V_1^2}{2g} + \dfrac{p_1}{\gamma} = \dfrac{V_2^2}{2g} + \dfrac{p_2}{\gamma}$. $\quad \therefore p_2 = 500\,000 + \dfrac{12^2 - 8.446^2}{2 \times 9.81} \times 9810$

Energy — ① → ③: $\dfrac{V_1^2}{2g} + \dfrac{p_1}{\gamma} = \dfrac{V_3^2}{2g} + \dfrac{p_3}{\gamma}$. $\qquad$ =536 300 Pa.

$$\therefore p_3 = 500\,000 + \frac{12^2 - 8^2}{2 \times 9.81} 9810 = 540\,000 \text{ Pa.}$$

$p_1 A_1 - p_2 A_2 - R_x = \dot{m}_2 V_{2x} + \dot{m}_3 V_{3x} - \dot{m}_1 V_{1x}$.

$\therefore R_x = 500\,000\pi \times .03^2 - 536\,300\pi \times .03^2 + 33.93 \times 12 - 23.88 \times 8.446 = \underline{103 \text{ N}}$.

$p_3 A_3 - R_y = \dot{m}_3 V_{3y} + \dot{m}_2 V_{2y} - \dot{m}_1 V_{1y}$.

$\therefore R_y = 540\,000\pi \times .02^2 - 10.05 \times (-8) = \underline{759 \text{ N}}$.

4.114 $\Sigma F_x = \dot{m}(V_{2x} - V_{1x})$. $\qquad -F = -\dot{m} V_1$. $\qquad V_1 = \dfrac{\dot{m}}{\rho A_1}$

$\therefore F = 300 \times 38.2 = \underline{11\,460 \text{ N}}$. $\qquad = \dfrac{300}{1000\pi \times .05^2} = 38.2$ m/s

b) $-F = \dot{m}_r (V_1 - V_B)(\cos\alpha - 1)$.

$\therefore F = 300 \times \dfrac{28.2}{38.2}(38.2 - 10) = \underline{6250 \text{ N}}$.

c) $-F = \dot{m}_r (V_1 - V_B)(\cos\alpha - 1)$.

$\therefore F = 300 \times \dfrac{48.2}{38.2}(38.2 - (-10)) = \underline{18\,250 \text{ N}}$.

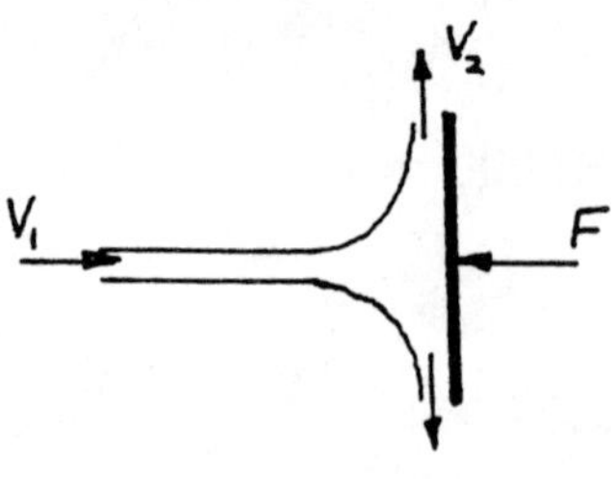

4.115 a) $-F = \dot{m}(V_{2x} - V_{1x})$. $\qquad 900 = 1000\pi \times .03^2 \times V_1^2$. $\qquad \therefore V_1 = \underline{17.84 \text{ m / s}}$.

b) $-F = \dot{m}_r (V_1 - V_B)(\cos\alpha - 1)$. $\quad 900 = 1000\pi \times .03^2 (V_1 - 10)^2$. $\quad \therefore V_1 = \underline{27.84 \text{ m / s}}$.

c) $-F = \dot{m}_r (V_1 - V_B)(\cos\alpha - 1)$. $\quad 900 = 1000\pi \times .03^2 (V_1 + 10)^2$. $\quad \therefore V_1 = \underline{7.84 \text{ m / s}}$.

4.115E a) $-F = \dot{m}(V_{2x} - V_{1x})$. $200 = 1.94\pi\left(\frac{1.25}{12}\right)^2 \times V_1^2$. $\therefore V_1 = \underline{55 \text{ fps.}}$

b) $-F = \dot{m}_r(V_1 - V_B)(\cos\alpha - 1)$. $200 = 1.94\pi\left(\frac{1.25}{12}\right)^2 (V_1 - 30)^2$. $\therefore V_1 = \underline{85 \text{ fps.}}$

c) $-F = \dot{m}_r(V_1 - V_B)(\cos\alpha - 1)$. $200 = 1.94\pi\left(\frac{1.25}{12}\right)^2 (V_1 + 30)^2$. $\therefore V_1 = \underline{25 \text{ fps.}}$

4.116 a) $-F = \dot{m}(V_{2x} - V_{1x})$. $-700 = 1000\pi\times.04^2 V_1(V_1 \times \cos 30^\circ - V_1)$. $\therefore V_1 = 32.24 \text{ m/s}$.

$\therefore \dot{m} = \rho A_1 V_1 = 1000\pi\times.04^2 \times 32.24 = \underline{162.1 \text{ kg/s}}$.

b) $-F = \dot{m}_r(V_1 - V_B)(\cos\alpha - 1)$. $-700 = 1000\pi\times.04^2(V_1 - 8)^2(.866 - 1)$. $\therefore V_1 = 40.24 \text{ m/s}$.

$\therefore \dot{m} = \rho A_1 V_1 = 1000\pi\times.04^2 \times 40.24 = \underline{202 \text{ kg/s}}$.

c) $-F = \dot{m}_r(V_1 - V_B)(\cos\alpha - 1)$. $-700 = 1000\pi\times.04^2(V_1 + 8)^2(.866 - 1)$. $\therefore V_1 = 24.24 \text{ m/s}$.

$\therefore \dot{m} = \rho A_1 V_1 = 1000\pi\times.04^2 \times 24.24 = \underline{121.8 \text{ kg/s}}$.

4.117 a) $-R_x = \dot{m}(V_{2x} - V_{1x}) = 1000\pi\times.025^2 \times 40(40\cos 60^\circ - 40)$. $\therefore R_x = \underline{1571 \text{ N}}$.

b) $-R_x = \dot{m}_r(V_1 - V_B)(\cos\alpha - 1) = 1000\pi\times.025^2 \times 20 \times 20(.5 - 1)$. $\therefore R_x = \underline{393 \text{ N}}$.

$R_y = \dot{m}_r(V_1 - V_B)\sin\alpha = 1000\pi\times.025^2 \times 20 \times (20\times.866)$. $\therefore R_y = \underline{680 \text{ N}}$.

c) $-R_x = \dot{m}_r(V_1 - V_B)(\cos\alpha - 1) = 1000\pi\times.025^2 \times 60 \times 60(.5 - 1)$. $\therefore R_x = \underline{3530 \text{ N}}$.

$R_y = \dot{m}_r(V_1 - V_B)\sin\alpha = 1000\pi\times.025^2 \times 60 \times (60\times.866)$. $\therefore R_y = \underline{6120 \text{ N}}$.

4.117E a) $-R_x = \dot{m}(V_{2x} - V_{1x}) = 1.94\pi \times \left(\frac{1}{12}\right)^2 \times 120(120\cos 60^\circ - 120)$. $\therefore R_x = \underline{305 \text{ lb}}$.

$R_y = \dot{m}(V_{2y} - V_{1y}) = 1.94\pi\left(\frac{1}{12}\right)^2 \times 120 \times (120\times.866)$. $\therefore R_y = \underline{528 \text{ lb}}$.

b) $-R_x = \dot{m}_r(V_1 - V_B)(\cos\alpha - 1) = 1.94\pi \times \left(\frac{1}{12}\right)^2 \times 60 \times 60(.5 - 1)$. $\therefore R_x = \underline{76.2 \text{ lb}}$.

$R_y = \dot{m}_r(V_1 - V_B)\sin\alpha = 1.94\pi\left(\frac{1}{12}\right)^2 \times 60 \times (60\times.866)$. $\therefore R_y = \underline{132 \text{ lb}}$.

c) $-R_x = \dot{m}_r(V_1 - V_B)(\cos\alpha - 1) = 1.94\pi \times \left(\frac{1}{12}\right)^2 \times 180 \times 180(.5 - 1)$. $\therefore R_x = \underline{686 \text{ lb}}$.

$R_y = \dot{m}_r(V_1 - V_B)\sin\alpha = 1.94\pi\left(\frac{1}{12}\right)^2 \times 180 \times (180\times.866)$. $\therefore R_y = \underline{1188 \text{ lb}}$.

4.118 $V_B = R\omega = 0.5\times 30 = 15$ m / s.

$-R_x = \dot{m}(V_1 - V_B)(\cos\alpha - 1) = 1000\pi\times.025^2 \times 40\times 25(.5-1)$. $\therefore R_x = 982$ N.

$\therefore \dot{W} = 10R_xV_B = 10\times 982\times 15 = \underline{147\ 300\ \text{W}}$.

4.119 a) $-R_x = \dot{m}(V_{2x} - V_{1x}) = 4\pi.02^2 \times 400(-400\cos 60^\circ - 400)$. $\therefore R_x = \underline{1206\text{ N}}$.

$R_y = \dot{m}(V_{2y} - V_{1y}) = 4\pi\times.02^2\times 400(400\sin 60^\circ)$. $\therefore R_y = \underline{696\text{ N}}$.

b) $-R_x = \dot{m}_r(V_1 - V_B)(\cos 120^\circ - 1) = 4\pi.02^2\times 300^2(-.5-1)$. $\therefore R_x = \underline{679\text{ N}}$.

$R_y = \dot{m}_r(V_1 - V_B)\sin\alpha = 4\pi\times.02^2\times 300^2\times.866$. $\therefore R_y = \underline{392\text{ N}}$.

c) $-R_x = \dot{m}_r(V_1 - V_B)(\cos 120^\circ - 1) = 4\pi.02^2\times 500^2(-.5-1)$. $\therefore R_x = \underline{1885\text{ N}}$.

$R_y = \dot{m}_r(V_1 - V_B)\sin\alpha = 4\pi\times.02^2\times 500^2\times.866$. $\therefore R_y = \underline{1088\text{ N}}$.

4.120 $-F_x = \dot{m}(V_1 - V_B)(\cos 120^\circ - 1) = 4\pi\times.02^2\times(400-180)^2(-.5-1)$. $\therefore R_x = 365$ N.

$V_B = 1.2\times 150 = 180$ m / s. $\dot{W} = 15\times 365\times 180 = \underline{986\ 000\ \text{W}}$.

The y-component force does no work.

4.121 Refer to Fig. 4.16: $\left.\begin{array}{r} 250\sin\beta_1 = V_{r1}\sin 45^\circ \\ 250\cos\beta_1 - 100 = V_{r1}\cos 45^\circ \end{array}\right\} \therefore \begin{array}{l} V_{r1} = 169.1 \text{ m / s } = V_{r2}. \\ \beta_1 = 28.6^\circ. \end{array}$

Note: $V_{2x} - V_{1x} = -V_{r2}\cos\alpha_2 + V_B - (V_{r1}\cos\alpha_1 + V_B) = -V_{r1}(\cos\alpha_2 + \cos\alpha_1)$.

$\therefore R_x = \dot{m}V_{r1}(\cos\alpha_2 + \cos\alpha_1) = 10\times\pi\times.01^2\times 250\times 169.1(\cos 30^\circ + \cos 45^\circ) = 209$ N.

$\therefore \dot{W} = 15V_BR_x = 15\times 100\times 209 = \underline{313\ 000\ \text{W}}$.

b) $\left.\begin{array}{r} 250\sin\beta_1 = V_{r1}\sin 60^\circ \\ 250\cos\beta_1 - 100 = V_{r1}\cos 60^\circ \end{array}\right\} 250^2 = V_{r1}^2\times.866^2 + (V_{r1}\times.5 + 100)^2.\ \therefore V_{r1} = 184.5 \text{ m / s}.$

$R_x = \dot{m}V_{r1}(\cos\alpha_2 + \cos\alpha_1) = 10\pi\times.01^2\times 250\times 184.5(\cos 30^\circ + \cos 60^\circ) = 197.9$ N.

$\therefore \dot{W} = 15V_BR_x = 15\times 100\times 197.9 = \underline{297\ 000\ \text{W}}$.

c) $\left.\begin{array}{r} 250\sin\beta_1 = V_{r1}\sin 90^\circ \\ 250\cos\beta_1 - 100 = V_{r1}\cos 90^\circ \end{array}\right\} 250^2 = V_{r1}^2 + 100^2.\ \therefore V_{r1} = 229 \text{ m / s}.$

$R_x = \dot{m}V_{r1}(\cos\alpha_2 + \cos\alpha_1) = 10\pi\times.01^2\times 250\times 229(\cos 30^\circ + 0) = 155.8$ N.

$\therefore \dot{W} = 15V_BR_x = 15\times 100\times 155.8 = \underline{234\ 000\ \text{W}}$.

4.121E a) Refer to Fig. 4.16: $\left.\begin{array}{r} 750\sin\beta_1 = V_{r1}\sin 45^\circ \\ 750\cos\beta_1 - 300 = V_{r1}\cos 45^\circ \end{array}\right\} \therefore \begin{array}{l} V_{r1} = 507 \text{ fps}. \\ = V_{r2} \end{array}$

Note: $V_{2x} - V_{1x} = -V_{r2}\cos\alpha_2 + V_B - V_{r1}\cos\alpha_1 - V_B = -V_{r1}(\cos\alpha_2 + \cos\alpha_1)$.

$\therefore R_x = \dot{m}V_{r1}(\cos\alpha_2 + \cos\alpha_1) = .015\pi\left(\frac{.5}{12}\right)^2\times 750\times 507(\cos 30^\circ + \cos 45^\circ) = 48.9$ lb.

$\therefore \dot{W} = 15R_xV_B = 15\times 48.9\times 300 = 220{,}000\dfrac{\text{ft - lb}}{\text{sec}}$ or $\underline{400\text{ Hp}}$.

b) $\left.\begin{array}{r}750\sin\beta_1 = V_{r1}\sin 60^\circ \\ 750\cos\beta_1 - 300 = V_{r1}\cos 60^\circ\end{array}\right\}\quad V_{r1} = 554 \text{ fps} = V_{r2}.$

$$\therefore R_x = \dot{m}V_{r1}(\cos\alpha_2 + \cos\alpha_1) = .015\pi \times \left(\frac{.5}{12}\right)^2 \times 750 \times 554(\cos 30^\circ + \cos 60^\circ) = 46.4 \text{ lb.}$$

$$\therefore \dot{W} = 15R_xV_B = 15 \times 46.4 \times 300 = 209{,}000\frac{\text{ft - lb}}{\text{sec}} \quad \text{or} \quad \underline{380 \text{ Hp.}}$$

c) $\left.\begin{array}{r}750\sin\beta_1 = V_{r1}\sin 90^\circ \\ 750\cos\beta_1 - 300 = V_{r1}\cos 90^\circ\end{array}\right\}\quad V_{r1} = 687 \text{ fps} = V_{r2}.$

$$\therefore R_x = \dot{m}V_{r1}(\cos\alpha_2 + \cos\alpha_1) = .015\pi \times \left(\frac{.5}{12}\right)^2 \times 750 \times 687(\cos 30^\circ + 0) = 36.5 \text{ lb.}$$

$$\therefore \dot{W} = 15R_xV_B = 15 \times 36.5 \times 300 = 164{,}300\frac{\text{ft - lb}}{\text{sec}} \quad \text{or} \quad \underline{299 \text{ Hp.}}$$

4.122 a) Refer to Fig. 4.16: $\left.\begin{array}{r}100\sin 30^\circ = V_{r1}\sin\alpha_1 \\ 100\cos 30^\circ - 20 = V_{r1}\cos\alpha_1\end{array}\right\}\quad \therefore \alpha_1 = \underline{36.9^\circ},\ V_{r1} = 83.3 \text{ m/s.}$

$$\left.\begin{array}{r}V_2\sin 60^\circ = 83.3\sin\alpha_2 \\ V_2\cos 60^\circ = 83.3\cos\alpha_2 - 20\end{array}\right\}\quad V_2 = 71.5,\quad \alpha_2 = \underline{48^\circ}.$$

$$-R_x = \dot{m}(V_{2x} - V_{1x}) = 1000\pi \times .015^2 \times 100(-71.5\cos 60^\circ - 100\cos 30^\circ). \quad \therefore R_x = 8650 \text{ N.}$$

$$\therefore \dot{W} = 12V_BR_x = 12 \times 20 \times 8650 = \underline{2.08 \times 10^6 \text{ W}}.$$

b) $\left.\begin{array}{r}100\sin 30^\circ = V_{r1}\sin\alpha_1 \\ 100\cos 30^\circ - 40 = V_{r1}\cos\alpha_1\end{array}\right\}\quad \therefore \alpha_1 = \underline{47^\circ},\ V_{r1} = V_{r2} = 68.35 \text{ m/s.}$

$$\left.\begin{array}{r}V_2\sin 60^\circ = 68.35\sin\alpha_2 \\ V_2\cos 60^\circ = 68.35\cos\alpha_2 - 40\end{array}\right\}\quad V_2 = 38.9 \text{ m/s},\quad \alpha_2 = \underline{29.5^\circ}.$$

$$-R_x = \dot{m}(V_{2x} - V_{1x}) = 1000\pi \times .015^2 \times 100(-38.9\cos 60^\circ - 100\cos 30^\circ). \quad \therefore R_x = 7500 \text{ N.}$$

$$\therefore \dot{W} = 12V_BR_x = 12 \times 40 \times 7500 = \underline{3.60 \times 10^6 \text{ W}}.$$

c) $\left.\begin{array}{r}100\sin 30^\circ = V_{r1}\sin\alpha_1 \\ 100\cos 30^\circ - 50 = V_{r1}\cos\alpha_1\end{array}\right\}\quad \therefore \alpha_1 = \underline{53.8^\circ},\ V_{r1} = V_{r2} = 61.96 \text{ m/s.}$

$$\left.\begin{array}{r}V_2\sin 60^\circ = 61.76\sin\alpha_2 \\ V_2\cos 60^\circ = 61.96\cos\alpha_2 - 50\end{array}\right\}\quad V_2 = 19.32 \text{ m/s},\quad \alpha_2 = \underline{15.66^\circ}.$$

$$-R_x = \dot{m}(V_{2x} - V_{1x}) = 1000\pi \times .015^2 \times 100(-19.32\cos 60^\circ - 100\cos 30^\circ). \quad \therefore R_x = 6800 \text{ N.}$$

$$\therefore \dot{W} = 12R_xV_B = 12 \times 6800 \times 50 = \underline{4.08 \times 10^6 \text{ W}}.$$

4.123 a) Refer to Fig. 4.16: $\left.\begin{array}{l} 50\sin 30^\circ = V_{r1}\sin\alpha_1 \\ 50\cos 30^\circ - V_B = V_{r1}\cos\alpha_1 \end{array}\right\} \quad \therefore V_{r1}^2 = 2500 - 86.6V_B + V_B^2$

$$\left.\begin{array}{l} 30\sin 60^\circ = V_{r2}\sin\alpha_2 \\ 30\cos 60^\circ - V_{r2}\cos\alpha_2 - V_B \end{array}\right\} \quad \therefore V_{r2}^2 = V_{r1}^2 = 900 + 30V_B + V_B^2.$$

Combine the above: $V_B = 13.72$ m/s. Then, $\alpha_1 = \underline{59.4^\circ}$, $\alpha_2 = \underline{42.1^\circ}$.

$$-R_x = \dot{m}(V_{2x} - V_{1x}) = 1000\pi\times.01^2 \times 50(-30\cos 60^\circ - 50\cos 30^\circ). \qquad \therefore R_x = 916 \text{ N}.$$

$$\therefore \dot{W} = 15V_B R_x = 15\times 13.72\times 916 = \underline{188\ 500\text{ W}}.$$

b) $\left.\begin{array}{l} 50\sin 30^\circ = V_{r1}\sin\alpha_1 \\ 50\cos 30^\circ - V_B = V_{r1}\cos\alpha_1 \end{array}\right\} \quad \therefore V_{r1}^2 = 2500 - 86.6V_B + V_B^2 \qquad \therefore V_B = 14.94 \text{ m/s}.$

$$\left.\begin{array}{l} 30\sin 70^\circ = V_{r2}\sin\alpha_2 \\ 30\cos 70^\circ - V_{r2}\cos\alpha_2 - V_B \end{array}\right\} \quad \therefore V_{r2}^2 = 900 + 20.52V_B + V_B^2. \qquad \alpha_1 = \underline{41.4^\circ}$$

$$\alpha_2 = \underline{48.2^\circ}$$

$$-R_x = \dot{m}(V_{2x} - V_{1x}) = 1000\pi\times.01^2 \times 50(-30\cos 70^\circ - 50\cos 30^\circ). \qquad \therefore R_x = 841 \text{ N}.$$

$$\therefore \dot{W} = 15V_B R_x = 15\times 14.94\times 841 = \underline{188\ 500\text{ W}}.$$

c) $\left.\begin{array}{l} 50\sin 30^\circ = V_{r1}\sin\alpha_1 \\ 50\cos 30^\circ - V_B = V_{r1}\cos\alpha_1 \end{array}\right\} \therefore V_{r1}^2 = 2500 - 86.6V_B + V_B^2 \quad \therefore V_B = 16.49 \text{ m/s}$

$$\left.\begin{array}{l} 30\sin 80^\circ = V_{r2}\sin\alpha_2 \\ 30\cos 80^\circ - V_{r2}\cos\alpha_2 - V_B \end{array}\right\} \quad \therefore V_{r2}^2 = 900 + 10.42V_B + V_B^2. \qquad \alpha_1 = \underline{43^\circ}$$

$$\alpha_2 = \underline{53.7^\circ}$$

$$-R_x = \dot{m}(V_{2x} - V_{1x}) = 1000\pi\times.01^2 \times 50(-30\cos 80^\circ - 50\cos 30^\circ). \qquad \therefore R_x = 762 \text{ N}.$$

$$\therefore \dot{W} = 15V_B R_x = 15\times 16.49\times 762 = \underline{188\ 500\text{ W}}.$$

4.124 To find F, sum forces normal to the plate: $\Sigma F_n = \dot{m}\left[(V_{out})_n^{\,0} - V_{1n}\right]$.

$\therefore F = 1000\times.02\times.4\times 40\left[-(-40\sin 60^\circ)\right] = \underline{11\ 080\text{ N}}$. (We have neglected friction)

a) $\Sigma F_t = 0 = \dot{m}_2V_2 + \dot{m}_3(-V_3) - \dot{m}_1\times 40\sin 30^\circ$. Bernoulli: $V_1 = V_2 = V_3$.

$$\left.\begin{array}{l} \therefore \quad 0 = \dot{m}_2 - \dot{m}_3 - .5\dot{m}_1 \\ \text{Continuity: } \dot{m}_1 = \dot{m}_2 + \dot{m}_3 \end{array}\right\} \quad \begin{array}{l} \therefore \dot{m}_2 = .75\dot{m}_1 = .75\times 320 = \underline{240\text{ kg/s}}. \\ \dot{m}_3 = \underline{80\text{ kg/s}}. \end{array}$$

b) $\Sigma F_n = \dot{m}_2V_{2n}^{\,0} + \dot{m}_3V_{3n}^{\,0} - \dot{m}_1V_{1n}$. $\quad \therefore F = -1000\times.02\times.4\times 40(-40\sin 45^\circ) = \underline{9050\text{ N}}$.

$\Sigma F_t = 0 = \dot{m}_2V_2 + \dot{m}_3(-V_3) - \dot{m}_1\times 40\sin 45^\circ$. Bernoulli: $V_1 = V_2 = V_3$.

$$\left.\begin{array}{l} \therefore \quad 0 = \dot{m}_2 - \dot{m}_3 - .707\dot{m}_1 \\ \text{Continuity: } \dot{m}_1 = \dot{m}_2 + \dot{m}_3 \end{array}\right\} \quad \begin{array}{l} \therefore \dot{m}_2 = .854\dot{m}_1 = .854\times 320 = \underline{273\text{ kg/s}}. \\ \dot{m}_3 = \underline{47\text{ kg/s}}. \end{array}$$

c) $\Sigma F_n = \dot{m}_2 V_{2n} + \dot{m}_3 V_{3n} - \dot{m}_1 V_{1n}$. $\quad \therefore F = -1000 \times .02 \times .4 \times 40(-40 \sin 30^\circ) = \underline{6400 \text{ N}}$.

$\Sigma F_t = 0 = \dot{m}_2 V_2 + \dot{m}_3(-V_3) - \dot{m}_1 \times 40 \sin 60^\circ$. Bernoulli: $V_1 = V_2 = V_3$.

$$\left.\begin{array}{r} \therefore \quad 0 = \dot{m}_2 - \dot{m}_3 - .866\dot{m}_1 \\ \text{Continuity: } \dot{m}_1 = \dot{m}_2 + \dot{m}_3 \end{array}\right\} \quad \therefore \dot{m}_2 = .933\dot{m}_1 = .933 \times 320 = \underline{298.6 \text{ kg/s}}.$$

$$\dot{m}_3 = \underline{21.4 \text{ kg/s}}.$$

4.124E To find F, sum forces normal to the plate: $\Sigma F_n = \dot{m}_2 \overset{0}{\cancel{V_{2n}}} + \dot{m}_3 \overset{0}{\cancel{V_{3n}}} - \dot{m}_1 V_{1n}$.

$\therefore F = -1.94 \times \dfrac{1}{12} \times \dfrac{20}{12} \times 120(-120 \sin 60^\circ) = \underline{3360 \text{ lb}}$. (We have neglected friction)

a) $\Sigma F_t = 0 = \dot{m}_2 V_2 + \dot{m}_3(-V_3) - \dot{m}_1 \times 120 \sin 30^\circ$. Bernoulli: $V_1 = V_2 = V_3$.

$$\left.\begin{array}{r} \therefore \quad 0 = \dot{m}_2 - \dot{m}_3 - 0.5\dot{m}_1 \\ \text{Continuity: } \dot{m}_1 = \dot{m}_2 + \dot{m}_3 \end{array}\right\} \quad \therefore \dot{m}_2 = .75\dot{m}_1 = .75 \times 1.94 \times \frac{20}{144} \times 120$$

$$= \underline{22.6 \text{ slug/sec.}}$$

$$\dot{m}_3 = \dot{m}_1 - \dot{m}_2 = \underline{9.7 \text{ slug/sec.}}$$

b) $\Sigma F_n = -\dot{m}_1 V_{1n} = 1.94 \times \dfrac{20}{144} \times 120(-120 \sin 45^\circ)$. $\quad \therefore F = \underline{2743 \text{ lb}}$.

$\Sigma F_t = 0 = \dot{m}_2 V_2 + \dot{m}_3(-V_3) - \dot{m}_1 V_1 \sin 45^\circ$. Bernoulli: $V_1 = V_2 = V_3$.

$$\left.\begin{array}{r} \therefore \quad 0 = \dot{m}_2 - \dot{m}_3 - .707\dot{m}_1 \\ \text{Continuity: } \dot{m}_1 = \dot{m}_2 + \dot{m}_3 \end{array}\right\} \quad \begin{array}{l} \therefore \dot{m}_2 = .854\dot{m}_1 = \underline{27.6 \text{ slug/sec.}} \\ \dot{m}_3 = \dot{m}_1 - \dot{m}_2 = \underline{4.7 \text{ slug/sec.}} \end{array}$$

c) $\Sigma F_n = -\dot{m}_1 V_{1n} = 1.94 \times \dfrac{20}{144} \times 120(-120 \sin 30^\circ)$. $\quad \therefore F = \underline{1940 \text{ lb}}$.

$\Sigma F_t = 0 = \dot{m}_2 V_2 + \dot{m}_3(-V_3) - \dot{m}_1 V_1 \sin 60^\circ$. Bernoulli: $V_1 = V_2 = V_3$.

$$\left.\begin{array}{r} \therefore \quad 0 = \dot{m}_2 - \dot{m}_3 - .866\dot{m}_1 \\ \text{Continuity: } \dot{m}_1 = \dot{m}_2 + \dot{m}_3 \end{array}\right\} \quad \begin{array}{l} \therefore \dot{m}_2 = .933\dot{m}_1 = \underline{30.2 \text{ slug/sec.}} \\ \dot{m}_3 = \underline{2.2 \text{ slug/sec.}} \end{array}$$

4.125 $F = \dot{m}_r (V_{1r})_n = 1000 \times .02 \times .4 \times (40 + 20)^2 \sin 60^\circ = 24\,940 \text{ N}$.

$F_x = 24\,940 \cos 30^\circ = 21\,600 \text{ N}$. $\quad \therefore \dot{W} = 21\,600 \times 20 = \underline{432\,000 \text{ W}}$.

4.126 $F = \dot{m}_r (V_{1r})_n = 1000 \times .02 \times .4(40 - V_B)^2 \sin 60^\circ$. $\quad F_x = 8(40 - V_B^2)\sin^2 60^\circ$.

$\dot{W} = V_B F_x = 8V_B(40 - V_B)^2 \times .75 = 6(1600V_B - 80V_B^2 + V_B^3)$.

$\dfrac{d\dot{W}}{dV_B} = 6(1600 - 160V_B + 3V_B^2) = 0$. $\quad \therefore V_B = \underline{13.33 \text{ m/s}}$.

4.127 $F = \dot{m}_r(\overset{0}{\cancel{V_1}} - V_B)(\cos\alpha - 1) = 1000 \times .1 \times .6 V_B(-V_B)(-2) = 120 V_B^2.$

At $t = 0$: $F = 120 \times \left(\dfrac{120 \times 1000}{3600}\right)^2 = 133\,300$ N. $a_o = \dfrac{133\,300}{100\,000} = \underline{1.33 \text{ m/s}^2}$

$$\frac{-F}{m} = \frac{dV_B}{dt} = \frac{-120V_B^2}{100\,000}. \quad \therefore \int_{33.33}^{16.67} -\frac{dV_B}{V_B^2} = .0012\int_0^t dt.$$

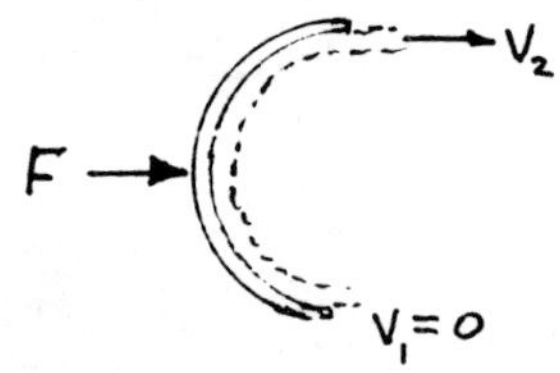

$$\therefore \left[\frac{1}{16.67} - \frac{1}{33.33}\right] = .0012 \text{ t}. \quad \therefore t = \underline{26.6 \text{ sec}}.$$

4.127E $F = \dot{m}_r(\overset{0}{\cancel{V_1}} - V_B)(\cos\alpha - 1) = 1.94 \times \dfrac{4}{12} \times \dfrac{24}{12} \times V_B(-V_B)(-2) = 2.587 V_B^2.$

At $t = 0$: $F = 2.587 \times \left(\dfrac{90 \times 5280}{3600}\right)^2 = 45{,}080$ lb. $\therefore a_o = \dfrac{F}{m} = \dfrac{45{,}080}{6000} = \underline{7.51 \text{ ft/sec}^2}$

$$\frac{-F}{m} = \frac{dV_B}{dt}. \quad \therefore \frac{dV_B}{dt} = \frac{-2.587V_B^2}{6000}. \quad \therefore \int_{132}^{58.67} \frac{-dV_B}{V_B^2} = \frac{2.587}{6000}\int_0^t dt.$$

$$\therefore \left[\frac{1}{58.67} - \frac{1}{132}\right] = \frac{2.587}{6000} t. \quad \therefore t = \underline{22.0 \text{ sec}}.$$

4.128 $F = \dot{m}_r(V_1 - V_B)(\cos\alpha - 1) = 90 \times .8 \times 2.5 \times 13.89 \times (-13.89)(-1) = 34725$ N.

$\left(V_B = \dfrac{50 \times 1000}{3600} = 13.89 \text{ m/s}\right)$ $\therefore \dot{W} = 34725 \times 13.89 = 482\,000$ W or $\underline{647 \text{ Hp}}$.

4.129 $F = \dot{m}_r(\overset{0}{\cancel{V_1}} - V_B)(\cos\alpha - 1) = 1000 \times .06 \times .2 \times V_B(-V_B)(-2) = 24\ V_B^2.$

$$-F = mV_B\frac{dV_B}{dx}. \quad \therefore -24V_B^2 = 5000V_B\frac{dV_B}{dx}.$$

(See figure in No. 4.127)

$$-\int_0^x \frac{24\,dx}{5000} = \int_{250}^{27.78} \frac{dV_B}{V_B}. \quad -\frac{24}{5000}x = \ell n 27.78 - \ell n 250. \quad \therefore x = \underline{458 \text{ m}}.$$

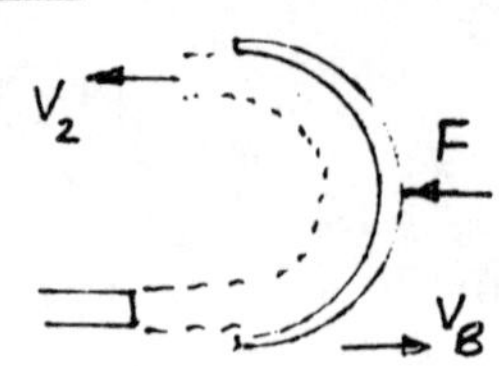

4.130 $-F = \dot{m}_r(V_1 - V_B)(\cos\alpha - 1) = 1000\pi \times .03^2 (V_1 - V_B)^2(-2).$

$$\therefore F = 5.655(V_1 - V_B)^2 = 300\frac{dV_B}{dt}.$$

At $t = 0$, $V_B = 0$. $300\dfrac{dV_B}{dt} = 5.655V_1^2$. With $\dfrac{dV_B}{dt} = 2$, $V_1 = \underline{10.3 \text{ m/s}}$.

$$5.655(10.3 - V_B)^2 = 300\frac{dV_B}{dt} \quad \text{or} \quad \frac{5.655}{300}\int_0^2 dt = \int_0^{V_B} \frac{dV_B}{(10.3 - V_B)^2}.$$

$$0.0377 = \left[\frac{-1}{V_B - 10.3} - \frac{1}{10.3}\right]. \quad \therefore V_B = \underline{2.88 \text{ m/s}}.$$

4.130E $-F = \dot{m}_r(V_1 - V_B)(\cos\alpha - 1) = 1.94\pi \times \left(\frac{1.25}{12}\right)^2 (V_1 - V_B)^2(-2)$.

$\therefore F = 0.1323(V_1 - V_B)^2 = 20\frac{dV_B}{dt}$. (See figure above)

At $t = 0$, $V_B = 0$. Then $20\frac{dV_B}{dt} = 0.1323V_1^2$. With $\frac{dV_B}{dt} = 6$, $V_1 = \underline{30.1 \text{ fps}}$.

For $t > 0$, $\int_0^{V_B} \frac{dV_B}{(30.1 - V_B)^2} = 0.006615\int_0^2 dt$. $0.01323 = \frac{1}{30.1 - V_B} - \frac{1}{30.1}$. $\therefore V_B = \underline{8.57 \text{ fps}}$.

4.131 For this steady-state flow, we fix the boat and move the upstream air. This provides us with the steady-state flow of Fig. 4.17. This is the same as observing the flow while standing on the boat.

$\dot{W} = FV_1$. $20\,000 = F\frac{50 \times 1000}{3600}$. $\therefore F = \underline{1440 \text{ N}}$. $(V_1 = 13.89 \text{ m/s})$

$F = \dot{m}(V_2 - V_1)$. $1440 = 1.23\pi \times 1^2 \frac{V_2 + 13.89}{2}(V_2 - 13.89)$. $\therefore V_2 = 30.6 \text{ m/s}$.

$\therefore Q = A_3V_3 = \pi \times 1^2 \frac{30.6 + 13.89}{2} = \underline{69.9 \text{ m}^3/\text{s}}$.

$\eta_p = \frac{V_1}{V_3} = \frac{13.89}{22.24} = 0.625$ or $\underline{62.5\%}$.

4.132 Fix the reference frame to the aircraft so that $V_1 = \frac{200 \times 1000}{3600} = 55.56 \text{ m/s}$.

$V_2 = \frac{320 \times 1000}{3600} = 88.89 \text{ m/s}$. $\therefore \dot{m} = 1.2 \times \pi \times 1.1^2 \frac{55.56 + 88.89}{2} = 329.5 \text{ kg/s}$.

$F = 329.5(88.89 - 55.56) = 10\,980 \text{ N}$.

$= \Delta p\pi \times 1.1^2$. $\therefore \Delta p = \underline{2890 \text{ Pa}}$.

$\dot{W} = F \times V_1 = 10\,980 \times 55.56 = 610\,000 \text{ W}$ or $\underline{818 \text{ Hp}}$.

4.133 Fix the reference frame to the boat so that $V_1 = \frac{30 \times 1000}{3600} = 8.333 \text{ m/s}$.

$V_2 = \frac{60 \times 1000}{3600} = 16.67 \text{ m/s}$. $\therefore F = \dot{m}(V_2 - V_1) = \rho A_3 \frac{V_1 + V_2}{2}(V_2 - V_1)$

$F = \Delta pA$ $\qquad = 1000\pi \times .25^2 \frac{8.33 + 16.67}{2}(16.67 - 8.33)$

$40\,900 = \Delta p\pi \times .25^2$. $\qquad = 20\,400 \text{ N}$.

$\therefore \Delta p = 208\,000 \text{ Pa}$ or 208 kPa.

$$\dot{W} = FV_1 = 20\,400 \times 8.333 = 170\,000 \text{ W} \text{ or } \underline{310 \text{ Hp}}.$$

$$\dot{m} = 1000\pi \times .25^2 \frac{8.33 + 16.67}{2} = \underline{2454 \text{ kg/s}}.$$

4.133E Fix the reference frame to the boat so that $V_1 = 20 \times \frac{88}{60} = 29.33$ fps.

$$V_2 = 40 \times \frac{88}{60} = 58.67 \text{ fps.} \quad \therefore F = \dot{m}(V_2 - V_1) = 1.94\pi \times \left(\frac{10}{12}\right)^2 \frac{29.33 + 58.67}{2}(58.67 - 29.33)$$

$$= 5460 \text{ lb.}$$

$$\dot{W} = F \times V_1 = 5460 \times 29.33 = 160{,}000 \frac{\text{ft - lb}}{\text{sec}} \text{ or } \underline{291 \text{ Hp}}.$$

$$\dot{m} = 1.94 \times \pi \times \left(\frac{10}{12}\right)^2 \frac{29.33 + 58.67}{2} = \underline{186.2 \text{ slug/sec}}.$$

4.134 Fix the reference frame to the boat: $V_1 = 10$ m/s, $V_2 = 20$ m/s.

$\therefore$ Thrust $= \dot{m}(V_2 - V_1) = 1000 \times 0.2(20 - 10) = \underline{2000 \text{ N}}$.

$\dot{W} = F \times V_1 = 2000 \times 10 = 20\,000$ W or $\underline{26.8 \text{ Hp}}$.

4.135 $0.2 = \overline{V}_1 A_1 = \overline{V}_1 \times .2 \times 1.0. \quad \therefore \overline{V}_1 = 1 \text{ m/s}. \quad \therefore V_{1\,max} = 2 \text{ m/s}. \quad \therefore V_1(y) = 20(0.1 - y).$

$$\text{flux in} = 2\int_0^{.1} \rho V^2 dy = 2\int_0^{.1} 1000 \times 20^2 (.1 - y)^2 dy = 800\,000 \frac{.1^3}{3} = 267 \text{ N}.$$

The slope at section ① is -20. $\therefore V_2(y) = -20y + A.$

$$\text{Continuity: } A_1\overline{V}_1 = A_2\overline{V}_2. \quad \therefore \overline{V}_2 = 2\overline{V}_1 = 2 \text{ m/s}. \qquad \left.\begin{aligned} V_2(0) &= A \\ V_2(.05) &= A - 1 \end{aligned}\right\} \therefore \overline{V}_2 = A - 1/2.$$

$2 = A - 1/2. \quad \therefore A = 2.5. \quad \therefore V_2(y) = 2.5 - 20y.$

$$\text{flux out} = 2\int_0^{.05} 1000(2.5 - 20y)^2 dy = 800\,000 \left[\frac{(y - .125)^3}{3}\right]_0^{.05} = \frac{800\,000}{3}[0.00153]$$

$= 408.3$ N. $\qquad \therefore$ change = 408 - 267 = $\underline{141 \text{ N}}$.

$$\text{4.136 a) } \beta = \frac{\int V^2 dA}{\overline{V}^2 A} = \frac{2\int_0^{.1} 20^2 (.1 - y)^2 dy}{1^2 \times .2 \times 1.0} = 4000 \frac{.1^3}{3} = \underline{\frac{4}{3}}.$$

b) See Problem 4.136: $V_2(y) = 20(0.125 - y),\ .05 \ge y \ge 0.\ \ \overline{V}_2 = 2\text{ m/s}.$

$$\beta = \frac{\int V^2 dA}{\overline{V}^2 A} = \frac{2\int_0^{.05} 20^2 (y-.125)^2 dy}{2^2 \times .1 \times 1.0} = 2000 \frac{(y-.125)^3}{3}\Big|_0^{.05} = \underline{1.021}.$$

4.137 From the c.v. shown: $(p_1 - p_2)\pi r_0^2 = \tau_w 2\pi r_o L.$

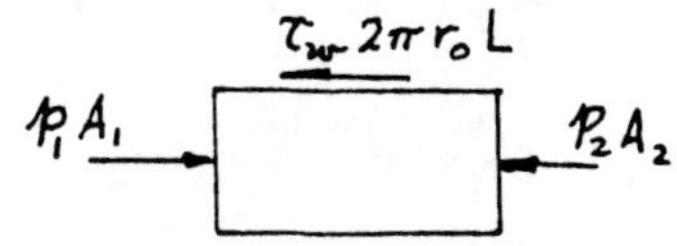

$$\therefore \tau_w = \frac{\Delta p\, r_o}{2L} = \frac{200 \times .02}{2 \times 10} = 0.2 = \mu \frac{du}{dr}\Big|_w = 10^{-3} \frac{du}{dr}\Big|_w .$$

$$\therefore \frac{du}{dr}\Big|_w = \underline{200 \frac{\text{m/s}}{\text{m}}}.$$

4.137E From the c.v. shown: $(p_1 - p_2)\pi r_0^2 = \tau_w 2\pi r_o L.$ (see above c.v.)

$$\therefore \tau_w = \frac{\Delta p\, r_o}{2L} = \mu \frac{du}{dr}\Big|_w . \quad \therefore \frac{du}{dr}\Big|_w = \frac{0.03 \times 144 \times .75/12}{2 \times 30 \times 2.36 \times 10^{-5}}$$

$$= \underline{191 \frac{\text{ft/sec}}{\text{ft}}}.$$

4.138 Write the equation of the parabola: $V(r) = V_{\max}\left(1 - \frac{r^2}{r_0^2}\right).$

Continuity: $\pi \times .006^2 \times 8 = \int_0^{.006} V_{\max}\left(1 - \frac{r^2}{.006^2}\right) 2\pi r dr. \quad \therefore V_{\max} = 16\text{ m/s}.$

Momentum: $p_1 A_1 - p_2 A_2 - F_{\text{Drag}} = \int \rho V^2 dA - \dot{m} V_1.$

$$40\,000\pi \times .006^2 - F_{\text{Drag}} = \int_0^{.006} 1000 \times 16^2 \left(1 - \frac{r^2}{.006^2}\right)^2 2\pi r dr - 1000 \times \pi \times .006^2 \times 8 \times 8$$

$$4.524 - F_{\text{Drag}} = 9.651 - 7.238. \qquad \therefore F_{\text{Drag}} = \underline{2.11\text{ N}}.$$

4.139 $\dot{m}_{top} = \rho A_1 V_1 - \rho \int V_2(y) dA = 1.23\left[2 \times 10 \times 32 - \int_0^2 (28 + y^2) 10 dy\right] = 65.6\text{ kg/s}.$

$$-\frac{F}{2} = \int \rho V^2 dA + \dot{m}_{top} V_1 - \dot{m}_1 V_1 = 1.23 \int_0^2 (28 + y^2)^2 10 dy + 65.6 \times 32 - 1.23 \times 20 \times 32^2.$$

$$\therefore F = \underline{3780\text{ N}}.$$

4.139E The value of y for $u(y) = 100$ is $y = 10'$. $\therefore u(y) = 96 + \frac{y^2}{25} \quad 0 \le y < 10.$

$$\dot{m}_{top} = \dot{m}_1 - \dot{m}_2 = \rho A_1 V_1 - \rho \int u(y) dA = .0024 \left[300 \times 100 - \int_0^{10} \left(96 + \frac{y^2}{25} \right) 30 dy \right] = 1.92 \text{ slug / sec.}$$

Apply the momentum equation to the top half to provide $F/2$:

$$-\frac{F}{2} = \int \rho V^2 dA + \dot{m}_{top} V_1 - \dot{m}_1 V_1 = .0024 \int_0^{10} \left(96 + \frac{y^2}{25} \right)^2 30 dy + 1.92 \times 100 - .0024 \times 300 \times 100^2$$

$= -186.$ $\therefore F = \underline{372 \text{ lb.}}$

4.140 a) $\dot{m}_{top} = \dot{m}_1 - \dot{m}_2 = \rho A_1 V_1 - \int \rho u(y) dA = 1.23 \left[.1 \times 2 \times 8 - \int_0^{.1} (20y - 100y^2) 8 \times 2 dy \right]$

$= 0.656$ kg / s. (Note: $y = 0.1$ for $u(y) = 8$).

Momentum: $-F_{\text{Drag}} = \rho \int_0^{.1} 64(20y - 100y^2)^2 2dy + .656 \times 8 - \rho \times .1 \times 2 \times 8^2$

$= 1.23 \times 6.83 + 5.25 - 1.23 \times 12.8.$ $\therefore F_{\text{Drag}} = \underline{2.1 \text{ N}}$

b) To find h: $8h = \int_0^{.1} 8(20y - 100y^2) dy.$

$$\therefore h = \frac{20 \times .1^2}{2} - \frac{100 \times .001}{3} = 0.0667 \text{ m.}$$

Momentum: $-F_{\text{Drag}} = 1.23 \int_0^{.1} 64(20y - 100y^2)^2 2dy - 1.23 \times .0667 \times 2 \times 8^2.$

$= 1.23 \times 6.83 - 10.50.$ $\therefore F_{\text{Drag}} = \underline{2.1 \text{ N.}}$

4.141 Continuity: $V_1 D^2 = V d^2$. $\therefore V_1 = \frac{d^2}{D^2} V.$

Energy: $\frac{V_1^2}{2g} + H(t) = \frac{V^2}{2g}$. $\therefore V = \sqrt{2gH(t)}.$

Momentum: $\Sigma F_x - (F_I)_x = \frac{d}{dt} \int_{c.v.} \rho V_x d\mathcal{V} + \dot{m}(V_{2x} - V_{1x}).$ $\left(\frac{d^2 \bar{s}}{dt^2} \right)_x = a_x.$

$$\therefore -a_x m(t) = \rho \frac{\pi d^2}{4} V(V). \quad m(t) = m_o - \rho \int_0^t \frac{\pi d^2}{4} V(t) dt.$$

But, $V_1 = -\frac{dH}{dt}$. $\therefore -\frac{dH}{dt} = \frac{d^2}{D^2} \sqrt{2gH}$. $\therefore -\frac{dH}{H^{1/2}} = \frac{d^2}{D^2} \sqrt{2g} dt$. $\therefore H^{1/2} = \frac{\sqrt{2g} d^2}{2D^2} t + \sqrt{H_o}.$

$$\therefore a_x = \frac{\rho \pi d^2}{4} 2g \left(\frac{\sqrt{2g} d^2}{2D^2} t + \sqrt{H_o} \right)^2 \left[\rho \frac{\pi d^2}{4} \int_0^t \sqrt{2g} \left(\frac{\sqrt{2g} d^2}{2D^2} t + \sqrt{H_o} \right) dt - m_o \right]$$

4.142 a) Energy: $\frac{V_1^2}{2g} + z_1 = \frac{V_2^2}{2g} + z_2 + h_L$. See Problem 4.107(a).

$$\frac{8^2}{2 \times 9.81} + 0.6 = \frac{1.912^2}{2 \times 9.81} + 2.51 + h_L. \qquad \therefore h_L = 1.166 \text{ m}.$$

$\therefore$ losses $= \gamma A_1 V_1 h_L = 9810 \times (.6 \times 1) \times 8 \times 1.166 = \underline{54\ 900 \text{ W / m}}$ of width.

b) See Problem 4.109: $\frac{V_1^2}{2g} + z_1 = \frac{V_2^2}{2g} + z_2 + h_L$.

$$\frac{7.19^2}{2 \times 9.81} + .417 = \frac{1.58^2}{2 \times 9.81} + 1.9 + h_L. \qquad \therefore h_L = 1.025 \text{ m}.$$

$\therefore$ losses $= \gamma A_1 V_1 h_L = 9810 \times .417 \times 3 \times 7.19 \times 1.025 = \underline{90\ 300 \text{ W}}$

c) See Problem 4.110: $\frac{5.17^2}{2 \times 9.81} + 1.16 = \frac{3^2}{2 \times 9.81} + 2 + h_L$. $\therefore h_L = 0.0636$ m.

$\therefore$ losses $= \gamma A_1 V_1 h_L = 9810 \times 1.16 \times 5.17 \times 0.0636 = \underline{3740 \text{ W / m}}$ of width.

d) See Problem 4.110E: $\frac{15.8^2}{2 \times 32.2} + 3.8 = \frac{10^2}{2 \times 32.2} + 6 + h_L$. $\therefore h_L = 0.124'$.

$\therefore$ losses $= \gamma A_1 V_1 h_L = 62.4 \times 3.8 \times 15.8 \times 0.124 = \underline{465 \frac{\text{ft - lb}}{\text{sec}}}$ / ft. of width.

4.143 See Problem 4.111: $V_1 = 20$ m / s, $V_2 = 5$ m / s, $p_1 = 60$ kPa, $p_2 = 135$ kPa.

Then, $\frac{V_1^2}{2g} + \frac{p_1}{\gamma} = \frac{V_2^2}{2g} + \frac{p_2}{\gamma} + h_L$. $\quad \frac{20^2}{2 \times 9.81} + \frac{60\ 000}{9810} = \frac{5^2}{2 \times 9.81} + \frac{135\ 000}{9810} + h_L$.

$$\therefore h_L = 11.47 \text{ m} = K\frac{V_1^2}{2g} = K\frac{20^2}{2 \times 9.81}. \qquad \therefore K = \underline{0.562}.$$

4.144 $V_e = \frac{\dot{m}}{\rho A_e} = \frac{4}{1000 \times 4 \times \pi \times .004^2} = 19.89$ m / s. Velocity in arm $= V$.

$$\vec{M}_I = \int_{c.v.} \vec{r} \times (2\Omega \times \vec{V})\rho d\forall = 4\int_0^{.3} r\hat{i} \times (-2\Omega\hat{k} \times V\hat{i})\rho A dr$$

$$= -8\rho A V \Omega \hat{k} \int_0^{.3} r dr = -0.36 \rho A V \Omega \hat{k}.$$

$\Sigma\vec{M} = 0$ and $\dfrac{d}{dt}\int_{c.v.}(\vec{r}\times\vec{V})\rho d\mathcal{V} = 0.$ $\int_{c.s.}\vec{r}\times\vec{V}(\vec{V}\cdot\hat{n})\rho dA = .3\hat{i}\times(.707V_e\hat{j}+.707V_e\hat{k})V_e\rho A_e.$

The z-component of $\int_{c.s.}\vec{r}\times\vec{V}(\vec{V}\cdot\hat{n})\rho dA = .3\times.707V_e^2A_e\rho.$

Finally, $-(M_I)_z = 0.36\rho AV\Omega = 4\times.3\times.707V_e^2A_e\rho.$ Using $AV = A_eV_e$,

$0.36\Omega = 4\times.3\times.707\times 19.89.$ $\therefore\ \Omega = \underline{46.9\text{ rad/s}}.$

4.145 A moment $\vec{M}$ resists the motion thereby producing power. One of the arms is shown.

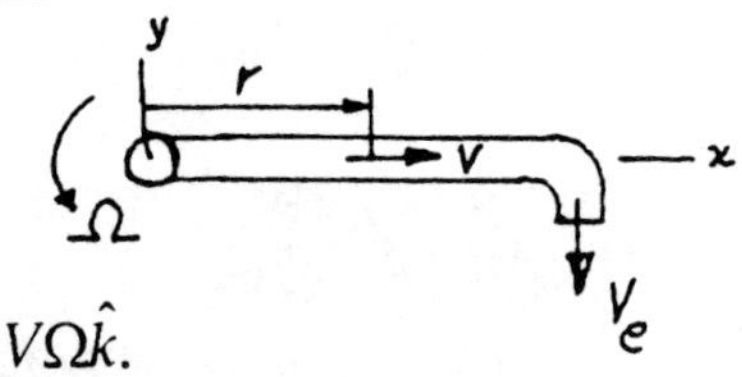

$\vec{M}_I = 4\int_0^{.25} r\hat{i}\times(-2\Omega\hat{k}\times V\hat{i})\rho A dr = -8\rho AV\Omega\hat{k}\int_0^{.25} r dr = -\dfrac{1}{4}\rho AV\Omega\hat{k}.$

$\Sigma\vec{M} = M\hat{k},$ $\dfrac{d}{dt}\int_{c.v.}(\vec{r}\times\vec{V})\rho d\mathcal{V} = 0,$ and $\int_{c.s.}\vec{r}\times\vec{V}(\vec{V}\cdot\hat{n})\rho dA = .25V_e^2\rho A_e\times 4\hat{k}.$

Thus, $M + \dfrac{1}{4}\times 1000\pi\times.015^2\times\dfrac{70}{9}\times 30 = 70^2\times 1000\pi\times.005^2.$ $\therefore\ M = 344\text{ N}\cdot\text{m}.$ $\dot{W} = M\Omega$

$= \underline{10\ 300\text{ W}}.$

4.145E A moment $\vec{M}$ resists the motion thereby producing power. One of the arms is shown. (see arm above)

$\vec{M}_I = \int_0^{.25} 4r\hat{i}\times(-2\Omega\hat{k}\times V\hat{i})\rho A dr = -8\rho AV\Omega\hat{k}\int_0^{10/12} r dr = -2.778\rho AV\Omega\hat{k}.$

$\Sigma\vec{M} = M\hat{k},$ $\dfrac{d}{dt}\int_{c.v.}(\vec{r}\times\vec{V})d\mathcal{V} = 0,$ and $\int_{c.s.}\vec{r}\times\vec{V}(\vec{V}\cdot\hat{n})\rho dA = \dfrac{10}{12}\times V_e^2\rho A_e\times 4\hat{k}.$

Thus, $M + 2.778\times 1.94\pi\left(\dfrac{.75}{12}\right)^2\times\dfrac{200}{9}\times 30 = 200^2\times\dfrac{10}{12}\times 1.94\pi\left(\dfrac{1/4}{12}\right)^2\times 4.$

$\therefore\ M = 309\text{ ft-lb}.$

$\dot{W} = M\Omega = 309\times 30 = \underline{9270\text{ ft-lb/sec}}.$

4.146 $\dot{m} = 10 = \rho AV = 1000\pi\times.01^2\,V_0.$ $\therefore\ V_0 = 31.8\text{ m/s}.$

Continuity: $V_0\pi\times.01^2 = V\pi\times.01^2 + V_e\times.006(r-.05).$

$V_0\pi\times.01^2 = V_e\times.006\times.15.$ $\therefore\ V_e = 11.1\text{ m/s}.$

$\therefore\ V = V_0 - 19.1(r-.05)V_e = 42.4 - 212r.$

$$\vec{M}_I = \int_0^{.05} 2r\hat{i} \times (+2\Omega\hat{k} \times V_0\hat{i})\rho A dr + \int_{.05}^{.2} 2r\hat{i} \times [+2\Omega\hat{k} \times (42.4 - 212r)\hat{i}]\rho A dr$$

$$= 4\Omega V_0 \rho A\hat{k} \int_0^{.05} r dr + 4\Omega\rho A\hat{k} \int_{.05}^{.2} (42.4r - 212r^2) dr$$

$$= 4\Omega \times 31.8 \times 1000\pi \times .01^2 \times \frac{.05^2}{2}\hat{k} + 4\Omega \times 1000\pi \times .01^2 \left[\frac{42.4}{2}(.2^2 - .05^2) - \frac{212}{3}(.2^3 - .05^3)\right]\hat{k}$$

$$= (0.05\Omega + 0.3\Omega)\hat{k} = 0.35\Omega\hat{k}.$$

$$\int_{.05}^{.2} r\hat{i} \times (-V_e\hat{j})V_e\rho \times .006 dr = -11.1^2 \times 1000 \times .006 \int_{.05}^{.2} r dr\, \hat{k} = -13.86\hat{k}.$$

$\therefore -0.35\Omega = -13.86.$ $\qquad \therefore \Omega = \underline{39.6 \text{ rad/s}}.$

4.147 $1000 = M\Omega.$ $\quad \therefore M = \dfrac{1000}{500} = 2 \text{ N}\cdot\text{m}.$

$$\vec{M}_I = \int r\hat{i}_r \times (-2\Omega\hat{k} \times V(r)\hat{i}_r)\rho 2\pi r \times .02 dr$$

$$= 0.08\pi\Omega \int_0^R r^2 V(r) dr\hat{k}.$$

Continuity: $V(r)2\pi r \times .02 = V_r \cos 30^\circ 2\pi R \times .02.$ $\quad \therefore V(r) = 0.866RV_r / r.$

$$\int_{c.s.} \vec{r} \times \vec{V}(\vec{V}\cdot\hat{n})\rho dA = -R(R\Omega + V_r \sin 30^\circ)V_r \cos 30^\circ \rho 2\pi R \times .02\hat{k} = -.00301V_r(35 + .5V_r)\hat{k}.$$

$$\therefore -2 - 16.32V_r \int_0^{.15} r\, dr = -.00301V_r(35 + .5V_r). \quad \therefore V_r^2 - 52.1V_r - 1333 = 0.$$

$$\therefore V_r = \frac{1}{2}(52.1 \pm \sqrt{52.1^2 + 4 \times 1333}) = 70.9 \text{ m/s}.$$

The flow rate is $Q = A_e V_r \cos 30^\circ = 2\pi \times .15 \times .02 \times 70.9 \times .866 = \underline{1.16 \text{ m}^3/\text{s}}.$

4.148 See Problem 4.144. $V_e = 19.89$ m/s. $\quad V = \dfrac{.008^2}{.02^2} \times 19.89 = 3.18$ m/s.

$$\vec{M}_I = 4\int_0^{.3} r\hat{i} \times \left[(-2\Omega\hat{k} \times V\hat{i}) + \left(-\frac{d\Omega}{dt}\hat{k}\right) \times r\hat{i}\right]\rho A dr. \quad A = \pi \times .01^2,\ A_e = \pi \times .004^2.$$

$$= -8\rho AV\Omega\hat{k}\int_0^{.3} r dr - 4\rho A\frac{d\Omega}{dt}\hat{k}\int_0^{.3} r^2 dr = -360AV\Omega\hat{k} - 36A\frac{d\Omega}{dt}\hat{k}.$$

$$\int_{c.s.} (\vec{r} \times \vec{V})_z (\vec{V} \cdot \hat{n}) \rho dA = 212 V_e^2 A_e \hat{k}.$$

Thus, $360 A V \Omega + 36 A \frac{d\Omega}{dt} = 212 V_e^2 A_e$ or $\frac{d\Omega}{dt} + 31.8\Omega = 373.$

The solution is $\Omega = Ce^{-31.8t} + 11.73.$

The initial condition is $\Omega(0) = 0.$ $\therefore C = -11.73.$

Finally, $\Omega = \underline{11.73(1 - e^{-31.8t}) \text{ rad/s}}.$

FIVE

The Differential Forms of the Fundamental Laws

5.1 $0 = \int_{c.v.} \frac{\partial \rho}{\partial t} d\mathcal{V} + \int_{c.s.} \rho \vec{V} \cdot \hat{n} dA$. Using Gauss' theorem:

$$0 = \int_{c.v.} \frac{\partial \rho}{\partial t} d\mathcal{V} + \int_{c.v.} \vec{\nabla} \cdot (\rho \vec{V}) d\mathcal{V} = \int_{c.v.} \left[\frac{\partial \rho}{\partial t} + \vec{\nabla} \cdot (\rho \vec{V}) \right] d\mathcal{V}.$$

Since this is true for all arbitrary control volumes (i.e., for all limits of integration), the integrand must be zero:

$$\frac{\partial \rho}{\partial t} + \vec{\nabla} \cdot (\rho \vec{V}) = 0.$$

This can be written in rectangular coordinates as

$$-\frac{\partial \rho}{\partial t} = \frac{\partial}{\partial x}(\rho u) + \frac{\partial}{\partial y}(\rho v) + \frac{\partial}{\partial z}(\rho w).$$

This is Eq. 5.2.2. The other forms of the continuity equation follow.

5.2 $\dot{m}_{in} - \dot{m}_{out} = \frac{\partial m_{element}}{\partial t}$.

$$\rho v_r (r d\theta dz) - \left[\rho v_r + \frac{\partial}{\partial r}(\rho v_r) dr \right] (r + dr) d\theta dz$$

$$+ \rho v_\theta dr dz - \left[\rho v_\theta + \frac{\partial}{\partial \theta}(\rho v_\theta) d\theta \right] dr dz$$

$$+ \rho v_z \left(r + \frac{dr}{2} \right) d\theta dr - \left[\rho v_z + \frac{\partial}{\partial z}(\rho v_z) dz \right] \left(r + \frac{dr}{2} \right) d\theta dr = \frac{\partial}{\partial t} \left[\rho \left(r + \frac{dr}{2} \right) d\theta dr dz \right].$$

Subtract terms and divide by $r d\theta dr dz$:

$$-\frac{\rho v_r}{r} - \frac{\partial}{\partial r}(\rho v_r) \frac{r + dr}{r} - \frac{\partial}{\partial \theta}(\rho v_\theta) \frac{1}{r} - \frac{\partial}{\partial z}(\rho v_z) \frac{r + dr/2}{r} = \frac{\partial}{\partial t} \rho \frac{r + dr/2}{r}.$$

Since dr is an infinitesimal, $(r + dr)/r = 1$ and $(r + dr/2)/r = 1$. Hence,

$\frac{\partial \rho}{\partial t} + \frac{\partial}{\partial r}(\rho v_r) + \frac{1}{r}\frac{\partial}{\partial \theta}(\rho v_\theta) + \frac{\partial}{\partial z}(\rho v_z) + \frac{1}{r}\rho v_r = 0$. This can be put in various forms.

5.3 $\dot{m}_{in} - \dot{m}_{out} = \frac{\partial m_{element}}{\partial t}$.

$$\rho v_r (rd\theta) r\sin\theta d\phi - \left[\rho v_r + \frac{\partial}{\partial r}(\rho v_r) dr\right](r+dr)d\theta(r+dr)\sin\theta d\phi$$

$$+\rho v_\theta dr\left(r+\frac{dr}{2}\right)\sin\theta d\phi - \left[\rho v_\theta + \frac{\partial}{\partial\theta}(\rho v_\theta)d\theta\right]dr\left(r+\frac{dr}{2}\right)\sin\theta d\phi$$

$$+\rho v_\phi dr\left(r+\frac{dr}{2}\right)d\theta - \left[\rho v_\phi + \frac{\partial}{\partial\phi}(\rho v_\phi)d\phi\right]dr\left(r+\frac{dr}{2}\right)d\theta$$

$$= \frac{\partial}{\partial t}\left[\rho\left(r+\frac{dr}{2}\right)^2 dr d\theta \sin\theta d\phi\right]$$

Because some areas are not rectangular, we used an average length $(r + dr/2)$. Now, subtract some terms and divide by $rd\theta d\phi dr$:

$$-\rho v_r \sin\theta - \rho v_r \sin\theta - \frac{\partial}{\partial r}(\rho v_r)\sin\theta\frac{(r+dr)^2}{r} - \frac{\partial}{\partial\theta}(\rho v_\theta)\frac{r+\frac{dr}{2}}{r}\sin\theta$$

$$-\frac{\partial}{\partial\phi}(\rho v_\phi)\frac{r+\frac{dr}{2}}{r} = \frac{\partial\rho}{\partial t}\frac{\left(r+\frac{dr}{2}\right)^2}{r}\sin\theta$$

Since dr is infinitesimal $(r+dr)^2/r = r$ and $(r+dr/2)/r = 1$. Divide by $r\sin\theta$ and there results

$$\frac{\partial\rho}{\partial t} + \frac{\partial}{\partial r}(\rho v_r) + \frac{1}{r}\frac{\partial}{\partial\theta}(\rho v_\theta) + \frac{1}{r\sin\theta}\frac{\partial}{\partial\phi}(\rho v_\phi) + \frac{2}{r}\rho v_r = 0$$

5.4 For a steady flow $\frac{\partial\rho}{\partial t} = 0$. Then, with $v = w = 0$ Eq. 5.2.2 yields

$$\frac{\partial}{\partial x}(\rho u) = 0 \qquad \text{or} \qquad \underline{\rho\frac{du}{dx} + u\frac{d\rho}{dx} = 0.}$$

Partial derivatives are not used since there is only one independent variable.

5.5 Since the flow is incompressible $\frac{D\rho}{Dt} = 0$. This gives

$$\underbrace{\frac{\partial\rho}{\partial t}}_{0} + u\frac{\partial\rho}{\partial x} + \underbrace{v}_{0}\frac{\partial\rho}{\partial y} + w\frac{\partial\rho}{\partial z} = 0 \qquad \text{or} \qquad \underline{u\frac{\partial\rho}{\partial x} + w\frac{\partial\rho}{\partial z} = 0.}$$

Also, $\vec{\nabla}\cdot\vec{V} = 0$, or $\underline{\frac{\partial u}{\partial x} + \frac{\partial w}{\partial z} = 0.}$

5.6 Given: $\frac{\partial}{\partial t} = 0,\ \frac{\partial \rho}{\partial z} \neq 0$. Since water can be considered to be incompressible, we demand that $\frac{D\rho}{Dt} = 0$. $\quad \therefore \underline{u\frac{\partial \rho}{\partial x} + w\frac{\partial \rho}{\partial z} = 0}$, assuming the x-direction to be in the direction of flow. Also, we demand that $\vec{\nabla} \cdot \vec{V} = 0$, or $\underline{\frac{\partial u}{\partial x} + \frac{\partial w}{\partial z} = 0}$.

5.7 We can use the ideal gas law, $\rho = \frac{p}{RT}$. Then, the continuity equation $\frac{D\rho}{Dt} = -\rho\vec{\nabla} \cdot \vec{V}$ becomes, assuming RT to be constant, $\frac{1}{RT}\frac{Dp}{Dt} = -\frac{p}{RT}\vec{\nabla} \cdot \vec{V}$ or $\frac{1}{p}\frac{Dp}{Dt} = -\vec{\nabla} \cdot \vec{V}$.

5.8 a) Use cylindrical coordinates with $v_\theta = v_z = 0$:

$$\frac{1}{r}\frac{\partial}{\partial r}(rv_r) = 0$$

Integrate:

$$rv_r = C. \qquad \therefore \underline{v_r = \frac{C}{r}}.$$

b) Use spherical coordinates with $v_\theta = v_\phi = 0$:

$$\frac{1}{r^2}\frac{\partial}{\partial r}(r^2 v_r) = 0$$

Integrate:

$$r^2 v_r = C. \qquad \therefore \underline{v_r = \frac{C}{r^2}}.$$

5.9 $\frac{D\rho}{Dt} = -\rho\vec{\nabla} \cdot \vec{V} = -\rho\left(\frac{\partial u}{\partial x} + \frac{\partial v}{\partial y}\right) = -2.3(200 \times 1 + 400 \times 1) = \underline{-1380\ \frac{\text{kg}}{\text{m}^3 \cdot \text{s}}}$.

5.10 In a plane flow, $u = u(x, y)$ and $v = v(x, y)$. Continuity demands that $\frac{\partial u}{\partial x} + \frac{\partial v}{\partial y} = 0$. If $u = \text{const}$, then $\frac{\partial u}{\partial x} = 0$ and hence $\frac{\partial v}{\partial y} = 0$. Thus, $\underline{v = \text{const}}$ also.

5.11 If $u = C_1$ and $v = C_2$, the continuity equation provides, for an incompressible flow,

$$\frac{\partial u}{\partial x}+\frac{\partial v}{\partial y}+\frac{\partial w}{\partial z}=0. \qquad \therefore \frac{\partial w}{\partial z}=0 \text{ and } w=C_3.$$

The z-component of velocity $\underline{w \text{ is also constant}}$.

We also have

$$\frac{D\rho}{Dt}=0=\frac{\partial \rho}{\partial t}+u\frac{\partial \rho}{\partial x}+v\frac{\partial \rho}{\partial y}+w\frac{\partial \rho}{\partial z}$$

The density may vary with x, y, z and t. It is not, necessarily, constant.

5.12 $\frac{\partial u}{\partial x}+\frac{\partial v}{\partial y}=0. \qquad \therefore A+\frac{\partial v}{\partial y}=0. \qquad \therefore v(x,y)=-Ay+f(x).$

But, $v(x,o)=0=f(x). \qquad \therefore \underline{v=-Ay}.$

5.13 $\frac{\partial u}{\partial x}+\frac{\partial v}{\partial y}=0. \qquad \therefore \frac{\partial v}{\partial y}=-\frac{\partial u}{\partial x}=-\frac{(x^2+y^2)5-5x(2x)}{(x^2+y^2)^2}=-\frac{5x^2-5y^2}{(x^2+y^2)^2}$

$$\therefore v(x,y)=\int\frac{5y^2-5x^2}{(x^2+y^2)^2}dy+f(x)=\frac{5y}{x^2+y^2}+f(x). \quad f(x)=0. \quad \therefore \underline{v=\frac{5y}{x^2+y^2}}.$$

5.14 From Table 5.1: $\frac{1}{r}\frac{\partial}{\partial r}(rv_r)=-\frac{1}{r}\frac{\partial v_\theta}{\partial \theta}=-\frac{1}{r}\left(10+\frac{.4}{r^2}\right)\sin\theta.$

$$\therefore rv_r=\int\left(10+\frac{.4}{r^2}\right)\sin\theta dr+f(\theta)=\left(10r-\frac{.4}{r}\right)\sin\theta+f(\theta).$$

$$.2v_r(.2,\theta)=\left(10\times.2-\frac{.4}{.2}\right)\sin\theta+f(\theta)=0. \qquad \therefore f(\theta)=0.$$

$$\therefore \underline{v_r=\left(10-\frac{0.4}{r^2}\right)\sin\theta.}$$

5.15 From Table 5.1: $\frac{1}{r}\frac{\partial}{\partial r}(rv_r)=-\frac{1}{r}\frac{\partial v_\theta}{\partial \theta}=\frac{-20}{r}\left(1+\frac{1}{r^2}\right)\cos\theta.$

$$\therefore rv_r=\int-20\left(1+\frac{1}{r^2}\right)\cos\theta dr+f(\theta)=-20\left(r-\frac{1}{r}\right)\cos\theta+f(\theta).$$

$$v_r(1,\theta)=-20(1-1)\cos\theta+f(\theta)=0. \qquad \therefore f(\theta)=0.$$

$$\therefore \underline{v_r=-20\left(1-\frac{1}{r^2}\right)\cos\theta.}$$

5.16 From Table 5.1, spherical coordinates: $\frac{1}{r^2}\frac{\partial}{\partial r}(r^2 v_r) = -\frac{1}{r\sin\theta}\frac{\partial}{\partial\theta}(v_\theta \sin\theta)$.

$$\therefore \frac{1}{r^2}\frac{\partial}{\partial r}(r^2 v_r) = \frac{1}{r\sin\theta}\left(10+\frac{40}{r^3}\right)2\sin\theta\cos\theta.$$

$$\therefore r^2 v_r = \int r\left(10+\frac{40}{r^3}\right)2\cos\theta dr + f(\theta) = \left(10r^2 - \frac{80}{r}\right)\cos\theta + f(\theta)$$

$$4v_r(2,\theta) = \left(10\times 2^2 - \frac{80}{2}\right)\cos\theta + f(\theta) = 0. \quad \therefore f(\theta) = 0.$$

$$\therefore v_r = \underline{\left(10 - \frac{80}{r^3}\right)\cos\theta.}$$

5.17 Continuity: $\frac{\partial}{\partial x}(\rho u) = 0. \quad \therefore \rho\frac{du}{dx} + u\frac{d\rho}{dx} = 0.$

$$\rho = \frac{p}{RT} = \frac{120}{.287\times 283} = 1.477\ \frac{\text{kg}}{\text{m}^3}. \quad \frac{du}{dx} = \frac{175-151}{2\times.05} = 240\frac{\text{m/s}}{\text{m}}.$$

$$\therefore \frac{d\rho}{dx} = -\frac{\rho}{u}\frac{du}{dx} = -\frac{1.477}{162}\times 240 = \underline{-2.19\ \text{kg/m}^4}.$$

5.17E Continuity: $\frac{\partial}{\partial x}(\rho u) = 0. \quad \therefore \rho\frac{du}{dx} + u\frac{d\rho}{dx} = 0.$

$$\rho = \frac{p}{RT} = \frac{18\times 144}{1716\times 500} = 0.00302\ \frac{\text{slug}}{\text{ft}^3}. \quad \frac{du}{dx} = \frac{526-453}{2\times 2/12} = 219\ \text{fps/ft}.$$

$$\therefore \frac{d\rho}{dx} = -\frac{\rho}{u}\frac{du}{dx} = -\frac{.00302}{486}\times 219 = \underline{-0.00136\ \text{slug/ft}^4}.$$

5.18 $\frac{\partial u}{\partial x} + \frac{\partial v}{\partial y} = 0. \quad \frac{\partial}{\partial x}\left[-20(1-e^{-x})\right] = -20e^{-x}$

Hence, in the vicinity of the x-axis:

$$\frac{\partial v}{\partial y} = 20e^{-x} \text{ and } v = 20ye^{-x} + C.$$

But $v = 0$ if $y = 0$. $\quad \therefore C = 0.$

$$v = 20ye^{-x} = 20(0.2)e^{-2} = \underline{0.541\ \text{m/s}}$$

5.19 $\frac{1}{r}\frac{\partial}{\partial r}(rv_r) + \frac{\partial v_z}{\partial z} = 0. \quad \frac{\partial}{\partial z}\left[-20(1-e^{-z})\right] = -20e^{-z}$

Hence, in the vicinity of the z-axis:

$$\frac{1}{r}\frac{\partial}{\partial r}(rv_r) = 20e^{-z} \text{ and } rv_r = \frac{r^2}{2}20e^{-z} + C.$$

But $v_r = 0$ if $r = 0$. $\quad \therefore C = 0.$

$$v_r = 10re^{-z} = 10(0.2)e^{-2} = \underline{0.271 \text{ m/s}}$$

5.20 The velocity is zero at the stagnation point. Hence,

$$0 = 10 - \frac{40}{R^2}. \qquad \therefore \underline{R = 2 \text{ m}}$$

The continuity equation for this plane flow is $\frac{\partial u}{\partial x} + \frac{\partial v}{\partial y} = 0$. Using $\frac{\partial u}{\partial x} = 80x^{-3}$,

we see that $\frac{\partial v}{\partial y} = -80x^{-3}$ near the x-axis. Consequently, for small Δy,

$$\Delta v = -80x^{-3}\Delta y$$

$$v = -80(-3)^{-3}(0.1) = \underline{0.296 \text{ m/s}}.$$

5.21 The velocity is zero at the stagnation point. Hence

$$0 = \frac{40}{R^2} - 10. \qquad \therefore \underline{R = 2 \text{ m}}$$

$$\frac{1}{r^2}\frac{\partial}{\partial r}\left(r^2 v_r\right) = \frac{1}{r^2}\frac{\partial}{\partial r}(40 - 10r^2) = -\frac{20}{r}.$$

Near the negative x-axis continuity provides us with

$$\frac{1}{r\sin\theta}\frac{\partial}{\partial\theta}\left(v_\theta \sin\theta\right) = \frac{20}{r}.$$

Integrate, letting $\theta = 0$ from the y-axis:

$$v_\theta \sin\theta = -20\cos\theta + C$$

Since $v_\theta = 0$ when $\theta = 90°$, $C = 0$. Then, with $\alpha = \tan^{-1}\frac{0.1}{3} = 1.909°$,

$$v_\theta = -20\frac{\cos\theta}{\sin\theta} = -20\frac{\cos 88.091}{\sin 88.091} = -20\frac{0.0333}{0.999} = \underline{0.667 \text{ m/s}}$$

5.22 Continuity: $\frac{\partial u}{\partial x} + \frac{\partial v}{\partial y} = 0. \quad \therefore \frac{\Delta v}{\Delta y} = -\frac{\Delta u}{\Delta x} = -\frac{13.5 - 11.3}{2 \times .005} = -220 \frac{\text{m/s}}{\text{m}}.$

$\therefore \Delta v = v - 0 = -220\Delta y. \quad \therefore v = -220 \times .004 = \underline{-0.88 \text{ m/s}}.$

b) $a_x = u\frac{\partial u}{\partial x} = 12.6 \times (+220) = \underline{2772 \text{ m/s}^2}.$

5.23 $\Sigma F_y = ma_y$. For the fluid particle occupying the volume of Fig. 5.3:

$$\left(\tau_{yy} + \frac{\partial\tau_{yy}}{\partial y}\frac{dy}{2}\right)dxdz + \left(\tau_{zy} + \frac{\partial\tau_{zy}}{\partial z}\frac{dz}{2}\right)dxdy + \left(\tau_{xy} + \frac{\partial\tau_{xy}}{\partial x}\frac{dx}{2}\right)dydz$$

$$-\left(\tau_{yy}-\frac{\partial\tau_{yy}}{\partial y}\frac{dy}{2}\right)dxdz-\left(\tau_{zy}-\frac{\partial\tau_{zy}}{\partial z}\frac{dz}{2}\right)dxdy-\left(\tau_{xy}-\frac{\partial\tau_{xy}}{\partial x}\frac{dx}{2}\right)dydz$$

$$+\rho g_y dx\,dy\,dz=\rho dx\,dy\,dz\frac{Dv}{Dt}$$

Dividing by $dx\,dy\,dz$, and adding and subtracting terms:

$$\frac{\partial\tau_{xy}}{\partial x}+\frac{\partial\tau_{yy}}{\partial y}+\frac{\partial\tau_{zy}}{\partial z}+\rho g_y=\rho\frac{Dv}{Dt}.$$

5.24 Check continuity:

$$\frac{\partial u}{\partial x}+\frac{\partial v}{\partial y}+\frac{\partial w}{\partial z}=\frac{(x^2+y^2)10-10x(2x)}{(x^2+y^2)^2}+\frac{(x^2+y^2)10-10y(2y)}{(x^2+y^2)^2}=0.$$

Thus, it is a possible flow. For a frictionless flow, Euler's Eqs. 5.3.7 give, with $g_x=g_y=0$:

$$\rho u\frac{\partial u}{\partial x}+\rho v\frac{\partial u}{\partial y}=-\frac{\partial p}{\partial x}.\quad\therefore\frac{\partial p}{\partial x}=-\rho\frac{10x}{x^2+y^2}\frac{10y^2-10x^2}{(x^2+y^2)^2}-\rho\frac{10y}{x^2+y^2}\frac{-20xy}{(x^2+y^2)^2}=\rho\frac{100(x^2+y^2)x}{(x^2+y^2)^3}$$

$$\rho u\frac{\partial v}{\partial x}+\rho v\frac{\partial v}{\partial y}=-\frac{\partial p}{\partial y}.\quad\therefore\frac{\partial p}{\partial y}=-\rho\frac{10x}{x^2+y^2}\frac{-20xy}{(x^2+y^2)^2}-\rho\frac{10y}{x^2+y^2}\frac{10x^2-10y^2}{(x^2+y^2)^2}=\rho\frac{100(x^2+y^2)y}{(x^2+y^2)^3}$$

$$\therefore\vec{\nabla}p=\frac{\partial p}{\partial x}\hat{i}+\frac{\partial p}{\partial y}\hat{j}=\frac{100x\rho}{(x^2+y^2)^2}\hat{i}+\frac{100y\rho}{(x^2+y^2)^2}\hat{j}=\underline{\frac{100\rho}{(x^2+y^2)^2}(x\hat{i}+y\hat{j})}.$$

5.25 Check continuity (cylindrical coord - Table 5.1):

$$\frac{1}{r}\frac{\partial}{\partial r}(rv_r)+\frac{1}{r}\frac{\partial v_\theta}{\partial\theta}=\frac{10}{r}\left(1+\frac{1}{r^2}\right)\cos\theta+\frac{-10}{r}\left(1+\frac{1}{r^2}\right)\cos\theta=0.$$ $\therefore$ It is a possible flow. For Euler's Eqs. (let $\nu=0$ in the momentum eqns of Table 5.1) in cylindrical coord:

$$\frac{\partial p}{\partial r}=\rho\frac{v_\theta^2}{r}-\rho v_r\frac{\partial v_r}{\partial r}-\rho\frac{v_\theta}{r}\frac{\partial v_r}{\partial\theta}=\frac{100\rho}{r}\left(1+\frac{1}{r^2}\right)^2\sin^2\theta-10\rho\left(1-\frac{1}{r^2}\right)\cos^2\theta\left(\frac{20}{r^3}\right)$$
$$-\frac{10\rho}{r}\left(1+\frac{1}{r^2}\right)\sin^2\theta\left(10-\frac{10}{r^2}\right).$$

$$\frac{1}{r}\frac{\partial p}{\partial\theta}=-\rho\frac{v_r v_\theta}{r}-\rho v_r\frac{\partial v_\theta}{\partial r}-\rho\frac{v_\theta}{r}\frac{\partial v_\theta}{\partial\theta}=\frac{100\rho}{r}\left(1-\frac{1}{r^4}\right)\sin\theta\cos\theta$$
$$-10\rho\left(1-\frac{1}{r^2}\right)\cos\theta\sin\theta\left(\frac{20}{r^3}\right)-\frac{100\rho}{r}\left(1+\frac{1}{r^2}\right)^2\sin\theta\cos\theta.$$

$$\therefore\vec{\nabla}p=\frac{\partial p}{\partial r}\hat{i}_r+\frac{1}{r}\frac{\partial p}{\partial\theta}\hat{i}_\theta=\underline{\frac{200\rho}{r^3}\left[\frac{1}{r^2}-\cos 2\theta\right]\hat{i}_r-\frac{200\rho}{r^3}\sin 2\theta\,\hat{i}_\theta}$$

5.26 This is an involved problem. Follow the steps of Problem 5.25. Good luck!

$$\frac{\partial p}{\partial r} = \rho\frac{\left(v_\theta^2 + v_\phi^2\right)}{r} - \rho v_r\frac{\partial v_r}{\partial r} - \rho\frac{v_\theta}{r}\frac{\partial v_r}{\partial \theta}$$

$$\frac{1}{r}\frac{\partial p}{\partial \theta} = -\rho\frac{(v_r v_\theta)}{r} - \rho v_r\frac{\partial v_\theta}{\partial r} - \rho\frac{v_\theta}{r}\frac{\partial v_\theta}{\partial \theta}$$

5.27 $$\frac{D\vec{V}}{Dt} = \frac{\partial}{\partial t}(V\hat{s}) + V\frac{\partial(V\hat{s})}{\partial s} + \overset{0}{\cancel{V_n}}\frac{\partial(V\hat{s})}{\partial n} = \hat{s}\frac{\partial V}{\partial t} + V\frac{\partial \hat{s}}{\partial t} + V\hat{s}\frac{\partial V}{\partial s} + V^2\frac{\partial \hat{s}}{\partial s}.$$

$$\frac{\partial \hat{s}}{\partial s} \cong \frac{\Delta\hat{s}}{\Delta s} = -\frac{\Delta\alpha\hat{n}}{R\Delta\alpha} = -\frac{\hat{n}}{R}.$$

$$\frac{\partial \hat{s}}{\partial t} \cong \frac{\Delta\hat{s}}{\Delta t} = \frac{\hat{n}\,\Delta\theta}{\Delta t} = \hat{n}\frac{\partial\theta}{\partial t}.$$

$$\therefore \frac{D\vec{V}}{Dt} = \left(\frac{\partial V}{\partial t} + V\frac{\partial V}{\partial s}\right)\hat{s} + \left(V\frac{\partial\theta}{\partial t} - \frac{V^2}{R}\right)\hat{n}.$$

For steady flow, the normal acc. is $\left(-\frac{V^2}{R}\right)$, the tangential acc. is $V\frac{\partial V}{\partial s}$.

5.28 For a rotating reference frame (see Eq. 3.2.15), we must add the terms due to $\vec{\Omega}$. Thus, Euler's equation becomes

$$\rho\left(\frac{D\vec{V}}{Dt} + 2\vec{\Omega}\times\vec{V} + \vec{\Omega}\times(\vec{\Omega}\times\vec{r}) + \frac{d\vec{\Omega}}{dt}\times\vec{r}\right) = -\vec{\nabla}p - \rho\vec{g}.$$

5.29 $$\tau_{xx} = -p + 2\mu\overset{0}{\cancel{\frac{\partial u}{\partial x}}} + \lambda\overset{0}{\cancel{\vec{\nabla}\cdot\vec{V}}} = -200 \text{ kPa}.$$

$$\tau_{yy} = \tau_{zz} = -p = -200 \text{ kPa}.$$

$$\tau_{xy} = \mu\left(\frac{\partial u}{\partial y} + \overset{0}{\cancel{\frac{\partial v}{\partial x}}}\right) = 10^{-3}[10 - 2000\times.002] = 0.006 \text{ Pa}.$$

$$\tau_{xz} = \tau_{yz} = 0. \qquad \frac{\tau_{xy}}{\tau_{xx}} = \frac{0.006}{200\,000} = 3\times10^{-8}.$$

5.29E $$\tau_{xx} = -p + 2\mu\overset{0}{\cancel{\frac{\partial u}{\partial x}}} + \lambda\overset{0}{\cancel{\vec{\nabla}\cdot\vec{V}}} = -30 \text{ psi}.$$

$$\tau_{yy} = \tau_{zz} = -p = -30 \text{ psi}.$$

$$\tau_{xy} = \mu\left(\frac{\partial u}{\partial y} + \overset{0}{\cancel{\frac{\partial v}{\partial x}}}\right) = 10^{-5}\left[30 - 1440\times\frac{.1}{12}\right] = 18\times10^{-5} \text{ psf}.$$

$$\tau_{xz} = \tau_{yz} = 0. \qquad \frac{\tau_{xy}}{\tau_{xx}} = \frac{18\times10^{-5}}{30\times144} = 4.17\times10^{-8}.$$

5.30 $\frac{\partial v}{\partial y}=-\frac{\partial u}{\partial x}=\frac{16y}{Cx^{9/5}}-\frac{16y^2}{C^2x^{13/5}}.$ $\quad\therefore v(x,y)=\frac{8y^2}{Cx^{9/5}}-\frac{16y^3}{3C^2x^{13/5}}+f(x).$

$v(x,o)=0.\quad \therefore f(x)=0.\quad 8=C\,1000^{4/5}.\quad \therefore C=0.0318.$

$\therefore u(x,y)=629yx^{-4/5}-9890y^2x^{-8/5}.$

$v(x,y)=252y^2x^{-9/5}-5270y^3x^{-13/5}.$

$\tau_{xx}=-p+2\mu\frac{\partial u}{\partial x}=-100+0=\underline{-100\text{ kPa}}.$

$\tau_{yy}=\tau_{zz}=-p=\underline{-100\text{ kPa}}.$

$\tau_{xy}=\mu\left(\frac{\partial u}{\partial y}+\frac{\partial v}{\partial x}\right)=2\times10^{-5}\left[629\times1000^{-4/5}\right]=\underline{5.01\times10^{-5}\text{ Pa}}.$

$\tau_{xz}=\tau_{yz}=0.$

5.31 $\frac{Du}{Dt}=\cancelto{0}{\frac{\partial u}{\partial t}}+\left(u\frac{\partial}{\partial x}+v\frac{\partial}{\partial y}+w\frac{\partial}{\partial z}\right)u=(\vec{V}\cdot\vec{\nabla})u.$

$\frac{Dv}{Dt}=\cancelto{0}{\frac{\partial v}{\partial t}}+\left(u\frac{\partial}{\partial x}+v\frac{\partial}{\partial y}+w\frac{\partial}{\partial z}\right)v=(\vec{V}\cdot\vec{\nabla})v.$

$\frac{Dw}{Dt}=\cancelto{0}{\frac{\partial w}{\partial t}}+\left(u\frac{\partial}{\partial x}+v\frac{\partial}{\partial y}+w\frac{\partial}{\partial z}\right)w=(\vec{V}\cdot\vec{\nabla})w$

$\therefore\frac{D\vec{V}}{Dt}=\frac{Du}{Dt}\hat{i}+\frac{Dv}{Dt}\hat{j}+\frac{Dw}{Dt}\hat{k}=\vec{V}\cdot\vec{\nabla}(u\hat{i}+v\hat{j}+w\hat{k})=(\vec{V}\cdot\vec{\nabla})\vec{V}.$

5.32 Follow the steps that lead to Eq. 5.3.17 and add the term due to compressible effects:

$$\rho\frac{D\vec{V}}{Dt}=-\vec{\nabla}p+\rho\vec{g}+\mu\nabla^2\vec{V}+\frac{\mu}{3}\frac{\partial}{\partial x}\vec{\nabla}\cdot\vec{V}\hat{i}+\frac{\mu}{3}\frac{\partial}{\partial y}\vec{\nabla}\cdot\vec{V}\hat{j}+\frac{\mu}{3}\frac{\partial}{\partial z}\vec{\nabla}\cdot\vec{V}\hat{k}$$

$$=-\vec{\nabla}p+\rho\vec{g}+\mu\nabla^2\vec{V}+\frac{\mu}{3}\left(\frac{\partial}{\partial x}\hat{i}+\frac{\partial}{\partial y}\hat{j}+\frac{\partial}{\partial z}\hat{k}\right)\vec{\nabla}\cdot\vec{V}$$

$$\therefore\underline{\rho\frac{D\vec{V}}{Dt}=-\vec{\nabla}p+\rho\vec{g}+\mu\nabla^2\vec{V}+\frac{\mu}{3}\vec{\nabla}(\vec{\nabla}\cdot\vec{V}).}$$

5.33 If $u=u(y)$, then continuity demands that $\frac{\partial v}{\partial y}=0.\quad\therefore v=C.$

But, at $y=0$ (the lower plate) $v=0$. $\therefore C=0$, and $v(x,y)=0.$

$$\therefore\rho\frac{Du}{Dt}=\rho\left(\frac{\partial u}{\partial t}+u\frac{\partial u}{\partial x}+v\frac{\partial u}{\partial y}+w\frac{\partial u}{\partial z}\right)=-\frac{\partial p}{\partial x}+\rho g_x+\mu\left(\frac{\partial^2u}{\partial x^2}+\frac{\partial^2u}{\partial y^2}+\frac{\partial^2u}{\partial z^2}\right).$$

$$\therefore 0 = -\frac{\partial p}{\partial x} + \mu \frac{\partial^2 u}{a y^2}.$$

$$\rho \frac{Dv}{Dt} = 0 = -\frac{\partial p}{\partial y}.$$

$$\rho \frac{Dw}{Dt} = 0 = -\frac{\partial p}{\partial z} + \rho(-g). \qquad \therefore 0 = -\frac{\partial p}{\partial z} - \rho g.$$

5.34 Continuity: $\frac{\partial}{\partial r}(r v_r) = 0.$ $\therefore r v_r = C.$ At $r = 0,\ v_r \neq \infty.$ $\therefore C = 0.$

$$\frac{Dv_r}{Dt} = 0 = -\frac{1}{\rho}\frac{\partial p}{\partial r}.$$

$$\frac{Dv_\theta}{Dt} = 0 = -\frac{1}{\rho r}\frac{\partial p}{\partial \theta}.$$

$$\rho \frac{Dv_z}{Dt} = \rho\left(\frac{\partial v_z}{\partial t} + v_r \frac{\partial v_z}{\partial r} + \frac{v_\theta}{r}\frac{\partial v_z}{\partial \theta} + v_z \frac{\partial v_z}{\partial z}\right) = -\frac{\partial p}{\partial z} + \mu\left(\frac{\partial^2 v_z}{\partial r^2} + \frac{1}{r}\frac{\partial v_z}{\partial r} + \frac{1}{r^2}\frac{\partial^2 v_z}{\partial \theta^2} + \frac{\partial^2 v_z}{\partial z^2}\right)$$

(terms marked 0: $\frac{\partial v_z}{\partial t}$, v_r, v_θ, $\frac{\partial v_z}{\partial z}$, $\frac{\partial^2 v_z}{\partial \theta^2}$, $\frac{\partial^2 v_z}{\partial z^2}$)

$$\therefore 0 = -\frac{\partial p}{\partial z} + \mu\left(\frac{\partial^2 v_z}{\partial r^2} + \frac{1}{r}\frac{\partial v_z}{\partial r}\right).$$

5.35 Continuity: $\frac{1}{r^2}\frac{\partial}{\partial r}(r^2 v_r) = 0.$ $\therefore r^2 v_r = C.$ At $r = r_1,\ v_r = 0.$ $\therefore C = 0.$

$$-\frac{v_\theta^2}{r}\rho = -\frac{\partial p}{\partial r} + \mu\left(-\frac{2 v_\theta}{r^2}\cot\theta\right).$$

$$0 = -\frac{1}{r}\frac{\partial p}{\partial \theta} + \mu\left[\frac{1}{r^2}\frac{\partial}{\partial r}\left(r^2 \frac{\partial v_\theta}{\partial r}\right) - \frac{v_\theta}{r^2 \sin^2\theta}\right]$$

$$0 = -\frac{1}{r\sin\theta}\frac{\partial p}{\partial \phi}.$$

5.36 For an incompressible flow $\vec{\nabla}\cdot\vec{V} = 0$. Substitute Eqs. 5.3.10 into Eq. 5.3.2 and 5.3.3:

$$\rho \frac{Du}{Dt} = \frac{\partial}{\partial x}\left(-p + 2\mu\frac{\partial u}{\partial x}\right) + \frac{\partial}{\partial y}\mu\left(\frac{\partial u}{\partial y} + \frac{\partial v}{\partial x}\right) + \frac{\partial}{\partial z}\mu\left(\frac{\partial u}{\partial z} + \frac{\partial w}{\partial x}\right) + \rho g_x.$$

$$= -\frac{\partial p}{\partial x} + \mu\frac{\partial^2 u}{\partial x^2} + \mu\frac{\partial^2 u}{\partial y^2} + \mu\frac{\partial^2 u}{\partial z^2} + \mu\frac{\partial}{\partial x}\left(\frac{\partial u}{\partial x} + \frac{\partial v}{\partial y} + \frac{\partial w}{\partial z}\right) + \rho g_x$$

(the last bracket marked 0)

$$\therefore \rho \frac{Du}{Dt} = -\frac{\partial p}{\partial x} + \mu\left(\frac{\partial^2 u}{\partial x^2} + \frac{\partial^2 u}{\partial y^2} + \frac{\partial^2 u}{\partial z^2}\right) + \rho g_x.$$

$$\rho\frac{Dv}{Dt}=\frac{\partial}{\partial x}\mu\left(\frac{\partial u}{\partial y}+\frac{\partial v}{\partial x}\right)+\frac{\partial}{\partial y}\left(-p+2\mu\frac{\partial v}{\partial y}\right)+\frac{\partial}{\partial z}\mu\left(\frac{\partial v}{\partial z}+\frac{\partial w}{\partial y}\right)+\rho g_y.$$

$$=-\frac{\partial p}{\partial y}+\mu\frac{\partial^2 v}{\partial x^2}+\mu\frac{\partial^2 v}{\partial y^2}+\mu\frac{\partial^2 v}{\partial z^2}+\mu\frac{\partial}{\partial y}\underbrace{\left(\frac{\partial u}{\partial x}+\frac{\partial v}{\partial y}+\frac{\partial w}{\partial z}\right)}_{0}+\rho g_y$$

$$\therefore \rho\frac{Dv}{Dt}=-\frac{\partial p}{\partial y}+\mu\left(\frac{\partial^2 v}{\partial x^2}+\frac{\partial^2 v}{\partial y^2}+\frac{\partial^2 v}{\partial z^2}\right)+\rho g_y.$$

$$\rho\frac{Dw}{Dt}=\frac{\partial}{\partial x}\mu\left(\frac{\partial u}{\partial z}+\frac{\partial w}{\partial x}\right)+\frac{\partial}{\partial y}\mu\left(\frac{\partial v}{\partial z}+\frac{\partial w}{\partial y}\right)+\frac{\partial}{\partial z}\left(-p+2\mu\frac{\partial w}{\partial z}\right)+\rho g_z$$

$$=-\frac{\partial p}{\partial z}+\mu\frac{\partial^2 w}{\partial x^2}+\mu\frac{\partial^2 w}{\partial y^2}+\mu\frac{\partial^2 w}{\partial z^2}+\mu\frac{\partial}{\partial z}\underbrace{\left(\frac{\partial u}{\partial x}+\frac{\partial v}{\partial y}+\frac{\partial w}{\partial z}\right)}_{0}+\rho g_z$$

$$\therefore \rho\frac{Dw}{Dt}=-\frac{\partial p}{\partial z}+\mu\left(\frac{\partial^2 w}{\partial x^2}+\frac{\partial^2 w}{\partial y^2}+\frac{\partial^2 w}{\partial z^2}\right)+\rho g_z.$$

5.37 If we substitute the constitutive equations (5.3.10) into Eqs. 5.3.2 and 5.3.3., with $\mu=\mu(x,y,z)$ we arrive at

$$\rho\frac{Du}{Dt}=-\frac{\partial p}{\partial x}+\rho g_x+\mu\left(\frac{\partial^2 u}{\partial x^2}+\frac{\partial^2 u}{\partial y^2}+\frac{\partial^2 u}{\partial z^2}\right)+2\frac{\partial\mu}{\partial x}\frac{\partial u}{\partial x}+\frac{\partial\mu}{\partial y}\left(\frac{\partial u}{\partial y}+\frac{\partial v}{\partial x}\right)+\frac{\partial\mu}{\partial z}\left(\frac{\partial u}{\partial z}+\frac{\partial w}{\partial x}\right)$$

5.38 If plane flow is only parallel to the plate, $v=w=0$. Continuity then demands that $\partial u/\partial x=0$. The first equation of (5.3.14) simplifies to

$$\rho\left(\frac{\partial u}{\partial t}+\underbrace{u\frac{\partial u}{\partial x}}_{0}+\underbrace{v\frac{\partial u}{\partial y}}_{0}+\underbrace{w\frac{\partial u}{\partial z}}_{0}\right)=-\underbrace{\frac{\partial p}{\partial x}}_{0}+\underbrace{\rho g_x}_{0}+\mu\left(\underbrace{\frac{\partial^2 u}{\partial x^2}}_{0}+\frac{\partial^2 u}{\partial y^2}+\underbrace{\frac{\partial^2 u}{\partial z^2}}_{0}\right)$$

$$\rho\frac{\partial u}{\partial t}=\mu\frac{\partial^2 u}{\partial y^2}$$

We assumed g to be in the y-direction, and since no forcing occurs other than due to the motion of the plate, we let $\partial p/\partial x=0$.

5.39 From Eqs. 5.3.10, $-\dfrac{\tau_{xx}+\tau_{yy}+\tau_{zz}}{3}=p-\dfrac{2\mu}{3}\left(\dfrac{\partial u}{\partial x}+\dfrac{\partial v}{\partial y}+\dfrac{\partial w}{\partial z}\right)-\lambda\,\vec{\nabla}\cdot\vec{V}.$

$$\therefore \bar{p}=p-\left(\frac{2\mu}{3}+\lambda\right)\vec{\nabla}\cdot\vec{V}. \qquad \therefore \bar{p}-p=-\left(\frac{2\mu}{3}+\lambda\right)\vec{\nabla}\cdot\vec{V}.$$

5.40 Studying the vorticity components of Eq. 3.2.21, we see that $\omega_z=-\partial u/\partial y$ is the component of interest. The third equation of 5.3.24 then simplifies to

$$\frac{D\omega_z}{Dt} = \nu\nabla^2\omega_z$$

$$= \nu\frac{\partial^2\omega_z}{\partial y^2}$$

since $\omega = 0$, and changes normal to the plate are much larger than changes along the plate, i.e., $\frac{\partial\omega_z}{\partial y} >> \frac{\partial\omega_z}{\partial x}$.

5.41 If viscous effects are negligible, Eq. 5.3.25 reduces to

$$\frac{D\omega_z}{Dt} = 0$$

that is, there is no change in vorticity (along a streamline) between sections 1 and 2. Since (see Eq. 3.2.21), at section 1,

$$\omega_z = \frac{\partial v}{\partial x} - \frac{\partial u}{\partial y} = -10$$

we conclude that, for the lower half of the flow at section 2,

$$\frac{\partial u}{\partial y} = 10.$$

This means the velocity profile at section 2 is a straight line with the same slope of the profile at section 1. Since we are neglecting viscosity, the flow can slip at the wall with a slip velocity u_0; hence, the velocity distribution at section 2 is $u_2(y) = u_0 + 10y$. Continuity then allows us to calculate the profile:

$$V_1A_1 = V_2A_2$$

$$\frac{1}{2}(10 \times 0.04)(0.04w) = (u_0 + 10 \times 0.02/2)(0.02w). \qquad \therefore u_0 = 0.3 \text{ m/s}.$$

Finally,

$$u_2(y) = \underline{0.3 + 10y}$$

5.42 No. The first of Eqs. 5.3.24 shows that, neglecting viscous effects,

$$\frac{D\omega_x}{Dt} = \omega_x\frac{\partial u}{\partial x} + \omega_y\frac{\partial u}{\partial y} + \omega_z\frac{\partial u}{\partial z}$$

so that ω_y, which is nonzero near the snow surface, creates ω_x through the term $\omega_y\partial u/\partial y$, since there would be a nonzero $\partial u/\partial y$ near the tree.

5.43 $$\int_{c.s.} k\vec{\nabla}T\cdot\hat{n}dA = \int_{c.v.}\frac{\partial}{\partial t}\left(\frac{V^2}{2} + gz + \tilde{u}\right)\rho d\forall + \int_{c.s.}\left(\frac{V^2}{2} + gz + \tilde{u} + \frac{p}{\rho}\right)\rho\vec{V}\cdot\hat{n}dA$$

$$\int_{c.v.}\vec{\nabla}\cdot(k\vec{\nabla}T)d\forall = \int_{c.v.}\frac{\partial}{\partial t}\left(\frac{V^2}{2} + gz + \tilde{u}\right)\rho d\forall + \int_{c.v.}\vec{\nabla}\cdot\rho\vec{V}\left(\frac{V^2}{2} + gz + \tilde{u} + \frac{p}{\rho}\right)d\forall$$

$$\therefore \int_{c.v.} \left[-k\nabla^2 T + \frac{\partial}{\partial t}\left(\rho\frac{V^2}{2} + \rho g z + \rho\tilde{u} \right) + \bar{\nabla}\cdot\rho\bar{V}\left(\frac{V^2}{2} + gz + \tilde{u} + \frac{p}{\rho} \right) \right] dV = 0.$$

$$\frac{\partial}{\partial t}\rho\frac{V^2}{2} + \bar{\nabla}\cdot\rho\bar{V}\left(\frac{V^2}{2} + gz + \frac{p}{\rho} \right) = \frac{V^2}{2}\left(\frac{\partial\rho}{\partial t} + \bar{\nabla}\cdot\rho\bar{V} \right) + \rho\bar{V}\cdot\left[\frac{\partial\bar{V}}{\partial t} + \bar{V}\cdot\nabla\bar{V} + \frac{\bar{\nabla}p}{\rho} + g\bar{\nabla}z \right] = 0.$$

continuity (= 0) momentum (= 0)

$$\therefore -k\nabla^2 T + \frac{\partial}{\partial t}\rho\tilde{u} + \rho\bar{V}\cdot\bar{\nabla}\tilde{u} = 0. \qquad \therefore \underline{\rho\frac{D\tilde{u}}{Dt} = k\nabla^2 T.}$$

5.44 See Eq. 5.4.9: $\tilde{u} = cT$. $\quad \therefore \rho c\left(\frac{\partial T}{\partial t} + u\frac{\partial T}{\partial x} + v\frac{\partial T}{\partial y} + w\frac{\partial T}{\partial z} \right) = k\nabla^2 T.$

Neglect terms with velocity: $\quad \underline{\rho c\frac{\partial T}{\partial t} = k\nabla^2 T.}$

5.45 The dissipation function Φ involves viscous effects. For flows with extremely large velocity gradients, it becomes quite large. Then

$$\rho c_p \frac{DT}{Dt} = \Phi$$

and $\frac{DT}{Dt}$ is large. This leads to very high temperatures on reentry vehicles.

5.46 $u = 10(1 - 10\ 000\, r^2)$. $\quad \therefore \frac{\partial u}{\partial r} = -2r\times 10^5$. $\quad$ (r takes the place of y)

From Eq. 5.4.17, $\Phi = 2\mu\left[\frac{1}{2}\left(\frac{\partial u}{\partial y} \right)^2 \right] = \mu 4 r^2 \times 10^{10}$.

At the wall where $r = 0.01$ m, $\quad \Phi = 1.8\times 10^{-5}\times 4 \times .01^2 \times 10^{10} = \underline{72\ \text{N}/\text{m}^2\cdot\text{s}}.$

At the centerline $\frac{\partial u}{\partial r} = 0$ so $\Phi = \underline{0}$.

At a point half-way: $\Phi = 1.8\times 10^{-5}\times 4\times .005^2\times 10^{10} = \underline{18\ \text{N}/\text{m}^2\cdot\text{s}}.$

5.47 (a) Momentum: $\frac{\partial u}{\partial t} = \nu\frac{\partial^2 u}{\partial y^2}$

Energy: $\rho c\frac{\partial T}{\partial t} = K\frac{\partial^2 T}{\partial y^2} + \mu\left(\frac{\partial u}{\partial y} \right)^2.$

(b) Momentum: $\rho\frac{\partial u}{\partial t} = \mu\frac{\partial^2 u}{\partial y^2} + \frac{\partial\mu}{\partial y}\frac{\partial u}{\partial y}$

Energy: $\rho c\frac{\partial T}{\partial t} = K\frac{\partial^2 T}{\partial y^2} + \mu\left(\frac{\partial u}{\partial y} \right)^2.$

SIX

Dimensional Analysis and Similitude

6.1 $$\frac{g}{V_1^2}\left[\frac{V_1^2}{2g}+\frac{p_1}{\rho g}+z_1\right]=\frac{g}{V_1^2}\left[\frac{V_2^2}{2g}+\frac{p_2}{\rho g}+z_2\right].$$

$$\frac{1}{2}+\frac{p_1}{\rho V_1^2}+\frac{gz_1}{V_1^2}=\frac{1}{2}\frac{V_2^2}{V_1^2}+\frac{p_2}{\rho V_1^2}+\frac{gz_2}{V_1^2}.$$

or $$\underline{\frac{1}{2}+\frac{p_1}{\rho V_1^2}+\frac{gz_1}{V_1^2}=\left(\frac{1}{2}+\frac{p_2}{\rho V_2^2}+\frac{gz_2}{V_2^2}\right)\frac{V_2^2}{V_1^2}}$$

6.2 a) $[\dot{m}]=\frac{\text{kg}}{\text{s}}=\frac{\text{N}\cdot\text{s}^2}{\text{m}\cdot\text{s}}=\frac{\text{N}\cdot\text{s}}{\text{m}}. \quad \therefore \frac{FT}{L}.$

b) $[p]=\frac{\text{N}}{\text{m}^2}. \quad \therefore \frac{F}{L^2}$

c) $[\rho]=\frac{\text{kg}}{\text{m}^3}=\frac{\text{N}\cdot\text{s}^2}{\text{m}\cdot\text{m}^3}=\frac{\text{N}\cdot\text{s}^2}{\text{m}^4}. \quad \therefore \frac{FT^2}{L^4}.$

d) $[\mu]=\frac{\text{N}\cdot\text{s}}{\text{m}^2}. \quad \therefore \frac{FT}{L^2}$

e) $[W]=\text{N}\cdot\text{m}. \quad \therefore FL$

f) $[\dot{W}]=\frac{\text{N}\cdot\text{m}}{\text{s}}. \quad \therefore \frac{FL}{T}$

g) $[\sigma]=\text{N}/\text{m}. \quad \therefore \frac{F}{L}$

6.3 $V=f(\ell,\rho,\mu). \quad [V]=\frac{L}{T},\ [\ell]=L,\ [\rho]=\frac{M}{L^3},\ [\mu]=\frac{M}{LT}.$

$\therefore$ There is one π – term: $\pi_1=\frac{\rho V\ell}{\mu}.$

$\therefore \pi_1=f_1(\pi_2^0)=\text{Const.} \quad \therefore \underline{\rho\frac{V\ell}{\mu}=C,}$ or Re = Const.

6.4 $V = f(\sigma, \rho, d)$. $[V] = \frac{L}{T}$, $[\sigma] = \frac{M}{T^2}$, $[\rho] = \frac{M}{L^3}$, $[d] = L$.

$$\therefore \pi_1 = \frac{\sigma}{\rho V^2 d}. \quad \therefore \pi_1 = f_1(\pi_2^0) = \text{Const.} \quad \therefore \underline{\frac{\sigma}{\rho V^2 d} = C}, \text{ or We = Const.}$$

6.5 $V = f(H, g, m)$. $[V] = \frac{L}{T}$, $[g] = \frac{L}{T^2}$, $[m] = M$, $[H] = L$.

$$\therefore \pi_1 = \frac{gHm^0}{V^2}. \quad \therefore \pi_1 = C. \quad \therefore \underline{V = \sqrt{gH/C}}.$$

6.6 $V = f(H, g, m, \rho, \mu)$. $[V] = \frac{L}{T}$, $[H] = L$, $[g] = \frac{L}{T^2}$, $[m] = M$, $[\rho] = \frac{M}{L^3}$, $[\mu] = \frac{M}{LT}$.

Choose repeating variables H, g, ρ (select ones with simple dimensions):

$$\pi_1 = VH^{a_1} g^{b_1} \rho^{c_1}, \quad \pi_2 = mH^{a_2} g^{b_2} \rho^{c_2}, \quad \pi_3 = \mu H^{a_3} g^{b_3} \rho^{c_3}.$$

$$\therefore \pi_1 = \frac{V\rho^0}{\sqrt{g}\sqrt{H}} = \frac{V}{\sqrt{gH}}. \quad \pi_2 = \frac{m}{\rho H^3}. \quad \pi_3 = \frac{\mu}{\rho\sqrt{g}H^{3/2}} = \frac{\mu}{\rho\sqrt{gH^3}}.$$

$$\therefore \underline{\frac{V}{\sqrt{gH}} = f_1\left(\frac{m}{\rho H^3}, \frac{\mu}{\rho\sqrt{gH^3}}\right)}.$$

Note: The above dimensionless groups are formed by observation: simply combine the dimensions so that the π – term is dimensionless.

6.7 $F_D = f(d, \ell, V, \mu, \rho)$. $[F_D] = \frac{ML}{T^2}$, $[d] = L$, $[V] = \frac{L}{T}$, $[\mu] = \frac{M}{LT}$, $[\rho] = \frac{M}{L^3}$.

$$\pi_1 = F_D \ell^{a_1} V^{b_1} \rho^{c_1}, \quad \pi_2 = dV^{b_2} \rho^{c_2} \ell^{a_2}, \quad \pi_3 = \mu \ell^{a_3} V^{b_3} \rho^{c_3}.$$

$$\therefore \pi_1 = \frac{F_D}{V^2 \rho \ell^2}, \quad \pi_2 = \frac{d}{\ell}, \quad \pi_3 = \frac{\mu}{\rho V \ell}.$$

$$\therefore \underline{\frac{F_D}{\rho \ell^2 V^2} = f_1\left(\frac{d}{\ell}, \frac{\mu}{\rho \ell V}\right)}.$$

We could write $\frac{\pi_1}{\pi_2^2} = f_2\left(\frac{1}{\pi_2}, \frac{\pi_3}{\pi_2}\right)$ or $\frac{F_D}{\rho d^2 V^2} = f_2\left(\frac{\ell}{d}, \frac{\mu}{\rho d V}\right)$. This is equivalent to the above. Either functional form must be determined by experimentation.

6.8 $F_D = f(d, \ell, V, \mu, \rho)$. $[F_D] = \frac{ML}{T^2}$, $[d] = L$, $[V] = \frac{L}{T}$, $[\mu] = \frac{M}{LT}$, $[\rho] = \frac{M}{L^3}$.

$\pi_1 = F_D d^{a_1} \mu^{b_1} V^{c_1}$, $\pi_2 = \ell d^{a_2} \mu^{b_2} V^{c_2}$, $\pi_3 = \rho d^{a_3} \mu^{b_3} V^{c_3}$.

$\therefore \pi_1 = \frac{F_D}{\mu V d}$, $\pi_2 = \frac{\ell}{d}$, $\pi_3 = \frac{\rho V d}{\mu}$.

$\therefore \underline{\frac{F_D}{\mu V d} = f_1\left(\frac{\ell}{d}, \frac{\rho V d}{\mu}\right)}$. Rather than $\pi_1 = f_1(\pi_2, \pi_3)$, we could write

$\frac{\pi_1}{\pi_3} = f_2\left(\pi_2, \frac{1}{\pi_3}\right)$, an acceptable form: $\frac{F_D}{\rho V^2 d^2} = f_2\left(\frac{\ell}{d}, \frac{\mu}{\rho V d}\right)$.

6.9 $h = f(\sigma, d, \gamma, \beta, g)$. $[h] = L$, $[\sigma] = \frac{M}{T^2}$, $[d] = L$, $[\gamma] = \frac{M}{L^2T^2}$, $[\beta] = 1$, $[g] = \frac{L}{T^2}$.

Select d, γ, g as repeating variables.

$\pi_1 = h d^{a_1} \gamma^{b_1} g^{c_1}$, $\pi_2 = \sigma d^{a_2} \gamma^{b_2} g^{c_2}$, $\pi_3 = \beta$.

$\therefore \pi_1 = \frac{h}{d}$, $\pi_2 = \frac{\sigma}{\gamma d^2}$, $\pi_3 = \beta$.

$\therefore \underline{\frac{h}{d} = f_1\left(\frac{\sigma}{\gamma d^2}, \beta\right)}$. Note: gravity does not enter the answer.

6.10 $F_C = f(m, \omega, R)$. $[F_C] = \frac{ML}{T^2}$, $[m] = M$, $[\omega] = \frac{1}{T}$, $[R] = L$.

$\therefore \pi_1 = F_C m^a \omega^b R^c = \frac{F_C}{m\omega^2 R}$. $\therefore \frac{F_C}{m\omega^2 R} = C$. $\therefore \underline{F_C = Cm\omega^2 R}$

6.11 $\sigma = f(M, y, I)$. $[\sigma] = \frac{M}{LT^2}$, $[M] = \frac{ML^2}{T^2}$, $[y] = L$, $[I] = L^4$.

$\therefore \pi_1 = \sigma M^a y^b I^c$. Given that $b = -1$. Hence,

$\pi_1 = \frac{\sigma I}{yM} = \text{Const.}$ $\therefore \underline{\sigma = \text{Const.} \frac{My}{I}}$.

6.12 $V = f\left((\mu, d, \frac{dp}{dx}\right)$. $[V] = \frac{L}{T}$, $[\mu] = \frac{M}{LT}$, $[d] = L$, $\left[\frac{dp}{dx}\right] = \frac{M}{L^2T^2}$.

$\therefore \pi_1 = V\mu^a d^b \left(\frac{dp}{dx}\right)^c$. Let's start with the ratio $\frac{\mu}{dp/dx}$ so that "M" is accounted for. Then the π_1 – term is $\frac{\mu V}{dp/dx\, d^2}$. Hence,

$\pi_1 = \frac{V\mu}{dp/dx\, d^2} = \text{Const.}$ $\therefore \underline{V = \text{Const} \frac{d^2 dp/dx}{\mu}}$.

6.13 $V = f(H, g, \rho)$. $[V] = \frac{L}{T}$, $[H] = L$, $[g] = \frac{L}{T^2}$, $[\rho] = \frac{M}{L^3}$.

$$\therefore \pi_1 = VH^a g^b \rho^c = V\frac{\rho^0}{\sqrt{g}\sqrt{H}} = \text{Const.} \quad \therefore V = \underline{\text{Const.}\ \sqrt{gH}}.$$

Density does not enter the expression.

6.14 $V = f(H, \mu, \rho, g, d)$. $[V] = \frac{L}{T}, [H] = L, [\mu] = \frac{M}{LT}, [\rho] = \frac{M}{L^3}, [g] = \frac{L}{T^2}, [d] = L$.

$$\pi_1 = VH^{a_1}\rho^{b_1}g^{c_1},\ \pi_2 = \mu H^{a_2}\rho^{b_2}g^{c_2},\ \pi_3 = dH^{a_3}\rho^{b_3}g^{c_3}. \quad \left.\begin{matrix}\text{Repeating}\\ \text{variables}\end{matrix}\right\} H, \rho, g.$$

$$\pi_1 = \frac{V}{\sqrt{gH}},\ \pi_2 = \frac{\mu}{\rho\sqrt{g}H^{3/2}},\ \pi_3 = \frac{d}{H}.$$

$$\therefore \pi_1 = f_1(\pi_2, \pi_3), \quad \text{or} \quad \underline{\frac{V}{\sqrt{gH}} = f_1\left(\frac{\mu}{\rho\sqrt{gH^3}}, \frac{d}{H}\right)}.$$

6.15 $\Delta p = f(V, d, \nu, L, \varepsilon, \rho)$.

$$[\Delta p] = \frac{M}{LT^2}, [V] = \frac{L}{T}, [d] = L, [\nu] = \frac{L^2}{T}, [L] = L, [\varepsilon] = L, [\rho] = \frac{M}{L^3}.$$

Repeating variables: V, d, ρ.

$$\pi_1 = \Delta p V^{a_1} d^{b_1} \rho^{c_1},\ \pi_2 = \nu V^{a_2} d^{b_2} \rho^{c_2},\ \pi_3 = L\, V^{a_3} d^{b_3} \rho^{c_3},\ \pi_4 = \varepsilon\, V^{a_4} d^{b_4} \rho^{c_4}.$$

$$\therefore \pi_1 = \frac{\Delta p}{\rho V^2},\ \pi_2 = \frac{\nu}{Vd},\ \pi_3 = \frac{L}{d},\ \pi_4 = \frac{\varepsilon}{d}.$$

$$\pi_1 = f_1(\pi_2, \pi_3, \pi_4). \quad \therefore \underline{\frac{\Delta p}{\rho V^2} = f_1\left(\frac{\nu}{Vd}, \frac{L}{d}, \frac{\varepsilon}{d}\right)}.$$

6.16 $F_D = f(V, \rho, \mu, c, h, r, \phi, w, \alpha)$

where c is the chord length, h is the maximum thickness, r is the nose radius, ϕ is the trailing edge angle, and α is the angle of attack. Repeating variables: V, c, ρ. The π – terms are

$$\pi_1 = \frac{F_D}{\rho V^2 c^2},\ \pi_2 = \frac{V\rho c}{\mu},\ \pi_3 = \frac{c}{h},\ \pi_4 = \frac{c}{r},\ \pi_5 = \phi,\ \pi_6 = \frac{c}{w},\ \pi_7 = \alpha.$$

Then,

$$\underline{\frac{F_D}{\rho V^2 c^2} = f_1\left(\frac{V\rho c}{\mu}, \frac{c}{h}, \frac{c}{r}, \phi, \frac{c}{w}, \alpha\right)}$$

6.17 $Q = f(R, A, e, S, g)$. $[Q] = \frac{L^3}{T}, [R] = L, [A] = L^2, [e] = L, [s] = 1, [g] = \frac{L}{T^2}$.

There are only two basic dimensions. Choose two repeating variables, R and g. Then,

$\pi_1 = QR^{a_1}g^{b_1}$, $\pi_2 = AR^{a_2}g^{b_2}$, $\pi_3 = eR^{a_3}g^{b_3}$, $\pi_4 = sR^{a_4}g^{b_4}$.

$\therefore \pi_1 = \frac{Q}{\sqrt{g}R^{5/2}}$, $\pi_2 = \frac{A}{R^2}$, $\pi_3 = \frac{e}{R}$, $\pi_4 = s$.

$\therefore \pi_1 = f_1(\pi_2, \pi_3, \pi_4)$. $\therefore \frac{Q}{\sqrt{gR^5}} = f_1\left(\frac{A}{R^2}, \frac{e}{R}, s\right)$.

6.18 $V_p = f(h, g, \sigma, \rho)$. $[V_p] = \frac{L}{T}, [h] = L, [g] = \frac{L}{T^2}, [\sigma] = \frac{M}{T^2}, [\rho] = \frac{M}{L^3}$.

Repeating variables: h, ρ, g. $\therefore \pi_1 = V_p h^{a_1}\rho^{b_1}g^{c_1}$, $\pi_2 = \sigma h^{a_2}\rho^{b_2}g^{c_2}$.

$\therefore \pi_1 = \frac{V_p}{\sqrt{hg}}$, $\pi_2 = \frac{\sigma}{\rho g h^2}$. $\therefore \frac{V_p}{\sqrt{gh}} = f_1\left(\frac{\sigma}{\rho g h^2}\right)$.

6.19 $F_D = f(V, \mu, \rho, e, I, d)$. Repeating variables: V, ρ, d.

$[F_D] = \frac{ML}{T^2}, [V] = \frac{L}{T}, [\mu] = \frac{M}{LT}, [\rho] = \frac{M}{L^3}, [e] = L, [I] = 1, [d] = L$.

$\pi_1 = F_D V^{a_1}\rho^{b_1}d^{c_1}$, $\pi_2 = \mu V^{a_2}\rho^{b_2}d^{c_2}$, $\pi_3 = e\,V^{a_3}\rho^{b_3}d^{c_3}$, $\pi_4 = I\,V^{a_4}\rho^{b_4}d^{c_4}$.

$\therefore \pi_1 = \frac{F_D}{\rho V^2 d^2}$, $\pi_2 = \frac{\mu}{V\rho d}$, $\pi_3 = \frac{e}{d}$, $\pi_4 = I$.

$\therefore \frac{F_D}{\rho V^2 d^2} = f_1\left(\frac{\mu}{V\rho d}, \frac{e}{d}, I\right)$.

6.20 $F_D = f(V, \rho_s, \rho, \mu, D, g)$. Repeating variables: V, ρ, D.

$[F_D] = \frac{ML}{T^2}, [V] = \frac{L}{T}, [\rho_s] = \frac{M}{L^3}, [\rho] = \frac{M}{L^3}, [\mu] = \frac{M}{LT}, [D] = L, [g] = \frac{L}{T^2}$.

$\pi_1 = F_D V^{a_1}\rho^{b_1}D^{c_1}$, $\pi_2 = \rho_s V^{a_2}\rho^{b_2}D^{c_2}$, $\pi_3 = \mu V^{a_3}\rho^{b_3}D^{c_3}$, $\pi_4 = gV^{a_4}\rho^{b_4}D^{c_4}$.

$\therefore \pi_1 = \frac{F_D}{\rho V^2 D^2}$, $\pi_2 = \frac{\rho_s}{\rho}$, $\pi_3 = \frac{\mu}{\rho V D}$, $\pi_4 = \frac{gD}{V^2}$.

$\therefore \frac{F_D}{\rho V^2 D^2} = f_1\left(\frac{\rho_s}{\rho}, \frac{\mu}{\rho V D}, \frac{gD}{V^2}\right)$.

6.21 $F_D = f(V,\mu,\rho,d,e,r,c)$. Repeating variables: V,ρ,d.

$$[F_D]=\frac{ML}{T^2},\ [V]=\frac{L}{T},\ [\mu]=\frac{M}{LT},\ [\rho]=\frac{M}{L^3},\ [d]=L,\ [e]=L,\ [r]=L,\ [c]=\frac{1}{L^2}.$$

$$\pi_1=F_DV^{a_1}\rho^{b_1}d^{c_1},\ \pi_2=\mu V^{a_2}\rho^{b_2}d^{c_2},\ \pi_3=eV^{a_3}\rho^{b_3}d^{c_3},\ \pi_4=rV^{a_4}\rho^{b_4}d^{c_4},\ \pi_5=cV^{a_5}\rho^{b_5}d^{c_5}.$$

$$\therefore \pi_1=\frac{F_D}{\rho V^2d^2},\ \pi_2=\frac{\mu}{\rho Vd},\ \pi_3=\frac{e}{d},\ \pi_4=\frac{r}{d},\ \pi_5=cd^2.$$

$$\therefore \frac{F_D}{\rho V^2d^2}=f_1\left(\frac{\mu}{\rho Vd},\frac{e}{d},\frac{r}{d},cd^2\right).$$

6.22 $f=g(\mu,\rho,V,d)$. $[f]=\frac{1}{T},\ [\mu]=\frac{M}{LT},\ [\rho]=\frac{M}{L^3},\ [V]=\frac{L}{T},\ [d]=L.$

Repeating variables, V, d, ρ. $\pi_1=fV^{a_1}d^{b_1}\rho^{c_1},\ \pi_2=\mu V^{a_2}d^{b_2}\rho^{c_2}$

$$\therefore \pi_1=\frac{fd}{V},\ \pi_2=\frac{\mu}{\rho Vd}. \qquad \therefore \frac{fd}{V}=g_1\left(\frac{\mu}{\rho Vd}\right).$$

6.23 $F_L=f(V,\ c,\ \rho,\ \ell_c,\ t,\ \alpha)$. Repeating variables: V, ρ, ℓ_c.

$$[F_L]=\frac{ML}{T^2},\ [V]=\frac{L}{T},\ [c]=\frac{L}{T},\ [\rho]=\frac{M}{L^3},\ [\ell_c]=L,\ [t]=L,\ [\alpha]=1.$$

$$\pi_1=F_LV^{a_1}\rho^{b_1}\ell_c^{\,c_1},\ \pi_2=cV^{a_2}\rho^{b_2}\ell_c^{\,c_2},\ \pi_3=tV^{a_3}\rho^{b_3}\ell_c^{\,c_3},\ \pi_4=\alpha V^{a_4}\rho^{b_4}\ell_c^{\,c_4}.$$

$$\therefore \pi_1=\frac{F_L}{\rho V^2\ell_c^{\,2}},\ \pi_2=\frac{c}{V},\ \pi_3=\frac{t}{\ell_c},\ \pi_4=\alpha.$$

$$\therefore \frac{F_L}{\rho V^2\ell_c^{\,2}}=f_1\left(\frac{c}{V},\frac{t}{\ell_c},\alpha\right).$$

6.24 $T=f(d,\ \omega,\ \rho,\ \mu,\ t)$. $[T]=\frac{ML^2}{T^2},\ [d]=L, [\omega]=\frac{1}{T}, [\rho]=\frac{M}{L^3}, [\mu]=\frac{M}{LT}, [t]=L.$

Repeating variables: d, ω, ρ. $\pi_1=Td^{a_1}\omega^{b_1}\rho^{c_1},\ \pi_2=\mu d^{a_2}\omega^{b_2}\rho^{c_2},\ \pi_3=t\,d^{a_3}\omega^{b_3}\rho^{c_3}.$

$$\therefore \pi_1=\frac{T}{\rho\omega^2d^5},\ \pi_2=\frac{\mu}{\rho\omega d^2},\ \pi_3=\frac{t}{d}. \qquad \therefore \frac{T}{\rho\omega^2d^5}=f_1\left(\frac{\mu}{\rho\omega d^2},\frac{t}{d}\right). \qquad \dot{W}=\rho\omega^3d^5f_1(\pi_2,\pi_3).$$

6.25 $F_D=f(V,\ \rho,\ \mu,\ d,\ L,\ \rho_c,\ \omega)$ where d is the cable diameter, L the cable length, ρ_c the cable density, and ω the vibration frequency.

Repeating variables: V, d, ρ. The π – terms are

$$\pi_1=\frac{F_D}{\rho V^2d^2},\ \pi_2=\frac{Vd\rho}{\mu},\ \pi_3=\frac{d}{L},\ \pi_4=\frac{\rho}{\rho_c},\ \pi_5=\frac{V}{\omega d}$$

We then have

$$\underline{\frac{F_D}{\rho V^2 d^2} = f_1\left(\frac{Vd\rho}{\mu},\ \frac{d}{L},\ \frac{\rho}{\rho_c},\ \frac{V}{\omega d}\right)}$$

6.26 $\Delta p = f(D,\ h,\ \omega,\ \rho,\ d_1,\ d_0)$. Repeating variables: $D,\ \omega,\ \rho$.

$$[\Delta p] = \frac{M}{LT^2},\ [D] = L,\ [h] = L,\ [\omega] = \frac{1}{T},\ [\rho] = \frac{M}{L^3},\ [d_1] = L,\ [d_0] = L$$

$$\pi_1 = \frac{\Delta p}{\rho\omega^2 D^2},\ \pi_2 = \frac{h}{D},\ \pi_3 = \frac{d_1}{D},\ \pi_4 = \frac{d_0}{D}.$$

$$\therefore \frac{\Delta p}{\rho\omega^2 D^2} = f_1\left(\frac{h}{D},\ \frac{d_1}{D},\ \frac{d_0}{D}\right).\quad \dot{W} = \text{force} \times \text{velocity}\ \Delta p D^2 \times \omega D.$$

$$\underline{\therefore \dot{W} = \rho\omega^3 D^5 f_1\left(\frac{h}{D},\ \frac{d_1}{D},\ \frac{d_0}{D}\right).}$$

6.27 $T = g(f,\ \omega,\ d,\ H,\ \ell,\ N,\ h,\ \rho)$. Repeating variables: $\omega,\ d,\ \rho$.

$$[T] = \frac{ML^2}{T^2},\ [f] = \frac{1}{T},\ [\omega] = \frac{1}{T},\ [d] = L,\ [H] = L,\ [\ell] = L,\ [N] = 1,\ [h] = L,\ [\rho] = \frac{M}{L^3}.$$

$$\pi_1 = \frac{T}{\rho\omega^2 d^5},\ \pi_2 = \frac{f}{\omega},\ \pi_3 = \frac{H}{d},\ \pi_4 = \frac{\ell}{d},\ \pi_5 = N,\ \pi_6 = \frac{h}{d}.$$

$$\underline{\therefore \frac{T}{\rho\omega^2 d^5} = g_1\left(\frac{f}{\omega},\ \frac{H}{d},\ \frac{\ell}{\mathrm{d}},\ N,\ \frac{h}{d}\right).}$$

6.28 $Q = f(H,\ w,\ g,\ \mu,\ \rho,\ \sigma)$. Repeating variables: $H,\ g,\ \rho$.

$$[Q] = \frac{L^3}{T},\ [H] = L,\ [w] = L,\ [g] = \frac{L}{T^2},\ [\mu] = \frac{M}{LT},\ [\rho] = \frac{M}{L^3},\ [\sigma] = \frac{M}{T^2}.$$

$$\therefore \pi_1 = \frac{Q}{\sqrt{gH^5}},\ \pi_2 = \frac{w}{H},\ \pi_3 = \frac{\mu}{\rho\sqrt{gH^3}},\ \pi_4 = \frac{\sigma}{\rho g H^2}.$$

$$\underline{\therefore \frac{Q}{\sqrt{gH^5}} = f_1\left(\frac{w}{H},\ \frac{\mu}{\rho\sqrt{gH^3}},\ \frac{\sigma}{\rho g H^2}\right).}$$

6.29 $d = f(V,\ V_j,\ D,\ \sigma,\ \rho,\ \mu,\ \rho_a)$. Repeating variables: $V_j,\ D,\ \rho$.

$$[d] = L,\ [V] = \frac{L}{T},\ [V_j] = \frac{L}{T},\ [D] = L,\ [\sigma] = \frac{M}{T^2},\ [\rho] = \frac{M}{L^3},\ [\mu] = \frac{M}{LT},\ [\rho_a] = \frac{M}{L^3}.$$

$$\pi_1 = \frac{d}{D},\ \pi_2 = \frac{V}{V_j},\ \pi_3 = \frac{\sigma}{\rho V_j^2 D},\ \pi_4 = \frac{\mu}{\rho V_j D},\ \pi_5 = \frac{\rho_a}{\rho}.$$

$$\underline{\therefore \frac{d}{D} = f_1\left(\frac{V}{V_j},\ \frac{\sigma}{\rho V_j^2 D},\ \frac{\mu}{\rho V_j D},\ \frac{\rho_a}{\rho}\right).}$$

6.30 $T = f(\omega, H, h, R, t, \mu, \rho)$. Repeating variables: ω, h, ρ.

$$[T] = \frac{ML^2}{T^2},\ [\omega] = \frac{1}{T},\ [H] = L,\ [h] = L,\ [R] = L,\ [t] = L,\ [\mu] = \frac{M}{LT},\ [\rho] = \frac{M}{L^3}.$$

$$\pi_1 = \frac{T}{\rho\omega^2 d^5},\ \pi_2 = \frac{H}{h},\ \pi_3 = \frac{R}{h},\ \pi_4 = \frac{t}{h},\ \pi_5 = \frac{\mu}{\rho\omega h^2}$$

$$\therefore \frac{T}{\rho\omega^2 d^5} = f_1\left(\frac{H}{h}, \frac{R}{h}, \frac{t}{h}, \frac{\mu}{\rho\omega h^2}\right).$$

6.31 $\mu = f(D, H, \ell, g, \rho, V)$ D = tube dia., H = head above outlet, ℓ = tube length.

Repeating variables: D, V, ρ. $\pi_1 = \frac{\mu}{\rho VD},\ \pi_2 = \frac{H}{D},\ \pi_3 = \frac{\ell}{D},\ \pi_4 = \frac{gD}{V^2}$

$$\therefore \frac{\mu}{\rho VD} = f_1\left(\frac{H}{D}, \frac{\ell}{D}, \frac{gD}{V^2}\right).$$

6.32 $T = f(R, \omega, \rho, e, r, \mu, \ell)$. Repeating variables: R, ω, ρ.

$$[T] = \frac{ML^2}{T^2},\ [R] = L,\ [\omega] = \frac{1}{T},\ [\rho] = \frac{M}{L^3},\ [e] = L,\ [r] = L,\ [\mu] = \frac{M}{LT},\ [\ell] = L.$$

$$\pi_1 = \frac{T}{\rho\omega^2 R^5},\ \pi_2 = \frac{e}{R},\ \pi_3 = \frac{r}{R},\ \pi_4 = \frac{\ell}{R},\ \pi_5 = \frac{\mu}{\rho\omega R^2}$$

$$\therefore \frac{T}{\rho\omega^2 R^5} = f_1\left(\frac{e}{R}, \frac{r}{R}, \frac{\ell}{R}, \frac{\mu}{\rho\omega R^2}\right).$$

6.33 $y_2 = f(V_1, y_1, \rho, g)$. Neglect viscous wall shear.

$[y_2] = L,\ [V_1] = \frac{L}{T},\ [y_1] = L,\ [\rho] = \frac{M}{L^3},\ [g] = \frac{L}{T^2}$. Repeating variables: V_1, y_1, ρ.

$\pi_1 = \frac{y_2}{y_1},\ \pi_2 = \frac{gy_1}{V_1^2}$. ($\rho$ does not enter the problem).

$$\therefore \frac{y_2}{y_1} = f\left(\frac{gy_1}{V_1^2}\right).$$

6.34 $f = g(d, \ell, \rho, \mu, V)$. $[f] = \frac{1}{T},\ [d] = L,\ [\ell] = L,\ [\rho] = \frac{M}{L^3},\ [\mu] = \frac{M}{LT},\ [V] = \frac{L}{T}$.

Repeating variables: d, ρ, V. (ℓ = length of cylinder).

$$\pi_1 = \frac{fd}{V},\ \pi_2 = \frac{\ell}{d},\ \pi_3 = \frac{\mu}{\rho Vd}. \qquad \therefore \frac{fd}{V} = f_1\left(\frac{\ell}{d}, \frac{\mu}{\rho Vd}\right).$$

6.35 $\dfrac{Q_m}{Q_p} = \dfrac{V_m \ell_m^2}{V_p \ell_p^2}, \quad \dfrac{\Delta p_m}{\Delta p_p} = \dfrac{\rho_m V_m^2}{\rho_p V_p^2}, \quad \dfrac{(F_p)_m}{(F_p)_p} = \dfrac{\rho_m V_m^2 \ell_m^2}{\rho_p V_p^2 \ell_p^2}$

$\dfrac{\tau_m}{\tau_p} = \dfrac{\rho_m V_m^2}{\rho_p V_p^2}, \quad \dfrac{T_m}{T_p} = \dfrac{\rho_m V_m^2 \ell_m^3}{\rho_p V_p^2 \ell_p^3}, \quad \dfrac{\dot{Q}_m}{\dot{Q}_p} = \dfrac{\rho_m V_m^3 \ell_m^2}{\rho_p V_p^3 \ell_p^2}$ ($\dot{Q}$ has same dimensions as $\dot{W}$)

6.36 a) $\text{Re}_m = \text{Re}_p. \quad \dfrac{V_m d_m}{\nu_m} = \dfrac{V_p d_p}{\nu_p}. \quad \therefore \dfrac{V_m}{V_p} = \dfrac{d_p}{d_m} = 7.$

$\dfrac{Q_m}{Q_p} = \dfrac{V_m \ell_m^2}{V_p \ell_p^2}. \quad \therefore Q_m = Q_p \dfrac{V_m}{V_p}\dfrac{\ell_m^2}{\ell_p^2} = 1.5 \times 7 \times \dfrac{1}{7^2} = \underline{0.214 \text{ m}^3/\text{s}}.$

$\dfrac{\dot{W}_m}{\dot{W}_p} = \dfrac{\rho_m V_m^3 \ell_m^2}{\rho_p V_p^3 \ell_p^2} = 7^3 \times \dfrac{1}{7^2} = 7. \quad \therefore \dot{W}_m = 7 \times 200 = \underline{1400 \text{ kW}}.$

b) $\text{Re}_m = \text{Re}_p. \quad \therefore \dfrac{V_m}{V_p} = \dfrac{d_p}{d_m}\dfrac{\nu_m}{\nu_p} = 7 \times \dfrac{.9}{1.3} = 4.85.$

$Q_m = 1.5 \times 4.85 \times \dfrac{1}{7^2} = \underline{0.148 \text{ m}^3/\text{s}}. \qquad \dot{W}_m = 4.85^3 \times \dfrac{1}{7^2} \times 200 = \underline{466 \text{ kW}}$

6.37 a) $\text{Re}_m = \text{Re}_p. \quad \dfrac{V_m d_m}{\nu_m} = \dfrac{V_p d_p}{\nu_p}. \quad \therefore \dfrac{V_m}{V_p} = \dfrac{d_p}{d_m} = 5.$

$\dfrac{\dot{m}_m}{\dot{m}_p} = \dfrac{\rho_m \ell_m^2 V_m}{\rho_p \ell_p^2 V_p} = \dfrac{1}{5^2} \times 5. \quad \therefore \dot{m}_m = \dot{m}_p \dfrac{1}{5} = 800/5 = \underline{160 \text{ kg/s}}.$

$\dfrac{\Delta p_m}{\Delta p_p} = \dfrac{\rho_m V_m^2}{\rho_p V_p^2} = 5^2. \quad \therefore \Delta p_m = 25 \Delta p_p = 25 \times 600 = \underline{15\,000 \text{ kPa}}.$

b) $\text{Re}_m = \text{Re}_p. \quad \therefore \dfrac{V_m}{V_p} = \dfrac{d_p}{d_m}\dfrac{\nu_m}{\nu_p} = 5 \times \dfrac{.8}{1.14} = 3.51.$

$\dot{m}_m = 800 \times \dfrac{1}{5^2} \times 3.51 = \underline{112 \text{ kg/s}}. \qquad \Delta p_m = 600 \times 3.51^2 = \underline{7390 \text{ kPa}}.$

6.38 a) $\text{Re}_m = \text{Re}_p. \quad \dfrac{V_m d_m}{\nu_m} = \dfrac{V_p d_p}{\nu_p}. \quad \therefore \dfrac{V_m}{V_p} = 10.$

$\dfrac{F_m}{F_p} = \dfrac{\rho_m V_m^2 \ell_m^2}{\rho_p V_p^2 \ell_p^2} = 10^2 \times \dfrac{1}{10^2} = 1. \quad \therefore F_m = F_p = \underline{45 \text{ N}}.$

b) $\text{Re}_m = \text{Re}_p. \quad \therefore \dfrac{V_m}{V_p} = \dfrac{d_p}{d_m}\dfrac{\nu_m}{\nu_p} = 10 \times \dfrac{0.9}{1.25} = 7.2.$

$F_m = 45 \times 7.2^2 \times \dfrac{1}{10^2} = \underline{23.3 \text{ N}}.$

6.38E a) $\text{Re}_m = \text{Re}_p$. $\frac{V_m d_m}{\nu_m} = \frac{V_p d_p}{\nu_p}$. $\therefore \frac{V_m}{V_p} = \frac{d_p}{d_m} = 10.$

$$\frac{F_m}{F_p} = \frac{\rho_m V_m^2 \ell_m^2}{\rho_p V_p^2 \ell_p^2} = 10^2 \times \frac{1}{10^2} = 1. \quad \therefore F_m = F_p = \underline{10 \text{ lb}}.$$

b) $\text{Re}_m = \text{Re}_p$. $\therefore \frac{V_m}{V_p} = \frac{d_p}{d_m}\frac{\nu_m}{\nu_p} = 10 \times \frac{1.06}{1.41} = 7.52.$

$$F_m = 10 \times 7.52^2 \times \frac{1}{10^2} = \underline{5.65 \text{ lb}}.$$

6.39 $\text{Re}_m = \text{Re}_p$. $\frac{V_m \ell_m}{\nu_m} = \frac{V_p \ell_p}{\nu_p}$. $\therefore \frac{V_m}{V_p} = \frac{\ell_p}{\ell_m}\frac{\nu_m}{\nu_p} = 10$ assuming $\frac{\nu_m}{\nu_p} = 1.$

$\therefore V_m = 10 V_p = 1000 \text{ km/hr}.$

This velocity is much too high for a model test; it is in the compressibility region. Thus, small-scale models of autos are not used. Full-scale wind tunnels are common.

6.40 $\text{Re}_m = \text{Re}_p$. $\therefore \frac{V_m \ell_m}{\nu_m} = \frac{V_p \ell_p}{\nu_p}$. $\therefore \frac{V_m}{V_p} = \frac{\ell_p}{\ell_m}\frac{\nu_m}{\nu_p}.$

Water: $\frac{V_m}{V_p} = \frac{\ell_p}{\ell_m} = 10$ assuming $\nu_m = \nu_p$. $\therefore V_m = 10 V_p = 900 \text{ km/hr}.$

Air: $V_m = V_p \frac{\ell_p}{\ell_m}\frac{\nu_m}{\nu_p} = 90 \times 10 \frac{1.5 \times 10^{-5}}{1 \times 10^{-6}} = 13\,500 \text{ km/hr}.$

Neither a water channel or a wind tunnel is recommended. Full-scale testing in a water channel is suggested.

6.41 $\text{Re}_m = \text{Re}_p$. $\frac{V_m \ell_m}{\nu_m} = \frac{V_p \ell_p}{\nu_p}$. $\therefore V_m / V_p = \ell_p / \ell_m = 10$ if $\nu_m = \nu_p$.

$\therefore V_m = 10 \times 50 = \underline{500 \text{ m/s}}$. This is in the compressibility range so is not recommended. Try a water channel for the model study. Then

$$\frac{V_m}{V_p} = \frac{\ell_p}{\ell_m}\frac{\nu_m}{\nu_p} = 10 \times \frac{1 \times 10^{-6}}{1.5 \times 10^{-5}} = 0.662 \qquad \therefore V_m = \underline{33.1 \text{ m/s}}.$$

This is a possibility, although 27.8 m/s is still quite large.

$$\frac{(F_D)_m}{(F_D)_p} = \frac{\rho_m V_m^2 \ell_m^2}{\rho_p V_p^2 \ell_p^2} = \frac{1000}{1.23} \times .662^2 \times \frac{1}{10^2} = \underline{3.56}$$

6.42 $\text{Re}_m = \text{Re}_p. \quad \frac{V_m d_m}{\nu_m} = \frac{V_p d_p}{\nu_p}. \quad \therefore d_m = d_p \frac{V_p}{V_m}\frac{\nu_m}{\nu_p} = 0.8 \times 1 \times \frac{1 \times 10^{-6}}{0.3/917} = \underline{0.00245 \text{ m}}.$

Find ν_{oil} using Fig. 1.9 and Table B.6.

$$\frac{\Delta p_m}{\Delta p_p} = \frac{\rho_m V_m^2}{\rho_p V_p^2} = \frac{1000}{917} \times 1^2 = \underline{1.09}.$$

6.42E $\text{Re}_m = \text{Re}_p. \quad \frac{V_m d_m}{\nu_m} = \frac{V_p d_p}{\nu_p}. \quad \therefore d_m = d_p \frac{V_p}{V_m}\frac{\nu_m}{\nu_p} = 2.5 \times 1 \times \frac{1.06 \times 10^{-5}}{7 \times 10^{-3}/1.78} = \underline{0.00674 \text{ ft}}.$

Find $\nu_{\text{oil}} = \frac{\mu_{\text{oil}}}{\rho_{\text{oil}}}$ using Fig. 1.9 and Table B.6

$$\frac{\Delta p_m}{\Delta p_p} = \frac{\rho_m V_m^2}{\rho_p V_p^2} = \frac{1.94}{1.78} \times 1^2 = \underline{1.09}.$$

6.43 $\text{Re}_m = \text{Re}_p. \quad \frac{V_m \ell_m}{\nu_m} = \frac{V_p \ell_p}{\nu_p}. \quad \therefore V_m = V_p \frac{\ell_p}{\ell_m}\frac{\nu_m}{\nu_p} = 0.1 \times .025 \times 10^{-3} \times \frac{\ell_p}{\ell_m}.$

If $\ell_p \cong 5$ cm, then $\frac{\ell_p}{\ell_m} = \frac{5}{.0025} = 2000$ and $V_m = 0.005 \text{ m/s}$.

We could try $\ell_p \cong 50$ cm, but $V_m = 0.05$ m/s. Each of these V_m's is quite small — too small for easy measurements. Let's try a wind tunnel. Then,

$V_m = V_p \frac{\ell_p}{\ell_m}\frac{\nu_m}{\nu_p} = 0.1 \times .025 \times 10^{-3} \frac{\ell_p}{\ell_m} \times \frac{1 \times 10^{-3}}{1.8 \times 10^{-5}} = 0.28 \text{ m/s if } \ell_p = 5 \text{ cm}.$ Or,

if $\ell_p = 50$ cm, $V_m = 2.8$ m/s. This is a much better velocity to work with in the lab. Thus, choose a wind tunnel.

6.44 $\text{Re}_m = \text{Re}_p. \therefore \frac{V_m \ell_m}{\nu_m} = \frac{V_p \ell_p}{\nu_p}. \quad \text{Fr}_m = \text{Fr}_p. \therefore \frac{V_m^2}{\ell_m g_m} = \frac{V_p^2}{\ell_p g_p}. \quad \therefore \frac{V_m}{V_p} = \sqrt{\frac{1}{30}}.$

$$\frac{V_m}{V_p} = \frac{\ell_p}{\ell_m}\frac{\nu_m}{\nu_p} = 30\frac{\nu_m}{\nu_p} = \sqrt{\frac{1}{30}}. \quad \therefore \frac{\nu_m}{\nu_p} = \frac{1}{164}. \quad \therefore \nu_m = \underline{6.1 \times 10^{-9} \text{ m}^2/\text{s}}. \text{ Impossible!}$$

6.45 $\text{Fr}_m = \text{Fr}_p. \quad \therefore \frac{V_m^2}{\ell_m g_m} = \frac{V_p^2}{\ell_p g_p}. \quad \therefore V_m = V_p\sqrt{\frac{\ell_m}{\ell_p}} = 10\sqrt{\frac{1}{60}} = \underline{1.29 \text{ m/s}}.$

$$\frac{(F_D)_m}{(F_D)_p} = \frac{\rho_m V_m^2 \ell_m^2}{\rho_p V_p^2 \ell_p^2}. \quad \therefore (F_D)_p = \frac{V_p^2}{V_m^2} \times \frac{\ell_p^2}{\ell_m^2}(F_D)_m = 60 \times 60^2 \times 10 = \underline{2.16 \times 10^6 \text{ N}}.$$

6.45E $\text{Fr}_m = \text{Fr}_p$. $\therefore \dfrac{V_m^2}{\ell_m g_m} = \dfrac{V_p^2}{\ell_p g_p}$. $\therefore V_m = V_p\sqrt{\dfrac{\ell_m}{\ell_p}} = 30\sqrt{\dfrac{1}{60}} = \underline{3.87 \text{ fps}}.$

$$\frac{(F_D)_m}{(F_D)_p} = \frac{\rho_m V_m^2 \ell_m^2}{\rho_p V_p^2 \ell_p^2}. \quad \therefore (F_D)_p = \frac{V_p^2}{V_m^2} \times \frac{\ell_p^2}{\ell_m^2}(F_D)_m = 60 \times 60^2 \times 3 = \underline{684{,}000 \text{ lb}}.$$

6.46 $\text{Fr}_m = \text{Fr}_p$. $\dfrac{V_m^2}{\ell_m g_m} = \dfrac{V_p^2}{\ell_p g_p}$. $\therefore \dfrac{V_m}{V_p} = \sqrt{\dfrac{\ell_m}{\ell_p}}$.

a) $\dfrac{Q_m}{Q_p} = \dfrac{V_m \ell_m^2}{V_p \ell_p^2}$. $\therefore Q_m = Q_p \dfrac{V_m}{V_p}\dfrac{\ell_m^2}{\ell_p^2} = 2 \times \dfrac{1}{\sqrt{10}} \times \dfrac{1}{10^2} = \underline{0.00632 \text{ m}^3/\text{s}}.$

b) $\dfrac{F_m}{F_p} = \dfrac{\rho_m V_m^2 \ell_m^2}{\rho_p V_p^2 \ell_p^2}$. $\therefore F_p = F_m \dfrac{V_p^2}{V_m^2}\dfrac{\ell_p^2}{\ell_m^2} = 12 \times 10 \times 10^2 = \underline{12\,000 \text{ N}}.$

6.47 Neglect viscous effects. $\text{Fr}_m = \text{Fr}_p$. $\therefore \dfrac{V_m}{V_p} = \sqrt{\dfrac{\ell_m}{\ell_p}} = \sqrt{\dfrac{1}{10}}$. $\therefore V_p = \underline{19 \text{ m/s}}.$

$$\frac{F_m}{F_p} = \frac{\rho_m V_m^2 \ell_m^2}{\rho_p V_p^2 \ell_p^2}. \quad \therefore F_p = F_m \frac{V_p^2}{V_m^2}\frac{\ell_p^2}{\ell_m^2} = 2.5 \times 10 \times 10^2 = \underline{2500 \text{ N}}.$$

6.47E Neglect viscous effects. $\text{Fr}_m = \text{Fr}_p$. $\therefore \dfrac{V_m}{V_p} = \sqrt{\dfrac{\ell_m}{\ell_p}} = \sqrt{\dfrac{1}{10}}$. $\therefore V_p = \underline{63.2 \text{ fps}}.$

$$\frac{F_m}{F_p} = \frac{\rho_m V_m^2 \ell_m^2}{\rho_p V_p^2 \ell_p^2}. \quad \therefore F_p = F_m \frac{V_p^2}{V_m^2}\frac{\ell_p^2}{\ell_m^2} = 0.8 \times 10 \times 10^2 = \underline{800 \text{ lb}}.$$

6.48 Neglect viscous effects, and account for wave (gravity) effects.

$\text{Fr}_m = \text{Fr}_p$. $\therefore \dfrac{V_m^2}{\ell_m g_m} = \dfrac{V_p^2}{\ell_p g_p}$. $\therefore \dfrac{V_m}{V_p} = \sqrt{\dfrac{\ell_m}{\ell_p}}$. $\dfrac{\omega_m}{\omega_p} = \dfrac{V_m/\ell_m}{V_p/\ell_p}$.

$$\therefore \omega_m = \omega_p \frac{V_m}{V_p}\frac{\ell_p}{\ell_m} = 600 \times \sqrt{\frac{1}{10}} \times 10 = 1897 \text{ rpm}.$$

$$\frac{T_m}{T_p} = \frac{\rho_m V_m^2 \ell_m^3}{\rho_p V_p^2 \ell_p^3}. \quad \therefore T_p = T_m \frac{V_p^2}{V_m^2}\frac{\ell_p^3}{\ell_m^3} = 1.2 \times 10 \times 10^3 = \underline{12000 \text{ N}\cdot\text{m}}.$$

6.49 $\text{Fr}_m = \text{Fr}_p$. $\therefore \dfrac{V_m^2}{\ell_m g_m} = \dfrac{V_p^2}{\ell_p g_p}$. $\therefore \dfrac{V_m}{V_p} = \sqrt{\dfrac{\ell_m}{\ell_p}}$. $\dfrac{6}{100} = \sqrt{\dfrac{\ell_m}{\ell_p}}$. $\therefore \dfrac{\ell_p}{\ell_m} = \underline{278}.$

6.50 Check the Reynolds number: $\text{Re}_p = \frac{V_p d_p}{\nu_p} = \frac{15 \times 2}{10^{-6}} = 30 \times 10^6$. This is a high-Reynolds-number flow. $\text{Re}_m = \frac{2 \times 2/30}{10^{-6}} = 1.33 \times 10^5$. This may be sufficiently large for similarity. $\frac{\dot{W}_m}{\dot{W}_p} = \frac{\rho_m V_m^3 \ell_m^2}{\rho_p V_p^3 \ell_p^2} = \frac{2^3}{15^3} \times \frac{1}{30^2} = 2.63 \times 10^{-6}$.

$\therefore \dot{W}_p = (2 \times 2.15)/2.63 \times 10^{-6} = \underline{1633 \text{ kW}}$.

6.50E Check Re: $\text{Re}_p = \frac{6 \times 45}{10^{-5}} = 27 \times 10^6$. This is a high-Reynolds-number flow.

$\text{Re}_m = \frac{6/30 \times 6}{10^{-5}} = 1.2 \times 10^5$. This may be sufficiently large for similarity.

$$\frac{\dot{W}_m}{\dot{W}_p} = \frac{\rho_m V_m^3 \ell_m^2}{\rho_p V_p^3 \ell_p^2} = \frac{6^3}{45^3} \times \frac{1}{30^2} = 2.63 \times 10^{-6}.$$

$$\therefore \dot{W}_p = 15 \times 0.5/2.63 \times 10^{-6} = \underline{2.85 \times 10^6 \ \frac{\text{ft - lb}}{\text{sec}}}.$$

6.51 This is due to the separated flow downwind of the stacks, a viscous effect. $\therefore$ Re is the significant parameter. $\text{Re}_p = \frac{10 \times 4}{1.5 \times 10^{-5}} = 26.7 \times 10^5$. This is a high-Reynolds-number flow. Let's assume the flow to be Reynolds number independent above $\text{Re} = 5 \times 10^5$ (see Fig. 6.4). Then

$\text{Re}_m = 5 \times 10^5 = \frac{V_m 4/20}{1.5 \times 10^{-5}}$. $\quad \therefore V_m \geq \underline{37.5 \text{ m/s}}$.

6.52 $\text{Re}_p = \frac{20 \times 10}{1.5 \times 10^{-5}} = 13.3 \times 10^6$. This is a high-Reynolds-number flow.

Let $\text{Re}_m = 10^5 = \frac{V_m \times .4}{1.5 \times 10^{-5}}$. $\quad \therefore V_m \geq 3.75 \text{ m/s}$ for the wind tunnel.

$\text{Re}_m = 10^5 = \frac{V_m \times .1}{1 \times 10^{-6}}$. $\quad \therefore V_m \geq 1.0 \text{ m/s}$ for the water channel.

Either could be selected. The better facility would be chosen.

$$\frac{F_{m_1}}{F_{m_2}} = \frac{\rho_{m_1} V_{m_1}^2 \ell_{m_1}^2}{\rho_{m_2} V_{m_2}^2 \ell_{m_2}^2} = \frac{3.2}{F_{m_2}}. \quad \therefore F_{m_2} = 3.2 \frac{1000}{1.23} \frac{2.4^2}{15^2} \times \frac{.1^2}{.4^2} = \underline{4.16 \text{ N}}.$$

$$\frac{\dot{W}_m}{\dot{W}_p} = \frac{\rho_m V_m^3 \ell_m^2}{\rho_p V_p^3 \ell_p^2} = \frac{15^3 \times .4^2}{20^3 \times 10^2}. \quad \therefore \dot{W}_p = (15 \times 3.2) \frac{20^3}{15^3} \times \frac{10^2}{.4^2} = \underline{71\,100 \text{ W}}.$$

6.53 Re is the significant parameter. This is undoubtedly a high-Reynolds-number flow. If the model is 1.2 m high then $\frac{\ell_p}{\ell_m} = \frac{300}{1.2} = 250$ and the model's diameter is 15/250 = 0.06 m. For $\text{Re}_m = 3 \times 10^5$, we have

$$\text{Re}_m = 3 \times 10^5 = \frac{V_m \times .06}{1.5 \times 10^{-5}}. \qquad \therefore V_m \geq 75 \text{ m/s, and } \underline{\text{a study is possible}}.$$

6.53E Re is the significant parameter. If the model is 4' high then $\frac{\ell_p}{\ell_m} = 250$, and the model's diameter is $\frac{45}{250} = 0.18'$. For $\text{Re}_m = 3 \times 10^5$, we have

$$\text{Re}_m = 3 \times 10^5 = \frac{V_m \times .18}{1.5 \times 10^{-4}}. \qquad \therefore V_m \geq 250 \text{ fps, and } \underline{\text{a study is possible}}.$$

6.54 Mach No. is the significant parameter. $\text{M}_m = \text{M}_p$.

a) $\text{M}_m = \text{M}_p. \qquad \therefore \frac{V_m}{C_m} = \frac{V_p}{C_p}. \qquad \therefore V_m = V_p = \underline{200 \text{ m/s}}.$

$$\frac{F_m}{F_p} = \frac{\rho_m V_m^2 \ell_m^2}{\rho_p V_p^2 \ell_p^2}. \qquad \therefore F_p = 10 \times 1^2 \times 20^2 = \underline{4000 \text{ N}}.$$

b) $V_p = V_m \frac{C_p}{C_m} = V_m \sqrt{\frac{T_p}{T_m}} = 200\sqrt{\frac{255.7}{296}} = \underline{186 \text{ m/s}}.$

$$F_p = F_m \frac{\rho_m V_p^2 \ell_p^2}{\rho_m V_m^2 \ell_m^2} = 10 \times .601 \times \frac{186^2}{200^2} \times 20^2 = \underline{2080 \text{ N}}.$$

c) $V_p = V_m \frac{C_p}{C_m} = V_m \sqrt{\frac{T_p}{T_m}} = 200\sqrt{\frac{223.3}{296}} = \underline{174 \text{ m/s}}.$

$$F_p = F_m \frac{\rho_p}{\rho_m} \frac{V_p^2}{V_m^2} \frac{\ell_p^2}{\ell_m^2} = 10 \times .338 \times \frac{174^2}{200^2} \times 20^2 = \underline{1023 \text{ N}}.$$

6.55 $\text{M}_m = \text{M}_p. \qquad \therefore \frac{V_m}{C_m} = \frac{V_p}{C_p}. \qquad \therefore V_m = 250\sqrt{\frac{273}{223.3}} = \underline{276 \text{ m/s}}.$

$$\frac{V_m}{V_p} = \frac{C_m}{C_p} = \sqrt{\frac{T_m}{T_p}}. \qquad \therefore V_p = 290\sqrt{\frac{223.3}{273}} = \underline{262 \text{ m/s}}.$$

$$\frac{p_m}{p_p} = \frac{\rho_m V_m^2}{\rho_p V_p^2}. \qquad \therefore p_p = p_m \frac{\rho_p}{\rho_m}\frac{V_p^2}{V_m^2} = 80 \frac{.338\rho_o}{.8\rho_o}\frac{262^2}{290^2} = \underline{34.6 \text{ kPa, abs}}.$$

$\alpha_p = \underline{5^\circ}$ for similarity. (Note: we use ρ_m at 2700 m where T = 0°C.)

6.56 a) $\text{Fr}_m = \text{Fr}_p$. $\dfrac{V_m^2}{\ell_m g_m} = \dfrac{V_p^2}{\ell_p g_p}$. $\therefore \dfrac{V_m}{V_p} = \sqrt{\dfrac{\ell_m}{\ell_p}}$.

$$\frac{\omega_m}{\omega_p} = \frac{V_m}{V_p}\frac{\ell_p}{\ell_m} = \frac{1}{\sqrt{10}} \times 10. \quad \therefore \omega_m = 2000 \times \frac{10}{\sqrt{10}} = \underline{6320 \text{ rpm}}.$$

b) $\text{Re}_m = \text{Re}_p$. $\dfrac{V_m \ell_m}{\nu_m} = \dfrac{V_p \ell_p}{\nu_p}$. $\therefore \dfrac{V_m}{V_p} = \dfrac{\ell_p}{\ell_m} = 10$.

$$\frac{\omega_m}{\omega_p} = \frac{V_m}{V_p}\frac{\ell_p}{\ell_m} = 10 \times \frac{1}{10} = 1. \quad \therefore \omega_m = \underline{2000 \text{ rpm}}.$$

6.57 There are no gravity effects nor compressibility effects. It is a high-Re flow. $\dfrac{T_m}{T_p} = \dfrac{\rho_m V_m^2 \ell_m^3}{\rho_p V_p^2 \ell_m^3}$. $\therefore T_p = T_m \dfrac{V_p^2}{V_m^2}\dfrac{\ell_p^3}{\ell_m^3} = 12 \times \dfrac{15^2}{60^2} \times 10^3 = \underline{750 \text{ N}\cdot\text{m}}$.

$$\frac{\omega_m}{\omega_p} = \frac{V_m}{V_p}\frac{\ell_p}{\ell_m}. \quad \therefore \omega_p = \omega_m \frac{V_p}{V_m}\frac{\ell_m}{\ell_p} = 500 \times \frac{15}{60} \times \frac{1}{10} = \underline{12.5 \text{ rpm}}.$$

6.57E There are no gravity effects nor compressibility effects. It is a high-Re flow. $\dfrac{T_m}{T_p} = \dfrac{\rho_m V_m^2 \ell_m^3}{\rho_p V_p^2 \ell_m^3}$. $\therefore T_p = T_m \dfrac{V_p^2}{V_m^2}\dfrac{\ell_p^3}{\ell_m^3} = 10 \times \dfrac{45^2}{180^2} \times 10^3 = \underline{625 \text{ ft-lb}}$.

$$\frac{\omega_m}{\omega_p} = \frac{V_m}{V_p}\frac{\ell_p}{\ell_m}. \quad \therefore \omega_p = \omega_m \frac{V_p}{V_m}\frac{\ell_m}{\ell_p} = 500 \times \frac{45}{180} \times \frac{1}{10} = \underline{12.5 \text{ rpm}}.$$

6.58 $\text{Re}_m = \text{Re}_p$. $\therefore \dfrac{V_m \ell_m}{\nu_m} = \dfrac{V_p \ell_p}{\nu_p}$. $\therefore V_m = V_p \dfrac{\ell_p}{\ell_m} = 10 \times 10 = 100 \text{ m/s}$.

This is too large for a water channel. Undoubtedly this is a high-Re flow. Select a speed of $\underline{5 \text{ m/s}}$. For this speed, $\text{Re}_m = \dfrac{5 \times 0.1}{1 \times 10^{-6}} = 5 \times 10^5$, where we used $\ell_m = 0.1$ ($\ell_p = 1$ m — i.e., the dia. of the porpoise). $\omega_m = \omega_p \dfrac{V_m}{V_p}\dfrac{\ell_p}{\ell_m} = 1 \times \dfrac{5}{10} \times 10 = \underline{5 \text{ motions / second}}$.

6.59 $\rho^* = \dfrac{\rho}{\rho_o}$, $t^* = tf$, $u^* = \dfrac{u}{V}$, $v^* = \dfrac{v}{V}$, $x^* = \dfrac{x}{\ell}$, $y^* = \dfrac{y}{\ell}$.

$$f\rho_o \frac{\partial \rho^*}{\partial t^*} + \rho_o \frac{V}{\ell}\frac{\partial(\rho^* u^*)}{\partial x^*} + \rho_o \frac{V}{\ell}\frac{\partial(\rho^* v^*)}{\partial y^*} = 0.$$

Divide by $\rho_o V/\ell$:

$$\therefore \frac{f\ell}{V}\frac{\partial \rho^*}{\partial t^*} + \frac{\partial}{\partial x^*}(\rho^* u^*) + \frac{\partial}{\partial y^*}(\rho^* v^*) = 0. \qquad \text{parameter} = \underline{\frac{f\ell}{V}}.$$

6.60 $\vec{V}^* = \frac{\vec{V}}{U}, u^* = \frac{u}{U}, v^* = \frac{v}{U}, w^* = \frac{w}{U}, x^* = \frac{x}{\ell}, y^* = \frac{y}{\ell}, z^* = \frac{z}{\ell}, p^* = \frac{p}{\rho U^2}, t^* = tf.$

$$Uf\frac{\partial \vec{V}^*}{\partial t^*} + \frac{U^2}{\ell}u^*\frac{\partial \vec{V}^*}{\partial x^*} + \frac{U^2}{\ell}v^*\frac{\partial \vec{V}^*}{\partial y^*} + \frac{U^2}{\ell}w^*\frac{\partial \vec{V}^*}{\partial z^*} = -\frac{\rho U^2}{\ell}\frac{\vec{\nabla}^* p^*}{\rho}.$$

Divide by U^2/ℓ:

$$\frac{f\ell}{U}\frac{\partial \vec{V}^*}{\partial t^*} + u^*\frac{\partial \vec{V}^*}{\partial x^*} + v^*\frac{\partial \vec{V}^*}{\partial y^*} + w^*\frac{\partial \vec{V}^*}{\partial z^*} = -\nabla^* p^*. \qquad \text{parameter} = \underline{\frac{f\ell}{U}}$$

6.61 $\vec{V}^* = \frac{\vec{V}}{U},\ t^* = \frac{tU}{\ell},\ \vec{\nabla}^* = \ell\vec{\nabla},\ p^* = \frac{p}{\rho U^2},\ h^* = \frac{h}{\ell}.$

$$\rho\frac{U^2}{\ell}\frac{D\vec{V}^*}{Dt^*} = -\rho\frac{U^2}{\ell}\vec{\nabla}^* p^* - \rho g\frac{\ell}{\ell}\vec{\nabla}^* h^*.$$

Divide by $\rho U^2/\ell$:

$$\frac{D\vec{V}^*}{Dt^*} = -\vec{\nabla}^* p^* - \frac{g\ell}{U^2}\vec{\nabla}^* h^*. \qquad \text{parameter} = \underline{\frac{g\ell}{U^2}}.$$

6.62 There is no y- or z-component velocity so continuity requires that $\partial u/\partial x = 0$. There is no initial pressure distribution tending to cause motion so $\partial p/\partial x = 0$. The x-component Navier-Stokes equation is then

$$\frac{\partial u}{\partial t} + u\frac{\partial u}{\partial x} + v\frac{\partial u}{\partial y} + w\frac{\partial u}{\partial z} = -\frac{1}{\rho}\frac{\partial p}{\partial x} + g_x + \nu\left(\frac{\partial^2 u}{\partial x^2} + \frac{\partial^2 u}{\partial y^2} + \frac{\partial^2 u}{\partial z^2}\right)$$

(wide plates)

This simplifies to

$$\frac{\partial u}{\partial t} = \nu\frac{\partial^2 u}{\partial y^2}.$$

a) Let $u^* = u/U$, $y^* = y/h$ and $t^* = tU/h$. Then

$$\frac{U^2}{h}\frac{\partial u^*}{\partial t^*} = \frac{\nu U}{h^2}\frac{\partial^2 u^*}{\partial y^{*2}}$$

The normalized equation is

$$\frac{\partial u^*}{\partial t^*} = \frac{1}{\text{Re}}\frac{\partial^2 u^*}{\partial y^{*2}} \qquad \text{where Re} = \frac{Uh}{\nu}$$

b) Let $u^* = u/U$, $y^* = y/h$ and $t^* = t\nu/h^2$. Then

$$\frac{\nu U}{h^2}\frac{\partial u^*}{\partial t^*} = \nu\frac{U}{h^2}\frac{\partial^2 u^*}{\partial y^{*2}}$$

The normalized equation is

$$\frac{\partial u^*}{\partial t^*} = \frac{\partial^2 u^*}{\partial y^{*2}}$$

6.63 The only velocity component is u. Continuity then requires that $\partial u / \partial x = 0$ (replace z with x and v_z with u in the equations written using cylindrical coordinates). The x-component Navier-Stokes equation is

$$\frac{\partial u}{\partial t}+v_r\frac{\partial u}{\partial r}+\frac{v_\theta}{r}\frac{\partial u}{\partial \theta}+u\frac{\partial u}{\partial x}=-\frac{1}{\rho}\frac{\partial p}{\partial x}+g_x+\nu\left(\frac{\partial^2 u}{\partial r^2}+\frac{1}{r}\frac{\partial u}{\partial r}+\frac{1}{r^2}\frac{\partial^2 u}{\partial \theta^2}+\frac{\partial^2 u}{\partial x^2}\right)$$

This simplifies to

$$\frac{\partial u}{\partial t}=-\frac{1}{\rho}\frac{\partial p}{\partial x}+\nu\left(\frac{\partial^2 u}{\partial r^2}+\frac{1}{r}\frac{\partial u}{\partial r}\right)$$

a) Let $u^*=u/V, x^*=x/d, t^*=tV/d, p^*=p/\rho V^2$ and $r^*=r/d$:

$$\frac{V^2}{d}\frac{\partial u^*}{\partial t^*}=-\frac{\rho V^2}{\rho d}\frac{\partial p^*}{\partial x^*}+\frac{\nu V}{d^2}\left(\frac{\partial^2 u^*}{\partial r^{*2}}+\frac{1}{r^*}\frac{\partial u^*}{\partial r^*}\right)$$

The normalized equation is

$$\frac{\partial u^*}{\partial t^*}=-\frac{\partial p^*}{\partial x^*}+\frac{1}{\text{Re}}\left(\frac{\partial^2 u^*}{\partial r^{*2}}+\frac{1}{r^*}\frac{\partial u^*}{\partial r^*}\right) \quad \text{where } \text{Re}=\frac{Vd}{\nu}$$

b) Let $u^*=u/V, x^*=x/d, t^*=t\nu/d^2, p^*=p/\rho V^2$ and $r^*=r/d$:

$$\frac{\nu V}{d^2}\frac{\partial u^*}{\partial t^*}=-\frac{\rho V^2}{\rho d}\frac{\partial p^*}{\partial x^*}+\frac{\nu V}{d^2}\left(\frac{\partial^2 u^*}{\partial r^{*2}}+\frac{1}{r^*}\frac{\partial u^*}{\partial r^*}\right)$$

The normalized equation is

$$\frac{\partial u^*}{\partial t^*}=-\text{Re}\frac{\partial p^*}{\partial x^*}+\frac{\partial^2 u^*}{\partial r^{*2}}+\frac{1}{r^*}\frac{\partial u^*}{\partial r^*} \quad \text{where } \text{Re}=\frac{Vd}{\nu}$$

6.64 Assume $w=0$ and $\frac{\partial}{\partial z}=0$. The x-component Navier-Stokes equation is then

$$\frac{\partial u}{\partial t}+u\frac{\partial u}{\partial x}+v\frac{\partial u}{\partial y}+w\frac{\partial u}{\partial z}=-\frac{1}{\rho}\frac{\partial p}{\partial x}+g_x+\nu\left(\frac{\partial^2 u}{\partial x^2}+\frac{\partial^2 u}{\partial y^2}+\frac{\partial^2 u}{\partial z^2}\right)$$

steady small

With $g_x=g$ the simplified equation is

$$u\frac{\partial u}{\partial x}=g+\nu\left(\frac{\partial^2 u}{\partial x^2}+\frac{\partial^2 u}{\partial y^2}\right)$$

Let $u^*=u/V$, $x^*=x/h$ and $y^*=y/h$. Then

$$\frac{V^2}{h}u^*\frac{\partial u^*}{\partial x^*}=g+\nu\frac{V}{h^2}\left(\frac{\partial^2 u^*}{\partial x^{*2}}+\frac{\partial^2 u^*}{\partial y^{*2}}\right)$$

The normalized equation is

$$u^*\frac{\partial u^*}{\partial x^*}=\frac{1}{\text{Fr}^2}+\frac{1}{\text{Re}}\left(\frac{\partial^2 u^*}{\partial x^{*2}}+\frac{\partial^2 u^*}{\partial y^{*2}}\right) \quad \text{where } \text{Fr}=\frac{V}{\sqrt{hg}} \text{ and } \text{Re}=\frac{Vh}{\nu}$$

6.65 $u^* = \frac{u}{U}, v^* = \frac{v}{U}, T^* = \frac{T}{T_o}, x^* = \frac{x}{\ell}, y^* = \frac{y}{\ell}, \nabla^{*2} = \ell^2 \nabla^2.$

$$\rho c_p \left[\frac{UT_o}{\ell} \frac{\partial T^*}{\partial x^*} + \frac{UT_o}{\ell} \frac{\partial T^*}{\partial y^*} \right] = \frac{K}{\ell^2} T_o \nabla^{*2} T^*.$$

Divide by $\rho c_p U T_o / \ell$:

$$\frac{\partial T^*}{\partial x^*} + \frac{\partial T^*}{\partial y^*} = \frac{K}{\rho c_p U \ell} \nabla^{*2} T^*. \qquad \text{Parameter} = \frac{K}{\mu c_p} \frac{\mu}{\rho U \ell} = \underline{\frac{1}{\text{Pr}} \frac{1}{\text{Re}}}.$$

6.66 $\rho^* = \frac{\rho}{\rho_o}, \vec{V}^* = \frac{\vec{V}}{U}, t^* = \frac{tU}{\ell}, \vec{\nabla}^* = \frac{1}{\ell} \vec{\nabla}, \nabla^{*2} = \frac{1}{\ell^2} \nabla^2, p^* = \frac{p}{p_o}, T^* = \frac{T}{T_o}.$

momentum: $\rho_o \rho^* \frac{U^2}{\ell} \frac{D\vec{V}^*}{Dt^*} = -\frac{p_o}{\ell} \vec{\nabla}^* p^* + \frac{\mu U}{\ell^2} \nabla^{*2} \vec{V}^* + \frac{\mu U}{3\ell^2} \vec{\nabla}^* (\vec{\nabla}^* \cdot \vec{V}^*).$

Divide by $\rho_o U^2 / \ell$:

$$\rho^* \frac{D\vec{V}^*}{Dt^*} = -\frac{p_o}{\rho_o U^2} \vec{\nabla}^* p^* + \frac{\mu}{\rho_o U \ell} \left[\nabla^{*2} \vec{V}^* + \vec{\nabla}^* (\vec{\nabla}^* \cdot \vec{V}^*) \right].$$

energy: $\rho^* c_v \rho_o T_o \frac{U}{\ell} \frac{DT^*}{Dt^*} = \frac{K}{\ell^2} T_o \nabla^{*2} T^* - p_o \frac{U}{\ell} p^* \vec{\nabla}^* \cdot \vec{V}^*.$

Divide by $\rho_o c_v T_o U / \ell$:

$$\rho^* \frac{DT^*}{Dt^*} = \frac{K}{\rho_o c_v U \ell} \nabla^{*2} T^* - \frac{p_o}{\rho_o c_v T_o} p^* \vec{\nabla}^* \cdot \vec{V}^*.$$

The parameters are $\frac{p_o}{\rho_o U^2} = \frac{RT_o}{U^2} = \frac{kRT_o}{kU^2} = \frac{c^2}{kU^2} = \frac{1}{kM^2}.$

$$\frac{\mu}{\rho_o U \ell} = \frac{1}{\text{Re}}.$$

$$\frac{K}{\rho_o c_v U \ell} = \frac{K}{\mu c_p} \frac{c_p}{c_v} \frac{\mu}{\rho_o U \ell} = \frac{K}{\text{Pr Re}}.$$

$$\frac{p_o}{\rho_o c_v T_o} = \frac{RT_o}{c_v T_o} = \frac{c_p - c_v}{c_v} = K - 1.$$

The significant parameters are $\underline{K, \text{M}, \text{Re}, \text{Pr}}$.

SEVEN

Internal Flows

7.1 $\text{Re} = \frac{VD}{\nu} = \frac{VD}{1\times10^{-6}}$. a) $2000 = \frac{V\times2}{1\times10^{-6}}$. $\therefore V = \underline{0.001 \text{ m/s}}$.

b) $2000 = \frac{V\times.02}{1\times10^{-6}}$. $\therefore V = \underline{0.1 \text{ m/s}}$.

c) $2000 = \frac{V\times.002}{1\times10^{-6}}$. $\therefore V = \underline{1.0 \text{ m/s}}$.

7.2 $\text{Re} = \frac{Vh}{\nu} = \frac{Vh}{1\times10^{-6}}$. a) $1500 = \frac{V\times4}{1\times10^{-6}}$. $\therefore V = \underline{0.000375 \text{ m/s}}$.

b) $1500 = \frac{V\times1}{1\times10^{-6}}$. $\therefore V = \underline{0.0015 \text{ m/s}}$.

c) $1500 = \frac{V\times0.3}{1\times10^{-6}}$. $\therefore V = \underline{0.005 \text{ m/s}}$.

7.3 $\text{Re} = \frac{Vh}{\nu} = \frac{0.5\times.005}{1.3\times10^{-6}} = \underline{1920}$. Using $R_{\text{crit}} = 1500$, the flow is <u>turbulent</u>.

7.3E $\text{Re} = \frac{Vh}{\nu} = \frac{1.5\times.2/12}{1.4\times10^{-5}} = \underline{1790}$. Using $R_{\text{crit}} = 1500$, the flow is <u>turbulent</u>.

7.4 $\text{Re} = \frac{Vh}{\nu} = \frac{(1/2)\times1.4}{10^{-6}} = 700\ 000$. $\therefore$ <u>Very turbulent</u>

7.5 $\text{Re} = \frac{VD}{\nu}$. a) $V = \frac{\text{Re}\times\nu}{D} = \frac{2000\times10^{-6}}{0.02} = \underline{0.1 \text{ m/s}}$.

b) $V = \frac{\text{Re}\times\nu}{D} = \frac{40\ 000\times10^{-6}}{0.02} = \underline{2 \text{ m/s}}$.

7.6 $\frac{L_E}{D} = 0.065\,\text{Re}$ $\text{Re} = \frac{VD}{\nu}$. $V = \frac{.0002}{\pi\times.02^2} = .1592 \text{ m/s}$.

a) $L_E = .065\frac{.1592\times.04}{1.31\times10^{-6}}\times.04 = \underline{12.6 \text{ m}}$.

b) $L_E = .065\dfrac{.1592 \times .04}{1.007 \times 10^{-6}} \times .04 = \underline{16.4 \text{ m}}.$

c) $L_E = .065\dfrac{.1592 \times .04}{.661 \times 10^{-6}} \times .04 = \underline{25.0 \text{ m}}.$

d) $L_E = .065\dfrac{.1592 \times .04}{.367 \times 10^{-6}} \times .04 = \underline{45.1 \text{ m}}.$

7.7 a) $V = \dfrac{\text{Re} \times \nu}{D} = \dfrac{1000 \times 1.51 \times 10^{-5}}{0.04} = \underline{0.378 \text{ m/s}}.$

$L_E = 0.065 \,\text{Re} \times D = 0.065 \times 1000 \times 0.04 = \underline{2.6 \text{ m}}.$

$L_i \cong L_E / 4 = 2.6/4 = \underline{0.65 \text{ m}}.$

b) $V = \dfrac{\text{Re} \times \nu}{D} = \dfrac{80\,000 \times 1.51 \times 10^{-5}}{0.04} = \underline{30.2 \text{ m/s}}.$

$L_E \cong 120D = 120 \times 0.04 = \underline{4.8 \text{ m}}.$

$L_i \cong 10D = 10 \times 0.04 = \underline{0.4 \text{ m}}.$

7.8 $V = \dfrac{.025}{\pi \times .03^2} = 8.84.$ $\text{Re} = \dfrac{8.84 \times .06}{1.007 \times 10^{-6}} = 5.3 \times 10^5.$ $\therefore$ Turbulent.

$\therefore L_E = 120 \times .06 = 7.2 \text{ m}.$

$\therefore$ Developed.

7.9 $V = \dfrac{Q}{A} = \dfrac{(18/1000)/2 \times 3600}{\pi \times 0.001^2} = 0.796 \text{ m/s}$

$\text{Re} = \dfrac{VD}{\nu} = \dfrac{0.796 \times 0.002}{1.14 \times 10^{-5}} = 139.6.$ $\therefore$ laminar.

$L_E = 0.065 \,\text{Re} \times D = 0.065 \times 139.6 \times 0.002 = 0.0181 \text{ m}.$ $\therefore$ negligible

7.10 $L_E = 0.04 \,\text{Re}\, h = 0.04 \times 7700 \times .012 = \underline{3.7 \text{ m}}.$

$(L_E)_{\min} = .04 \times 1500 \times .012 = \underline{0.72 \text{ m}}.$

7.11 $(L_E)_{\text{lam}} = 0.065 \,\text{Re}\, D = 0.065\dfrac{5 \times .06}{1.55 \times 10^{-5}} \times .06 = \underline{75.5 \text{ m}}.$

$(L_E)_{\text{turb}} = 120 \times .06 = \underline{7.2 \text{ m}}.$ (Re = 32 300)

7.12 $\text{Re} = \dfrac{VD}{\nu} = \dfrac{0.2 \times 0.04}{10^{-6}} = 8000.$

a) If laminar, $L_E = 0.065 \times \text{Re} \times D = 0.065 \times 8000 \times 0.04 = \underline{20.8 \text{ m}}$

$L_i = L_E / 4 = 20.8/4 = \underline{5.2 \text{ m}}$

b) This is a low Reynolds number turbulent flow. A minimum entrance length would be $L_E = 120D = 120 \times 0.04 = \underline{4.8 \text{ m}}$ with a minimum inviscid core length of $L_i = 10D = 10 \times 0.04 = \underline{0.4 \text{ m}}$.

7.13 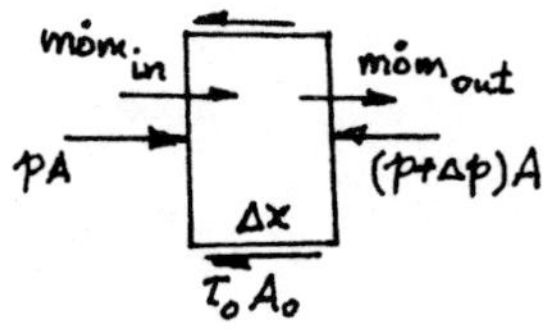

$$\Sigma F_x = \Delta p \pi r_0^2 - \tau_0 2\pi r_0 \Delta x = \dot{m}\text{om}_{\text{out}} - \dot{m}\text{om}_{\text{in}}.$$

$$\therefore \frac{\Delta p}{\Delta x} = \frac{2\tau_0}{r_0} + \frac{\Delta\, \dot{m}\text{om}}{\Delta x}$$

For developed flow $\frac{\Delta p}{\Delta x} = \frac{2\tau_0}{r_0}$ since $\dot{m}\text{om}$ = Const. From the velocity distribution in an entrance (see Fig. 7.1) it is obvious that $(\tau_0)_{\text{entrance}} > (\tau_0)_{\text{developed}}$ since $\left.\frac{\partial u}{\partial y}\right|_{\text{wall}}$ is greater in the entrance. Also, $\frac{\Delta\, \dot{m}\text{om}}{\Delta x} > 0$ since the momentum flux increases from the inlet to the developed flow. Hence,

$$\left(\frac{\Delta p}{\Delta x}\right)_{\text{entrance}} > \left(\frac{\Delta p}{\Delta x}\right)_{\text{developed}}.$$

7.14 a) For a high Re flow transition to turbulence occurs near the origin. In the entrance region the velocity gradient $\partial u / \partial y$ at the wall is very large resulting in a large wall shear. This large wall shear requires a large pressure gradient. In addition, the momentum flux is increasing in the x-direction, also requiring an increased pressure gradient (see the solution to 7.13 for more detail).

b) For a low Re turbulent flow, the flow is laminar through much of the entrance region, up to about L_d (see Fig. 7.2). The laminar flow results in a much smaller velocity gradient at the wall compared with that of the turbulent flow of part (a) requiring a much smaller pressure gradient. This results in the lower distribution of Fig. 7.3.

c) The pressure distribution must move from the lower distribution to the higher distribution of Fig. 7.3 as the Re increases. This occurs at an intermediate Re when transition occurs near L_i. Research that gives accurate data for such low Re transition does not exist.

7.15 $1600 = \frac{VD}{\nu} = \frac{V \times .02}{1 \times 10^{-6}}$. $\quad \therefore V = 0.08 \text{ m/s}.$

Using Eq. 7.3.14: $\quad L = \frac{\Delta p\, r_0^2}{8\mu V} = \frac{400 \times .01^2}{8 \times 1 \times 10^{-3} \times .08} = \underline{62.5 \text{ m}}.$

Using Eq. 7.3.18: $\tau_0 = \frac{r_0 \Delta p}{2L} = \frac{.01 \times 400}{2 \times 62.5} = \underline{0.032 \text{ Pa}}.$

$$\therefore f = \frac{\tau_0}{\frac{1}{8}\rho V^2} = \frac{.032}{\frac{1}{8} \times 1000 \times .08^2} = \underline{0.04}.$$

7.15E $1600 = \frac{VD}{\nu} = \frac{V \times .8/12}{1.06 \times 10^{-5}}.$ $\therefore V = 0.254 \text{ fps}.$

Using Eq. 7.3.14: $L = \frac{\Delta p\, r_0^2}{8\mu V} = \frac{0.07 \times 144 \times .4^2/144}{8 \times 2.06 \times 10^{-5} \times .254} = \underline{268 \text{ ft}}.$

Using Eq. 7.3.18: $\tau_0 = \frac{r_0 \Delta p}{2L} = \frac{.4/12 \times .07 \times 144}{2 \times 268} = \underline{6.27 \times 10^{-4} \text{ psf}}.$

$$\therefore f = \frac{\tau_0}{\frac{1}{8}\rho V^2} = \frac{6.27 \times 10^{-4}}{\frac{1}{8} \times 1.94 \times .254^2} = \underline{0.04}.$$

7.16 $1500 = \frac{VD}{\nu} = \frac{V \times .01}{6.61 \times 10^{-7}}.$ $\therefore V = 0.0992 \text{ m/s}.$

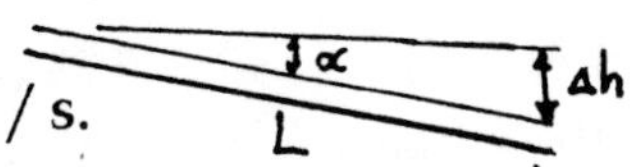

$\alpha = \frac{\Delta h}{L}$

Eq. 7.3.14: $\cancel{\frac{\Delta p}{L}}^{\,0} + \gamma \frac{\Delta h}{L} = \frac{8\mu V}{r_0^2}.$ $\therefore \frac{\Delta h}{L} = \frac{8\mu V}{r_0^2 \rho g}.$

$$\therefore \alpha = \frac{8\nu V}{r_0^2 g} = \frac{8 \times 6.61 \times 10^{-7} \times .0992}{.005^2 \times 9.81} = 0.00214 \text{ rad or } \underline{0.123^\circ}$$

$$Q = AV = \pi \times .005^2 \times 0.0992 = \underline{7.79 \times 10^{-6} \text{ m}^3/\text{s}}.$$

7.17 $V = \frac{Q}{A} = \frac{0.0002}{\pi \times .01^2} = 0.637 \text{ m/s}.$ Use Eq. 7.3.14.

a) $\Delta p = \frac{8\mu VL}{r_0^2} = \frac{8 \times 0.1 \times .637 \times 10}{.01^2} = \underline{51\,000 \text{ Pa}}.$

b) $\Delta p = \frac{8\mu VL}{r_0^2} = \frac{8 \times 1 \times 10^{-3} \times .637 \times 10}{.01^2} = \underline{510 \text{ Pa}}.$

c) $\Delta p = \frac{8\mu VL}{r_0^2} = \frac{8 \times 1.5 \times .637 \times 10}{.01^2} = \underline{764\,000 \text{ Pa}}.$

7.18 Eq. 7.3.14: $\cancel{\frac{\Delta p}{L}}^{\,0} + \gamma \frac{\Delta h}{L} = \frac{8\mu V}{r_0^2}.$ For a vertical pipe $\Delta h = L$. Thus,

$$V = \frac{\rho g r_0^2}{8\mu} = \frac{g r_0^2}{8\nu} = \frac{9.81 \times .01^2}{8\nu} = \frac{1.226 \times 10^{-4}}{\nu}.$$

a) $V = \dfrac{1.226 \times 10^{-4}}{1.52 \times 10^{-6}} = 80.7\ \dfrac{\text{m}}{\text{s}}$. $\therefore Q = \pi \times .01^2 \times 80.7 = \underline{0.0254\ \dfrac{\text{m}^3}{\text{s}}}$.

$\text{Re} = \dfrac{80.7 \times .02}{1.52 \times 10^{-6}} = 1.06 \times 10^6$. $\therefore$ not laminar

b) $V = \dfrac{1.226 \times 10^{-4}}{0.34 / 917} = 0.33\ \dfrac{\text{m}}{\text{s}}$. $\therefore Q = \underline{1.04 \times 10^{-4}\ \dfrac{\text{m}^3}{\text{s}}}$. Re = 17.8. $\therefore$ laminar.

c) $V = \dfrac{1.226 \times 10^{-4}}{1.5 / 1258} = 0.103\ \dfrac{\text{m}}{\text{s}}$. $\therefore Q = \underline{3.24 \times 10^{-5}\ \dfrac{\text{m}^3}{\text{s}}}$. Re = 1.73. $\therefore$ laminar.

7.19 $2000 = \dfrac{VD}{\nu} = \dfrac{4QD\rho}{\pi D^2 \mu} = \dfrac{4 \times .004 \times 917}{\pi \times 0.1 \times D} = \dfrac{46.7}{D}$. $\therefore D = \underline{0.0234\ \text{m}}$.

a) $\Delta p = \dfrac{8\mu VL}{r_0^2} = \dfrac{8 \times .1 \times (.004 / \pi \times .0117^2) \times 10}{0.0117^2} = \underline{544\ 000\ \text{Pa}}$.

b) $\Delta p + \gamma \Delta h = 544\ 000$. $\therefore \Delta p = 544\ 000 - 9800 \times 10 \sin 30^\circ = \underline{495\ 000\ \text{Pa}}$

c) $\Delta p + \gamma \Delta h = 544\ 000$. $\therefore \Delta p = 544\ 000 - 9800(-10 \sin 30^\circ) = \underline{593\ 000\ \text{Pa}}$

7.20 Neglect the effects of the entrance region and assume developed flow for the whole length. Also, assume $p_{\text{inlet}} = \gamma h$ (neglect $V^2/2g$ compared to 4 m).

$\therefore \Delta p = \gamma h - 0$. $\therefore \rho g h = \dfrac{8\mu VL}{r_0^2}$. $\therefore V = \dfrac{g h r_0^2}{8 \nu L} = \dfrac{9.81 \times 4 \times .0025^2}{8 \times 1 \times 10^{-6} \times 40} = 0.766\ \text{m/s}$.

$\therefore Q = AV = \pi \times .0025^2 \times .766 = \underline{1.5 \times 10^{-5}\ \dfrac{\text{m}^3}{\text{s}}}$. $L_E = .065 \times \dfrac{.766 \times .005}{1 \times 10^{-6}} \times .005 = \underline{1.2\ \text{m}}$.

7.21 Neglect the effects of the entrance region and assume developed flow for the whole length. Also, assume $p_{\text{inlet}} = \gamma h$ (neglect $V^2/2g$ compared to 4 m).

$\therefore \Delta p = \gamma h = \dfrac{8\mu VL}{r_0^2}$ and $r_0^2 = \dfrac{8V\mu L}{\gamma h} = \dfrac{8(0.0034/60 \times 60) \times 10^{-3} \times 4}{9800 \times 4 \times \pi r_0^2}$

where $V = Q/A = Q/\pi r_0^2$. This gives $r_0 = 7.04 \times 10^{-4}$ m or $\underline{0.704\ \text{mm}}$.

The velocity is then $V = \dfrac{0.0034/60 \times 60}{\pi \times (0.000704)^2} = 0.6066\ \text{m/s}$.

$\dfrac{V^2}{2g} = \dfrac{0.6066^2}{2 \times 9.81} = 0.0188$ m. This is negligible compared to 4 m.

$L_E = 0.065\, \text{Re} \times D = 0.065 \dfrac{0.6066 \times 14.08 \times 10^{-4}}{10^{-6}} \times 14.08 \times 10^{-4} = 1.2$ m

This entrance region may be significant; to include its effects would be quite difficult, certainly beyond the scope of an introductory course.

7.22 $\text{Re} = 2000 = \frac{VD}{\nu} = \frac{V \times .02}{1.51 \times 10^{-5}}. \quad \therefore V = 1.51 \text{ m/s.}$

$$\Delta p = \frac{8\mu VL}{r_0^2} = \frac{8 \times 1.81 \times 10^{-5} \times 1.51 \times 10}{.01^2} = \underline{21.9 \text{ Pa}}.$$

7.22E $\text{Re} = 2000 = \frac{VD}{\nu} = \frac{V \times .8/12}{1.6 \times 10^{-4}}. \quad \therefore V = 4.8 \text{ fps.}$

$$\Delta p = \frac{8\mu VL}{r_0^2} = \frac{8 \times 3.82 \times 10^{-7} \times 4.8 \times 30}{(.4/12)^2} = \underline{0.396 \text{ psf}}.$$

7.23 $\Delta p + \gamma \Delta h = \frac{8\mu VL}{r_0^2}. \quad -6000 + 9810 \times 10 \sin 10^\circ = \frac{8 \times 1 \times 10^{-3} V \times 10}{.002^2}.$

$$\therefore V = 0.552 \text{ m/s.} \quad \text{Re} = \frac{VD}{\nu} = \frac{.552 \times .004}{1 \times 10^{-6}} = \underline{2210}.$$

$$\Delta p + \gamma \Delta h = \frac{2\tau_0 L}{r_0}. \quad -6000 + 9810 \times 10 \sin 10^\circ = \frac{2 \times \tau_0 \times 10}{.002}. \quad \therefore \tau_0 = \underline{1.1 \text{ Pa}}.$$

7.24 $\text{Re} = 40\,000 = \frac{VD}{\nu} = \frac{V \times 0.1}{1.51 \times 10^{-5}}. \quad \therefore V = 6.04 \text{ m/s.} \quad \therefore V_{max} = 2V = \underline{12.1 \text{ m/s}}$

$$\Delta p = \frac{8\mu VL}{r_0^2} = \frac{8 \times 1.81 \times 10^{-5} \times 6.04 \times 10}{.05^2} = \underline{3.5 \text{ Pa}}. \quad L_E = .065 \times 40\,000 \times .1 = \underline{260 \text{ m}}.$$

7.25 If $Q = \pi R^2 \sqrt{2gH}$, then $V = \sqrt{2gH}$. The parabolic velocity profile is

$$u(r) = u_{max}\left(1 - \frac{r^2}{R^2}\right) = 2V\left(1 - \frac{r^2}{R^2}\right) = 2\sqrt{2gH}\left(1 - \frac{r^2}{R^2}\right).$$

But, the manometer requires

$$p + \frac{u^2}{2}\rho - p = \gamma H. \quad \therefore H = \frac{u^2}{2g}.$$

Substituting yields

$$u = 2u\left(1 - \frac{r^2}{R^2}\right). \quad \therefore r^2 = \frac{R^2}{2} \text{ or } \underline{r = R/\sqrt{2}}.$$

7.26 $V = -\frac{R^2}{8\mu}\frac{\Delta p + \gamma \Delta h}{L} = -\frac{R^2}{8\mu}\frac{\gamma(-L)}{L} = \frac{\gamma R^2}{8\mu} = \frac{9800 \times 0.001^2}{8 \times 10^{-3}} = 1.225 \text{ m/s}$

$$Q = AV = \pi \times 0.001^2 \times 1.225 = \underline{3.85 \times 10^{-6} \text{ m}^3/\text{s}}$$

$$\text{Re} = \frac{VD}{\nu} = \frac{1.225 \times 0.002}{10^{-6}} = 2450$$

It probably is not a laminar flow. Since Re > 2000 it would most likely be turbulent. If the pipe were smooth, rigidly held, with a well-rounded entrance and disturbance-free water entering the pipe it could be laminar.

7.27 a) Combine Eq. 7.3.12 and 7.3.15: $u = u_{max}\left(1-\frac{r^2}{r_0^2}\right)$.

$$V = 2V\left(1-\frac{r^2}{r_0^2}\right). \qquad \therefore r^2 = \frac{r_0^2}{2} \text{ and } \underline{r = 0.707\, r_0}$$

b) From Eq. 7.3.17: $\tau = Cr$ where C is a constant. Then $\tau_w = Cr_0$.

If $\tau = \tau_w / 2$, then $\tau_w / 2 = Cr$ and

$$r = \frac{1}{2}\left(\frac{\tau_w}{C}\right) = \frac{r_0}{2}$$

7.28 $$\frac{Q_{\text{pipe}}}{Q_{\text{annulus}}} = \frac{\pi r_0^4}{8\mu}\frac{\Delta p}{L} \Big/ \frac{\pi}{8\mu}\frac{\Delta p}{L}\left\{r_0^4 - \left(\frac{r_0}{2}\right)^4 - \frac{\left[r_0^2 - (r_0/2)^2\right]^2}{\ell n(2r_0/r_0)}\right\}$$

$$= \frac{r_0^4}{r_0^4 - \frac{r_0^4}{16} - \frac{9r_0^4/16}{\ell n 2}} = \underline{7.938}$$

7.29 $\text{Re} = 20\,000 = \frac{VD}{\nu} = \frac{V \times .04}{1 \times 10^{-6}}. \qquad \therefore V = .5 \text{ m/s}.$

$$h_L = \frac{\Delta p}{\gamma} = \frac{8\mu VL}{\gamma r_0^2} = \frac{8 \times 1 \times 10^{-3} \times .5 \times 10}{9810 \times .02^2} = \underline{0.0102 \text{ m}}.$$

$$\tau_0 = \frac{r_0 \Delta p}{2L} = \frac{.02 \times .0102 \times 9810}{2 \times 10} = \underline{0.100 \text{ Pa}}. \quad L_E = .065 \times 20\,000 \times .04 = \underline{52 \text{ m}}.$$

7.29E $\text{Re} = 20\,000 = \frac{VD}{\nu} = \frac{V \times 2/12}{1.2 \times 10^{-5}}. \qquad \therefore V = 1.44 \text{ fps}.$

$$h_L = \frac{\Delta p}{\gamma} = \frac{8\mu VL}{\gamma r_0^2} = \frac{8 \times 2.36 \times 10^{-5} \times 1.44 \times 30}{62.4 \times (1/12)^2} = \underline{0.0188 \text{ ft}}.$$

$$\tau_0 = \frac{r_0 \Delta p}{2L} = \frac{(1/12).0188 \times 62.4}{2 \times 30} = \underline{0.00163 \text{ psf}}. \quad L_E = .065 \times 20\,000 \times \frac{2}{12} = \underline{217 \text{ ft}}.$$

7.30 See Example 7.2: $Q = -\frac{\pi}{8\mu}\frac{(-\Delta p)}{L}\left[r_2^4 - r_1^4 - \frac{(r_2^2 - r_1^2)^2}{\ell n(r_2 / r_1)}\right]$

$$\therefore Q = -\frac{\pi}{8 \times 1 \times 10^{-3}}\frac{(-100)}{10}\left[.03^4 - .02^4 - \frac{(.03^2 - .02^2)^2}{\ell n(.03/.02)}\right] = \underline{1.31 \times 10^{-4}\ \text{m}^3/\text{s}}.$$

$$\tau_{r_1} = \mu\frac{\partial u}{\partial r}\bigg|_{r=r_1} = \frac{1}{4}\frac{(-\Delta p)}{L}\left[2r_1 - \frac{r_2^2 - r_1^2}{\ell n(r_2/r_1)}\frac{1}{r_1}\right]$$

$$= -\frac{100}{4 \times 10}\left[2 \times .02 - \frac{.03^2 - .02^2}{\ell n(.03/.02)}\frac{1}{.02}\right] = \underline{0.054\ \text{Pa}}.$$

7.31 See Example 7.2: $Q = -\frac{\pi}{8\mu}\frac{(-\Delta p)}{L}\left[r_2^4 - r_1^4 - \frac{(r_2^2 - r_1^2)^2}{\ell n(r_2 / r_1)}\right]$

$$\therefore Q = \frac{\pi \times 10}{8 \times 1.81 \times 10^{-5} \times 10}\left[.03^4 - .02^4 - \frac{(.03^2 - .02^2)^2}{\ell n(.03/.02)}\right] = 7.25 \times 10^{-4}\ \text{m}^3/\text{s}.$$

$$\therefore V = \frac{Q}{A} = \frac{7.25 \times 10^{-4}}{\pi(.03^2 - .02^2)} = \underline{0.462\ \text{m/s}}.$$

$$\tau_{r_1} = \mu\frac{\partial u}{\partial r}\bigg|_{r=r_1} = -\frac{\Delta p}{4L}\left[2r_1 - \frac{r_2^2 - r_1^2}{\ell n(r_2/r_1)}\frac{1}{r_1}\right]$$

$$= -\frac{10}{4 \times 10}\left[2 \times .02 - \frac{.03^2 - .02^2}{\ell n(.03/.02)}\frac{1}{.02}\right] = \underline{0.0054\ \text{Pa}}.$$

7.32 See Example 7.2: $\frac{dp}{dx} = \frac{\mu(T)}{r}\frac{d}{dr}\left(r\frac{du}{dr}\right)$.

The function $\mu(T)$ requires that $T(r)$ be known. The energy equation is needed to find $T(r)$. But, as Eq. 5.5.1 shows, we must know $u(r)$ to find $T(r)$. Hence, the above momentum equation and the energy equation are coupled. A simultaneous solution (a numerical approach is needed) would provide $u(r)$ and $T(r)$.

7.33 From Example 7.2 $u(r) = \frac{1}{4\mu}\frac{dp}{dx}\left[r^2 - r_2^2 + \frac{r_2^2 - r_1^2}{\ell n(r_1/r_2)}\ell n\frac{r}{r_2}\right]$

As $r_1 \to 0$, $\ell n(r_1/r_2) \to -\infty$ so that $\frac{r_2^2 - r_1^2}{\ell n(r_1/r_2)}\ell n\frac{r}{r_2} \to 0.$

Thus, $u(r) = \frac{1}{4\mu}\frac{dp}{dx}\left(r^2 - r_0^2\right)$ where $r_2 = r_0$. See Eq. 7.3.11.

As $r_1 \to r_2$, $\frac{r_2^2 - r_1^2}{\ell n(r_1/r_2)} = \frac{0}{0}$. $\therefore$ differentiate *w.r.t.* r_1:

$$\frac{-2r_1}{1/r_1} = -2r_1^2. \qquad \text{Also, } \ell n\frac{r}{r_2} = \ell n\left(1-\frac{y}{r_2}\right) \cong -\frac{y}{r_2}, \text{ where } y = r_2 - r.$$

$$\therefore u(r) = \frac{1}{4\mu}\frac{dp}{dx}\left(r^2 - r_2^2 + 2r_1^2 y / r_2\right) = \frac{1}{4\mu}\frac{dp}{dx}\left[y^2 - 2r_2 y + \frac{2r_1^2}{r_2}y\right] = \frac{1}{4\mu}\frac{dp}{dx}(y^2 - ay).$$

7.34 $2000 = \frac{Vh}{\nu}$. a) $V = \frac{2000\nu}{h} = \frac{2000 \times 1 \times 10^{-6}}{.012} = 0.167 \text{ m/s}.$

$\therefore Q = AV = .012 \times .5 \times .167 = \underline{0.001 \text{ m}^3/\text{s}}.$

b) $V = \frac{2000\nu}{h} = \frac{2000 \times 1.51 \times 10^{-5}}{.012} = 0.252 \text{ m/s}.$

$\therefore Q = AV = .012 \times .5 \times .252 = \underline{0.00151 \text{ m}^3/\text{s}}.$

7.34E a) $2000 = \frac{Vh}{\nu}$. $\therefore V = \frac{2000 \times 1.2 \times 10^{-5}}{1/24} = 0.576 \text{ fps}.$ $\therefore Q = AV = \underline{0.04 \text{ ft}^3/\text{sec}}.$

b) $V = \frac{2000 \times 1.6 \times 10^{-5}}{1/24} = 0.768 \text{ fps}.$ $\therefore Q = \frac{1}{24} \times \frac{20}{12} \times .768 = \underline{0.053 \text{ cfs}}.$

7.35 There is no pressure gradient. $\therefore$ Eq. 7.4.13 gives $u = \frac{V}{a}y.$

The friction balances the weight component.

$\tau A = W\sin\theta.$ $\quad \tau = \mu\frac{\partial u}{\partial y} = \mu\frac{V}{a} = \mu\frac{.2}{.0004}.$

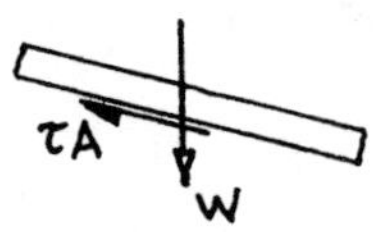

a) $\mu\frac{.2}{.0004} \times 1 \times 1 = 40\sin 20°.$ $\quad \therefore \mu = \underline{0.0274 \frac{\text{N}\cdot\text{s}}{\text{m}^2}}.$

b) $\mu\frac{.2}{.0004} \times 1 \times 1 = 40\sin 30°.$ $\quad \therefore \mu = \underline{0.04 \frac{\text{N}\cdot\text{s}}{\text{m}^2}}.$

7.36 With the pressure gradient zero, Eq. 7.4.13 gives $u = \frac{U}{a}y.$ Thus,

$\tau A = W\sin\theta.$ $\quad \tau = \mu\frac{\partial u}{\partial y} = \mu\frac{V}{a} = \frac{1 \times 10^{-3}}{.0004}V.$

a) $\frac{1 \times 10^{-3}V}{.0004} \times 1 \times 1 = 40\sin 20°.$ $\quad \therefore V = \underline{5.47 \text{ m/s}}.$

b) $\frac{1 \times 10^{-3}V}{.0004} \times 1 \times 1 = 40\sin 30°.$ $\quad \therefore V = \underline{8 \text{ m/s}}.$

7.37 From Eq. 7.4.13 $u(y) = \frac{\gamma}{2\mu}\frac{dh}{dx}(y^2 - ay)$ since $U = 0$.

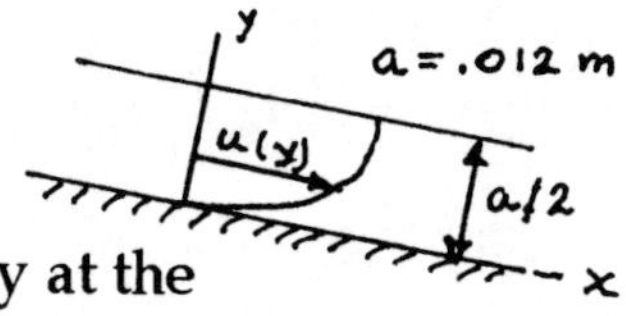

The depth of water is $a/2$, with the maximum velocity at the surface.

$$-\frac{dh}{dx} = \sin\theta. \quad \text{Hence, } u(y) = -\frac{\gamma\sin\theta}{2\mu}(y^2 - ay).$$

$$Q = -\int_0^{a/2} \frac{\gamma\sin\theta}{2\mu}(y^2 - ay)50dy = -\frac{50\gamma\sin\theta}{2\mu}\left(\frac{a^3}{24} - \frac{a^3}{8}\right) = \frac{25}{12\mu}\gamma\sin\theta a^3$$

$$= \frac{25}{12\times10^{-3}} \times 9810 \times \sin 20^\circ \times .012^3 = \underline{12.1 \text{ m}^3/\text{s}}. \quad \therefore V = \frac{12.1}{.006\times 50} = 40.3 \text{ m/s}.$$

$\therefore \text{Re} = \frac{Va/2}{\nu} = \frac{40.3\times.006}{10^{-6}} = 241\ 000.$ $\therefore$ The assumption of laminar flow was not a good one! $u_{max} = \frac{9810\times\sin 20^\circ}{2\times10^{-3}}\left(\frac{.012^2}{4}\right) = \underline{60.4 \text{ m/s}}.$

$$\tau_0 = \mu\left.\frac{\partial u}{\partial y}\right|_{y=0} = -\frac{\gamma\sin\theta}{2}(-a) = \frac{9810\sin 20^\circ}{2}\times.012 = \underline{20.1 \text{ Pa}}.$$

Obviously the flow would be turbulent and the above analysis would have to be modified for an actual flow.

7.38 $u(y) = \frac{\gamma}{2\mu}\frac{dh}{dx}(y^2 - ay) = \frac{9810}{2\times10^{-3}}\times(-.00015)(y^2 - .02y)$

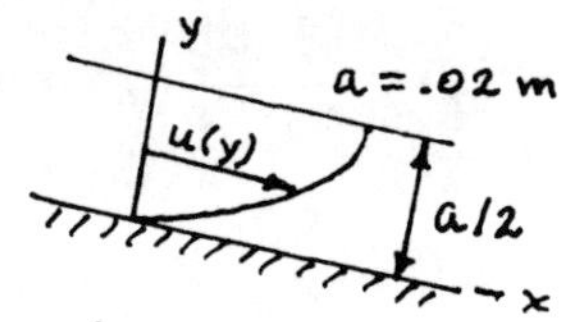

$$Q = \int_0^{.01} \frac{9810(.00015)}{2\times10^{-3}}(.02y - y^2)100dy = 73\ 600(.01\times.01^2 - \frac{.01^3}{3}) = \underline{0.049 \text{ m}^3/\text{s}}.$$

$$V = \frac{.049}{100\times.01} = 0.049 \text{ m/s}. \quad \tau_0 = \mu\left.\frac{du}{dy}\right|_0 = \frac{9810\times.00015\times.02}{2} = \underline{0.015 \text{ Pa}}.$$

$$f = \frac{8\tau_0}{\rho V^2} = \underline{0.050} \qquad \text{Re} = \underline{490}.$$

7.39 Eq. 7.4.17: $\Delta p = \frac{12\mu VL}{a^2}$. $\therefore V = \frac{50\times.02^2}{12\times1.81\times10^{-5}\times60} = 1.53 \text{ m/s}.$

$\therefore Q = AV = .02\times.9\times1.53 = \underline{0.028 \text{ m}^3/\text{s}}$. This is maximum since laminar flow is assumed. $\text{Re} = \frac{Va}{\nu} = \frac{1.53\times.02}{1.51\times10^{-5}} = \underline{2030}.$

7.40 $(p_A - p_B)_{\text{static}} = \gamma h = 9810 \times 20 \sin 30^\circ = 98\ 100$ or 98.1 kPa. $\therefore$ flow is down.

Eq. 7.4.17: $\Delta p + \gamma \Delta h = \dfrac{12\mu LV}{a^2}$. $\therefore -96\ 000 + 98\ 100 = \dfrac{12 \times 10^{-3} \times 20V}{.008^2}$.

$\therefore V = 0.56$ m/s. $\tau_0 = \dfrac{a\Delta p}{2L} + \dfrac{a\gamma\Delta h}{2L} = \dfrac{.008}{2 \times 20}[-96\ 000 + 98\ 100] = 0.42$ Pa.

$\therefore f = \dfrac{8\tau_0}{\rho V^2} = \dfrac{8 \times .42}{1000 \times .56^2} = \underline{0.0107}.$

7.41 Assume laminar flow:

$$\Delta p = \frac{12\mu VL}{a^2}. \quad \therefore 4\ 000\ 000 = \frac{12 \times .06V \times .04}{.0005^2}. \quad \therefore V = 34.7 \text{ m/s}.$$

$\therefore Q = AV = .0005 \times 0.1 \times 34.7 = \underline{1.74 \times 10^{-3}\ \text{m}^3/\text{s}}$

7.41E Assume laminar flow:

$$\Delta p = \frac{12\mu VL}{a^2}. \quad \therefore 600 \times 144 = \frac{12 \times 1.2 \times 10^{-3} V \times 2/12}{(.02/12)^2}. \quad \therefore V = 100 \text{ fps}.$$

$\therefore Q = AV = (.02 \times 4/144) \times 100 = \underline{0.0556\ \text{ft}^3/\text{sec}}$

7.42 $u(y) = \dfrac{1}{2\mu}\dfrac{dp}{dx}(y^2 - ay) + \dfrac{U}{a}y$. $\tau = \mu\dfrac{du}{dy} = \dfrac{1}{2}\dfrac{dp}{dx}(2y - a) + \mu\dfrac{U}{a}$.

a) $\tau_{y=a} = 0 = \dfrac{1}{2}\dfrac{dp}{dx}(2a - a) + \mu\dfrac{U}{a}$. $\therefore \dfrac{dp}{dx} = -\dfrac{2\mu U}{a^2} = -\dfrac{2 \times 1.81 \times 10^{-5} \times 6}{.004^2} = \underline{-13.6 \text{ Pa/m}}.$

b) $\tau_{y=o} = 0 = -\dfrac{a}{2}\dfrac{dp}{dx} + \mu\dfrac{U}{a}$. $\therefore \dfrac{dp}{dx} = \dfrac{2\mu U}{a^2} = \dfrac{2 \times 1.81 \times 10^{-5} \times 6}{.004^2} = \underline{+13.6 \text{ Pa/m}}.$

c) $Q = \displaystyle\int_0^{.004}\left[\frac{1}{2\mu}\frac{dp}{dx}(y^2 - ay) + \frac{U}{a}y\right]dy = \frac{1}{2 \times 1.81 \times 10^{-5}}\frac{dp}{dx}\left(\frac{.004^3}{3} - \frac{.004^3}{2}\right) + \frac{6}{.004} \times \frac{.004^2}{2} = 0.$

$\therefore \dfrac{dp}{dx} = \underline{40.7 \text{ Pa/m}}.$

d) $u(.002) = 4 = \dfrac{1}{2 \times 1.81 \times 10^{-5}}\dfrac{dp}{dx}(.002^2 - .004 \times .002) + \dfrac{6}{.004} \times .002.$ $\therefore \dfrac{dp}{dx} = \underline{9.05 \text{ Pa/m}}.$

7.43 $u = \dfrac{1}{2\mu}\dfrac{dp}{dx}(y^2 - ay) + \dfrac{U}{a}y$. $\tau = \mu\dfrac{du}{dy} = \dfrac{1}{2}\dfrac{dp}{dx}(2y - a) + \mu\dfrac{U}{a}$.

a) $\tau_{y=.006} = \dfrac{1}{2}(-20)(.006) + 1.95 \times 10^{-5}\dfrac{U}{.006} = 0.$ $\therefore U = \underline{18.5 \text{ m/s}}.$

b) $\tau_{y=0} = \dfrac{1}{2}(-20)(.006) + 1.95 \times 10^{-5}\dfrac{U}{.006} = 0.$ $\therefore U = \underline{-18.5 \text{ m/s}}.$

c) $Q = \int_0^{.006} \left[\frac{1}{2 \times 1.95 \times 10^{-5}}(-20)(y^2 - .006y) + \frac{U}{.006} y \right] dy \quad \therefore U = \underline{-6.15 \text{ m/s}}.$

d) $u(.002) = \frac{1}{2 \times 1.95 \times 10^{-5}}(-20)(.002^2 - .006 \times .002) + \frac{U}{.006} \times .002. \quad \therefore U = \underline{-12.3 \text{ m/s}}.$

7.44 $u = \frac{U}{a} y = \frac{15}{.0002} y = 75\,000y. \quad \tau = \mu \frac{du}{dy} = .01 \times 75\,000 = 750 \text{ Pa}.$

$\therefore F = \tau A = 750 \times 2\pi \times .02 \times .3 = \underline{28.3 \text{ N}}.$

7.44E $u = \frac{U}{a} y = \frac{45}{.008/12} y = 67{,}500y. \quad \tau = \mu \frac{du}{dy} = 10^{-4} \times 67{,}500 = 6.75 \text{ psf}.$

$\therefore F = \tau A = 6.75 \times (2\pi \times 10/144) = \underline{2.95 \text{ lb}}.$

7.45 $v_\theta = \frac{U}{a} y. \quad \therefore \tau = \mu \frac{dv_\theta}{dy} = 0.1 \times \frac{.2 \times 30}{.0008} = 750 \text{ Pa}.$

$T = F \times R = \tau \pi DL \times R = 750\pi \times .4 \times .8 \times .2 = \underline{151 \text{ N} \cdot \text{m}}.$

7.46 Assume a linear velocity profile between the rotating disc and the wall.

$\therefore v_\theta = \frac{U}{a} y = \frac{r\omega}{a} y. \quad \therefore \tau = \mu \frac{dv_\theta}{dy} = \frac{\mu\omega}{a} r = \frac{.01 \times 60}{.0012} r = 500 \text{ r}.$

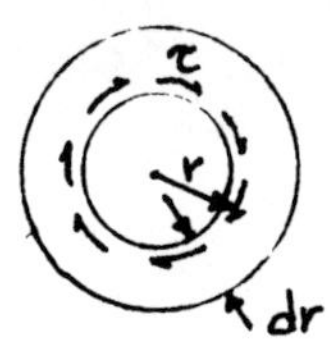

$T = \int \tau dA \times r = \int_0^{.2} 500r \times r 2\pi r dr = 1000\pi \int_0^{.2} r^3 dr = \underline{1.26 \text{ N} \cdot \text{m}}.$

The largest Re occurs at $r = .2$ m: $(\text{Re})_{\max} = \frac{R\omega \times a}{\nu} = \frac{.2 \times 60 \times .0012}{.01/(.86 \times 1000)} = \underline{1240}.$

The laminar flow assumption is valid.

7.47 Neglect the shear on the cylinder bottom; assume a linear velocity profile:

$v_\theta = \frac{U}{a} y = \frac{.1 \times 30}{.001} y = 3000y. \quad \therefore \tau = \mu \frac{dv_\theta}{dy} = 0.42 \times 3000 = 1260 \text{ Pa}.$

$\therefore T = F \times R = \pi DL \times \tau \times R = \pi \times .2 \times .1 \times 1260 \times .1 = \underline{7.9 \text{ N} \cdot \text{m}}.$

7.48 Assume a linear velocity profile: $v_\theta = \frac{U}{a} y = \frac{r\omega}{a} y = \frac{50r}{.002} y.$

$\tau = \mu \frac{dv_\theta}{dy} = 25\,000\, r. \quad T = \int_0^{.0707} \tau \times dA \times r = \int_0^{.0707} (25\,000\, r)(2\pi r \frac{dr}{.707}) \times r.$

$\therefore T = 25\,000 \times \frac{2\pi}{.707} \int_0^{.0707} r^3 dr = \underline{0.694 \text{ N} \cdot \text{m}}.$

7.49 Assume that all losses occur in the 8-m-long channel. The velocity through the straws and screens is so low that the associated losses will be neglected. Assume a developed flow in the channel:

$$V = \frac{\text{Re} \times \nu}{h} = \frac{7000 \times 1.5 \times 10^{-5}}{0.012} = 8.75 \text{ m/s}$$

$$\Delta p = \frac{12 \times 1.8 \times 10^{-5} \times 8.75 \times 8}{0.012^2} = 105 \text{ Pa}$$

Energy: $$\dot{W}_{\text{fan}} = \frac{\Delta p}{\rho \eta} \dot{m} = \frac{\Delta p}{\rho \eta}(\rho A V) = \Delta p A V / \eta$$

$$= 105 \times 1.2 \times 0.012 \times 8.75 / 0.7 = \underline{18.9 \text{ W}}$$

7.50 The solution for $v_\theta(r)$ is (see Eq. 7.5.15) $v_\theta = \frac{A}{2} r + \frac{B}{r}$.

If $v_\theta = 0$ as $r \to \infty$, then $A = 0$ and $v_\theta = \frac{B}{r}$. Also, $v_\theta = R\omega$

at $r = R$. $\therefore R\omega = \frac{B}{R}$, and $B = \omega R^2$. $\therefore v_\theta = \underline{\omega R^2 / \text{r}}$.

The shear stress at $r = R$ is $\tau_1 = -\left[\mu r \frac{d v_\theta / r}{dr}\right] = \mu \omega \times 2.$

The torque is $T = \tau_1 \pi D L \times R = .001 \times \left(\frac{1000 \times 2\pi}{60}\right) \times 2\pi \times .04 \times 1.0 \times .02$

$$= \underline{5.26 \times 10^{-4} \text{ N} \cdot \text{m}}.$$

7.50E The solution for $v_\theta(r)$ is (see Eq. 7.5.15) $v_\theta = \frac{A}{2} r + \frac{B}{r}$.

If $v_\theta = 0$ as $r \to \infty$, $A = 0$. $\therefore v_\theta = \frac{B}{r}$. Also, $v_\theta = R\omega$ at $r = R$.

$\therefore R\omega = \frac{B}{R}$. $\therefore B = \omega R^2$. $\therefore v_\theta = \underline{\omega R^2 / \text{r}}$. The shear stress at

$r = R$ is $\tau_1 = -\left[\mu r \frac{d v_\theta / r}{dr}\right] = 2\mu\omega$. $\therefore T = \tau_1 2\pi R L \times R = 4\pi\mu\omega R^2 L.$

$$\therefore T = 4\pi \times 2.36 \times 10^{-5} \times \left(\frac{1000 \times 2\pi}{60}\right) \times \left(\frac{1}{12}\right)^2 \times \frac{40}{12} = \underline{7.19 \times 10^{-4} \text{ ft-lb}}.$$

7.51 Use Eq. 7.5.19: $T = \frac{4\pi\mu r_1^2 r_2^2 L \omega_1}{r_2^2 - r_1^2} = \frac{4\pi \times .035 \times .02^2 \times .03^2 \times .4 \times (3000 \times 2\pi / 60)}{.03^2 - .02^2}$

$\therefore T = \underline{0.040 \text{ N} \cdot \text{m}}$ $\therefore \dot{W} = T\omega = .04 \times (3000 \times 2\pi / 60) = \underline{12.6 \text{ W}}.$

$\text{Re} = \frac{\omega r_1 (r_2 - r_1) \rho}{\mu} = \frac{3000 \times 2\pi \times .02 \times (.01) \times 917}{60 \times .035} = \underline{1650}.$ $\therefore$ Eq. 7.5.15 is OK.

7.52 Use Eq. 7.5.19: $T = \dfrac{4\mu\pi r_1^2 r_2^2 L\omega_1}{r_2^2 - r_1^2} = .015 = \dfrac{4\pi\mu .04^2 \times .05^2 \times .5 \times 40}{.05^2 - .04^2}$

$\therefore \mu = \underline{0.0134 \dfrac{\text{N}\cdot\text{s}}{\text{m}^2}}$. $\text{Re} = \dfrac{40 \times .04(.05 - .04) \times 1000 \times .9}{.0134} = \underline{1070}$. Eq. 7.5.17 is OK.

7.53 With $\omega_1 = 0$, Eq. 7.5.15 is $v_\theta = \dfrac{r_2^2\omega_2}{r_2^2 - r_1^2}\left(r - \dfrac{r_1^2}{r}\right)$. $\tau_2 = \mu r \dfrac{d}{dr}\left(\dfrac{v_\theta}{r}\right) = \dfrac{2\mu r_1^2 \omega_2}{r_2^2 - r_1^2}$.

$\therefore T_2 = \tau_2 A_2 r_2 = \dfrac{2\mu r_1^2\omega_2}{r_2^2 - r_1^2} 2\pi r_2 L r_2 = \dfrac{4\pi\mu r_1^2 r_2^2 L\omega_2}{r_2^2 - r_1^2}$.

7.54 $T = \dfrac{4\pi\mu r_1^2 r_2^2 L\omega_1}{r_2^2 - r_1^2} = \dfrac{4\pi \times 0.1 \times 0.2^2 \times 0.2008^2 \times 0.8 \times 30}{0.2008^2 - 0.2^2} = 152\ \text{N}\cdot\text{m}$

$\%\ \text{error} = \dfrac{152 - 151}{151} \times 100 = \underline{0.66\%}$

7.55 Let $u = \bar{u} + u'$, $v = \bar{v} + v'$, $w = \bar{w} + w'$. The continuity equation becomes

$\dfrac{\partial}{\partial x}(\bar{u} + u') + \dfrac{\partial}{\partial y}(\bar{v} + v') + \dfrac{\partial}{\partial z}(\bar{w} + w') = \dfrac{\partial \bar{u}}{\partial x} + \dfrac{\partial \bar{v}}{\partial y} + \dfrac{\partial \bar{w}}{\partial z} + \dfrac{\partial u'}{\partial x} + \dfrac{\partial v'}{\partial y} + \dfrac{\partial w'}{\partial z}$.

Now, time-average the above equation recognizing that $\overline{\dfrac{\partial \bar{u}}{\partial x}} = \dfrac{\partial \bar{u}}{\partial x}$ and

$\overline{\dfrac{\partial u'}{\partial x}} = \dfrac{\partial}{\partial x}\overline{u'} = 0$. Then, $\dfrac{\partial \bar{u}}{\partial x} + \dfrac{\partial \bar{v}}{\partial y} + \dfrac{\partial \bar{w}}{\partial z} = 0$. Substitute this back into the

continuity equation, so that $\dfrac{\partial u'}{\partial x} + \dfrac{\partial v'}{\partial y} + \dfrac{\partial w'}{\partial z} = 0$.

7.56 $\overline{\dfrac{Du}{Dt}} = \overline{(\bar{u} + u')\dfrac{\partial}{\partial x}(\bar{u} + u')} + \overline{(\bar{v} + v')\dfrac{\partial}{\partial y}(\bar{u} + u')} + \overline{(\bar{w} + w')\dfrac{\partial}{\partial z}(\bar{u} + u')} + \overline{\dfrac{\partial}{\partial t}(\bar{u} + u')}$

$= \bar{u}\overline{\dfrac{\partial \bar{u}}{\partial x}} + \bar{u}\overline{\dfrac{\partial u'}{\partial x}}^{\,0} + \overline{u'}^{\,0}\overline{\dfrac{\partial \bar{u}}{\partial x}} + \overline{u'\dfrac{\partial u'}{\partial x}} + \bar{v}\overline{\dfrac{\partial \bar{u}}{\partial y}} + \bar{v}\overline{\dfrac{\partial u'}{\partial y}}^{\,0} + \overline{v'}^{\,0}\overline{\dfrac{\partial \bar{u}}{\partial y}} + \overline{v'\dfrac{\partial u'}{\partial y}} + \bar{w}\overline{\dfrac{\partial \bar{u}}{\partial z}} + \bar{w}\overline{\dfrac{\partial u'}{\partial z}}^{\,0}$

$+ \overline{w'}^{\,0}\overline{\dfrac{\partial \bar{u}}{\partial z}} + \overline{w'\dfrac{\partial u'}{\partial z}} + \overline{\dfrac{\partial \bar{u}}{\partial t}}$.

$= \bar{u}\dfrac{\partial \bar{u}}{\partial x} + \bar{v}\dfrac{\partial \bar{u}}{\partial y} + \bar{w}\dfrac{\partial \bar{u}}{\partial t} + \dfrac{\partial \bar{u}}{\partial t} + \overline{u'\dfrac{\partial u'}{\partial x}} + \overline{v'\dfrac{\partial u'}{\partial y}} + \overline{w'\dfrac{\partial u'}{\partial z}}$.

$\dfrac{\bar{D}\bar{u}}{Dt} = \bar{u}\dfrac{\partial \bar{u}}{\partial x} + \bar{v}\dfrac{\partial \bar{u}}{\partial z} + \bar{w}\dfrac{\partial \bar{u}}{\partial z} + \dfrac{\partial \bar{u}}{\partial t}$.

$\therefore \underline{\overline{\dfrac{Du}{Dt}} - \dfrac{\bar{D}\bar{u}}{Dt} = \dfrac{\partial}{\partial x}\overline{u'^2} + \dfrac{\partial}{\partial y}\overline{u'v'} + \dfrac{\partial}{\partial z}\overline{u'w'}}$. (We used continuity.)

7.57 Use the fact that $\frac{\partial}{\partial y}\overline{u'v'} = \overline{\frac{\partial}{\partial y}u'v'}$. See Eq. 7.6.2. This is equivalent to $\frac{\partial}{\partial y}\left(\frac{1}{T}\int_0^T u'v'dt\right) = \frac{1}{T}\int_0^T \frac{\partial}{\partial y}(u'v')dt$, which is obviously correct. Also,

$$\frac{\partial}{\partial x}u'u' + \frac{\partial}{\partial y}u'v' + \frac{\partial}{\partial z}u'w' = u'\underbrace{\left(\frac{\partial u'}{\partial x} + \frac{\partial v'}{\partial y} + \frac{\partial w'}{\partial z}\right)}_{0} + u'\frac{\partial u'}{\partial x} + v'\frac{\partial u'}{\partial y} + w'\frac{\partial u'}{\partial z}.$$

Time average both sides and obtain the result.

7.58 The x-component Navier-Stokes equation for a horizontal channel is

$$\rho\left(\frac{\partial u}{\partial t} + u\frac{\partial u}{\partial x} + v\frac{\partial u}{\partial y} + w\frac{\partial u}{\partial z}\right) = -\frac{\partial p}{\partial x} + \mu\left(\frac{\partial^2 u}{\partial x^2} + \frac{\partial^2 u}{\partial y^2} + \frac{\partial^2 u}{\partial z^2}\right)$$

Substitute $u = \bar{u} + u'$, $v = v'$, $w = w'$ into the N-S eq and time-average:

$$\rho\left[\frac{\partial \bar{u}}{\partial t} + \frac{\partial \overline{u'}}{\partial t} + \frac{\partial}{\partial x}\overline{u'^2} + \frac{\partial}{\partial y}\overline{u'v'} + \frac{\partial}{\partial z}\overline{u'w'}\right] = -\frac{\partial \bar{p}}{\partial x} - \frac{\partial \overline{p'}}{\partial x} + \mu\left(\frac{\partial^2 \bar{u}}{\partial x^2} + \frac{\partial^2 \bar{u}}{\partial y^2} + \frac{\partial^2 \bar{u}}{\partial z^2}\right)$$

where we used the result written in Problem 7.56. Then

$$\rho\frac{\partial}{\partial y}\overline{u'v'} = -\frac{\partial \bar{p}}{\partial x} + \mu\frac{\partial^2 \bar{u}}{\partial y^2}$$

In terms of stresses

$$-\frac{\partial \bar{\tau}_{\text{turb}}}{\partial y} = -\frac{\partial \bar{p}}{\partial x} + \frac{\partial \bar{\tau}_{\text{lam}}}{\partial y} \quad \text{or} \quad \frac{\partial \bar{p}}{\partial x} = \frac{\partial}{\partial y}\left(\bar{\tau}_{\text{turb}} + \bar{\tau}_{\text{lam}}\right)$$

7.59

$\bar{u} = \Sigma u_i / 11 = \underline{16.2 \text{ m/s}}$. $\bar{v} = \Sigma v_i / 11 = \underline{-1.6 \text{ m/s}}$. $u' = u - \bar{u}$, $v' = v - \bar{v}$.

t	0	.01	.02	.03	.04	.05	.06	.07	.08	.09	.1	
u'	-.1	9.5	-5.6	1.1	-11	-6	0.9	12.4	-9.5	3	5.4	
u'^2	.01	90.2	31.4	1.2	121	36	.81	153.8	90.2	9	29.2	$\overline{u'^2} = \underline{51.2 \text{ m}^2/\text{s}^2}$
v'	3.2	-3.6	-7.0	5.1	5.7	-4.4	0.2	8.3	-3.6	-6.6	3.1	
v'^2	10.2	14.4	49	26	32.5	19.4	.04	68.9	13.0	43.6	9.6	$\overline{v'^2} = \underline{26.1 \text{ m}^2/\text{s}^2}$
$u'v'$	-.3	-36.1	39.2	5.6	-62.7	26.4	.2	102.9	34.2	-19.8	16.7	$\overline{u'v'} = \underline{9.7 \text{ m}^2/\text{s}^2}$.

7.60 $\overline{u'v'} = \nu\frac{\partial\bar{u}}{\partial y} - \frac{\bar{\tau}}{\rho} = \nu\frac{\partial\bar{u}}{\partial y} - \frac{r}{2}\frac{\Delta p}{L\rho} = 1.6\times10^{-5}\frac{25.5-20.2}{0.02} - \frac{.22}{2}\frac{400}{10\times1.23} = \underline{-3.57\text{ m}^2/\text{s}^2}$

7.60E $\overline{u'v'} = \nu\frac{\partial\bar{u}}{\partial y} - \frac{\bar{\tau}}{\rho} = \nu\frac{\partial\bar{u}}{\partial y} - \frac{r}{2}\frac{\Delta p}{L\rho} = 1.6\times10^{-4}\frac{76.9-60.7}{0.09} - \frac{.69}{2}\frac{8}{30\times.0035} = \underline{-26.3\text{ ft}^2/\text{sec}^2}$

7.61 $\eta = -\frac{\overline{u'v'}}{d\bar{u}/dy} = -\frac{-3.57}{(25.5-20.2)/.02} = \underline{0.0135\text{ m}^2/\text{s}}.$

$K_{uv} = \frac{\overline{u'v'}}{\sqrt{\overline{u'^2}}\sqrt{\overline{v'^2}}} = \frac{-3.57}{\sqrt{37}\sqrt{17.2}} = \underline{-0.142}$

$\ell_m = \sqrt{\eta/\partial\bar{u}/\partial y} = \sqrt{0.0135/(25.5-20.2)/0.02} = 0.00714\text{ m or }\underline{7.14\text{ mm}}$

7.61E $\eta = -\frac{\overline{u'v'}}{d\bar{u}/dy} = -\frac{-26.3}{(76.9-60.7)/.09} = \underline{0.146\text{ ft}^2/\text{s}}.$

$K_{uv} = \frac{\overline{u'v'}}{\sqrt{\overline{u'^2}}\sqrt{\overline{v'^2}}} = \frac{-26.3}{\sqrt{316}\sqrt{156}} = \underline{-0.118}$

$\ell_m = \sqrt{\eta/\partial\bar{u}/\partial y} = \sqrt{0.146/(76.9-60.7)/0.09} = 0.0285\text{ ft or }\underline{0.342\text{ in}}.$

7.62 $v' = \frac{1}{2}\sin\frac{2\pi}{.2}t.\quad \overline{u'v'} = \frac{1}{.2}\int_0^{.2}\left(\frac{1}{2}\sin10\pi t\right)\left(\frac{1}{2}\sin10\pi t\right)dt = \frac{1}{.8}\int_0^{.2}\sin^2 10\pi t dt$

$= \frac{1}{.8}\int_0^{.2}\left(\frac{1}{2}+\frac{1}{2}\cos20\pi t\right)dt = \frac{1}{.8}\left(\frac{.2}{2}\right) = \underline{0.125\text{ m}^2/\text{s}^2}.$

$\eta = -\frac{\overline{u'v'}}{d\bar{u}/dy} = -\frac{.125}{-10} = \underline{0.0125\text{ m}^2/\text{s}}.$

$\ell_m = \sqrt{\eta/|\partial\bar{u}/\partial y|} = \sqrt{0.0125/10} = 0.0354\text{ m or }\underline{3.54\text{ cm}}$

$u'^2 = \frac{1}{4}\sin^2 10\pi t = v'^2 \qquad \therefore \overline{u'^2} = \overline{v'^2} = .125.$

$K_{uv} = \frac{\overline{u'v'}}{\sqrt{\overline{u'^2}}\sqrt{\overline{v'^2}}} = \frac{.125}{\sqrt{.125}\sqrt{.125}} = \underline{1.0}$

7.63 a) $\text{Re} = \frac{VD}{\nu} = \frac{0.02 \times 0.2}{10^{-6}} = 4000, \quad \frac{e}{D} = \frac{0.26}{200} = 0.0013.$

From the Moody diagram, this is effectively a "smooth" pipe so we conclude that $\delta_\nu > e$.

b) $\text{Re} = \frac{0.2 \times 0.2}{10^{-6}} = 40\,000, \quad \frac{e}{D} = 0.0013.$

From the Moody diagram, this is in the transition zone where δ_ν may be near, in magnitude, to e. The pipe is rough.

c) $\text{Re} = 400\,000, \quad \frac{e}{D} = 0.0013$ and the Moody diagram indicates a rough pipe.

7.64 $\text{Re} = \frac{VD}{\nu} = \frac{6 \times 0.1}{1.1 \times 10^{-4}} = 5455.$ From the Moody diagram, if $\frac{e}{D} = 0.001$, then the pipe is "smooth". Thus, $e = 0.001\,D = 0.001 \times 100 = 0.1 \text{ mm}$

7.65 a) Using $\text{Re} = 4000$ and $e/D = 0.0013$, the Moody diagram provides $f = 0.04$.

Then $$\tau_0 = \frac{f}{8}\rho V^2 = \frac{0.04}{8}1000 \times 0.02^2 = 0.002 \text{ Pa}$$

$$\therefore u_\tau = \sqrt{\tau_0/\rho} = \sqrt{0.002/1000} = 0.001414 \text{ m/s}$$

Eq. 7.6.16 gives

$$u_{\max} = u_\tau\left(2.44\ell n\frac{u_\tau r_o}{\nu} + 5.7\right)$$

$$= 0.001414\left(2.44\ell n\frac{0.001414 \times 0.1}{10} + 5.7\right) = 0.0251 \text{ m/s}$$

b) With $\text{Re} = 40\,000$, $e/D = 0.0013$ the Moody diagram gives $f = 0.026$. Then

$$\tau_0 = \frac{0.026}{8}1000 \times 0.2^2 = 0.130 \text{ Pa}$$

$$u_\tau = \sqrt{0.130/1000} = 0.01140 \text{ m/s}$$

Eq. 7.6.17 gives

$$u_{\max} = u_\tau\left(2.44\ell n\frac{r_o}{e} + 8.5\right)$$

$$= 0.0114\left(2.44\ell n\frac{0.1}{0.00026} + 8.5\right) = 0.262 \text{ m/s}$$

c) With Re = 400 000, $e/D = 0.0013$ we find $f = 0.022$:

$$\tau_0 = \frac{0.022}{8} 1000 \times 2^2 = 11 \text{ Pa}$$

$$u_\tau = \sqrt{11/1000} = 0.105 \text{ m/s}$$

$$u_{\max} = u_\tau \left(2.44 \ell n \frac{r_o}{e} + 8.5 \right)$$

$$= 0.105 \left(2.44 \ell n \frac{0.1}{0.00026} + 8.5 \right) = \underline{2.41 \text{ m/s}}$$

7.66 Here $n = 7$ so that, from Eq. 7.6.21, $f = \frac{1}{n^2} = \frac{1}{49} = 0.0204.$ $V = 4.9$ m / s. (from Eq. 7.6.20)

a) $\therefore \tau_0 = \frac{1}{8} \rho V^2 f = \frac{1}{8} \times 1000 \times 4.9^2 \times .0204 = \underline{61.2 \text{ Pa.}}$

b) $\frac{d\bar{u}}{dy} = 9.2 \times \frac{1}{7} y^{-6/7} \Big|_{y=0} = \underline{\infty}.$ (The profile is not too good near the wall!)

c) $\frac{\partial p}{\partial x} = -\frac{\Delta p}{L} = -\frac{2\tau_0}{r_o} = -\frac{2 \times 61.2}{.05} = \underline{-2450 \text{ Pa/m.}}$

d) τ varies linearly with r. $\therefore \tau_{r=2.5 \text{ cm}} = \frac{1}{2} \tau_0 = 30.6$ Pa. ($r_o = 5$ cm)

$$\tau = \rho \eta \frac{d\bar{u}}{dy}. \quad \therefore 30.6 = 1000 \eta \left(9.2 \times \frac{1}{7} \times .025^{-6/7} \right). \quad \therefore \eta = \underline{9.85 \times 10^{-4} \text{ m}^2 \text{/s.}}$$

7.67 $V = \frac{Q}{A} = \frac{.08}{\pi \times .06^2} = 7.074$ m / s. $\text{Re} = \frac{VD}{\nu} = \frac{7.074 \times .12}{10^{-6}} = 8.49 \times 10^5.$

From Table 7.1, $\underline{n \cong 9}$. From Eq. 7.6.20

$$u_{\max} = \frac{(n+1)(2n+1)}{2n^2} \times V = \frac{10 \times 19}{2 \times 81} \times 7.074 = \underline{8.30 \text{ m/s.}}$$

7.67E $V = \frac{Q}{A} = \frac{2.5}{\pi \times (2.5/12)^2} = 18.33$ fps. $\text{Re} = \frac{VD}{\nu} = \frac{18.33 \times 5/12}{1.08 \times 10^{-5}} = 7.09 \times 10^5.$

From Table 7.1, $\underline{n \cong 8.5}$. From Eq. 7.6.20

$$u_{\max} = \frac{(n+1)(2n+1)}{2n^2} \times V = \frac{9.5 \times 18}{2 \times 8.5^2} \times 18.33 = \underline{21.7 \text{ fps.}}$$

7.68 $n = 5$: $V = \frac{2n^2}{(n+1)(2n+1)}u_{max} = \frac{2\times 5^2}{6\times 11}u_{max} = 0.758u_{max}$.

$$\alpha = \frac{\int u^3 dA}{V^3 A} = \frac{\int_0^{r_o} u_{max}^3 y^{3/n} 2\pi(r_o - y)(dy)}{.758^3 u_{max}^3 \pi r_o^2 r_o^{3/n}} = \frac{2}{.758^3 r_o^{2+3/n}}\int_0^{r_o}\left(r_o y^{3/n} - y^{1+3/n}\right)dy$$

$$= \frac{4.59}{r_o^{2+3/n}}\left[\frac{r_o^{2+3/n}}{1+3/n} - \frac{r_o^{2+3/n}}{2+3/n}\right] = 4.59\left(\frac{1}{1.6} - \frac{1}{2.6}\right) = \underline{1.10}$$

With $n = 10$, $V = \frac{2\times 10^2}{11\times 21}u_{max} = .866u_{max}$.

$$\alpha = \frac{2}{.866^3}\left(\frac{1}{1.3} - \frac{1}{2.3}\right) = \underline{1.03}.$$

7.69 With $n = 7$, Eq. 7.6.21 gives $f = \frac{1}{n^2} = \frac{1}{49} = 0.0204$.

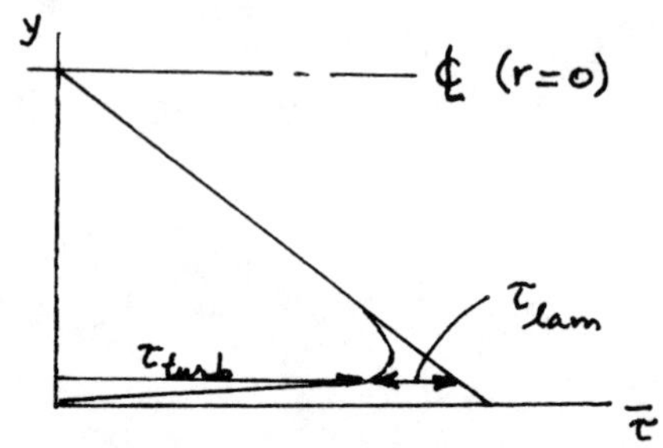

$\therefore \tau_0 = \frac{1}{8}\rho V^2 \times f = \frac{1}{2}\times 1000\times 10^2 \times .0204 = 1020$ Pa.

Since τ varies linearly with r and is zero at $r = 0$,

$$\bar{\tau} = \tau_0 \frac{r}{r_o} = \frac{1020}{.05}r = 20\,400\,r.$$

$$\tau_{lam} = \mu\frac{\partial\bar{u}}{\partial y} = 10^{-3}\left[u_{max}\frac{1}{7}\frac{1}{r_0^{1/7}}y^{-6/7}\right] = \frac{10^{-3}\times 12.24}{7\times .05^{1/7}}y^{-6/7} = 0.00268y^{-6/7}.$$

$\tau_{turb} = \bar{\tau} - \tau_{lam} = 20\,400\,r - 0.00268y^{-6/7}$ where $y + r = .05$.

$\tau_{lam}(y)$ is good away from y=0 (the wall).

$\tau_{lam}(.00625) = .21$, $\tau_{lam}(.003125) = .38$, $\tau_{lam}(.00156) = .68$, $\tau_{lam}(.00078) = 1.24$

$$\frac{d\bar{p}}{dx} = -\frac{2\tau_0}{r_0} = -\frac{2\times 1020}{.05} = \underline{-40\,800 \text{ Pa/m}}.$$

7.70 a) $V = \frac{Q}{A} = \frac{1.2}{\pi\times .4^2} = 19.63$ m/s. $\quad \text{Re} = \frac{VD}{\nu} = \frac{19.63\times 0.8}{2.2\times 10^{-4}} = \underline{71\,400}.$

b) $\frac{e}{D} = 0$. $\therefore$ From Fig. 7.13 $f = \underline{0.019}$. $\quad \tau_0 = \frac{1}{8}\times 917\times 19.63^2 \times .019 = 840$ Pa.

c) $V = \frac{2n^2}{(n+1)(2n+1)}u_{max}$. $\quad \therefore u_{max} = \frac{19.63\times 8\times 15}{2\times 49} = \underline{24.0 \text{ m/s}}$

d) $\delta_\nu = \frac{5\nu}{u_\tau} = \frac{5\times 2.2\times 10^{-4}}{0.957} = \underline{0.0015 \text{ m}}$ where $u_\tau = \sqrt{840/917} = 0.957$ m/s.

e) $u_{max} = 0.957\left[2.44\ell n\dfrac{0.957\times.4}{2.2\times10^{-4}} + 5.7\right] = \underline{22.9 \text{ m/s.}}$

Note that u_{max} is nearly the same using either form of the velocity profile.

7.71 We must find u_τ:

$$V = \frac{Q}{A} = \frac{1.2}{\pi\times0.4^2} = 19.63 \text{ m/s.} \qquad \text{Re} = \frac{19.63\times0.8}{2.2\times10^{-4}} = 71\,400$$

$$\frac{e}{D} = \frac{0.26}{800} = 0.000325. \quad \therefore \text{ Moody diagram} \Rightarrow f = 0.021$$

$$\tau_0 = \frac{1}{8}f\rho V^2 = \frac{1}{8}\times0.021\times917\times19.63^2 = 928 \text{ Pa}$$

$$u_\tau = \sqrt{\tau_0/\rho} = \sqrt{928/917} = 1.006 \text{ m/s.}$$

$$\therefore u_{max} = 1.006\left[2.44\ell n\frac{1.006\times0.4}{2.2\times10^{-4}} + 5.7\right] = \underline{24.2 \text{ m/s}}$$

7.72 a) $\dfrac{\Delta p}{L} = \dfrac{10\,000}{5} = \dfrac{2\tau_0}{r_0} = \dfrac{2\tau_0}{.05}. \quad \therefore \tau_0 = \underline{50 \text{ Pa}}. \quad \therefore u_\tau = \sqrt{\dfrac{50}{1000}} = 0.224 \text{ m/s.}$

b) $u_{max} = u_\tau\left[2.44\ell n\dfrac{u_\tau r_0}{\nu} + 5.7\right] = .224\left[2.44\ell n\dfrac{.224\times.05}{.661\times10^{-6}} + 5.7\right] = \underline{6.60 \text{ m/s.}}$

c) Assume $n = 8$: $V = \dfrac{2n^2}{(n+1)(2n+1)}u_{max} = \dfrac{2\times64}{9\times17}\times6.6 = \underline{5.52 \text{ m/s.}}$

d) Check Re: $\text{Re} = \dfrac{VD}{\nu} = \dfrac{5.52\times.05\times2}{.661\times10^{-6}} = \underline{8.35\times10^5}. \quad \therefore n \cong 8$ is OK.

e) $Q = AV = \pi\times.05^2\times5.52 = \underline{0.0433 \text{ m}^3\text{/s.}}$

7.72E a) $\dfrac{\Delta p}{L} = \dfrac{1.5\times144}{15} = 14.4 = \dfrac{2\tau_0}{r_0} = \dfrac{2\tau_0}{2/12}. \quad \therefore \tau_0 = \underline{1.2 \text{ psf}}. \quad \therefore u_\tau = \sqrt{\dfrac{1.2}{1.94}} = 0.786 \text{ fps.}$

b) $u_{max} = u_\tau\left[2.44\ell n\dfrac{u_\tau r_0}{\nu} + 5.7\right] = .786\left[2.44\ell n\dfrac{.786\times2/12}{.736\times10^{-5}} + 5.7\right] = \underline{23.2 \text{ fps.}}$

c) Assume $n = 8$: $V = \dfrac{2n^2}{(n+1)(2n+1)}u_{max} = \dfrac{2\times64}{9\times17}\times23.2 = \underline{19.41 \text{ fps.}}$

d) Check Re: $\text{Re} = \dfrac{VD}{\nu} = \dfrac{19.41\times4/12}{.736\times10^{-5}} = \underline{8.79\times10^5}. \quad \therefore n \cong 8$ is OK.

e) $Q = AV = \pi\times(2/12)^2\times19.41 = \underline{1.69 \text{ cfs.}}$

7.73 a) From a control volume of the 10-m section of pipe

$$\Delta p\frac{\pi D^2}{4} = \tau_0\pi DL. \quad \therefore \tau_0 = \frac{0.12\times5000}{4\times10} = \underline{15 \text{ Pa}}$$

Assume n = 7. Then Eq. 7.6.21 gives

$$f = \frac{1}{n^2} = \frac{1}{49} = 0.0204$$

and, from Eq. 7.3.19,

$$V^2 = \frac{8\tau_0}{\rho f} = \frac{8 \times 15}{917 \times 0.0204} = 6.41 \quad \text{or} \quad V = 2.53 \text{ m / s}$$

Check: $\text{Re} = \dfrac{2.53 \times 0.12}{2.2 \times 10^{-4}} = 1381.$ This suggests laminar flow. Use Eq. 7.3.14:

$$V = \frac{r_0^2 \Delta p}{8\mu L} = \frac{0.06^2 \times 5000}{8 \times (917 \times 2.2 \times 10^{-4}) \times 10} = \underline{1.12 \text{ m / s}}.$$

Check: $\text{Re} = \dfrac{1.12 \times 0.12}{2.2 \times 10^{-4}} = 611.$ OK

Finally, $Q = AV = \pi \times 0.06^2 \times 1.12 = \underline{0.0127 \text{ m}^3 / \text{s}}$

b) From a control volume or Eq. 7.6.18:

$$\tau_0 = -\frac{0.06(-20\ 000)}{2 \times 10} = 60 \text{ Pa}$$

Assume n = 6. Then Eq. 7.6.21 gives

$$f = \frac{1}{36} = 0.0278$$

and

$$V^2 = \frac{8\tau_0}{\rho f} = \frac{8 \times 60}{917 \times 0.0278} = 18.83 \text{ or } V = 4.34 \text{ m / s}.$$

Check: $\text{Re} = \dfrac{4.34 \times 0.12}{2.2 \times 10^{-4}} = 2370.$ OK

$$Q = \pi \times 0.06^2 \times 4.34 = \underline{0.0491 \text{ m}^3 / \text{s}}$$

c) $\tau_0 = -\dfrac{0.06(-200\ 000)}{2 \times 10} = 600 \text{ Pa}$

Assume n = 7. Then

$$f = \frac{1}{49} = 0.0204$$

and

$$V = \sqrt{\frac{8\tau_0}{\rho f}} = \sqrt{\frac{8 \times 600}{917 \times 0.0204}} = \underline{16.0 \text{ m / s}}$$

Check: $\text{Re} = \dfrac{16.0 \times 0.12}{2.2 \times 10^{-4}} = 8740.$ OK

$$Q = \pi \times 0.06^2 \times 16 = \underline{0.181 \text{ m}^3 / \text{s}}$$

7.74 $\frac{u}{u_{max}} = \left(\frac{y}{r_0}\right)^{1/8}$

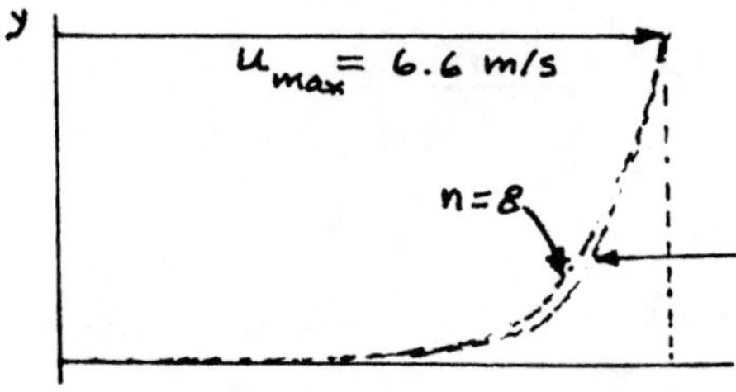

Eq. 7.6.15 for $\frac{y}{r_0} < .15$.

7.75 a) $Re = \frac{0.025 \times 0.04}{10^{-6}} = 1000.$ $\therefore$ laminar. $\therefore f = \frac{64\nu}{VD} = \frac{64 \times 10^{-6}}{0.025 \times 0.04} = \underline{0.064}$

b) $Re = \frac{0.25 \times 0.04}{10^{-6}} = 10\ 000,\ \frac{e}{D} = \frac{0.26}{40} = 0.0065.$ $\therefore f = \underline{0.04}$

c) $Re = \frac{2.5 \times 0.04}{10^{-6}} = 100\ 000,\ \frac{e}{D} = \frac{0.26}{40} = 0.0065.$ $\therefore f = \underline{0.034}$

d) $Re = 10^6,\ \frac{e}{D} = 0.0065.$ $\therefore f = \underline{0.033}$

7.76 $V = \frac{Q}{A} = \frac{.02}{\pi \times .05^2} = 2.55 \text{ m/s}.$ $\Delta p = \gamma f \frac{L}{D}\frac{V^2}{2g} = \rho f \frac{100}{.10}\frac{2.55^2}{2} = 3251\rho f.$

a) $Re = \frac{2.55 \times .1}{10^{-6}} = 2.55 \times 10^5.$ $\frac{e}{D} = \frac{.046}{100} = .00046.$ $\therefore f = .0185$

$\therefore \Delta p = 1000 \times .0185 \times 3251 = \underline{60\ 100 \text{ Pa}}.$

$$\Delta p = \gamma h_L = 9800 \times 1.07 \frac{.02^2 \times 100}{9.8 \times 0.1^5}\left\{\ell n\left[\frac{.046}{3.7 \times 100} + 4.62\left(\frac{10^{-6} \times .1}{.02}\right)^{.9}\right]\right\}^{-2} = \underline{59\ 200 \text{ Pa}}$$

b) $Re = \frac{2.55 \times .1}{8.1 \times 10^{-5}} = 3.1 \times 10^3.$ $\frac{e}{D} = .00046.$ $\therefore f = .042.$

$\therefore \Delta p = 1260 \times .042 \times 3251 = \underline{172\ 000 \text{ Pa}}.$

$$\Delta p = \gamma h_L = 12\ 360 \times 1.07 \frac{.02^2 \times 100}{9.8 \times .1^5}\left\{\ell n\left[\frac{.046}{3.7 \times 100} + 4.62\left(\frac{8.1 \times 10^{-5} \times .1}{.02}\right)^{.9}\right]\right\}^{-2} = \underline{180\ 400 \text{ Pa}}$$

c) $Re = \frac{2.55 \times .1}{2.4 \times 10^{-4}} = 1.06 \times 10^3.$ $\therefore$ laminar. $\therefore f = \frac{64}{Re} = .06.$

$\therefore \Delta p = 917 \times .06 \times 3251 = \underline{179\ 000 \text{ Pa}}.$

Eq. 7.6.31 is not applicable if Re < 3000.

d) $\text{Re} = \dfrac{2.55\times.1}{2.5\times10^{-6}} = 1.02\times10^5.$ $\dfrac{\text{e}}{\text{D}} = .00046.$ $\therefore f = .02.$

$$\therefore \Delta p = 809\times.02\times3251 = \underline{52\ 600\ \text{Pa}}.$$

$$\Delta p = 7940\times1.07\frac{.02^2\times100}{9.8\times.1^5}\left\{\ell n\left[\frac{.046}{3.7\times100}+4.62\left(\frac{2.5\times10^{-6}\times.1}{.02}\right)^{.9}\right]\right\}^{-2} = \underline{52\ 800\ \text{Pa}}$$

7.77 $V = \dfrac{.002}{\pi\times.02^2} = 1.59\ \text{m/s}.$ $\text{Re} = \dfrac{1.59\times.04}{10^{-6}} = 6.4\times10^4.$ $h_L = f\dfrac{200}{.04}\times\dfrac{1.59^2}{2\times9.81} = 644\ \text{f}.$

a) $\dfrac{e}{D} = .0065.$ $\therefore f = .035.$ $\therefore h_L = \underline{22.5\ \text{m}}.$

b) $\dfrac{e}{D} = .00375.$ $\therefore f = .029.$ $\therefore h_L = \underline{18.7\ \text{m}}.$

c) $\dfrac{e}{D} = .0012.$ $\therefore f = .024.$ $\therefore h_L = \underline{15.5\ \text{m}}.$

d) $\dfrac{e}{D} = 0.$ $\therefore f = .0195.$ $\therefore h_L = \underline{12.6\ \text{m}}.$

7.77E $V = \dfrac{Q}{A} = \dfrac{.06}{\pi(.75/12)^2} = 4.89\ \text{fps}.$ $\text{Re} = \dfrac{VD}{\nu} = \dfrac{4.89\times1.5/12}{1.22\times10^{-5}} = 5\times10^4.$

$$h_L = f\frac{L}{D}\frac{V^2}{2g} = f\frac{600}{1.5/12}\frac{4.89^2}{2\times32.2} = 1782\ \text{f}.$$

a) $\dfrac{e}{D} = \dfrac{.00085}{1.5/12} = .0068$ $\therefore f = .035.$ $\therefore h_L = 1782\times.035 = \underline{62\ \text{ft}}.$

b) $\dfrac{e}{D} = \dfrac{.0005}{1.5/12} = .004$ $\therefore f = .033.$ $\therefore h_L = 1782\times.033 = \underline{59\ \text{ft}}.$

c) $\dfrac{e}{D} = \dfrac{.00015}{1.5/12} = .0012$ $\therefore f = .0245.$ $\therefore h_L = 1782\times.0245 = \underline{42\ \text{ft}}.$

d) $\dfrac{e}{D} = 0$ $\therefore f = .0205.$ $\therefore h_L = 1782\times.0205 = \underline{36.5\ \text{ft}}.$

7.78 a) $\rho = \dfrac{p}{RT} = \dfrac{500}{.287\times293} = 5.95\ \text{kg/m}^3.$ $\nu = \dfrac{\mu}{\rho} = \dfrac{1.81\times10^{-5}}{5.95} = 3.04\times10^{-4}\ \text{m}^2/\text{s}.$

$\therefore V = \dfrac{\dot{m}}{\rho A} = \dfrac{1.2}{5.95\pi\times.05^2} = 25.7\ \text{m/s}.$ $\text{Re} = \dfrac{VD}{\nu} = \dfrac{25.7\times.1}{3.04\times10^{-4}} = 8450.$

$$\therefore f = .032. \quad \Delta p = \gamma f \frac{L}{D}\frac{V^2}{2g} = (5.95 \times 9.81) \times .032 \frac{10}{.1}\frac{25.7^2}{2 \times 9.81} = \underline{6290 \text{ Pa}}.$$

$$\Delta p = \gamma \Delta h = 5.95 \times 9.8 \times 1.07 \frac{.202^2 \times 10}{9.8 \times .1^5}\left\{ \ell n \left[0 + 4.62 \left(\frac{3.04 \times 10^{-4} \times .1}{.202} \right)^{.9} \right] \right\}^{-2} = \underline{6360 \text{ Pa}}$$

b) $\rho = \dfrac{p}{RT} = \dfrac{500}{.189 \times 293} = 9.03 \text{ kg/m}^3. \quad \nu = \dfrac{\mu}{\rho} = \dfrac{1.2 \times 10^{-5}}{9.03} = 1.329 \times 10^{-4} \text{ m}^2/\text{s}.$

$$\therefore V = \frac{\dot{m}}{\rho A} = \frac{1.2}{9.03 \pi \times .05^2} = 16.92 \text{ m/s}. \quad \text{Re} = \frac{VD}{\nu} = \frac{16.92 \times .1}{1.329 \times 10^{-4}} = 12\,730.$$

$$\therefore f = .029. \quad \Delta p = \gamma f \frac{L}{D}\frac{V^2}{2g} = (9.03 \times 9.81) \times .029 \frac{10}{.1}\frac{16.92^2}{2 \times 9.81} = \underline{3750 \text{ Pa}}.$$

$$\Delta p = 9.03 \times 1.07 \frac{.133^2 \times 10}{.1^5}\left\{ \ell n \left[0 + 4.62 \left(\frac{1.33 \times 10^{-4} \times .1}{.133} \right)^{.9} \right] \right\}^{-2} = \underline{3740 \text{ Pa}}$$

c) $\rho = \dfrac{p}{RT} = \dfrac{500}{4.124 \times 293} = 0.414 \text{ kg/m}^3. \quad \nu = \dfrac{\mu}{\rho} = \dfrac{7 \times 10^{-6}}{.414} = 17 \times 10^{-6} \text{ m}^2/\text{s}.$

$$V = \frac{1.2}{.414 \pi \times .05^2} = 369 \text{ m/s}. \quad \text{Re} = \frac{369 \times .1}{17 \times 10^{-6}} = 2.2 \times 10^6.$$

$$\therefore f = .01. \quad \Delta p = \gamma f \frac{L}{D}\frac{V^2}{2g} = (.414 \times 9.81) \times .01 \frac{10}{.1}\frac{369^2}{2 \times 9.81} = \underline{28\,000 \text{ Pa}}.$$

$$\Delta p = 0.414 \times 1.07 \frac{2.9^2 \times 10}{.1^5}\left\{ \ell n \left[0 + 4.62 \left(\frac{17 \times 10^{-6} \times .1}{2.9} \right)^{.9} \right] \right\}^{-2} = \underline{28\,700 \text{ Pa}}$$

7.79 $\quad V = \dfrac{Q}{A} = \dfrac{.08}{\pi \times .075^2} = 4.73 \text{ m/s}. \quad \dfrac{e}{D} = \dfrac{.15}{150} = .001.$

$$\Delta p = \gamma f \frac{L}{D}\frac{V^2}{2g} = (917 \times 9.81) f \frac{10}{.15}\frac{4.73^2}{2 \times 9.81} = 6.84 \times 10^5 f.$$

a) $\text{Re} = \dfrac{4.73 \times .15}{.0022} = 322. \quad \therefore \text{laminar}. \quad \Delta p = 6.84 \times 10^5 \times \dfrac{64}{\text{Re}} = \underline{1.36 \times 10^5 \text{ Pa}}.$

Eq. 7.6.31 is not applicable to a laminar flow.

b) $\text{Re} = \dfrac{4.73 \times .15}{.00022} = 3220. \quad \therefore f = .042. \quad \Delta p = 6.84 \times 10^5 \times .042 = \underline{29\,400 \text{ Pa}}.$

$$\Delta p = 9.8 \times 917 \times 1.07 \frac{.08^2 \times 10}{9.8 \times .15^5}\left\{ \ell n\left[\frac{.15}{3.7 \times 50} + 4.62\left(\frac{.00022 \times .15}{.08}\right)^{.9}\right]\right\}^{-2} = 29\,400 \text{ Pa.}$$

c) $\text{Re} = \frac{4.73 \times .15}{.000044} = 16\,300.$ $\therefore f = .029.$ $\Delta p = 6.84 \times 10^5 \times .029 = \underline{20\,000 \text{ Pa}}.$

$$\Delta p = 9.8 \times 917 \times 1.07 \frac{.08^2 \times 10}{9.8 \times .15^5}\left\{ \ell n\left[\frac{.15}{3.7 \times 50} + 4.62\left(\frac{4.4 \times 10^{-5} \times .15}{.08}\right)^{.9}\right]\right\}^{-2} = 20\,700 \text{ Pa.}$$

d) $\text{Re} = \frac{4.73 \times .15}{1.5 \times 10^{-5}} = 46\,000.$ $\therefore f = .0245.$ $\Delta p = 6.84 \times 10^5 \times .0245 = \underline{17\,000 \text{ Pa}}.$

$$\Delta p = 9.8 \times 917 \times 1.07 \frac{.08^2 \times 10}{9.8 \times .15^2}\left\{ \ell n\left[\frac{.15}{3.7 \times 50} + 4.62\left(\frac{1.5 \times 10^{-5} \times .15}{.08}\right)^{.9}\right]\right\}^{-2} = 16\,800 \text{ Pa.}$$

7.80 $V = \frac{Q}{A} = \frac{.4/60}{\pi \times .025^2} = 3.395 \text{ m/s}.$ $\text{Re} = \frac{3.395 \times .05}{10^{-6}} = 1.7 \times 10^5.$

a) $\Delta p = \gamma f \frac{L}{D}\frac{V^2}{2g} = 9800 f \frac{10}{.05}\frac{3.395^2}{2 \times 9.8} = 36\,000.$ $\therefore f = 0.031.$

$\therefore \frac{e}{D} = .005$ and $e = .005 \times 50 = .25$ mm. $\therefore$ Cast iron

b) $\Delta p = 9800 f \frac{10}{.05}\frac{3.395^2}{2 \times 9.8} = 24\,000.$ $\therefore f = 0.021.$

$\therefore \frac{e}{D} = .0009$ and $e = .0009 \times 50 = .045$ mm. $\therefore$ Wrought iron

c) $\Delta p = 9800 f \frac{10}{.05}\frac{3.395^2}{2 \times 9.8} = 19\,000.$ $\therefore f = 0.0165.$

$\therefore \frac{e}{D} = .0003$ and $e = .0015.$ $\therefore$ Drawn tubing

7.81 $V = \frac{Q}{A} = \frac{.01}{\pi \times .03^2} = 3.54 \text{ m/s}.$ $\text{Re} = \frac{3.54 \times .06}{10^{-6}} = 2.12 \times 10^5.$ $\frac{e}{D} = 0.$

$\therefore f = .0155.$ $\therefore \Delta p - \gamma \Delta h = \gamma f \frac{L}{D}\frac{V^2}{2g}.$ (Recall: $\Delta p = p_1 - p_2$, $\Delta h = h_2 - h_1$)

$\therefore \Delta p = 9810 \times .0155 \frac{100}{.06}\frac{3.54^2}{2 \times 9.81} + 9810 \times 100 \sin 30^\circ = \underline{652\,000 \text{ Pa}}.$

7.81E $V = \frac{Q}{A} = \frac{.3}{\pi(1.25/12)^2} = 8.8 \text{ fps}.$ $\text{Re} = \frac{8.8 \times 2.5/12}{1.41 \times 10^{-5}} = 1.3 \times 10^5.$ $\frac{e}{D} = 0.$

$\therefore f = .0167.$ $\therefore \Delta p - \gamma \Delta h = \gamma f \frac{L}{D}\frac{V^2}{2g}.$ (Recall: $\Delta p = p_1 - p_2$, $\Delta h = h_2 - h_1$)

$$\therefore \Delta p = 62.4 \times .0167 \frac{300}{2.5/12} \frac{8.8^2}{2 \times 32.2} + 62.4 \times 300 \sin 30^\circ = \underline{11\ 200 \text{ psf}}.$$

7.82 $V = \frac{Q}{A} = \frac{.02}{\pi \times .025^2} = 10.2 \text{ m/s}.$ $\text{Re} = \frac{10.2 \times .05}{.661 \times 10^{-6}} = 7.7 \times 10^5.$ $\frac{e}{D} = \frac{.046}{50} = .00092.$

For $\frac{e}{D} = 0$: $f = .013.$ For $\frac{e}{D} = .00092$: $f = .0197.$

This is a significant difference. $\therefore$ <u>The roughness is significant</u>.

7.83 $V = \frac{Q}{A} = \frac{5}{\pi \times .4^2} = 9.95 \text{ m/s}.$ $\text{Re} = \frac{9.95 \times .8}{10^{-6}} = 8 \times 10^6.$

$\frac{e}{D} = \frac{1.6}{800} = .002$ (using an average "e" value). $\therefore f = .0237.$

$$\therefore \Delta p = \gamma f \frac{L}{D} \frac{V^2}{2g} = 9810 \times .0237 \frac{100}{.8} \frac{9.95^2}{2 \times 9.81} = \underline{147\ 000 \text{ Pa}}.$$

7.84 Use Eq. 7.6.32: $Q = -.965 \left[g \frac{D^5 h_L}{L} \right]^{.5} \ell n \left[\frac{e}{3.7D} + \left(\frac{3.17 \nu^2 L}{g D^3 h_L} \right)^{.5} \right].$

a) $h_L = \frac{\Delta p}{\gamma} = \frac{500\ 000}{9810} = 51.0 \text{ m}.$ $\frac{e}{D} = \frac{.26}{100} = .0026.$ $\nu = 10^{-6} \text{ m}^2/\text{s}.$

$$\therefore Q = -.965 \left[\frac{9.81 \times .1^5 \times 51}{200} \right]^{.5} \ell n \left[\frac{.0026}{3.7} + \left(\frac{3.17 \times 10^{-12} \times 200}{9.81 \times .1^3 \times 51} \right)^{.5} \right] = \underline{0.035 \text{ m}^3/\text{s}}.$$

b) $h_L = \frac{\Delta p}{\gamma} = \frac{500\ 000}{12\ 350} = 40.5 \text{ m}.$ $\frac{e}{D} = .0026.$ $\nu = .0012 \text{ m}^2/\text{s}.$

$$\therefore Q = -.965 \left[\frac{9.81 \times .1^5 \times 40.5}{200} \right]^{.5} \ell n \left[\frac{.0026}{3.7} + \left(\frac{3.17 \times .0012^2 \times 200}{9.81 \times .1^3 \times 40.5} \right)^{.5} \right] = \underline{0.013 \text{ m}^3/\text{s}}.$$

c) $h_L = \frac{\Delta p}{\gamma} = \frac{500\ 000}{9000} = 55.6 \text{ m}.$ $\frac{e}{D} = .0026.$ $\nu = 1.1 \times 10^{-4} \text{ m}^2/\text{s}.$

$$\therefore Q = -.965 \left[\frac{9.81 \times .1^5 \times 55.6}{200} \right]^{1/2} \ell n \left[\frac{.0026}{3.7} + \left(\frac{3.17 \times 1.12^2 \times 10^{-8} \times 200}{9.81 \times .1^3 \times 55.6} \right)^{.5} \right] = \underline{0.027 \text{ m}^3/\text{s}}.$$

d) $h_L = \frac{\Delta p}{\gamma} = \frac{500\ 000}{7930} = 63.1 \text{ m}.$ $\frac{e}{D} = .0026.$ $\nu = 1.85 \times 10^{-6} \text{ m}^2/\text{s}.$

$$\therefore Q = -.965 \left[\frac{9.81 \times .1^5 \times 63.1}{200} \right]^{.5} \ell n \left[\frac{.0026}{3.7} + \left(\frac{3.17 \times 1.85^2 \times 10^{-12} \times 200}{9.81 \times .1^3 \times 63.1} \right)^{.5} \right] = \underline{0.039 \text{ m}^3/\text{s}}.$$

7.85 Use Eq. 7.6.32: $h_L = \frac{\Delta p}{\gamma} = \frac{200\ 000}{9810} = 20.4$ m. $\nu = 10^{-6}$ m^2/s.

a) $$Q = -.965\left[\frac{9.81 \times .04^5 \times 20.4}{100}\right]^{.5} \ell n\left[\frac{.26}{3.7 \times 40} + \left(\frac{3.17 \times 10^{-12} \times 100}{9.81 \times .04^3 \times 20.4}\right)^{.5}\right] = \underline{0.0027 \text{ m}^3/\text{s}}.$$

b) $$Q = -.965\left[\frac{9.81 \times .04^5 \times 20.4}{100}\right]^{.5} \ell n\left[\frac{.046}{3.7 \times 40} + \left(\frac{3.17 \times 10^{-12} \times 100}{9.81 \times .04^3 \times 20.4}\right)^{.5}\right] = \underline{0.0033 \text{ m}^3/\text{s}}.$$

c) $$Q = -.965\left[\frac{9.81 \times .04^5 \times 20.4}{100}\right]^{.5} \ell n\left[0 + \left(\frac{3.17 \times 10^{-12} \times 100}{9.81 \times .04^3 \times 20.4}\right)^{.5}\right] = \underline{0.0038 \text{ m}^3/\text{s}}.$$

7.86 Use Eq. 7.6.32:

$h_L = 40 - 10 = 30$ m, $\nu = 1.31 \times 10^{-6}$ m^2/s, $\frac{3.17 \times (1.31 \times 10^{-6})^2 \times 200}{9.8 \times 30} = 3.7 \times 10^{-12}$.

a) $$Q = -.965\left[\frac{9.8 \times .04^5 \times 30}{200}\right]^{.5} \ell n\left[\frac{.15}{3.7 \times 40} + \left(\frac{3.7 \times 10^{-12}}{.04^3}\right)^{.5}\right] = \underline{0.0025 \text{ m}^3/\text{s}}.$$

b) $$Q = -.965\left[\frac{9.8 \times .08^5 \times 30}{200}\right]^{.5} \ell n\left[\frac{.15}{3.7 \times 80} + \left(\frac{3.7 \times 10^{-12}}{.08^3}\right)^{.5}\right] = \underline{0.0157 \text{ m}^3/\text{s}}.$$

c) $$Q = -.965\left[\frac{9.8 \times .12^5 \times 30}{200}\right]^{.5} \ell n\left[\frac{.15}{3.7 \times 120} + \left(\frac{3.7 \times 10^{-12}}{.12^3}\right)^{.5}\right] = \underline{0.0459 \text{ m}^3/\text{s}}.$$

d) $$Q = -.965\left[\frac{9.8 \times .16^5 \times 30}{200}\right]^{.5} \ell n\left[\frac{.15}{3.7 \times 160} + \left(\frac{3.7 \times 10^{-12}}{.16^3}\right)^{.5}\right] = \underline{0.0979 \text{ m}^3/\text{s}}.$$

7.87 Use Eq. 7.6.32: $\dot{m} = \rho Q$. $\frac{e}{3.7D} = \frac{.046}{3.7 \times 120} = .000104$.

a) $\rho = \frac{p}{RT} = \frac{200}{.287 \times 313} = 2.23$ kg/m^3. $\therefore h_L = \frac{\Delta p}{\gamma} = \frac{400}{9.81 \times 2.23} = 18.3$ m.

$$\nu = \frac{2 \times 10^{-5}}{2.23} = 9 \times 10^{-6} \text{ m}^2/\text{s}.$$

$$Q = -.965\left[\frac{9.81 \times .12^5 \times 18.3}{400}\right]^{.5} \ell n\left[.000104 + \left(\frac{3.17 \times 9^2 \times 10^{-12} \times 400}{9.81 \times .12^3 \times 18.3}\right)^{.5}\right] = .0235.$$

$$\therefore \dot{m} = \underline{.052 \text{ kg/s}}.$$

b) $\rho = \frac{p}{RT} = \frac{200}{.189 \times 313} = 3.38$ kg/m^3. $\therefore h_L = \frac{400}{9.81 \times 3.38} = 12.1$ m.

$$\nu = \frac{\mu}{\rho} = \frac{1.3 \times 10^{-5}}{3.38} = 3.8 \times 10^{-6} \text{ m}^2/\text{s}.$$

$$Q=-.965\left[\frac{9.81\times.12^5\times12.1}{400}\right]^{.5}\ell n\left[.000104+\left(\frac{3.17\times3.8^2\times10^{-12}\times400}{9.81\times.12^3\times12.1}\right)^{.5}\right]=.0205.$$

$$\therefore\ \dot{m}=\underline{.069\ \text{kg}/\text{s}}.$$

c) $\rho=\frac{p}{RT}=\frac{200}{4.12\times313}=.155\ \text{kg}/\text{m}^3.\quad\therefore\ h_L=\frac{400}{9.81\times.155}=263\ \text{m}.$

$$\nu=\frac{\mu}{\rho}=\frac{7.2\times10^{-6}}{.155}=46\times10^{-6}\ \text{m}^2/\text{s}.$$

$$Q=-.965\left[\frac{9.81\times.12^5\times263}{400}\right]^{.5}\ell n\left[.000104+\left(\frac{3.17\times46^2\times10^{-12}\times400}{9.81\times.12^3\times263}\right)^{.5}\right]=.086.$$

$$\therefore\ \dot{m}=\underline{.0133\ \text{kg}/\text{s}}.$$

7.88 a) Use an intermediate value for e = 1.6 mm.

$$h_L=f\frac{L}{D}\frac{Q^2}{2A^2g}\quad\text{or}\quad\frac{200\ 000}{9800}=f\frac{200}{1.2}\frac{Q^2}{2\times(\pi\times.6^2)^2\times9.8}\quad\text{or } fQ^2=3.07.$$

Guess: f = 0.02. Then Q = 12.54 m³/s.

Check: $V=11.65/\pi\times.6^2=11.1,\ \text{Re}=\frac{11.1\times1.2}{10^{-6}}=13.3\times10^6,\ \frac{e}{D}=\frac{1.6}{1200}=.0013.$

$$\therefore\ f=0.021.$$

This is good. Use f = 0.021 and $\underline{Q=12.1\ \text{m}^3/\text{s}}$

b) Use Eq. 7.6.32: $h_L=\frac{200\ 000}{9810}=20.4.\quad\frac{e}{3.7D}=\frac{1.6}{3.7\times1200}=.00036.\quad\nu=10^{-6}\ \text{m}^2/\text{s}.$

$$Q=-.965\left[\frac{9.81\times1.2^5\times20.4}{200}\right]^{.5}\ell n\left[.00036+\left(\frac{3.17\times10^{-12}\times200}{9.81\times1.2^3\times20.4}\right)^{.5}\right]=\underline{12.1\ \text{m}^3/\text{s}}.$$

7.88E a) Use an intermediate value for e = 0.005 ft.

$$h_L=f\frac{L}{D}\frac{Q^2}{2A^2g}\quad\text{or}\quad\frac{30\times144}{62.4}=f\frac{600}{4}\frac{Q^2}{2(\pi\times2^2)^2\times32.2}\quad\text{or } fQ^2=4694.$$

Guess: f = 0.02. Then Q = 484 ft³/sec.

Check: $V=484/\pi\times2^2=38.6\ \text{fps},\ \text{Re}=\frac{38.6\times4}{1.22\times10^{-5}}=12.6\times10^6,\ \frac{e}{D}=\frac{.005}{4}=.00125.$

$$\therefore\ f=0.021.$$

This is good. Use f = 0.021 and $\underline{Q=473\ \text{ft}^3/\text{sec}}.$

b) Use Eq. 7.6.32: $h_L=\frac{30\times144}{62.4}=69.2'.\quad\frac{e}{3.7D}=\frac{.005}{3.7\times4}=.00034$ using an average value for e. $\nu=1.22\times10^{-5}\ \text{ft}^2/\text{sec}.$

$$Q = -.965\left[\frac{32.2\times 4^5\times 69.2}{600}\right]^{.5} \ell n\left[.00034+\left(\frac{3.17\times 1.22^2\times 10^{-10}\times 600}{32.2\times 4^3\times 69.2}\right)^{.5}\right] = \underline{475 \text{ cfs.}}$$

7.89 Use Eq. 7.6.33: $e = 0$ $D = .66\left[e^{1.25}\left(\frac{LQ^2}{gh_L}\right)^{4.75} + \nu Q^{9.4}\left(\frac{L}{gh_L}\right)^{5.2}\right]^{.04}$.

a) $h_L = \frac{\Delta p}{\gamma} = \frac{200\ 000}{9810} = 20.4 \text{ m.}$ $\nu = 10^{-6} \text{ m}^2/\text{s.}$

$$D = .66\left[10^{-6}\times .002^{9.4}\left(\frac{100}{9.81\times 20.4}\right)^{5.2}\right]^{.04} = \underline{0.032 \text{ m}}.$$

b) $h_L = \frac{\Delta p}{\gamma} = \frac{200\ 000}{12\ 350} = 16.2 \text{ m.}$ $\nu = 7.9\times 10^{-5} \text{ m}^2/\text{s.}$

$$D = .66\left[7.9\times 10^{-5}\times .002^{9.4}\left(\frac{100}{9.81\times 16.2}\right)^{5.2}\right]^{.04} = \underline{0.040 \text{ m}}.$$

c) $h_L = \frac{\Delta p}{\gamma} = \frac{200\ 000}{7930} = 25.2 \text{ m.}$ $\nu = 2.1\times 10^{-6} \text{ m}^2/\text{s.}$

$$D = .66\left[2.1\times 10^{-6}\times .002^{9.4}\left(\frac{100}{9.81\times 25.2}\right)^{5.2}\right]^{.04} = \underline{0.031 \text{ m}}.$$

d) $h_L = \frac{\Delta p}{\gamma} = \frac{200\ 000}{9020} = 22.2 \text{ m.}$ $\nu = 3.8\times 10^{-5} \text{ m}^2/\text{s.}$

$$D = .66\left[3.8\times 10^{-5}\times .002^{9.4}\left(\frac{100}{9.81\times 22.2}\right)^{5.2}\right]^{.04} = \underline{0.036 \text{ m}}.$$

7.90 a) Use an intermediate value for e: $e = 1.6$ mm.

$$h_L = f\frac{L}{D}\frac{V^2}{2g} \text{ or } 20 = f\frac{300}{D}\frac{5^2}{(\pi D^2/4)^2\times 2\times 9.8} \text{ or } 31.0f = D^5.$$

Guess: $f = 0.02$. Then $D = 0.91$ m.

Check: $V = \frac{5}{\pi\times .455} = 7.69 \text{ m/s}$, $\text{Re} = \frac{7.69\times .91}{10^{-6}} = 7.0\times 10^6$, $\frac{e}{D} = \frac{1.6}{910} = .00176.$

$\therefore f = 0.0225.$

This is good. Use $f = 0.0225$ and $D = 0.93$ m.
Choose a standard diameter, i.e., $\underline{D = 1.0 \text{ m}}$ (It cannot be smaller than 0.93 m).
b) Use Eq. 7.6.33: Select an average "e": $e = 1.6$ mm. $h_L = 20$ m.

$$D = .66\left[.0016^{1.25}\left(\frac{300\times 5^2}{9.81\times 20}\right)^{4.75} + 10^{-6}\times 5^{9.4}\left(\frac{300}{9.81\times 20}\right)^{5.2}\right]^{.04} = \underline{0.96 \text{ m}}.$$

A diameter of 1 m would be selected.

7.91 a) $V = \frac{.4/60}{\pi D^2/4} = .00849/D^2$, $h_L = f\frac{L}{D}\frac{V^2}{2g}$ or $3 = f\frac{1200}{D}\frac{(.00849/D^2)^2}{2 \times 9.8}$ or

$f = 680D^5$.

Guess: $f = 0.02$. Then $D = .124$ m.

Check: $V = .551$ m/s, $\text{Re} = \frac{.551 \times .25}{1.31 \times 10^{-6}} = 1.05 \times 10^5$, $\frac{e}{D} = \frac{.0015}{124} = 1.2 \times 10^{-5}$.

$\therefore f = 0.0175$

Use $f = .0175$ and $\underline{D = 0.121 \text{ m}}$.

b) Use Eq. 7.6.33:

$$D = 0.66\left[(1.5 \times 10^{-5})^{1.25}\left(\frac{1200 \times (.4/60)^2}{9.8 \times 3}\right)^{4.75} + 1.31 \times 10^{-6} \times \left(\frac{.4}{60}\right)^{9.4}\left(\frac{1200}{9.8 \times 3}\right)^{5.2}\right]^{.04}$$

$$= \underline{0.127 \text{ m}}.$$

Exiting K.E. $= \frac{V^2}{2g} = \frac{0.892^2}{2 \times 9.8} = 0.041$ m. $\therefore$ negligible.

7.92 $h_L = 10$ m. For a square duct, $R = \frac{w^2}{4w} = w/4$. But, $R = D/4$.

Sheet metal is relatively smooth like drawn tubing. $\therefore$ Use $e = .001$ mm.

$$\therefore D = .66\left[(10^{-6})^{1.25}\left(\frac{200 \times 4^2}{9.81 \times 10}\right)^{4.75} + 1.6 \times 10^{-5} \times 4^{9.4}\left(\frac{200}{9.81 \times 10}\right)^{5.2}\right]^{.04} = 0.829 \text{ m}.$$

$\therefore R = \frac{.829}{4} = .207$ m. $\therefore w = 4R = \underline{0.83 \text{ m}}$.

7.93 Use Eq. 7.6.32: $h_L = \frac{80}{9810}$, $\nu = 10^{-6}$ m^2/s, use $D = 4R = 4\frac{.02 \times .04}{2(.02+.04)} = .027$ m.

$$Q = -.965\left[\frac{9.81 \times .027^5}{2} \times \frac{80}{9810}\right]^{.5} \ell n\left[\frac{.0015}{3.7 \times 27} + \left(\frac{3.17 \times 10^{-12} \times 2 \times 9810}{9.81 \times .027^3 \times 80}\right)^{.5}\right] = \underline{0.000143 \text{ m}^3/\text{s}}.$$

7.94 $D = 4R = 4\frac{.04 \times .1}{2(.04+.1)} = 0.057$ m. Use Eq. 7.6.32 with $e = 0$. $\nu = 10^{-6}$ m^2/s.

$$Q = -.965\left[\frac{9.81 \times .057^5}{5} \times \frac{100}{9810}\right]^{.5} \ell n\left[0 + \left(\frac{3.17 \times 10^{-12} \times 5 \times 9810}{9.81 \times .057^3 \times 100}\right)^{.5}\right] = \underline{0.00074 \text{ m}^3/\text{s}}.$$

7.95 a) $R = \dfrac{A}{P_{\text{wet}}} = \dfrac{0.3 \times 1.2}{1.8} = 0.2$ m, $h_L = 10\ 000 \times 0.0015 = 15$ m (from energy eq.)

$$Q = -.965\left[\frac{9.8(4\times.2)^5 \times 15}{10\ 000}\right]^{.5} \ell n\left[\frac{1.5}{3.7\times 800} + \left(\frac{3.17\times 10^{-12}\times 10\ 000}{9.8\times(4\times.2)^3\times 15}\right)^{.5}\right] = \underline{0.0351\ \text{m}^3/\text{s}}.$$

b) $R = \dfrac{A}{P_{\text{wet}}} = \dfrac{0.6 \times 1.2}{2.4} = 0.3$ m, $h_L = 10\ 000 \times 0.0015 = 15$ m (from energy eq.)

$$Q = -.965\left[\frac{9.8\times 1.2^5 \times 15}{10\ 000}\right]^{.5} \ell n\left[\frac{1.5}{3.7\times 1200} + \left(\frac{3.17\times 10^{-12}\times 10\ 000}{9.8\times 1.2^3\times 15}\right)^{.5}\right] = \underline{0.257\ \text{m}^3/\text{s}}.$$

c) $R = \dfrac{A}{P_{\text{wet}}} = \dfrac{0.9 \times 1.2}{3.0} = 0.36$ m, $h_L = 10\ 000 \times 0.0015 = 15$ m (from energy eq.)

$$Q = -.965\left[\frac{9.8\times 1.44^5 \times 15}{10\ 000}\right]^{.5} \ell n\left[\frac{1.5}{3.7\times 1440} + \left(\frac{3.17\times 10^{-12}\times 10\ 000}{9.8\times 1.44^3\times 15}\right)^{.5}\right] = \underline{2.37\ \text{m}^3/\text{s}}.$$

7.96 a) $V_1 = \dfrac{.02}{\pi\times.01^2} = 63.7,\ V_2 = \dfrac{V_1}{4} = 15.9\ \text{m/s}.\ \rho_1 = \dfrac{100+50}{.287\times 293} = 1.78\ \text{kg/m}^3.$

$$\text{Energy: } 0 = \frac{15.9^2 - 63.7^2}{2\times 9.81} + \frac{p_2 - 50\ 000}{1.78\times 9.81} + \left(1-\frac{1}{4}\right)^2\frac{63.7^2}{2\times 9.81}. \quad \therefore p_2 = \underline{51\ 400\ \text{Pa}}.$$

b) $V_1 = 63.7,\ V_2 = \dfrac{V_1}{9} = 7.08\ \text{m/s}.\ \rho_1 = 1.78\ \text{kg/m}^3.$

$$\text{Energy: } 0 = \frac{7.08^2 - 63.7^2}{2\times 9.81} + \frac{p_2 - 50\ 000}{1.78\times 9.81} + \left(1-\frac{1}{9}\right)^2\frac{63.7^2}{2\times 9.81}. \quad \therefore p_2 = \underline{50\ 700\ \text{Pa}}.$$

7.97 a) $V_1 = 63.7,\ V_2 = 15.9,\ \rho_1 = 1.78\ \text{kg/m}^3.$

$$\text{Energy: } 0 = \frac{15.9^2 - 63.7^2}{2\times 9.81} + \frac{p_2 - 50\ 000}{1.78\times 9.81} + .4\frac{(63.7-15.9)^2}{2\times 9.81}. \quad \therefore p_2 = \underline{52\ 600\ \text{Pa}}.$$

b) $V_1 = 63.7,\ V_2 = 7.08,\ \rho_1 = 1.78\ \text{kg/m}^3.$

$$\text{Energy: } 0 = \frac{7.08^2 - 63.7^2}{2\times 9.81} + \frac{p_2 - 50\ 000}{1.78\times 9.81} + .4\frac{(63.7-7.08)^2}{2\times 9.81}. \quad \therefore p_2 = \underline{52\ 400\ \text{Pa}}.$$

7.98 a) $A_c = \left[.62 + .38\left(\dfrac{1}{4}\right)^3\right]A_2 = .626A_2$. Neglect losses from ① → ©.

$$\text{From © → ②: } h_L = \left(1-\frac{A_c}{A_2}\right)^2\frac{V_c^2}{2g} = (1-.626)^2\left(\frac{A_2}{A_c}\right)^2\frac{V_2^2}{2g}$$

$$= .140\times\frac{1}{.626^2}\frac{V_2^2}{2g}. \qquad \therefore K = \frac{.14}{.626^2} = \underline{0.36}.$$

b) $A_c = \left[.6+.4\left(\frac{1}{4}\right)^2\right] A_0 = .625 A_0$. Neglect losses from ① → ©.

From © → ②: $h_L = \left(1-\frac{A_c}{A_2}\right)^2 \frac{V_c^2}{2g} = \left(1-\frac{A_c}{A_0}\frac{A_0}{A_2}\right)^2 \left(\frac{A_2}{A_c}\right)^2 \frac{V_2^2}{2g}$.

$\therefore K = \left(1-.625\times\frac{1}{4}\right)^2 \frac{A_2^2}{A_0^2}\frac{A_0^2}{A_c^2} = .712\times 4^2 \times \frac{1}{.625^2} = \underline{29}.$

c) $A_c = \left[.6+.4\left(\frac{1}{4}\right)^2\right] A_0 = .625 A_0$. Neglect losses from ① → ©.

From © → ②: $h_L = \left(1-\frac{A_c}{A_2}\right)^2 \frac{V_c^2}{2g} = \left(1-\frac{A_c}{A_0}\frac{A_0}{A_2}\right)^2 \frac{A_2^2}{A_c^2} \frac{V_2^2}{2g}$.

$\therefore K = \left(1-.625\times\frac{1}{9}\right)^2 \frac{A_2^2}{A_0^2}\frac{A_0^2}{A_c^2} = .866\times 9^2 \times \frac{1}{.625^2} = \underline{180}.$

7.99 $V = \frac{Q}{A} = \frac{.004}{\pi\times .02^2} = 3.18$ m / s. Neglect pipe losses (i.e., $f\frac{L}{D}\frac{V^2}{2g}$ is negligible).

Energy: $0 = \frac{3.18^2 - 0^2}{2\times 9.81} + 0 - 2 + (K+.03)\frac{3.18^2}{2\times 9.81}$. $\therefore \underline{K = 2.85}$.

7.99E $V = \frac{Q}{A} = \frac{.12}{\pi(1/12)^2} = 5.5$ fps. Neglect pipe friction (i.e., $f\frac{L}{D}\frac{V^2}{2g} \cong 0$).

Energy: $0 = \frac{5.5^2 - 0^2}{2\times 32.2} + 0 - 6 + (K+.03)\frac{5.5^2}{2\times 32.2}$. $\therefore \underline{K = 11.7}$.

7.100 a) manometer: $.04\times 9810 + p_1 = .04\times(9810\times 13.6) + p_2$. $V = \frac{.006}{\pi\times .02^2} = 4.77$ m / s.

energy: $0 = \frac{V_2^2 - V_1^2}{2g} + \frac{p_2 - p_1}{\gamma} + K\frac{V^2}{2g}$. (with $\frac{V_2^2 - V_1^2}{2g}$ struck out, marked 0)

$\therefore K = \frac{.04\times 9810\times 12.6\times 2\times 9.81}{9810\times 4.77^2} = \underline{0.435}.$

b) $K = \frac{p_1 - p_2}{\gamma}\frac{2g}{V^2} = \frac{.08\times 9810\times 12.6\times 2\times 9.81}{9810\times 4.77^2} = \underline{0.869}.$

7.101 Assume completely turbulent regime. $\frac{e}{D}=\frac{.046}{40}=.0012.\quad \therefore f=.0205,\ \ \mathrm{Re}>10^6.$

Energy: $$0=\frac{V^2}{2g}-40+\left[f\frac{L}{D}+2K_{\text{elbow}}+K_{\text{entrance}}\right]\frac{V^2}{2g}$$

$$40=\left[1+.0205\frac{110}{.04}+2\times 1.0+.5\right]\frac{V^2}{2\times 9.81}.\quad \text{(Assume a screwed elbow.)}$$

$$\therefore V=3.62\ \text{m/s}.\quad \mathrm{Re}=\frac{3.62\times .04}{10^{-6}}=1.4\times 10^5.\quad \therefore \text{Try } f=.022.$$

$$\therefore V=3.50\ \text{m/s}.\quad \therefore Q=AV=\pi\times .02^2\times 3.5=\underline{0.0044\ \text{m}^3/\text{s}}.$$

The EGL and HGL have sudden drops at the elbows, and a gradual slope over the pipe length.

7.102 Energy: $$0=\frac{V^2}{2g}-5+\left(f\frac{400}{.1}+K_{\text{entrance}}+K_{\text{valve}}\right)\frac{V^2}{2g}.\quad \frac{e}{D}=.0026.$$

a) Assume $f=.0245$ $(\mathrm{Re}>4\times 10^5)$: $5=\left(1+.5+15+.0245\frac{400}{.1}\right)\frac{V^2}{2\times 9.81}.$

$$\therefore V=0.926\ \text{m/s}.\quad \mathrm{Re}=\frac{.926\times .1}{10^{-6}}=9.3\times 10^4.\quad \therefore f=.027.\quad \therefore V=0.888\ \text{m/s}.$$

$$\therefore Q=AV=\pi\times .05^2\times .888=\underline{0.0070\ \text{m}^3/\text{s}}.$$

b) Assume $f=.025$. $10=(16.5+.025\times 4000)\frac{V^2}{2\times 9.81}.\quad \therefore V=1.3\ \text{m/s}.$

$$\mathrm{Re}=\frac{1.3\times .1}{10^{-6}}=1.3\times 10^5.\quad \therefore f=.026.\quad \therefore V=1.28\ \text{m/s}.\quad \therefore Q=\pi\times .05^2\times 1.28=\underline{0.01\ \text{m}^3/\text{s}}.$$

c) Assume $f=.025$. $20=(16.5+.025\times 4000)\frac{V^2}{2\times 9.81}.\quad \therefore V=1.84\ \text{m/s}.$

$$\mathrm{Re}=\frac{1.84\times .1}{10^{-6}}=1.84\times 10^5.\quad \therefore f=.026.\quad \therefore V=1.82\ \text{m/s}.\quad \therefore Q=\pi\times .05^2\times 1.82=\underline{0.014\ \text{m}^3/\text{s}}.$$

7.102E Energy: $$0=\frac{V^2}{2g}-H+\left(f\frac{L}{D}+K_{\text{entrance}}+K_{\text{valve}}\right)\frac{V^2}{2g}.\quad \frac{e}{D}=\frac{.00085}{4/12}=.0026.$$

a) Assume $f=.027$ (slightly larger than the value of the completely turbulent flow):

$$15=\left(1+.5+15+.027\frac{1200}{4/12}\right)\frac{V^2}{2\times 32.2}.\quad \therefore V=2.91\ \text{ft/sec}.$$

$$\mathrm{Re}=\frac{2.91\times 4/12}{1.06\times 10^{-5}}=9.2\times 10^4.\quad \therefore f \text{ is OK}.\quad \therefore Q=\pi\times\left(\frac{2}{12}\right)^2\times 2.91=\underline{0.25\ \text{cfs}}.$$

b) Assume $f = .026$. $30 = \left(16.5 + .026\dfrac{1200}{4/12}\right)\dfrac{V^2}{64.4}$. $\therefore V = 4.19$ fps.

$\text{Re} = \dfrac{4.19 \times 4/12}{1.06 \times 10^{-5}} = 1.3 \times 10^5$. $\therefore f$ is OK. $\therefore Q = \pi \times \left(\dfrac{2}{12}\right)^2 \times 4.19 = \underline{0.37 \text{ cfs}}$.

c) Assume $f = .026$. $60 = \left(16.5 + .026\dfrac{1200}{4/12}\right)\dfrac{V^2}{2g}$. $\therefore V = 5.92$ fps.

$\text{Re} = 1.8 \times 10^5$. $\therefore f$ is OK. $\therefore Q = \pi \times \left(\dfrac{2}{12}\right)^2 \times 5.92 = \underline{0.52 \text{ cfs}}$.

7.103 Assume screwed elbows: $0 = \dfrac{V^2}{2g} - 2 + \left(K_{\text{entrance}} + 2K_{\text{elbows}} + f\dfrac{L}{D}\right)\dfrac{V^2}{2g}$.

a) Assume $f = .022$: $2 = \left(.8 + 1 + 2 \times 1 + .022\dfrac{26}{.04}\right)\dfrac{V^2}{2 \times 9.81}$. $\therefore V = 1.47$ m/s.

$\therefore \text{Re} = \dfrac{1.47 \times .04}{1.14 \times 10^{-6}} = 5.2 \times 10^4$. $\therefore f = .021$. $\therefore V = 1.50$, $Q = \pi \times .02^2 \times 1.50 = \underline{.0019 \text{ m}^3/\text{s}}$.

b) Assume $f = .018$: $2 = \left(.8 + 1 + 2 \times .8 + .018\dfrac{26}{.08}\right)\dfrac{V^2}{2 \times 9.81}$. $\therefore V = 2.06$ m/s.

$\therefore \text{Re} = \dfrac{2.06 \times .08}{1.14 \times 10^{-6}} = 1.5 \times 10^5$. $f = .0164$. $\therefore V = 2.12$, $Q = \pi \times .04^2 \times 2.12 = \underline{0.011 \text{ m}^3/\text{s}}$.

c) Assume $f = .013$: $2 = \left(.8 + 1 + 2 \times .6 + .01\dfrac{26}{.12}\right)\dfrac{V^2}{2 \times 9.81}$. $\therefore V = 2.76$ m/s.

$\therefore \text{Re} = \dfrac{2.76 \times .12}{1.14 \times 10^{-6}} = 3 \times 10^5$. $\therefore f = .014$. $\therefore V = 2.55$. $Q = \pi \times .06^2 \times 2.55 = \underline{0.029 \text{ m}^3/\text{s}}$.

7.104 a) $\dfrac{e}{D_1} = \dfrac{.26}{20} = .013$. $\therefore f_1 = .041$.

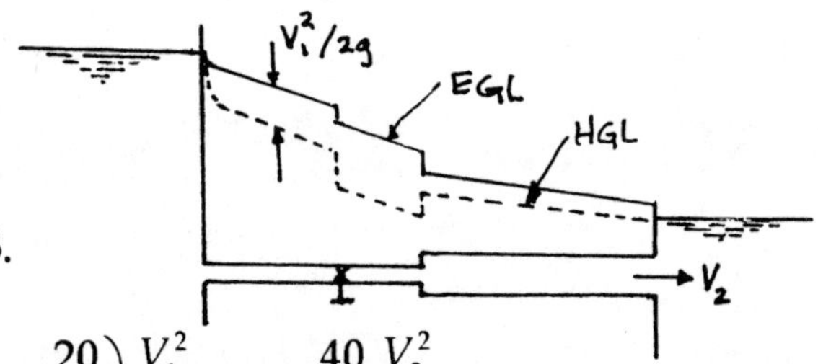

$\dfrac{e}{D_2} = \dfrac{.26}{40} = .0065$. $\therefore f_2 = .033$.

$0 = \dfrac{V_2^2}{2g} - 20 + \left((.5 + .56 + 5 + .041\dfrac{20}{.02}\right)\dfrac{V_1^2}{2g} + .033\dfrac{40}{.04}\dfrac{V_2^2}{2g}$.

But, $V_1 = 4V_2$. $\therefore V_2 = .706$, $V_1 = 2.82$ m/s.

$\therefore Q = A_1V_1 = \pi \times .01^2 \times 2.82 = \underline{0.00089 \text{ m}^3/\text{s}}$.

b) Lowest pressure occurs in the small pipe. This occurs at the enlargement.

$0 = \dfrac{V_1^2}{2g} + \dfrac{p_1}{9810} - H + \left((.5 + 5 + .041\dfrac{20}{.02}\right)\dfrac{V_1^2}{2g}$. With $H = 20$, $V_1 = 2.82$, then

$p_1 = \underline{7300 \text{ Pa}}$.

7.105 Assume constant water level, neglect the velocity head at the pipe exit and let $p_{exit} = 0$. Then the energy equation says $h_L = 0.8$ m.

$$0.8 = f\frac{10}{.025}\frac{V^2}{2\times 9.8} + 2\times 1.5\frac{V^2}{2\times 9.8} \quad \text{or} \quad 0.8 = 400fV^2 + .153V^2$$

Try $f = 0.02$. Then $V = 0.319$ m/s, $\text{Re} = \dfrac{.313\times .025}{.73\times 10^{-6}} = 1.1\times 10^4$. $\therefore f = 0.029$.

Try $f = 0.029$. Then $V = 0.262$ m/s.

$$\therefore Q = AV = \pi \times 0.0125^2 \times 0.262 = 1.28\times 10^{-4} \text{ m}^3/\text{s}.$$

Finally, $\Delta t = \dfrac{V\!\!\!\!-}{Q} = \dfrac{10}{.128} = 78.1$ sec or $\underline{1.3 \text{ min}}$.

7.106 The siphon operates unless $p_{min} = p_{vapor} = 1.176$ kPa abs or -98.8 kPa. Since pressure decreases along the pipe, we will let the pressure after the second elbow be -98.8 kPa. From there, the pressure increases to zero at the exit. If there are no losses, $\gamma H = -p_{elbow}$ or $H = 98\,800/9800 = 10.1$ m.

$$\text{Energy:} \qquad 0 = \frac{V^2}{2g} - 6 + \left(f\frac{8+2H}{.03} + 0.8 + 2\times 2.0\right)\frac{V^2}{2g}.$$

Let $H = 6$ m and assume $f = 0.02$. Then energy provides $V = 2.48$ m/s.

Check: $\text{Re} = \dfrac{2.48\times .03}{1.31\times 10^{-6}} = 5.7\times 10^4$. From the Moody diagram, $f = 0.02$. $\therefore$ Good.

Now, check the pressure after the elbow:

$$0 = \frac{0 - p_{elbow}}{9800} - (6+6) + .02\frac{12}{.03}\frac{2.48^2}{2\times 9.8}. \qquad \therefore p_{elbow} = -93\,000 \text{ Pa}.$$

Try $\underline{H = 6.5 \text{ m}}$ and the pressure after the second elbow is close to p_v.

7.107 Energy: $\qquad 0 = \dfrac{V_3^2}{2g} - \dfrac{p_0}{\gamma} - 1.2 + 0.8\dfrac{V_1^2}{2g} + f_1\dfrac{1.2}{.008}\dfrac{V_1^2}{2g} + f_2\dfrac{1.2}{.005}\dfrac{V_2^2}{2g}$

Continuity: $V_3 = \dfrac{25}{4}V_2$, $V_2 = \dfrac{64}{25}V_1$.

Assume $\qquad f_1 = f_2 = 0.02$. Then energy says $V_1 = 2.09$ m/s.

Check: $\qquad \text{Re}_1 = \dfrac{2.09\times .008}{10^{-6}} = 16\,700$. $\quad \therefore f = .026$. Then $V_1 = 2.88$.

Finally, $\qquad V_3 = 16V_1 = \underline{46.1 \text{ m/s}}$

7.108 Use an average value of e: $\dfrac{e}{D} = \dfrac{1.65}{800} = .0021$. $\quad \therefore f = .024$.

$$0 = \Delta z + \left(K_{entrance} + K_{valve} + K_{exit} + f\frac{L}{D}\right)\frac{V^2}{2g}.$$

$$-\Delta z=\left(0.8+1+1+.024\frac{2000}{.8}\right)\frac{V^2}{2\times 9.81}$$

a) $\Delta z=-80.\quad \therefore V=5.0,\ \ Q=\pi\times.4^2\times 5=\underline{2.5\text{ m}^3/\text{s.}}$

b) $\Delta z=-150.\quad \therefore V=6.85,\ \ Q=\pi\times.4^2\times 6.85=\underline{3.44\text{ m}^3/\text{s.}}$

c) $\Delta z=-200.\quad \therefore V=7.9,\ \ Q=\pi\times.4^2\times 7.9=\underline{3.97\text{ m}^3/\text{s.}}$

Check: $\text{Re}=\frac{5\times.8}{10^{-6}}=4\times 10^6$. This is sufficiently large.

7.109 Energy: $0=-3+f\frac{300}{.0094}\frac{V^2}{2\times 9.8}$.

Assume turbulent: $f=.04$. Then $V=0.215\text{ m/s}$.

Check: $\text{Re}=\frac{0.215\times.0094}{10^{-6}}=2020.\quad \therefore$ laminar.

Assume laminar flow with $f=64/\text{Re}$. Then

$$3=\frac{64\times 10^{-6}}{V\times.0094}\frac{300}{.0094}\frac{V^2}{2\times 9.8}.\qquad \therefore V=0.271\text{ m/s.}$$

Check: $\text{Re}=\frac{.271\times.0094}{10^{-6}}=2540.\quad \therefore$ turbulent.

The flow is neither laminar nor turbulent but may oscillate between the two. The wall friction is too low in the laminar state so it speeds up and becomes turbulent. The wall friction is too high in the turbulent state so it slows down and becomes laminar, etc., etc.

7.110 a) $V=\frac{Q}{A}=\frac{.02}{\pi\times.02^2}=15.92\text{ m/s.}\quad \text{Re}=\frac{15.92\times.04}{10^{-6}}=6.4\times 10^5.\quad \frac{e}{D}=\frac{.26}{40}=.0065$

$\therefore f=.033.$

$$\therefore H_P=\frac{V_2^2-V_1^2}{2g}+\frac{p_2-p_1}{\gamma}+z_2-z_1+\left(\Sigma K_i+f\frac{L}{D}\right)\frac{V^2}{2g}$$

(with V_1, p_1, z_1 struck out and marked 0)

$$=\frac{15.92^2}{2\times 9.81}+\frac{200\,000}{9810}+10+\left(.5+2\times 1.0+.033\frac{300}{.04}\right)\frac{15.92^2}{2\times 9.81}=3270\text{ m.}$$

$\therefore \dot{W}_P=\gamma QH_P/\eta_P=9810\times.02\times 3270/.8=\underline{802\,000\text{ W.}}$

b) $V=\frac{.02}{\pi\times.04^2}=3.98\text{ m/s.}\quad \text{Re}=\frac{3.98\times.08}{10^{-6}}=3.2\times 10^5.\quad \frac{e}{D}=\frac{.26}{80}=.0032.$

$\therefore f=.027.$

$$\therefore H_P=\frac{3.98^2}{2\times 9.81}+\frac{200\,000}{9810}+10+\left(.5+2\times.8+.027\frac{300}{.08}\right)\frac{3.98^2}{2\times 9.81}=115\text{ m.}$$

$\therefore \dot{W}_P=\gamma QH_P/\eta_P=9810\times.02\times 115/.8=\underline{28\,200\text{ W.}}$

c) $V=\frac{.02}{\pi\times.06^2}=1.77\text{ m/s.}\quad \text{Re}=\frac{1.17\times.12}{10^{-6}}=1.4\times 10^5.\quad \frac{e}{D}=\frac{.26}{120}=.0022.$

$$\therefore f = .026.$$

$$H_P = \frac{1.77^2}{2\times 9.81} + \frac{200\ 000}{9810} + 10 + \left(.5 + 2\times.6 + .026\frac{300}{.12}\right)\frac{1.77^2}{2\times 9.81} = 41.2\text{ m.}$$

$$\therefore \dot{W}_P = \gamma Q H_P / \eta_P = 9810\times.02\times 41.2/.8 = \underline{10\ 100\text{ W}}.$$

7.110E a) $V = \dfrac{.6}{\pi\times(.75/12)^2} = 48.9.$ $\text{Re} = \dfrac{48.9\times 1.5/12}{1.22\times 10^{-5}} = 5\times 10^5.$ $\dfrac{e}{D} = \dfrac{.00085}{1.5/12} = .0068$

$$\therefore f = .033.$$

$$H_P = \frac{48.9^2}{32.2\times 2} + \frac{30\times 144}{62.4} + 30 + \left(.5 + 2\times 1.2 + .033\frac{900}{1.5/12}\right)\frac{48.9^2}{2\times 32.2} = 9066'.$$

$$\therefore \dot{W}_P = \gamma Q H_P / \eta_P = 62.4\times.6\times 9066/.8 = 424{,}000\text{ ft-lb/sec or }\underline{771\text{ Hp}}.$$

b) $V = \dfrac{.6}{\pi\times(1.5/12)^2} = 12.2\text{ fps.}$ $\text{Re} = \dfrac{12.2\times 3/12}{1.22\times 10^{-5}} = 2.5\times 10^5.$ $\dfrac{e}{D} = .0034.$

$$\therefore f = .027.$$

$$\therefore H_P = \frac{12.2^2}{64.4} + \frac{30\times 144}{62.4} + 30 + \left(.5 + 2\times.8 + .027\frac{900}{3/12}\right)\frac{12.2^2}{64.4} = 331'.$$

$$\therefore \dot{W}_P = \gamma Q H_P / \eta_P = 62.4\times.6\times 331/.8 = 15{,}500\text{ ft-lb/sec or }\underline{28.2\text{ Hp}}.$$

c) $V = \dfrac{.6}{\pi\times(2.25/12)^2} = 5.43\text{ fps.}$ $\text{Re} = \dfrac{5.43\times 4.5/12}{1.22\times 10^{-5}} = 1.67\times 10^5.$ $\dfrac{e}{D} = .0023.$

$$\therefore f = .026.$$

$$\therefore H_P = \frac{5.43^2}{64.4} + \frac{30\times 144}{62.4} + 30 + \left(.5 + 2\times.6 + .026\frac{900}{4.5/12}\right)\frac{5.43^2}{64.4} = 129'.$$

$$\therefore \dot{W}_P = \gamma Q H_P / \eta_P = 62.4\times.6\times 129/.8 = 6040\text{ ft-lb/sec or }\underline{11.0\text{ Hp}}.$$

7.111 $V = \dfrac{Q}{A} = \dfrac{.01}{\pi\times.02^2} = 7.96\text{ m/s.}$ $\text{Re} = \dfrac{7.96\times.04}{1.14\times 10^{-6}} = 2.8\times 10^5.$ $\dfrac{e}{D} = \dfrac{.0015}{40} = .000038$

$$\therefore f = .0145.$$

$$H_P = 80 - 10 + \left(.5 + 1.0 + .0145\frac{800}{.04}\right)\frac{7.96^2}{2\times 9.81} = 1010\text{ m.}$$

$$\therefore \dot{W}_P = \gamma Q H_P / \eta_P = 9810\times.01\times 1010/.85 = \underline{117\ 000\text{ W}}.$$

$$0 = \frac{7.96^2}{2\times 9.81} + \frac{1670 - 100\ 000}{9810} + 0 - 10 + \left(.5 + .0145\frac{L}{.04}\right)\frac{7.96^2}{2\times 9.81}. \quad \therefore \underline{L = 13.0\text{ m}}.$$

7.112 $V = \frac{Q}{A} = \frac{2}{\pi \times .45^2} = 3.14 \text{ m/s}$. $\text{Re} = \frac{3.14 \times .9}{1.14 \times 10^{-6}} = 2.5 \times 10^6$. $\frac{e}{D} = \frac{.26}{900} = .000029$.
$\therefore f = .015$.

a) $-H_T = -20 + \left(.5 + 1 + .015\frac{400}{.9}\right)\frac{3.14^2}{2 \times 9.81} = -15.9 \text{ m}$.

$\therefore \dot{W}_T = \gamma H_T Q \eta_T = 9810 \times 2 \times 15.9 \times .85 = \underline{265\ 000 \text{ W}}$.

b) $-H_T = -60 + \left(.5 + 1 + .015\frac{400}{.9}\right)\frac{3.14^2}{2 \times 9.81} = -55.9 \text{ m}$.

$\therefore \dot{W}_T = \gamma Q H_T \eta_T = 9810 \times 2 \times 55.9 \times .85 = \underline{932\ 000 \text{ W}}$.

c) $-H_T = -100 + \left(.5 + 1 + .015\frac{400}{.9}\right)\frac{3.14^2}{2 \times 9.81} = -95.9 \text{ m}$.

$\therefore \dot{W}_T = \gamma Q H_T \eta_T = 9810 \times 2 \times 95.9 \times .85 = 1\ 600\ 000 \text{ W or } \underline{1.6 \text{ MW}}$.

7.113 Energy across nozzle (neglect losses):

$$\frac{V_1^2}{2g} + \frac{700\ 000}{9810} = \frac{V_2^2}{2g} = \frac{16V_1^2}{2g}. \quad \therefore V_1 = 9.66 \text{ m/s}.$$

$\therefore \text{Re} = \frac{9.66 \times .04}{1.14 \times 10^{-6}} = 3.4 \times 10^5$. $\frac{e}{D} = \frac{.046}{40} = .00115$. $\therefore f = .021$.

$$H_P = \frac{(4 \times 9.66)^2}{2 \times 9.81} - 20 + \left(.5 + .021\frac{400}{.04}\right)\frac{9.66^2}{2 \times 9.81} = 1060 \text{ m}.$$

$\therefore \dot{W}_P = \gamma Q H_P / \eta_P = 9810 \times (\pi \times .02^2 \times 9.66) \times 1060/.75 = \underline{168\ 000 \text{ W}}$.

$$0 = \frac{9.66^2}{2 \times 9.81} + \frac{1670 - 100\ 000}{9810} - 20 + \left(.5 + .021\frac{L}{.04}\right)\frac{9.66^2}{2 \times 9.81}. \quad \therefore \underline{L = 9.2 \text{ m}}.$$

$\therefore$ The pump must be located near the entrance.

7.113E Energy across nozzle (neglect losses): $\frac{V_1^2}{2g} + \frac{100 \times 144}{62.4} = \frac{V_2^2}{2g} = \frac{(4V_1)^2}{2g}$.

$\therefore V_1 = 31.5 \text{ fps}$. $\text{Re} = \frac{31.5 \times 2/12}{1.06 \times 10^{-5}} = 4.95 \times 10^5$. $\frac{e}{D} = \frac{.00015}{2/12} = .0009$. $\therefore f = .02$.

$$H_P = \frac{(4 \times 31.5)^2}{64.4} - 60 + \left(.5 + .02\frac{1200}{2/12}\right)\frac{31.5^2}{64.4} = 2410'.$$

$\therefore \dot{W}_P = \gamma Q H_P / \eta_P = 62.4 \times \left(\pi \times \left(\frac{1}{12}\right)^2 \times 31.5\right) \times 2410/.75 = 138{,}000 \text{ ft-lb/sec or } \underline{251 \text{ Hp}}$.

$$0 = \frac{31.5^2}{64.4} + \frac{(.34 - 14.7)144}{62.4} - 60 + \left(.5 + .02\frac{L}{2/12}\right)\frac{31.5^2}{64.4}. \quad \therefore \underline{L = 37.9'}.$$

7.114 $\frac{e}{D}=\frac{.26}{200}=.0013. \quad \therefore f=.021. \quad Q=AV=\pi\times.1^2V=.0314\text{ V}$

$$H_P=20+\left(.5+1.0+.021\frac{300}{.2}\right)\frac{V^2}{2g}=20+1.68V^2=20+1700Q^2.$$

Try $Q=.2\text{ m}^3/\text{s}$: $(H_P)_{\text{energy}}=88.$ $(H_P)_{\text{char.}}=74.$

Try $Q=.18\text{ m}^3/\text{s}$: $(H_P)_{\text{energy}}=75.$ $(H_P)_{\text{char.}}=77.$ $\therefore \underline{Q=.182\text{ m}^3/\text{s}}.$

$\therefore H_P=20+1700\times.182^2=76\text{ m}.$

$\therefore \dot{W}_P=9810\times.182\times76/.7=\underline{194\ 000\text{ W}}.$

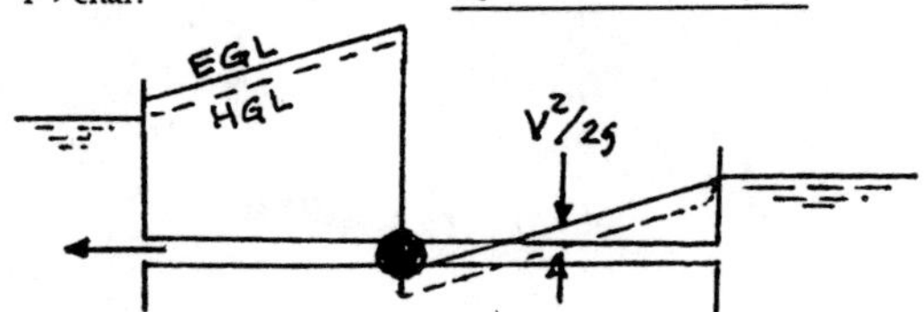

7.115 $\frac{e}{D}=.0013. \quad \therefore f=.021. \quad Q=.0314\ V.$

$$H_P=-20+\left(.5+1.0+.021\frac{300}{.2}\right)\frac{V^2}{2g}=-20+1700Q^2.$$

Try $\underline{Q=.23\text{ m}^3/\text{s}}$: $(H_P)_E=70.$ $(H_P)_C=70.$

$\therefore \dot{W}_P=9810\times.23\times70/.83=\underline{190\ 000\text{ W}}.$ (Re$=1.5\times10^6$. $\therefore$OK.)

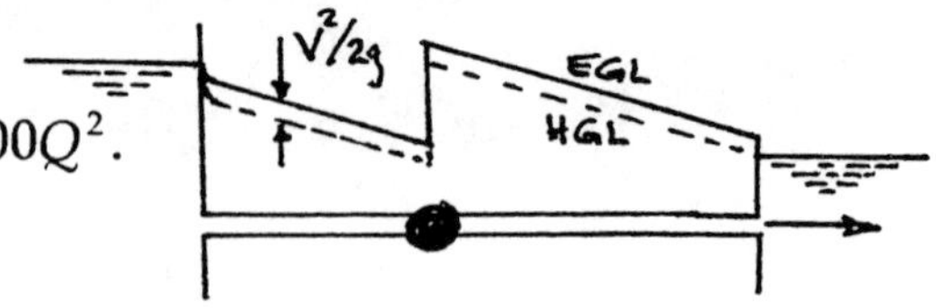

7.116 Energy: $$-H_T=-20+\left(.5+1.0+.021\frac{300}{.2}\right)\frac{V^2}{2g}$$

$$800Q=-20+1706Q^2. \quad \therefore Q=0.493\text{ m}^3/\text{s}.$$

$$\dot{W}_T=H_T\gamma Q\eta=800\times.493\times9800\times.88=3.4\times10^6\text{ W or }\underline{3.4\text{ MW}}.$$

7.117 $\frac{e}{D}=\frac{.046}{160}=.00029. \quad \therefore f=.015. \quad Q=\pi\times.08^2V=.0201\ V.$

$$H_P=25+\left(.5+2\times.4+1+.015\frac{50}{.16}\right)\frac{Q^2}{2g\times.0201^2}=25+882Q^2.$$

Try $Q=.23\text{ m}^3/\text{s}$. $(H_P)_E=72.$ $(H_P)_C=70.$

a) $\therefore \dot{W}_P=9810\times.23\times72/.83=\underline{195\ 000\text{ W}}.$

b) $0=\frac{11.44^2}{2\times9.81}+\frac{p_2}{9810}-8+\left(.5+.015\frac{10}{.16}\right)\frac{11.44^2}{2\times9.81}. \quad \therefore \underline{p_{\text{in}}=-81\ 000\text{ Pa}}.$

c) $72=\frac{11.44^2}{2\times9.81}+\frac{p_3}{9810}-8+\left(.5+.015\frac{10}{.16}\right)\frac{11.44^2}{2\times9.81}. \quad \therefore \underline{p_{\text{out}}=625\ 000\text{ Pa}}.$

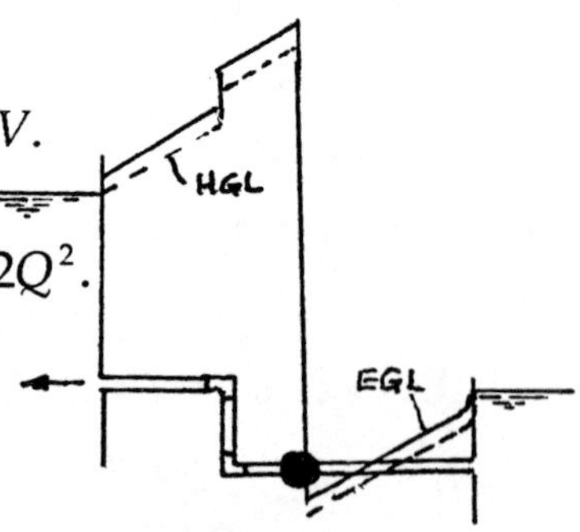

7.118 $\frac{e}{D}=.00029. \quad \therefore f=.015. \quad Q=.0201\ V.$

$H_P=-25+882Q^2.$

Try $Q=.31\text{ m}^3/\text{s}$. $(H_P)_E=60.$ $(H_P)_C=60.$

a) $\therefore \dot{W}_P=9810\times.31\times60/.82=223\ 000\text{ W}.$

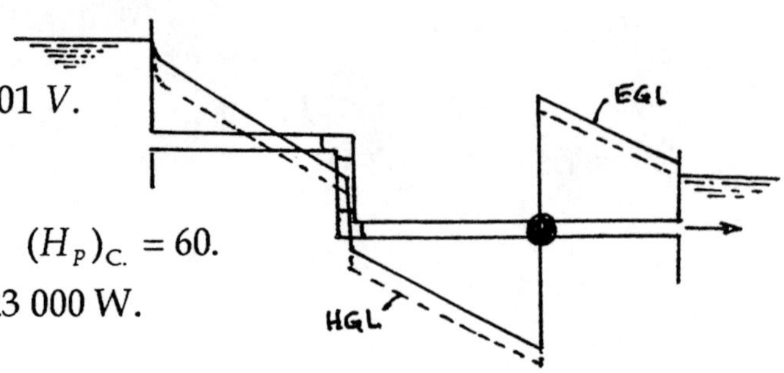

b) $0 = \frac{15.42^2}{2 \times 9.81} + \frac{p_2}{9810} - 33 + \left(.5 + .015\frac{40}{.16}\right)\frac{15.42^2}{2 \times 9.81}$.

$\therefore p_{in} = -300\ 000$ Pa. This is below absolute zero and the pump would not function according to the curves of Example 7.16. Part (a) is also not correct. The pump must be repositioned for reverse flow.

c) Since p_{in} is too low, the problem is not workable as posed.

7.119 $\frac{e}{D} = \frac{1.65}{1200} = .0014$. $\therefore f = .021$. $Q = \pi \times .6^2 V = 1.13\ V$.

$-H_T = -60 + \left(.8 + 1 + .021\frac{1000}{1.2}\right)\frac{Q^2}{1.13^2 \times 2 \times 9.81}$. $\therefore H_T = 60 - .77Q^2$.

$\dot{W}_T = \gamma Q H_T \eta_T = 9810 \times (60Q - .77Q^3) \times .9 = 530\ 000\ Q\ - 6800\ Q^3$.

But, the characteristic curve is $\dot{W}_T = 500\ 000\ Q$ (in Watts).

Hence, 500 000 Q = 530 000 Q - 6800 Q^3. $\therefore Q = 2.1\ m^3/s$.

$\therefore \dot{W}_T = \underline{1.05\ \text{MW}}$.

7.120 $\tau_w A_w = W \sin\theta$. $\sin\theta \cong .001$.

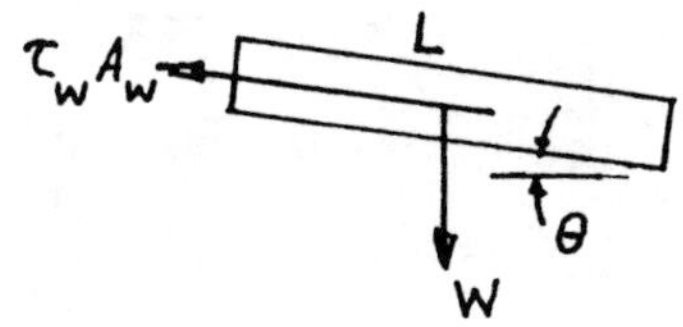

$\tau_w \times (3 + 2 \times 2)L = \gamma \times 3 \times 2 \times L \times .001$.

$\therefore \tau_w = 9810 \times 6 \times .001/7 = \underline{8.41\ \text{Pa}}$.

7.120E $\tau_w A_w = W \sin\theta$ (see figure above). $\sin\theta = .001$.

$\tau_w \times (10 + 2 \times 6)L = 62.4 \times 10 \times 6 \times L \times .001$. $\therefore \tau_w = \underline{0.170\ \text{psf}}$.

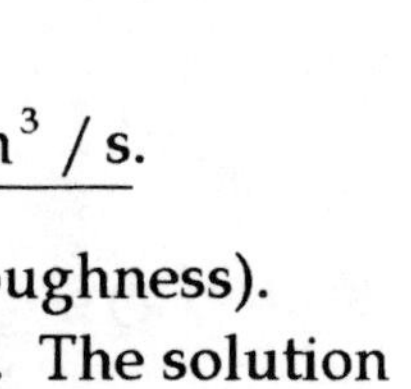

7.121 $\tau_w A_w = W \times .0016$(see fig. of Prob. 7.120). $\cos\alpha = \frac{10}{20}$. $\therefore \alpha = 60°$.

$\tau_w 2\pi \times .2 \times \frac{120}{360} L = 9810\left(\pi \times .2^2 \frac{120}{360} - .1 \times .1732\right) \times .0016L$. $\therefore \tau_w = \underline{0.921\ \text{Pa}}$.

7.122 a) $Q = \frac{1.0}{n} A R^{2/3} S^{1/2} = \frac{1.0}{.012} \times (2 \times .6)\left(\frac{2 \times .6}{2 + 2 \times .6}\right)^{2/3} \times .001^{1/2} = \underline{1.64\ m^3/s}$.

b) Planed wood and finished concrete have the same "n" (i.e., roughness).

$\therefore$ Use a value for "e" which is relatively small, say e =0.6 mm. The solution is not too sensitive to this choice. Then, with R = .375 m,

$\frac{e}{D} = \frac{e}{4R} = \frac{.6}{4 \times 375} = .0004$ $\therefore f = .016$

$h_L = f\frac{L}{D}\frac{V^2}{2g} = LS$. $\therefore V^2 = 2 \times 9.81 \times .001 \times (4 \times .375)/.016$. $\therefore V = 1.36\ m/s$.

$$\therefore Q = 1.36 \times 2 \times .6 = \underline{1.63 \text{ m}^3/\text{s.}} \quad \left(\text{Check: } \text{Re} = \frac{1.36 \times (4 \times .375)}{10^{-6}} = 2.0 \times 10^6. \quad \therefore OK\right)$$

7.123 a) $R = \dfrac{\pi \times 1^2}{\pi \times 2} = \dfrac{1}{2}$ m. $\therefore Q = \dfrac{1.0}{.012} \times \pi \times 1^2 \times .5^{2/3} \times .0012^{1/2} = \underline{5.71 \text{ m}^3/\text{s.}}$

b) $\cos\alpha = \dfrac{.9}{1.0}$. $\therefore \alpha = 25.84°$. $A_1 = \dfrac{51.68}{360}\pi \times 1^2 - .4359 \times .9 = .0587 \text{ m}^2$.

$\therefore A = \pi - .0587 = 3.083$. $R = \dfrac{3.083}{\pi \times 2}\dfrac{360}{308.3} = .573$ m.

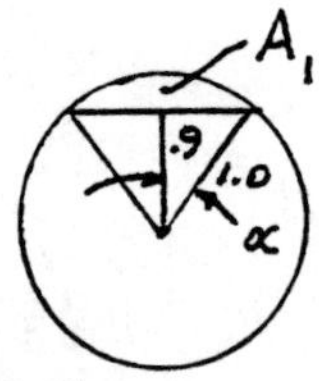

$\therefore Q = \dfrac{1.0}{.012} \times 3.083 \times .573^{2/3} \times .0012^{1/2} = \underline{6.14 \text{ m}^3/\text{s.}}$

c) $R = \dfrac{\pi/2}{\pi} = \dfrac{1}{2}$ m. $\therefore Q = \dfrac{1.0}{.012} \times \dfrac{\pi}{2} \times .5^{2/3} \times .0012^{1/2} = \underline{2.86 \text{ m}^3/\text{s.}}$

d) $\cos\alpha = \dfrac{.5}{1.0}$. $\therefore \alpha = 60°$. $A = \pi \times \dfrac{120}{360} - .5 \times .866 = .6142 \text{ m}^2$.

$$\therefore R = \frac{.6142}{2\pi \times 120/360} = .293 \text{ m.}$$

$\therefore Q = \dfrac{1.0}{.012} \times .6142 \times .293^{2/3} \times .0012^{1/2} = \underline{0.783 \text{ m}^3/\text{s.}}$

e) $\cos\alpha = \dfrac{.8}{1.0}$. $\therefore \alpha = 36.87°$. $A = \pi \times \dfrac{73.74}{360} - .8 \times .6 = .1635 \text{ m}^2$.

$$\therefore R = \frac{.1635}{2\pi \times 73.74/360} = .127 \text{ m.}$$

$\therefore Q = \dfrac{1.0}{.012} \times .1635 \times .127^{2/3} \times .0012^{1/2} = \underline{0.119 \text{ m}^3/\text{s.}}$

7.123E a) $R = \dfrac{\pi \times 3^2}{6\pi} = 1.5'$. $\therefore Q = \dfrac{1.49}{.012} \times \pi \times 3^2 \times 1.5^{2/3} \times .0012^{1/2} = \underline{159 \text{ cfs.}}$

b) $\cos\alpha = \dfrac{2.7}{3}$. $\alpha = 25.84°$. $\therefore A_1 = \dfrac{51.68}{360}\pi \times 3^2 - 1.308 \times 2.7 = 0.5285 \text{ ft}^2$.

$\therefore A = 9\pi - .5285 = 27.75 \text{ ft}^2$. $R = \dfrac{27.75 \times 360}{6\pi \times 308.3} = 1.719'$.

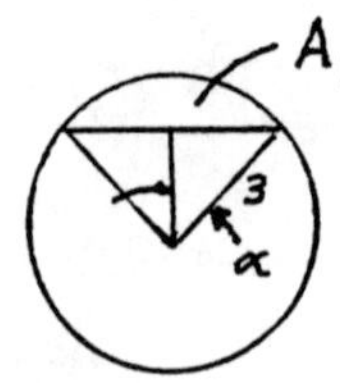

$\therefore Q = \dfrac{1.49}{.012} \times 27.75 \times 1.719^{2/3} \times .0012^{1/2} = \underline{171 \text{ cfs.}}$

c) $R = \dfrac{9\pi/2}{3\pi} = 1.5'$. $\therefore Q = \dfrac{1.49}{.012} \times \dfrac{9\pi}{2} \times 1.5^{2/3} \times .0012^{1/2} = \underline{79.7 \text{ cfs.}}$

d) $\alpha = 60°$. $A = \pi \times 9 \times \dfrac{120}{360} - 1.5 \times 2.6 = 5.52$. $R = \dfrac{5.52}{2\pi \times 3 \times 120/360} = .879'$.

$Q = \dfrac{1.49}{.012} \times 5.52 \times .879^{2/3} \times .0012^{1/2} = \underline{21.8 \text{ cfs.}}$

e) $\cos\alpha = \frac{2.5}{3}$. $\therefore \alpha = 33.56°$. $A = 9\pi \times \frac{67.11}{360} - 2.5 \times 1.658 = 1.125 \text{ ft}^2$.

$$R = \frac{1.125}{6\pi \times 67.11/360} = 0.32'. \quad \therefore Q = \frac{1.49}{.012} \times 1.125 \times .32^{2/3} \times .0012^{1/2} = \underline{2.26 \text{ cfs}}.$$

7.124 a) $R = \frac{1.2 \times .5 + .5 \times .5}{1.2 + 2 \times .5/.707} = .325 \text{ m} \quad \therefore Q = \frac{1.0}{.016} \times .85 \times .325^{2/3} \times .001^{1/2} = \underline{0.794 \text{ m}^3/\text{s}}.$

$$V = Q/A = \frac{.794}{.85} = \underline{.934 \text{ m/s}}$$

b) Use an "e" for rough concrete, e = 3 mm. $\therefore \frac{e}{D} = \frac{3}{4 \times 325} = .0023$. $\therefore f = .025$.

$$h_L = LS = f\frac{L}{4R}\frac{V^2}{2g}. \quad \therefore V^2 = \frac{4 \times .325 \times 2 \times 9.81 \times .001}{.025}. \quad \therefore V = \underline{1.01 \text{ m/s}}.$$

$$\therefore Q = 1.01 \times .85 = \underline{0.86 \text{ m}^3/\text{s}}.$$

7.125 a) $R = \frac{2y}{2+2y} = \frac{y}{1+y}$. $5 = \frac{1.0}{.016} 2y\left(\frac{y}{1+y}\right)^{2/3} \times .001^{1/2}$. $\therefore y\left(\frac{y}{1+y}\right)^{2/3} = 1.265$

Trial-and-error: $y = 1.5$: $1.07 \overset{?}{=} 1.265$

$y = 1.8$: $1.34 \overset{?}{=} 1.265$

$y = 1.7$: $1.25 \overset{?}{=} 1.265$ $\quad \therefore \underline{y = 1.71 \text{ m}}$.

b) Use an "e" value of rough concrete: e = 3 mm. $\therefore \frac{e}{4R} = \frac{3(1+y)}{4y \times 1000}$.

$$h_L = LS = f\frac{L}{4R}\frac{V^2}{2g}. \quad \therefore V^2 = 4RS \times 2g/f = \frac{4y \times .001 \times 2 \times 9.81}{(y+1)f} = 0.0785\frac{y}{(1+y)f}.$$

Try $y = 1.7$: $\frac{e}{4R} = .0012$. $\therefore f = .02$, $V = 1.57$. $\therefore Q = 5.35 \text{ m}^3/\text{s}$.

Try $y = 1.6$: $\frac{e}{4R} = .0012$. $\therefore f = .022$, $V = 1.48$. $Q = 4.74 \text{ m}^3/\text{s}$. $\quad \therefore y = \underline{1.67 \text{ m}}$.

7.126 Assume $y > 3$. $R = \frac{30 + 20(y-3)}{26 + 2(y-3)} = \frac{20y - 30}{2y + 20} = \frac{10y - 15}{y + 10}$.

$$100 = \frac{1.0}{.022}(20y - 30)\left(\frac{10y - 15}{y + 10}\right)^{2/3} \times .001^{1/2}$$

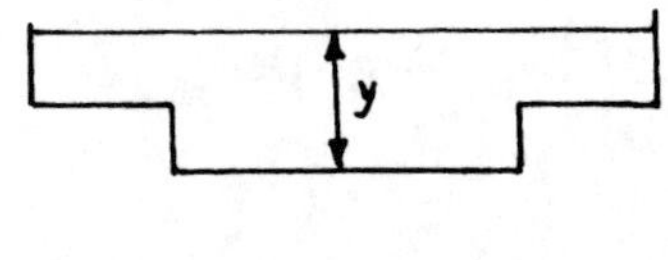

or $\quad 6.96 = (2y-3)\left(\dfrac{10y-15}{y+10}\right)^{2/3}$. $\quad$ Try $y=5$: $6.96 \overset{?}{=} 12.3$

$y=4$: $6.96 \overset{?}{=} 7.36$

$y=3.8$: $6.96 \overset{?}{=} 6.47$

$y=3.9$: $6.96 \overset{?}{=} 6.91$ $\quad \therefore y = \underline{3.91 \text{ m}}$

1.127 $R = \dfrac{2y+y^2}{2+2.83y}$. $10 = \dfrac{1.0}{.012}(2y+y^2)\left(\dfrac{2y+y^2}{2+2.83y}\right)^{2/3}(.0016)^{1/2}$ or $\dfrac{(2y+y^2)^{5/3}}{(2+2.83y)^{2/3}} = 3$

Try $y=1.2$: $3.06 \overset{?}{=} 3$ $\quad$ Try $y=1.15$: $2.83 \overset{?}{=} 3$ $\quad \therefore y = \underline{1.19 \text{ m}}$.

b) $10 = \dfrac{1.0}{.016}(2y+y^2)\left(\dfrac{2y+y^2}{2+2.83y}\right)^{2/3}(.0016)^{1/2}$ or $\dfrac{(2y+y^2)^{5/3}}{(2+2.83y)^{2/3}} = 4$

Try $y=1.5$: $4.68 \overset{?}{=} 4$ $\quad$ Try $y=1.4$: $4.10 \overset{?}{=} 4$ $\quad \therefore y = \underline{1.38 \text{ m}}$.

7.128 Try $y=.6$: $Q = \dfrac{1.0}{.013}\dfrac{\pi\times.6^2}{2}\left(\dfrac{\pi\times.6^2/2}{\pi\times.6}\right)^{2/3}\times.001^{1/2} = .62$. $\quad \cos\alpha = \dfrac{y-.6}{.6}$

$\therefore y > .6$. $R = \dfrac{A}{1.2\pi\alpha/180}$, $\quad A = \pi\times.6^2\dfrac{180-\alpha}{180} + (y-.6).6\sin\alpha$.

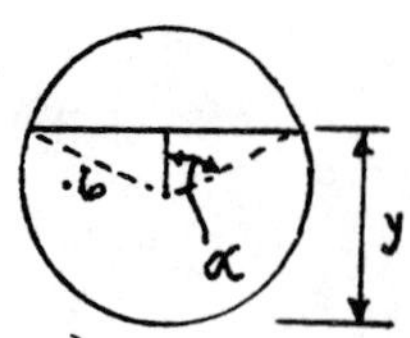

$Q = \dfrac{1.0}{.013}AR^{2/3}.001^{1/2} = 2.43AR^{2/3} = 0.8$. $\quad \therefore AR^{2/3} = .329$.

Try $y=.7$ m: $\alpha = 80.41°$, $A = .6849 \text{ m}^2$, $R = .4067$ m. $.376 \overset{?}{=} .329$

Try $y=.66$ m: $\alpha = 84.26°$, $A = .6374 \text{ m}^2$, $R = .3612$ m. $.323 \overset{?}{=} .329$ $\quad y = \underline{.662 \text{ m}}$.

1.28E Try $y = 2'$: $Q = \dfrac{1.49}{.013}\times 2\pi\left(\dfrac{2\pi}{2\pi}\right)^{2/3}\times.001^{1/2} = 22.8$. $\quad \cos\alpha = \dfrac{y-2}{2}$

$\therefore y > 2$. $R = \dfrac{A}{4\pi\times\alpha/180}$, $\quad A = 4\pi\dfrac{180-\alpha}{180} + (y-2)2\sin\alpha$.

$Q = \dfrac{1.49}{.013}AR^{2/3}.001^{1/2} = 24$. $\quad \therefore AR^{2/3} = 6.62$.

$$\left.\begin{array}{l}\text{Try } y = 2.1': \ \alpha = 87.13^\circ,\ A = 6.683 \text{ ft}^2,\ R = 1.099'.\ \ 7.12 \overset{?}{=} 6.62 \\ \text{Try } y = 2.04': \ \alpha = 88.85^\circ,\ A = 6.443 \text{ ft}^2,\ R = 1.039'.\ \ 6.61 \overset{?}{=} 6.62\end{array}\right\} y = \underline{2.04'}.$$

7.129 Try $y = .4$ m: $Q = \frac{1.0}{.013}\frac{\pi \times .4^2}{2}\left(\frac{.08\pi}{.4\pi}\right)^{2/3} \times .001^{1/2} = 0.209 \text{ m}^3/\text{s}. \quad \therefore y < .4 \text{ m}.$

$$\cos\alpha = \frac{.4-y}{.4},\quad R = \frac{A}{.8\pi \times \alpha / 180},\quad A = .4^2\pi\frac{\alpha}{180} - (.4-y).4\sin\alpha.$$

$$Q = \frac{1.0}{.013}AR^{2/3} \times .001^{1/2} = 0.2. \quad \therefore AR^{2/3} = 0.0822.$$

Try $y = .396$ m: $\alpha = 89.43^\circ$, $A = .2481 \text{ m}^2$, $R = .1987$ m. $\quad .0845 \overset{?}{=} .0822$

Try $y = .392$ m: $\alpha = 88.85^\circ$, $A = .2449 \text{ m}^2$, $R = .1974$ m. $\quad .0830 \overset{?}{=} .0822$

Try $y = .391$ m: $\alpha = 88.71^\circ$, $A = .2441 \text{ m}^2$, $R = .1971$ m. $\quad .0827 \overset{?}{=} .0822$

$\therefore y = \underline{0.389 \text{ m}}.$

EIGHT

External Flows

8.1

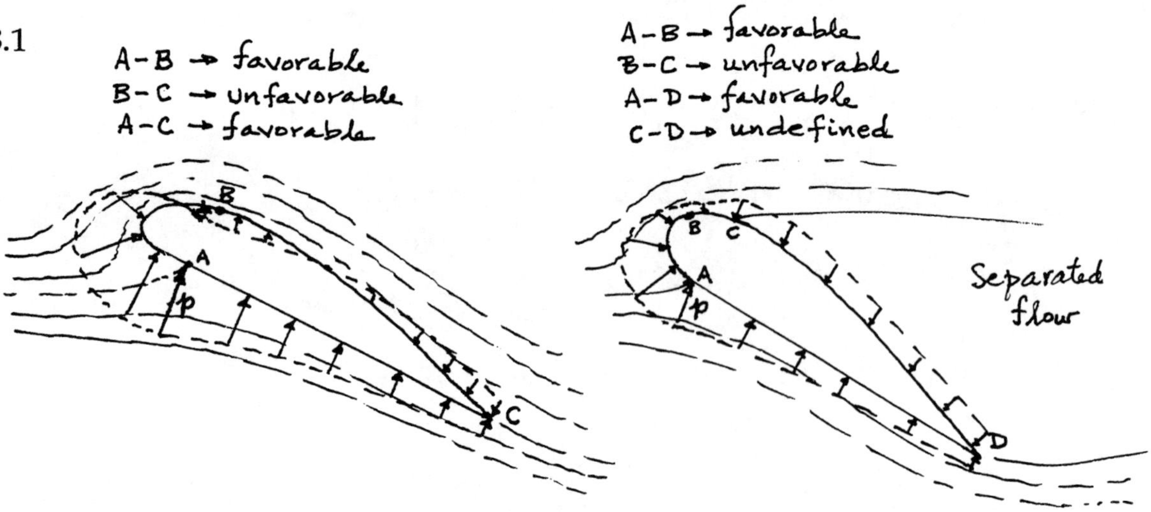

8.2 $\text{Re} = 5 = \frac{VD}{\nu}. \quad \therefore D = \frac{5 \times 1.51 \times 10^{-5}}{20} = \underline{3.78 \times 10^{-6} \text{ m}}.$

$\text{Re} = 10^5 = \frac{VD}{\nu}. \quad \therefore D = \frac{10^5 \times 1.51 \times 10^{-5}}{20} = \underline{0.755 \text{ mm}}.$

8.3

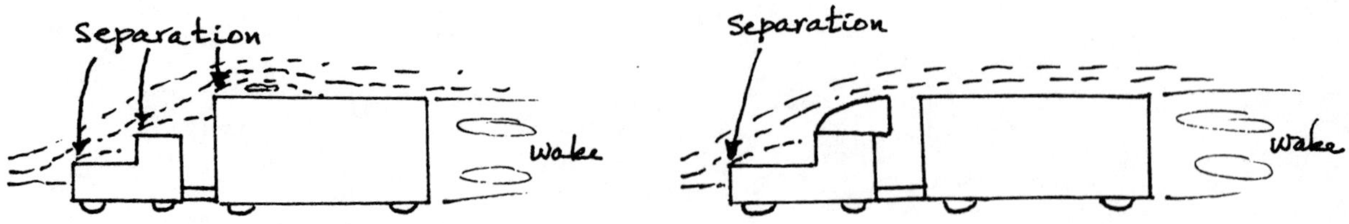

8.4

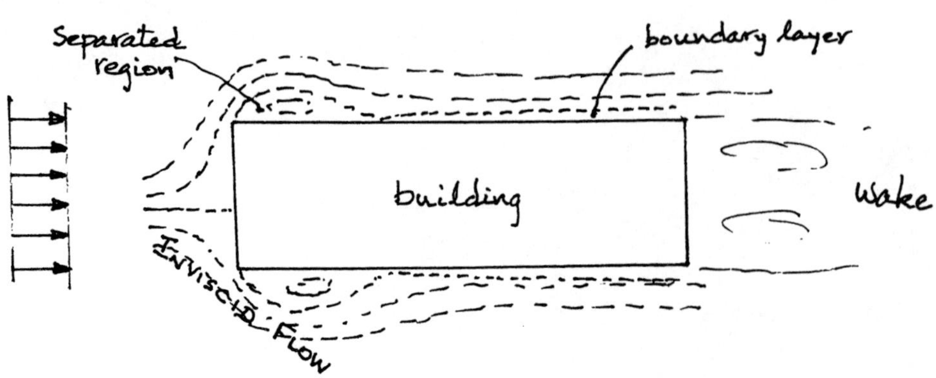

8.5 a) $5 = \frac{VD}{\nu} \quad \therefore V = \frac{5 \times 10^{-6}}{.02} = \underline{.00025 \text{ m/s.}}$

b) $5 = \frac{VD}{\nu} \quad \therefore V = \frac{5 \times .328 \times 10^{-6}}{.02} = \underline{.000082 \text{ m/s.}}$

c) $5 = \frac{VD}{\nu} \quad \therefore V = \frac{5 \times 1.51 \times 10^{-5}}{.02} = \underline{.00378 \text{ m/s.}}$

8.5E $5 = \frac{VD}{\nu} \quad \therefore V = 5\nu / D.$ a) $V = 5 \times 1.22 \times 10^{-5} / \frac{.8}{12} = \underline{.000915 \text{ fps.}}$

b) $V = 5 \times .388 \times 10^{-5} / \frac{.8}{12} = \underline{.000291 \text{ fps.}}$ c) $V = 5 \times 1.6 \times 10^{-4} / \frac{.8}{12} = \underline{0.012 \text{ fps.}}$

8.6 $\text{Re} = \frac{VD}{\nu} = \frac{20 \times D}{1.51 \times 10^{-5}} = 13.25 \times 10^5 D.$

a) $\text{Re} = 13.25 \times 10^5 \times 6 = \underline{7.9 \times 10^6}.$ $\therefore$Separated flow.

b) $\text{Re} = 13.25 \times 10^5 \times .06 = \underline{7.9 \times 10^4}.$ $\therefore$Separated flow.

c) $\text{Re} = 13.25 \times 10^5 \times .006 = \underline{7950}.$ $\therefore$Separated flow.

8.7 $F_D = \int_{A_{\text{front}}} p dA - p_{\text{back}} A_{\text{back}} = p_0 \int_0^1 (1 - r^2) 2\pi r dr = p_0 2\pi \left(\frac{1}{2} - \frac{1}{4} \right) = \frac{1}{2} \pi p_0$

Bernoulli: $p_\infty + \frac{1}{2} \rho V_\infty^2 = p_0. \quad \therefore p_0 = \frac{1}{2} \times 1.21 \times 20^2 = 242 \text{ Pa.}$

$\therefore F_D = \frac{1}{2} \pi (242) = \underline{380 \text{ N}}$

$C_D = \frac{F_D}{\frac{1}{2} \rho V^2 A} = \frac{2 \times 380}{1.21 \times 20^2 \times \pi \times 1^2} = \underline{0.5}$

8.8 $F_{\text{total}} = F_{\text{bottom}} + F_{\text{top}} = 20\,000 \times .3 \times .3 + 10\,000 \times .3 \times .3 = 2700 \text{ N.}$

$F_{\text{lift}} = 2700 \cos 10^\circ = \underline{2659 \text{ N}}$

$F_{\text{drag}} = 2700 \sin 10^\circ = \underline{469 \text{ N}}$

$C_L = \frac{F_L}{\frac{1}{2} \rho V^2 A} = \frac{2 \times 2659}{1000 \times 5^2 \times .3 \times .3} = \underline{2.36}$

$C_D = \frac{F_D}{\frac{1}{2} \rho V^2 A} = \frac{2 \times 469}{1000 \times 5^2 \times .3 \times .3} = \underline{0.417}$

8.9 $F_\ell = p_\ell A_\ell = 26\ 000 \times Lw$. $F_u = p_u A_u = 8000 \times \dfrac{Lw}{2\cos 5^\circ} = 4015\ Lw$

$F_L = F_\ell \cos 5^\circ - F_u \cos 10^\circ = 21\ 950\ Lw$

$F_D = F_\ell \sin 5^\circ - F_u \sin 10^\circ = 1569\ Lw$

$$C_L = \frac{F_L}{\frac{1}{2}\rho V^2 A} = \frac{2 \times 21\ 950\ Lw}{.3119 \times 750^2 Lw} = \underline{0.25}$$

$$C_D = \frac{F_D}{\frac{1}{2}\rho V^2 A} = \frac{2 \times 1569 Lw}{.3119 \times 750^2 Lw} = \underline{0.0179}$$

8.10 If $C_D = 1.0$ for a sphere, Re = 100 (see Fig. 8.8). $\therefore \dfrac{V \times .1}{\nu} = 100$, $V = 1000\nu$.

a) $V = 1000 \times 1.46 \times 10^{-5} = .0146 \text{ m/s}$. $\therefore F_D = \dfrac{1}{2} \times 1.22 \times .0146^2 \pi \times .05^2 \times 1.0$

$= \underline{3.25 \times 10^{-7}\ \text{N}}$.

b) $V = 1000 \times \dfrac{1.46 \times 10^{-5}}{.015 \times 1.22} = 0.798 \text{ m/s}$. $\therefore F_D = \dfrac{1}{2} \times (.015 \times 1.22) \times .798^2 \pi \times .05^2 \times 1.0$

$= \underline{4.58 \times 10^{-5}\ \text{N}}$.

c) $V = 1000 \times 1.31 \times 10^{-6} = .00131 \text{ m/s}$. $\therefore F_D = \dfrac{1}{2} \times 1000 \times .00131^2 \pi \times .05^2 \times 1.0$

$= \underline{6.74 \times 10^{-6}\ \text{N}}$.

8.11 a) $\text{Re} = \dfrac{VD}{\nu} = \dfrac{6 \times .5}{1.5 \times 10^{-5}} = 2 \times 10^5$. $\therefore C_D = 0.45$ from Fig. 8.8.

$\therefore F_D = \dfrac{1}{2}\rho V^2 A C_D = \dfrac{1}{2} \times 1.22 \times 6^2 \times \pi \times .25^2 \times .45 = 1.94$ N.

b) $\text{Re} = \dfrac{15 \times .5}{1.5 \times 10^{-5}} = 5 \times 10^5$. $\therefore C_D = 0.2$ from Fig. 8.8.

$\therefore F_D = \dfrac{1}{2}\rho V^2 A C_D = \dfrac{1}{2} \times 1.22 \times 15^2 \times \pi \times .25^2 \times .2 = 5.4$ N.

8.12 The velocities associated with the two Re's are

$$V_1 = \frac{\text{Re}_1\, \nu}{D} = \frac{3 \times 10^5 \times 1.5 \times 10^{-5}}{.0445} = 101 \text{ m/s},$$

$$V_2 = \frac{\text{Re}_2\, \nu}{D} = \frac{6 \times 10^4 \times 1.5 \times 10^{-5}}{.0445} = 20 \text{ m/s}.$$

The drag, between these two velocities, is reduced by a factor of 2.5 $\left[(C_D)_{\text{high}} = 0.5 \text{ and } (C_D)_{\text{low}} = 0.2\right]$. Thus, between 20 m/s and 100 m/s the drag is reduced by a factor of 2.5. This would significantly lengthen the flight of the ball.

8.13 a) $F_D = \frac{1}{2}\rho V^2 AC_D. \quad \therefore 2 = \frac{1}{2} \times 1.22 \times V^2 \times \pi \times .05^2 C_D. \quad \therefore V^2C_D = 417.$

$\text{Re} = \frac{VD}{\nu} = \frac{V \times .1}{1.5 \times 10^{-5}} = 6670V.$ Try $C_D = .5$: $V = 28.9 \text{ m/s}$, $\text{Re} = 1.9 \times 10^5$.

Try $C_D = .45$: $V = \underline{30.4 \text{ m/s}}$, $\text{Re} = 2.0 \times 10^5$.

b) $C_D = 0.2$: $2 = \frac{1}{2} \times 1.22 \times V^2 \pi \times .05^2 \times .2. \quad \therefore V = \underline{46 \text{ m/s}}.$

8.13E a) $F_D = \frac{1}{2}\rho V^2 AC_D. \quad \therefore 0.5 = \frac{1}{2} \times .00238 V^2 \pi \times \left(\frac{2}{12}\right)^2 C_D. \quad \therefore V^2C_D = 4810.$

$\text{Re} = \frac{VD}{\nu} = \frac{V \times 4/12}{1.6 \times 10^{-4}} = 2080V.$ Try $C_D = .5$: $V = 98$ fps, $\text{Re} = 2 \times 10^5$.

Try $C_D = .4$: $V = \underline{110 \text{ fps}}$, $\text{Re} = 2.3 \times 10^5$.

b) $C_D = 0.2$: $0.5 = \frac{1}{2} \times .00238 V^2 \pi \times \left(\frac{2}{12}\right)^2 \times .2. \quad \therefore V = \underline{155 \text{ fps}}.$

8.14 $4.2 = \frac{1}{2} \times 1000 V^2 \pi \times .1^2 C_D. \quad \therefore V^2C_D = 0.267. \quad \text{Re} = \frac{V \times .2}{10^{-6}} = 2 \times 10^5 V.$

Try $\underline{C_D = 0.5}$: $\therefore V = 0.73 \text{ m/s}$. $\text{Re} = \underline{1.46 \times 10^5}$. $\therefore$ OK.

8.15 $\text{Re} = \frac{VD}{\nu} = \frac{40 \times 2}{1.5 \times 10^{-5}} = 5.3 \times 10^6. \quad \therefore C_D = 0.7$ (This is extrapolated from Fig. 8.8).

$\therefore F_D = \frac{1}{2} \times 1.22 \times 40^2 \times (2 \times 60) \times .7 = \underline{81\,900 \text{ N}}.$

$M = 81\,900 \times 30 = \underline{2.46 \times 10^6 \text{ N} \cdot \text{m}}.$

8.16 a) $\text{Re}_1 = \frac{25 \times .05}{1.08 \times 10^{-5}} = 1.2 \times 10^5. \quad \text{Re}_2 = 1.8 \times 10^5. \quad \text{Re}_3 = 2.4 \times 10^5.$ Assume a rough cylinder (the air is highly turbulent).

$\therefore (C_D)_1 = 0.7, \ (C_D)_2 = 0.8, \ (C_D)_3 = 0.9.$

$\therefore F_D = \frac{1}{2} \times 1.45 \times 25^2 (.05 \times 10 \times .7 + .075 \times 15 \times .8 + .1 \times 20 \times .9) = \underline{1380 \text{ N}}.$

$$M = \frac{1}{2} \times 1.45 \times 25^2 (.05 \times 10 \times .7 \times 40 + .075 \times 15 \times .8 \times 27.5 + .1 \times 20 \times .9 \times 10) = \underline{25\ 700\ \text{N} \cdot \text{m}}.$$

b) $\text{Re}_1 = \frac{25 \times .05}{1.65 \times 10^{-5}} = 7.6 \times 10^4$. $\text{Re}_2 = 1.14 \times 10^5$, $\text{Re}_3 = 1.5 \times 10^5$.

$\therefore (C_D)_1 = .8,\ (C_D)_2 = .7,\ (C_D)_3 = .8.$ $\rho = \frac{101}{.287 \times 308} = 1.17\ \text{kg/m}^3$.

$\therefore F_D = \frac{1}{2} \times 1.17 \times 25^2 (.05 \times 10 \times .8 + .075 \times 15 \times .7 + .1 \times 20 \times .8) = \underline{1020\ \text{N}}.$

$$M = \frac{1}{2} \times 1.17 \times 25^2 (.05 \times 10 \times .8 \times 40 + .075 \times 15 \times .7 \times 27.5 + .1 \times 20 \times .8 \times 10) = \underline{19\ 600\ \text{N} \cdot \text{m}}.$$

8.17 It is assumed that an atmospheric air flow is highly turbulent. $\therefore$ Use the "rough" curve. $\therefore C_D = 0.7$. $F_D = 42 = \frac{1}{2} \times 1.22 V^2 \times 2D \times .7$. $\therefore 49 = V^2 D$.

$10^5 = \frac{VD}{1.5 \times 10^{-5}} = \frac{V \times 49 / V^2}{1.5 \times 10^{-5}}$. $\therefore V = \frac{49}{1.5} = 33\ \text{m/s}$. $\therefore D = .045\ \text{m}$.

$\therefore L/D = 2/.045 = 44$ $\therefore (C_D)_{\text{actual}} = .83$ (from Table 8.1) $\times .7 = .58$. $\therefore V^2 D = 59$.

$\therefore V = 59/1.5 = \underline{40\ \text{m/s}}$. $D = \underline{.037\ \text{m}}$.

8.17E Atmospheric air is turbulent. $\therefore$ Use the "rough" curve. $\therefore C_D = 0.7$.

$F_D = 10 = \frac{1}{2} \times .00238 V^2 \times 6D \times .7$. $\therefore 2000 = V^2 D$. $10^5 = \frac{V \times 2000 / V^2}{1.6 \times 10^{-4}}$.

$\therefore V = 125$ fps. $\therefore D = .128'$. $\therefore \frac{L}{D} = \frac{6}{.128} = 47$.

$\therefore (C_D)_{\text{actual}} = .84$ (from Table 8.1) $\times .7 = .59$.

$\therefore V^2 D = 2370$. $\therefore V = \underline{148\ \text{fps}}$. $D = \underline{0.108'}$.

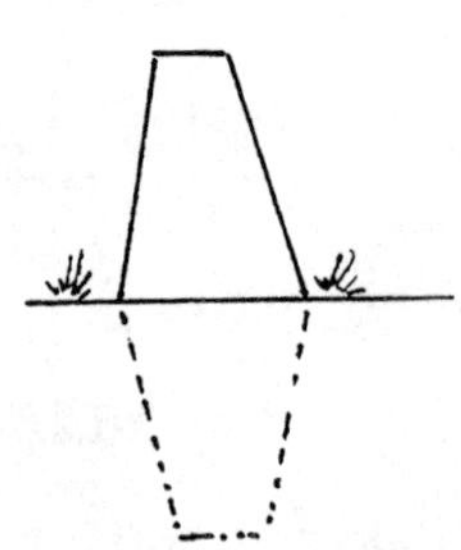

8.18 Since the air cannot flow around the bottom, we imagine the structure to be mirrored as shown. Then $L/D = 40/5 = 8$. $\therefore C_D = 0.66 C_{D_\infty}$.

$\text{Re}_{\text{min}} = \frac{VD_{\text{min}}}{\nu} = \frac{30 \times 2}{1.5 \times 10^{-5}} = 4 \times 10^6$. $\therefore C_D = 1.0 \times .66 = .66$.

$\therefore F_D = \frac{1}{2} \times 1.22 \times 30^2 \times \left(\frac{2+8}{2} \times 20 \right) \times .66 = \underline{36\ 000\ \text{N}}.$

8.19 $F_B + F_D = F_W$. $9810 \times \frac{4}{3} \pi r^3 + \frac{1}{2} \times 1000 V^2 \pi r^2 C_D = 9810 \times 7.82 \times \frac{4}{3} \pi r^3$.

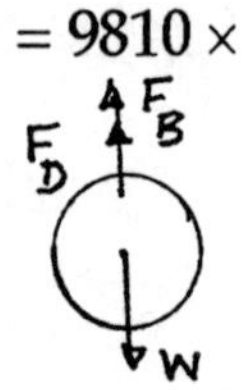

$\text{Re} = \frac{V \times 2r}{10^{-6}} = 2 \times 10^6 Vr$. $\therefore V^2 C_D = 178r$

a) $r = .05$ m. $\therefore \text{Re} = 10^5 V,\ V^2 C_D = 8.9$. Assume a smooth sphere.

Try $C_D = .5$: $\therefore V = 4.22$ m/s. $\text{Re} = 4.22 \times 10^5$. This is too large for Re.

Try $C_D = .2$: $\therefore V = \underline{6.67 \text{ m/s}}$. $\text{Re} = 6.67 \times 10^5$. OK.

b) $r = .025$ m. $\text{Re} = 5 \times 10^4 V,\ V^2 C_D = 4.45$.

Try $C_D = .2$: $V = \underline{4.72 \text{ m/s}}$. $\text{Re} = 2.4 \times 10^5$. OK.

c) $r = .005$ m. $\text{Re} = 10^4 V,\ V^2 C_D = 0.89$.

Try $C_D = .5$: $V = \underline{1.33 \text{ m/s}}$. $\text{Re} = 1.33 \times 10^4$. OK.

d) $r = .001$ m. $\text{Re} = 2 \times 10^3 V,\ V^2 C_D = 0.178$.

Try $C_D = .4$: $V = \underline{0.67 \text{ m/s}}$. $\text{Re} = 1.33 \times 10^3$. OK.

8.20 $F_B + F_D = F_W$. $1.22 \times 9.81 \times \frac{4}{3}\pi(.25)^3 + \frac{1}{2} \times 1.22 V^2 \pi (.25)^2 C_D = 9810 S \frac{4}{3}\pi(.25)^3$.

$$\text{Re} = \frac{V \times .5}{1.5 \times 10^{-5}} = 3.33 \times 10^4 V. \quad \therefore 1 + .153 V^2 C_D = 820 S.$$

a) $S = .005$. $V^2 C_D = 20.3$. Assume a rough sphere because of turbulence.

Try $C_D = .4$: $V = \underline{7.1 \text{ m/s}}$. $\text{Re} = 2.4 \times 10^5$. OK.

b) $S = .02$. $V^2 C_D = 101$. Try $C_D = .4$: $V = \underline{15.9 \text{ m/s}}$. $\text{Re} = 5 \times 10^5$. $\therefore$ OK.

c) $S = 1.0$. $V^2 C_D = 5350$. Try $C_D = .4$: $V = \underline{116 \text{ m/s}}$. $\text{Re} = 3.8 \times 10^6$. $\therefore$ OK.

8.20E $F_B + F_D = F_W$. $.077 \times \frac{4}{3}\pi\left(\frac{10}{12}\right)^3 + \frac{1}{2} \times .00238 V^2 \pi \left(\frac{10}{12}\right)^2 C_D = 62.4 S \frac{4}{3}\pi\left(\frac{10}{12}\right)^3$.

$$\text{Re} = \frac{V \times 10/12}{1.6 \times 10^{-4}} = 5.2 \times 10^3 V. \qquad \therefore 1 + .0139 V^2 C_D = 810 S$$

a) $S = .005$. $V^2 C_D = 219$. Assume atmospheric turbulence, i.e., rough.

Try $C_D = .4$: $V = 23.4$ fps. $\text{Re} = 1.2 \times 10^5$. $\therefore C_D = .3$. $V = \underline{27 \text{ fps}}$.

b) $S = .02$. $V^2 C_D = 1090$. Try $C_D = .4$: $V = \underline{52 \text{ fps}}$. $\text{Re} = 2.7 \times 10^5$. $\therefore$ OK.

c) $S = 1.0$. $V^2 C_D = 58\,200$. Try $C_D = .4$: $V = \underline{381 \text{ fps}}$.

8.21 Assume a 180 lb, 6' sky diver, with components as shown. If V is quite large, then $\text{Re} > 2 \times 10^5$. $F_D = F_W$.

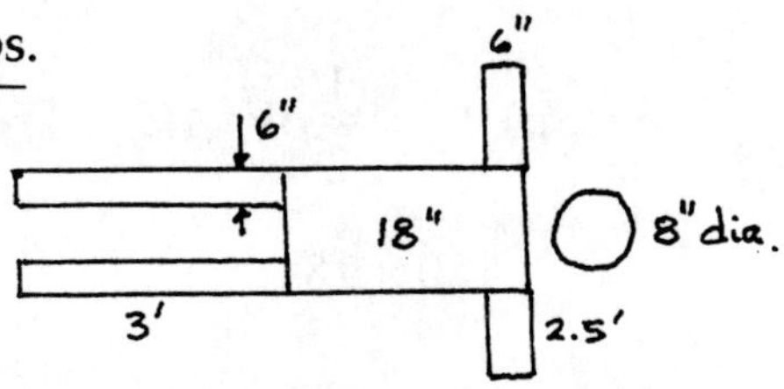

$$\frac{1}{2} \times .00238 V^2 \left[2 \times 3 \times \frac{1}{2} \times 1.0 \times .7 + 2 \times 2.5 \times \frac{1}{2} \times 1.0 \times .7 + \frac{18}{12} \times 2.5 \times 1.0 + \pi \times \left(\frac{4}{12}\right) \times .4 \right] = 180.$$

Table 8.1

$\therefore V = \underline{140 \text{ fps}}$.

8.22 From Table 8.2 $C_D = 0.35$. $F_D = \frac{1}{2} \times 1.22V^2 \times 3.2 \times 0.35 = .683V^2$.

a) $F_D = .683 \times \left(\frac{80 \times 1000}{3600}\right)^2 = 337 \text{ N}$. $\therefore \dot{W} = 337\frac{80 \times 1000}{3600} = 7500 \text{ W}$ or $\underline{10 \text{ Hp}}$.

b) $V = 25 \text{ m/s}$. $F_D = .683 \times 25^2 = 427 \text{ N}$. $\therefore \dot{W} = 427 \times 25 = 10\,700 \text{ W}$ or $\underline{14.3 \text{ Hp}}$.

c) $V = 27.8 \text{ m/s}$. $F_D = .683 \times 27.8^2 = 527 \text{ N}$. $\therefore \dot{W} = 527 \times 27.8 = 14\,700 \text{ W}$ or $\underline{19.6 \text{ Hp}}$.

8.23 $1.2F_D = 1.1 \times 400$. $F_D = \frac{1}{2}\rho V^2 AC_D$. $C_D = 1.1$

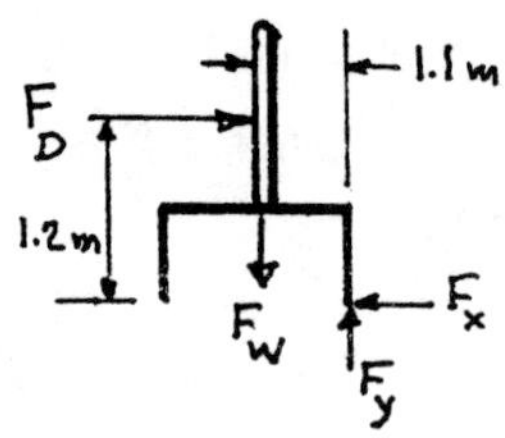

$1.2 \times \frac{1}{2} \times 1.22V^2 \times (2 \times 3) \times 1.1 = 1.1 \times 400$.

$\therefore V = \underline{9.5 \text{ m/s}}$.

8.24 $\text{Re} = \frac{VD}{\nu} = \frac{(40\,000/3600)0.6}{1.51 \times 10^{-5}} = 4.42 \times 10^5$. $\therefore C_D = 0.35$ from Fig. 8.8.

a) $F_D = \frac{1}{2}\rho V^2 AC_D = \frac{1}{2} \times 1.204 \times (40\,000/3600)^2 \times 0.6 \times 6 \times 0.35 = \underline{93.6 \text{ N}}$

b) $F_D = 93.6 \times 0.68 = \underline{63.7 \text{ N}}$ where $L/D = 6/0.6 = 10$.

c) $F_D = 93.6 \times 0.76 = \underline{71.1 \text{ N}}$ where we can use $L/D = 20$ since only one end is free. The ground acts like the mid-section of a 12-m-long cylinder.

8.25 a) Curled up, she makes an approximate sphere of about 1.2 m in diameter (just a guess!). Assume a rough sphere at large Re. From Fig. 8.8, $C_D = 0.4$:

$$F_D = \frac{1}{2}\rho V^2 AC_D$$

$$80 \times 9.8 = \frac{1}{2} \times 1.21 \times V^2 \pi \times 0.6^2 \times 0.4. \quad \therefore V = \underline{53.7 \text{ m/s}}.$$

Check Re: $\text{Re} = \frac{53.7 \times 1.2}{1.51 \times 10^{-5}} = 4.27 \times 10^6$. $\therefore$ OK.

b) $F_D = \frac{1}{2}\rho V^2 AC_D$. From Table 8.2, $C_D = 1.4$:

$$80 \times 9.8 = \frac{1}{2} \times 1.21 \times V^2 \pi \times 4^2 \times 1.4. \quad \therefore V = \underline{4.29 \text{ m/s}}.$$

Check Re: $\text{Re} = \frac{4.29 \times 8}{1.51 \times 10^{-5}} = 2.27 \times 10^6$. Should be larger but the velocity should be close.

c) $F_D = \frac{1}{2}\rho V^2 AC_D$

$$80 \times 9.8 = \frac{1}{2} \times 1.21 \times V^2 \pi \times 1^2 \times 1.4. \quad \therefore V = \underline{17.2 \text{ m/s}}.$$

Check Re: $\text{Re} = \dfrac{17.2 \times 1}{1.51 \times 10^{-5}} = 1.14 \times 10^6$. This should be greater than 10^7 for C_D to be acceptable. Hence, the velocity is approximate.

8.26 With the deflector the drag coefficient is 0.76 rather than 0.96. The required power (directly related to fuel consumed) is reduced by the ratio of 0.76/0.96. The cost per year without the deflector is

$$\text{Cost} = (200\,000/1.2) \times 0.25 = \$41{,}667.$$

With the deflector it is

$$\text{Cost} = 41{,}667 \times 0.76/0.96 = \$32{,}986.$$

The savings is

$$\text{Savings} = \underline{\$8{,}861}.$$

8.27 $F_D = \dfrac{1}{2}\rho V^2 A C_D = \dfrac{1}{2} \times 1.22 \times \left(\dfrac{100 \times 1000}{3600}\right)^2 \times (1.8 \times .6) \times 1.1 = 508 \text{ N}.$

$$\dot{W} = F_D \times V = 508 \frac{100 \times 1000}{3600} = 14\,100 \text{ W or } \underline{18.9 \text{ Hp}}.$$

8.27E $F_D = \dfrac{1}{2}\rho V^2 A C_D = \dfrac{1}{2} \times .00238 \times 88^2 \times (6 \times 2) \times 1.1 = 122 \text{ lb}.$

$$\dot{W} = F_D \times V = 122 \times 88 = 10{,}700 \ \frac{\text{ft - lb}}{\text{sec}} \text{ or } \underline{19.5 \text{ Hp}}.$$

8.28 $F_D = \dfrac{1}{2}\rho V^2 A C_D = \dfrac{1}{2} \times 1.22 \times (27.8 \times 1.6)^2 \times \pi \times .05^2 \times 1.1 = 10.43 \text{ N}.$

$$\dot{W} = F_D \times V \times 2 = 10.43 \times (27.8 \times 1.6) \times 2 = 226 \text{ W or } \underline{1.24 \text{ Hp}}.$$

8.29 The projected area is $\dfrac{(2 + 0.3)}{2} \times 4 = 4.6 \text{ m}^2$.

$$F_D = \frac{1}{2}\rho V^2 A C_D = \frac{1}{2} \times 1.18 \times 20^2 \times 4.6 \times 0.4 = 434 \text{ N}.$$

Since there are two free ends, we use Table 8.1 with $L/D = 4/1.15 = 3.47$, and approximate the force as

$$F_D = 434 \times 0.62 = \underline{269 \text{ N}}.$$

8.30 The net force acting up is (use absolute pressure)

$$F_{\text{up}} = \frac{4}{3}\pi \times 0.4^3 \times 1.21 \times 9.8 - 0.5 - \frac{4}{3}\pi \times 0.4^3 \frac{120}{2.077 \times 293} \times 9.8 = 2.16 \text{ N}$$

From a force triangle (2.16 up and F_D to the right), we see that

$$\tan\alpha = F_{\text{up}} / F_D.$$

a) $F_D = 2.16 / \tan 80^\circ = 0.381.$

$$0.381 = \frac{1}{2} \times 1.21V^2 \pi \times 0.4^2 \times 0.2. \quad \therefore V = 2.50 \text{ m/s}.$$

Check Re: $\text{Re} = \frac{2.5 \times 0.8}{1.51 \times 10^{-5}} = 1.33 \times 10^5.$ Too low. Use $C_D = 0.5$:

$$0.381 = \frac{1}{2} \times 1.21V^2 \pi \times 0.4^2 \times 0.5. \quad \therefore V = \underline{1.58 \text{ m/s}}$$

b) $F_D = 2.16 / \tan 70^\circ = 0.786.$

$$0.786 = \frac{1}{2} \times 1.21V^2 \pi \times 0.4^2 \times 0.2. \quad \therefore V = 3.60 \text{ m/s}.$$

Check Re: $\text{Re} = \frac{3.6 \times 0.8}{1.51 \times 10^{-5}} = 1.9 \times 10^5.$ Too low. Use $C_D = 0.5$:

$$0.786 = \frac{1}{2} \times 1.21V^2 \pi \times 0.4^2 \times 0.5. \quad \therefore V = \underline{2.27 \text{ m/s}}$$

c) $F_D = 2.16 / \tan 60^\circ = 1.25.$

$$1.25 = \frac{1}{2} \times 1.21V^2 \pi \times 0.4^2 \times 0.5. \quad \therefore V = \underline{2.86 \text{ m/s}}.$$

Check Re: $\text{Re} = \frac{2.86 \times 0.8}{1.51 \times 10^{-5}} = 1.5 \times 10^5.$ $\therefore$OK.

d) $F_D = 2.16 / \tan 50^\circ = 1.81.$

$$1.81 = \frac{1}{2} \times 1.21V^2 \pi \times 0.4^2 \times 0.5. \quad \therefore V = \underline{3.45 \text{ m/s}}.$$

Check Re: $\text{Re} = \frac{3.45 \times 0.8}{1.51 \times 10^{-5}} = 1.8 \times 10^5.$ Close, but OK.

8.31 Assume each section of the tree is a cylinder. The average diameter of the tree is 1 m. The top doesn't have a blunt end around which the air flows, however, the bottom does; so assume $L/D = (5/2) \times 2 = 5.$ So, use a factor of 0.62 from Table 8.1 to multiply the drag coefficient. The force acts near the centroid of the triangular area, one-third the way up. Finally,

$$F \times d = 5000$$

$$\left[\frac{1}{2} \times 1.21V^2(5) \times 0.4 \times 0.62\right] \times \left(\frac{5}{3} + 0.6\right) = 5000. \quad V = \underline{54.2 \text{ m/s}}.$$

8.32 Power to move the sign:

$$F_D V = \frac{1}{2}\rho V^2 A C_D \times V$$

$$= \frac{1}{2} \times 1.21 \times 11.11^2 \times 0.72 \times 1.1 \times 11.11 = 657 \text{ J/s}.$$

This power comes from the engine:

$$657 = (12\,000 \times 1000)\dot{m} \times 0.3. \quad \therefore \dot{m} = 1.825 \times 10^{-4} \text{ kg/s}.$$

Assuming the density of gas to be 900 kg/m³,

$$1.825\times10^{-4}\times10\times3600\times6\times52\times\frac{1000}{900}\times0.30=\underline{\$683}$$

8.33 The power expended is $F_D\times V$. $V=(25\times88/60)/3.281=11.18\text{ m/s}$

$$\frac{1}{2}\times1.21\times11.18^3\times0.56\times C_D=\frac{1}{2}\times1.21\times V^3\times0.4\times C_D\times0.8$$

$$\therefore V=13.47\text{ m/s or }\underline{30.1\text{ mph}}.$$

8.34 $\dot{W}=40\times746\eta=F_D\times V=\frac{1}{2}\rho V^2AC_D\times V=\frac{1}{2}\rho AC_DV^3$.

$\therefore 40\times746\times.9=\frac{1}{2}\times1.22\times3\times0.35V^3$. $\therefore V=34.7\text{ m/s or }\underline{125\text{ km/hr}}$.

8.35 $40<\text{Re}<10\,000$. $40<\frac{V\times.003}{1.5\times10^{-5}}<10\,000$. $\therefore \underline{0.2<V<50\text{ m/s}}$.

$\text{St}=.12=\frac{f\times.003}{.2}$. $\therefore f_{\text{low}}=8\text{ Hz}$.

$\text{St}=.21=\frac{f\times.003}{50}$. $\therefore f_{\text{lhigh}}=3500\text{ Hz}$.

The vortices could be heard over most of the range.

8.36 $40>\frac{VD}{\nu}=\frac{2D}{10^{-6}}$. $\therefore \underline{D<20\mu\text{m}}$.

$10\,000<\frac{VD}{\nu}=\frac{2D}{10^{-6}}$. $\therefore \underline{D>5\text{mm}}$.

8.36E $40>\frac{VD}{\nu}=\frac{6D}{1.22\times10^{-5}}$. $\therefore \underline{D<8.13\times10^{-5}\text{ ft}}$.

$10\,000<\frac{VD}{\nu}=\frac{6D}{1.22\times10^{-5}}$. $\therefore D>0.020\text{ ft or }\underline{0.24''}$.

8.37 From Fig. 8.9, Re is related to St. $\text{St}=\frac{f\times D}{V}=\frac{0.2\times.1}{V}$.

$\text{Re}=\frac{VD}{\nu}=\frac{V\times.1}{1.5\times10^{-5}}$. Try St =.21: $V=0.095\text{ m/s}$. $\therefore \text{Re}=630$.

This is acceptable. $\therefore V=\underline{0.095\text{ m/s}}$.

8.38 $\text{St}=\frac{fD}{V}=\frac{.002\times2}{V}$. $\text{Re}=\frac{VD}{\nu}=\frac{V\times2}{10^{-6}}$. Use Fig. 8.9.

Try St = 0.21: $V=\underline{.0191\text{ m/s}}$. $\text{Re}=38\times10^3$. $\therefore$ OK.

8.39 Let $S_t = 0.21$ for the wind imposed vorticies. When this frequency equals the natural frequency, or one of its odd harmonics, resonance occurs:

$$f = \sqrt{T / \rho L^2 d^2 \pi}$$

$$\frac{0.21 \times 10}{0.016} = \sqrt{30\,000 / 7850 L^2 \times 0.016^2 \times \pi}. \qquad \therefore L = \underline{0.525 \text{ m}}$$

Consider the third and fifth harmonics:

$$f = 3\sqrt{T / \rho L^2 d^2 \pi}. \qquad \therefore L = \underline{1.56 \text{ m}}.$$

$$f = 5\sqrt{T / \rho L^2 d^2 \pi}. \qquad \therefore L = \underline{2.62 \text{ m}}.$$

8.40 $F_D = \frac{1}{2}\rho V^2 A C_D = \frac{1}{2} \times 1.22 \times 25^2 \times 1.0 \times .8 \times (2 \times .15) = 91.5 \text{ N}$

$\therefore \dot{W} = F_D \times V = 91.5 \times 25 = 2290 \text{ W}$ or $\underline{3.1 \text{ Hp}}$.

$\text{Re} = \frac{VD}{\nu} = \frac{25 \times .15}{1.5 \times 10^{-5}} = 2.5 \times 10^5. \quad \therefore C_D = .035. \quad \therefore F_D = 3.2 \text{ N}.$

$\therefore \dot{W} = 3.2 \times 25 = 80 \text{ W}$ or $\underline{0.11 \text{ Hp}}$.

Fig. 8.8 Table 8.1

8.40E $\text{Re} = \frac{88 \times 6/12}{1.6 \times 10^{-4}} = 2.8 \times 10^5. \quad F_D = \frac{1}{2} \times .00238 \times 88^2 \times 1.0 \times .8 \times \left(6 \times \frac{6}{12}\right) = 22 \text{ lb}.$

$\dot{W} = F_D \times V = 22 \times 88 = 1946 \text{ ft-lb/sec}$ or $\underline{3.5 \text{ Hp}}$.

$(C_D)_{\text{streamlined}} = 0.035. \quad \therefore F_D = 0.77 \text{ lb}. \quad \dot{W} = 67.8 \frac{\text{ft-lb}}{\text{sec}}$ or $\underline{0.12 \text{ Hp}}$.

8.41 $\text{Re} = \frac{VD}{\nu} = \frac{3 \times .08}{1.5 \times 10^{-5}} = 16\,000.$

Table 8.1

$$\therefore F_D = \frac{1}{2} \times 1.22 \times 3^2 \times (0.08 \times 2) \times 1.2 \times .78 = \underline{0.822 \text{ N}}$$

$(C_D)_{\text{streamlined}} = .35. \quad \therefore F_D = 0.24 \text{ N}. \quad \therefore \%\text{ reduction} = \frac{0.822 - 0.24}{0.822} \times 100 = \underline{70.8\%}$

8.42 $\text{Re} = \frac{VD}{\nu} = \frac{2 \times 0.8}{10^{-6}} = 1.6 \times 10^6. \quad \therefore C_D = 0.45$ from Fig. 8.8.

$\frac{L}{D} = \frac{4}{0.8} = 5. \quad \therefore C_D = 0.62 \times 0.45 = 0.28.$

Because only one end is free, we double the length.

$F_D = \frac{1}{2}\rho V^2 A C_D = \frac{1}{2} \times 1000 \times 2^2 \times 0.8 \times 2 \times 0.28 = \underline{900 \text{ N}}.$

If streamlined, $C_D = 0.03 \times 0.62 = 0.0186.$

$$\therefore F_D = \frac{1}{2} \times 1000 \times 2^2 \times 0.8 \times 2 \times 0.0186 = \underline{60 \text{ N}}.$$

8.43 $V = 50 \times 1000 / 3600 = 13.9 \text{ m/s}.$

Assume the ends to not be free. $\therefore$ Use C_D from Fig. 8.8.

$$\text{Re} = \frac{13.9 \times 0.02}{1.5 \times 10^{-5}} = 1.85 \times 10^4. \quad \therefore C_D = 1.2. \quad (C_D)_{\text{streamlined}} = 0.3$$

$$\dot{W} = F_D \times V = \frac{1}{2}\rho V^3 A C_D = \frac{1}{2} \times 1.2 \times 13.9^3 \times 0.02 \times 20 \times 1.2 = 773 \text{ W or } \underline{1.04 \text{ Hp}}.$$

$$\dot{W}_{\text{streamlined}} = \frac{1}{2} \times 1.2 \times 13.9^3 \times 0.02 \times 20 \times 0.3 = 193 \text{ W or } \underline{0.26 \text{ Hp}}$$

8.44 $V = 50 \times 1000 / 3600 = 13.9 \text{ m/s}.$ $\quad \text{Re} = \frac{13.9 \times 0.3}{1.5 \times 10^{-5}} = 2.8 \times 10^5. \quad \therefore C_D = 0.4$

We assumed a head diameter of 0.3 m and used the rough sphere curve.

$$F_D = \frac{1}{2} \times 1.2 \times 13.9^2 (\pi \times 0.3^2 / 4) \times 0.4 = \underline{3.3 \text{ N}}.$$

$$F_D = \frac{1}{2} \times 1.2 \times 13.9^2 (\pi \times 0.3^2 / 4) \times 0.035 = \underline{0.29 \text{ N}}.$$

8.45 $\sigma = \dfrac{p_\infty - p_v}{\frac{1}{2}\rho V^2}. \quad 0.7 = \dfrac{150\,000 - 1670}{\frac{1}{2} \times 1000 V^2}$ where $p_\infty = \gamma h + p_{\text{atm}} = 150\,000$ Pa.

$\therefore V = \underline{20.6 \text{ m/s}}.$

8.46 $C_L = \dfrac{F_L}{\frac{1}{2}\rho V^2 A} = \dfrac{200\,000}{\frac{1}{2} \times 1000 \times 12^2 \times .4 \times 10} = 0.69. \quad \therefore \underline{\alpha \cong 3^\circ}.$

$$C_D = .0165 = \frac{F_D}{\frac{1}{2} \times 1000 \times 12^2 \times .4 \times 10}. \quad \therefore F_D = \underline{4800 \text{ N}}.$$

$$\sigma_{\text{crit}} = .75 \overset{?}{>} \frac{(9810 \times .4 + 101\,000) - 1670}{\frac{1}{2} \times 1000 \times 12^2} = 1.43. \quad \therefore \underline{\text{No cavitation.}}$$

8.46E $C_L = \dfrac{F_L}{\frac{1}{2}\rho V^2 A} = \dfrac{50\,000}{\frac{1}{2} \times 1.94 \times 35^2 \times \frac{16}{12} \times 30} = 1.05. \quad \therefore \alpha = 7.3^\circ.$

$$C_D = .027 = \frac{F_D}{\frac{1}{2} \times 1.94 \times 35^2 \times \frac{16}{12} \times 30}. \quad \therefore F_D = \underline{1280 \text{ lb}}.$$

$$\sigma_{\text{crit}} = 1.6 \overset{?}{>} \frac{62.4 \times 16/12 + 2117 - .25 \times 144}{\frac{1}{2} \times 1.94 \times 35^2} = 1.82. \qquad \therefore \underline{\text{No cavitation.}}$$

8.47 $p_\infty = 9810 \times 5 + 101\,000 = 150\,000$ Pa. $p_v = 1670$ Pa. $\text{Re} = \frac{20 \times .8}{10^{-6}} = 16 \times 10^6.$

$$\sigma = \frac{150\,000 - 1670}{\frac{1}{2} \times 1000 \times 20^2} = 0.74. \quad \therefore C_D = C_D(0)(1+\sigma) = .3(1+.74) = .52$$

$$\therefore F_D = \frac{1}{2}\rho V^2 A C_D = \frac{1}{2} \times 1000 \times 20^2 \times \pi \times .4^2 \times .52 = \underline{52\,000 \text{ N}}.$$

Note: We retain 2 sig. figures since C_D is known to only 2 sig. nos.

8.48 For a 6° angle of attack we find from Table 8.4 $C_L = 0.95$.

$$F_L = \frac{1}{2}\rho V^2 A C_L = \frac{1}{2} \times 1000 \times 15^2 \times 4 \times 0.4L \times .95 = 12\,000 \times 9.8.$$

$$\therefore L = \underline{0.69 \text{ m}}.$$

8.49 $\Sigma F = ma.$ a) $400 - 9810 \times \frac{4}{3}\pi \times .2^3 = \frac{400}{9.81}a. \quad \therefore a = \underline{1.75 \text{ m/s}^2}.$

b) $400 - 9810 \times \frac{4}{3}\pi \times .2^3 = \left(\frac{400}{9.81} + \frac{1}{2} \times 1000 \times \frac{4}{3}\pi \times .2^3\right)a. \quad \therefore a = \underline{1.24 \text{ m/s}^2}.$

8.50 $F = ma_1 = 1000 \times 1.2 \times \forall a_1. \quad \therefore a_1 = \frac{F}{1200\forall}. \quad m_a = 0.2 \times 1000\forall.$

$F = (m + m_a)a_2. \quad \therefore a_2 = \frac{F}{1200\forall + 200\forall} = \frac{F}{1400\forall}. \quad a_2$ is true acceleration.

$$\therefore \%\text{ error} = \left|\frac{a_2 - a_1}{a_2}\right| \times 100 = \left|\frac{\frac{F}{1400\forall} - \frac{F}{1200\forall}}{\frac{F}{1400\forall}}\right| \times 100 = \underline{16.7\%}.$$

8.51 $C_L = \frac{F_L}{\frac{1}{2}\rho V^2 A} = \frac{1000 \times 9.81}{\frac{1}{2} \times .412 \times 80^2 \times 15} = 0.496. \quad \therefore \alpha = 3.2°. \quad C_D = .0065.$

$$\dot{W} = F_D V = \left(\frac{1}{2} \times .412 \times 80^2 \times 15 \times .0065\right) \times 80 = 10\,300 \text{ W} \quad \text{or} \quad \underline{13.8 \text{ Hp}}.$$

8.52 a) $C_L = 1.22 = \dfrac{1500 \times 9.81 + 3000}{\frac{1}{2} \times 1.22 \times V^2 \times 20}$. $\therefore V = \underline{34.5 \text{ m/s}}$.

b) $(C_L)_{max} = 1.72 = \dfrac{1500 \times 9.81 + 3000}{\frac{1}{2} \times .412 \times V^2 \times 20}$. $\therefore V = \underline{50 \text{ m/s}}$. (at 10 000 m)

c) $\dot{W} = F_D V = \left(\frac{1}{2} \times .412 \times 80^2 \times 20 \times .0065\right) \times 80 = 13\,700 \text{ W}$ or $\underline{18.4 \text{ Hp}}$

where we found C_D as follows:

$$(C_L)_{cruise} = \frac{1500 \times 9.81 + 3000}{\frac{1}{2} \times .412 \times 80^2 \times 20} = .67. \quad \therefore C_D = .0065, \text{ from Fig. 8.12.}$$

$$\therefore \text{Power} = \frac{18.4}{0.45} = \underline{40.9 \text{ Hp}}.$$

8.53 $C_L = 1.22 = \dfrac{1500 \times 9.81 + 3000}{\frac{1}{2} \times 1.007 \times V^2 \times 20}$. $\therefore V = \underline{38.0 \text{ m/s}}$.

8.54 $(C_L)_{cruise} = \dfrac{1500 \times 9.81 + 3000}{\frac{1}{2} \times 1.007 \times 80^2 \times 20} = 0.275.$ $\therefore C_D = \dfrac{0.275}{48} = 0.0057.$

$$\therefore \dot{W} = F_D V = \frac{1}{2} \times 1.007 \times 80^3 \times 20 \times 0.0057 = 29\,400 \text{ W or } 39.4 \text{ Hp}$$

$$\% \text{ change} = \frac{39.4 - 18.4}{18.4} \times 100 = \underline{114\% \text{ increase}}$$

The increased power is due to the increase in air density.

8.55 $C_L = 1.22 = \dfrac{1500 \times 9.81 + 9000}{\frac{1}{2} \times 1.22 \times V^2 \times 20}$. $\therefore V = \underline{39.9 \text{ m/s}}$.

8.56 $C_L = 1.72 = \dfrac{250\,000 \times 9.81}{\frac{1}{2} \times 1.22 \times V^2 \times 60 \times 8}$. $\therefore V = \underline{69.8 \text{ m/s}}$.

8.57 a) $C_L = 1.72 = \dfrac{250\,000 \times 9.81}{\frac{1}{2} \times 1.05 \times V^2 \times 60 \times 8}$. $\therefore V = 75.2 \text{ m/s}$.

$$\% \text{ change} = \frac{75.2 - 69.8}{69.8} \times 100 = \underline{7.77\% \text{ increase}}$$

b) $C_L = 1.72 = \dfrac{250\ 000 \times 9.81}{\frac{1}{2} \times 1.515 V^2 \times 60 \times 8}$. $\therefore V = 62.6 \text{ m/s}$ $\left(\rho = \dfrac{101.3}{.287 \times 233} = 1.515 \text{ kg/m}^3\right)$

$$\% \text{ change} = \frac{62.6 - 69.8}{69.8} \times 100 = \underline{-10.3\%}$$

c) $C_L = 1.72 = \dfrac{250\ 000 \times 9.81}{\frac{1}{2} \times 1.093 V^2 \times 60 \times 8}$. $\therefore V = 73.7 \text{ m/s}$ $\left(\rho = \dfrac{101.3}{.287 \times 323} = 1.093 \text{ kg/m}^3\right)$

$$\% \text{ change} = \frac{73.7 - 69.8}{69.8} \times 100 = \underline{5.63\% \text{ increase}}$$

8.58 For a conventional airfoil assume $C_L / C_D = 47.6$ at $C_L = 0.3$.

$$0.3 = \frac{m \times 9.81}{\frac{1}{2} \times 0.526 \times 222^2 \times 200 \times 30}. \quad \therefore m = \underline{2.38 \times 10^6 \text{ kg}}$$

$$\dot{W} = F_D V = \frac{1}{2} \times 0.526 \times 222^3 \times 200 \times 30 \times \frac{0.3}{47.6} = 490\ 000 \text{ W or } \underline{657 \text{ Hp}}$$

8.59 $\vec{\nabla} \times \left[\dfrac{\partial \vec{V}}{\partial t} + (\vec{V} \cdot \vec{\nabla})\vec{V} + \dfrac{\vec{\nabla} p}{\rho} - \nu \nabla^2 \vec{V} \right] = 0.$

$$\vec{\nabla} \times \frac{\partial \vec{V}}{\partial t} = \frac{\partial}{\partial t}(\vec{\nabla} \times \vec{V}) = \frac{\partial \vec{\omega}}{\partial t}. \quad \vec{\nabla} \times \frac{\vec{\nabla} p}{\rho} = \frac{1}{\rho} \vec{\nabla} \times \vec{\nabla} p = 0.$$

$\vec{\nabla} \times (\nabla^2 \vec{V}) = \nabla^2 (\vec{\nabla} \times \vec{V}) = \nabla^2 \vec{\omega}$ (we have interchanged derivatives)

$$\vec{\nabla} \times \left[(\vec{V} \cdot \vec{\nabla})\vec{V} \right] = \vec{\nabla} \times \left[\frac{1}{2} \vec{\nabla} V^2 - \vec{V} \times (\vec{\nabla} \times \vec{V}) \right] = \frac{1}{2} (\underbrace{\vec{\nabla} \times \vec{\nabla} V^2}_{0}) - \vec{\nabla} \times (\vec{V} \times \vec{\omega})$$

$$= \vec{V}(\underbrace{\vec{\nabla} \cdot \vec{\omega}}_{0}) - \vec{\omega}(\underbrace{\vec{\nabla} \cdot \vec{V}}_{0}) + (\vec{V} \cdot \vec{\nabla})\vec{\omega} - (\vec{\omega} \cdot \vec{\nabla})\vec{V}$$

$$= (\vec{V} \cdot \vec{\nabla})\vec{\omega} - (\vec{\omega} \cdot \vec{\nabla})\vec{V} \quad \text{where } \vec{\nabla} \cdot \vec{\omega} = \vec{\nabla} \cdot (\vec{\nabla} \times \vec{V}) = 0.$$

There results: $\dfrac{\partial \vec{\omega}}{\partial t} + (\vec{V} \cdot \vec{\nabla})\vec{\omega} - (\vec{\omega} \cdot \vec{\nabla})\vec{V} - \nu \nabla^2 \vec{\omega} = 0.$

This is written as $\dfrac{D\vec{\omega}}{Dt} = (\vec{\omega} \cdot \vec{\nabla})\vec{V} + \nu \nabla^2 \vec{\omega}.$

8.60 x-comp: $\dfrac{\partial \omega_x}{\partial t} + u \dfrac{\partial \omega_x}{\partial x} + v \dfrac{\partial \omega_x}{\partial y} + w \dfrac{\partial \omega_x}{\partial z} = \omega_x \dfrac{\partial u}{\partial x} + \omega_y \dfrac{\partial u}{\partial y} + \omega_z \dfrac{\partial u}{\partial z} + \nu \nabla^2 \omega_x$

y-comp:

$$\frac{\partial \omega_y}{\partial t} + u \frac{\partial \omega_y}{\partial x} + v \frac{\partial \omega_y}{\partial y} + w \frac{\partial \omega_y}{\partial z} = \omega_x \frac{\partial v}{\partial x} + \omega_y \frac{\partial v}{\partial y} + \omega_z \frac{\partial v}{\partial z} + \nu \left(\frac{\partial^2 \omega_y}{\partial x^2} + \frac{\partial^2 \omega_y}{\partial y^2} + \frac{\partial^2 \omega_y}{\partial z^2} \right)$$

z-comp:

$$\frac{\partial\omega_z}{\partial t}+u\frac{\partial\omega_z}{\partial x}+v\frac{\partial\omega_z}{\partial y}+w\frac{\partial\omega_z}{\partial z}=\omega_x\frac{\partial w}{\partial x}+\omega_y\frac{\partial w}{\partial y}+\omega_z\frac{\partial w}{\partial z}+\nu\left(\frac{\partial^2\omega_z}{\partial x^2}+\frac{\partial^2\omega_z}{\partial y^2}+\frac{\partial^2\omega_z}{\partial z^2}\right)$$

8.61 $\omega_x=\cancelto{0}{\frac{\partial w}{\partial y}}-\cancelto{0}{\frac{\partial v}{\partial z}}=0. \quad \omega_y=\cancelto{0}{\frac{\partial u}{\partial z}}-\cancelto{0}{\frac{\partial w}{\partial x}}=0. \quad \omega_z=\frac{\partial v}{\partial x}-\frac{\partial u}{\partial y}\neq 0.$

$$\frac{D\omega_z}{DT}=(\vec{\omega}\cdot\vec{\nabla})\cancelto{0}{w}+\nu\nabla^2\omega_z \;;\quad \therefore \underline{\frac{D\omega_z}{Dt}=\nu\nabla^2\omega_z}.$$

If viscous effects are negligible, then $\frac{D\omega_z}{Dt}=0.$

Thus, for a planer flow, $\underline{\omega_z=\text{const}}$ if viscous effects are small.

8.62 a) $\vec{\nabla}\times\vec{V}=\left(\frac{\partial w}{\partial y}-\frac{\partial v}{\partial z}\right)\hat{i}+\left(\frac{\partial u}{\partial z}-\frac{\partial w}{\partial x}\right)\hat{j}+\left(\frac{\partial v}{\partial x}-\frac{\partial u}{\partial y}\right)\hat{k}=0.$ $\quad\therefore$ <u>irrotational</u>

$\frac{\partial\phi}{\partial x}=10x. \quad \therefore\phi=5x^2+f(y)$

$\frac{\partial\phi}{\partial y}=\frac{\partial f}{\partial y}=20y. \quad \therefore f=10y^2+C. \quad \text{Let } C=0.$

$\therefore \underline{\phi=5x^2+10y^2}$

b) $\vec{\nabla}\times\vec{V}=0\hat{i}+0\hat{j}+(8-8)\hat{k}=0.$ $\quad\therefore$ <u>irrotational</u>

$\frac{\partial\phi}{\partial x}=8y. \quad \therefore\phi=8xy+f(y,z)$

$\frac{\partial\phi}{\partial y}=8x+\frac{\partial f}{\partial y}=8x. \quad \therefore\frac{\partial f}{\partial y}=0 \text{ and } f=f(z).$

$\frac{\partial\phi}{\partial z}=\frac{df}{dz}=-6z. \quad \therefore f=-3z^2+C. \quad \text{Let } C=0.$

$\therefore \underline{\phi=8xy-3z^2}$

c) $\vec{\nabla}\times\vec{V}=0\hat{i}+0\hat{j}+\left(\frac{-y\frac{1}{2}(x^2+y^2)^{-1/2}2x}{x^2+y^2}-\frac{-x\frac{1}{2}(x^2+y^2)^{-1/2}2y}{x^2+y^2}\right)\hat{k}=0.$ $\quad\therefore$ <u>irrotational</u>

$\frac{\partial\phi}{\partial x}=\frac{x}{\sqrt{x^2+y^2}}. \quad \therefore\phi=\sqrt{x^2+y^2}+f(y)$

$$\frac{\partial \phi}{\partial y} = \frac{1}{2}(x^2+y^2)^{-1/2}2y + \frac{\partial f}{\partial y} = \frac{y}{\sqrt{x^2+y^2}}. \quad \therefore \frac{\partial f}{\partial y} = 0. \quad \therefore f = C. \text{ Let } C = 0.$$

$\therefore \underline{\phi = \sqrt{x^2+y^2}}$

d) $\vec{\nabla} \times \vec{V} = 0\hat{i} + 0\hat{j} + \left[\frac{-y(2x)}{(x^2+y^2)^2} - \frac{-x(2y)}{(x^2+y^2)^2}\right]\hat{k} = 0.$ $\therefore$ <u>irrotational</u>

$$\frac{\partial \phi}{\partial x} = \frac{x}{x^2+y^2}. \quad \therefore \phi = \frac{1}{2}\ell n(x^2+y^2) + f(y)$$

$$\frac{\partial \phi}{\partial y} = \frac{y}{x^2+y^2} = \frac{1}{2}\frac{2y}{x^2+y^2} + \frac{\partial f}{\partial y}. \quad \therefore \frac{\partial f}{\partial y} = 0. \quad \therefore f = C. \text{ Let } C = 0.$$

$\therefore \underline{\phi = \ell n\sqrt{x^2+y^2}}$

8.63 $\frac{\partial^2 \psi}{\partial x^2} + \frac{\partial^2 \psi}{\partial y^2} = 0.$ This requires two conditions on x and two on y.

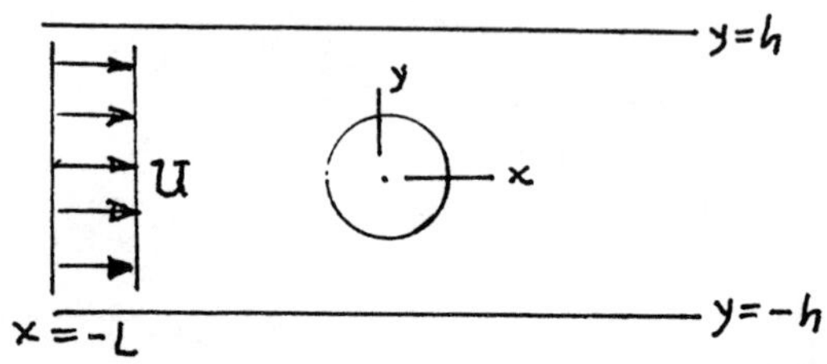

At $x = -L$, $u = U$. $\quad \therefore \frac{\partial \psi}{\partial y} = U.$

At $x = L$, $u = U$. $\quad \therefore \frac{\partial \psi}{\partial y} = U.$

At $y = -h$, $\psi = 0$.

At $y = h$, $\psi = \mathrm{U} \times \mathrm{h}$. (See Example 8.8).

The boundary conditions are stated as:

$\underline{\frac{\partial \psi}{\partial y}(-L,y) = U, \ \frac{\partial \psi}{\partial y}(L,y) = U, \ \psi(x,-h) = 0, \ \psi(x,h) = 2Uh.}$

8.64 $u = \frac{\partial \psi}{\partial y} = 100. \quad \therefore \psi = 100y + f(x). \quad v = -\frac{\partial \psi}{\partial x} = -\frac{df}{dx} = 50. \quad \therefore f = -50x + C.$

$\therefore \underline{\psi(x,y) = 100y - 50x.}$ (We usually let $C = 0$.)

$u = \frac{\partial \phi}{\partial x} = 100. \quad \therefore \phi = 100x + f(y). \quad v = \frac{\partial \phi}{\partial y} = \frac{df}{dy} = 50. \quad \therefore f = 50y + C.$

$\therefore \underline{\phi(x,y) = 100x + 50y.}$

8.65 a) $\psi = 40\theta$.

b) $\frac{1}{r}\frac{\partial}{\partial r}\left(\frac{\partial \psi}{\partial \theta}\right) + \frac{1}{r}\left(-\frac{\partial^2 \psi}{\partial \theta \partial r}\right) = \frac{1}{r}\frac{\partial}{\partial r}(40) + \frac{1}{r}\left(-\frac{\partial}{\partial \theta}(0)\right) = 0.$

$\therefore$ It is incompressible since the above continuity equation is satisfied.
Note: The continuity equation is found in Table 5.1.

c) $\dfrac{\partial\phi}{\partial r}=\dfrac{1}{r}\dfrac{\partial\psi}{\partial\theta}=\dfrac{40}{r}.\quad \therefore \phi = 40\ell n r + f(\theta)$

$\dfrac{\partial\phi}{\partial\theta}=\dfrac{\partial f}{\partial\theta}=-r\dfrac{\partial\psi}{\partial r}=0.\quad \therefore f = \text{C}.\ \ \text{Let C} = 0.$

$\therefore \underline{\phi = 40\ell n\, r}$

d) $v_r=\dfrac{40}{r},\quad v_\theta = 0.\quad a_r = v_r\dfrac{\partial v_r}{\partial r}=\dfrac{40}{r}\left(-\dfrac{40}{r^2}\right)=-10.$

$\therefore \underline{r = 5.43 \text{ m}}$

8.66 $u=\dfrac{\partial\psi}{\partial y}=20\dfrac{2y}{x^2+y^2}=\dfrac{\partial\phi}{\partial x}.\quad \therefore \phi = -40\tan^{-1}\dfrac{y}{x}+f(y).$

$v=\dfrac{\partial\phi}{\partial y}=-\dfrac{40/x}{1+y^2/x^2}+\dfrac{\partial f}{\partial y}=-\dfrac{40x}{x^2+y^2}+\dfrac{\partial f}{\partial y}=-20\dfrac{2x}{x^2+y^2}.\quad \therefore f = \text{C}.\ \ \text{Let C}=0.$

$\underline{\phi = -40\tan^{-1}\dfrac{y}{x}.}$

8.67 a) $\dfrac{\partial^2\psi}{\partial x^2}+\dfrac{\partial^2\psi}{\partial y^2}=0.\qquad \dfrac{\partial\psi}{\partial x}=10y(x^2+y^2)^{-2}(2x).$

$\dfrac{\partial^2\psi}{\partial x^2}=20y(x^2-y^2)^{-2}-80x^2y(x^2+y^2)^{-3}$

$\dfrac{\partial\psi}{\partial y}=10-10(x^2+y^2)^{-1}+10y(x^2+y^2)^{-2}(2y).$

$\dfrac{\partial^2\psi}{\partial y^2}=20y(x^2+y^2)^{-2}+40y(x^2+y^2)^{-2}-80y^3(x^2+y^2)^{-3}.$

$$\therefore \frac{\partial^2\psi}{\partial x^2}+\frac{\partial^2\psi}{\partial y^2}=\frac{20y}{(x^2+y^2)^2}-\frac{80x^2y}{(x^2+y^2)^3}+\frac{60y}{(x^2+y^2)^2}-\frac{80y^3}{(x^2+y^2)^3}$$

$$=\frac{80y(x^2+y^2)}{(x^2+y^2)^3}-\frac{80x^2y}{(x^2+y^2)^3}-\frac{80y^3}{(x^2+y^2)^3}=\frac{80x^2y+80y^3-80x^2y-80y^3}{(x^2+y^2)^3}=0.$$

b) In polar coord: $\psi(r,\theta)=10r\sin\theta-\dfrac{10r\sin\theta}{r^2}=10r\sin\theta-\dfrac{10}{r}\sin\theta.$

$\dfrac{1}{r}\dfrac{\partial\psi}{\partial\theta}=\left(10-\dfrac{10}{r^2}\right)\cos\theta=\dfrac{\partial\phi}{\partial r}.\quad \therefore \phi=\left(10+\dfrac{10}{r}\right)\cos\theta+f(\theta).$

$\dfrac{1}{r}\dfrac{\partial\phi}{\partial\theta}=\dfrac{1}{r}\dfrac{df}{d\theta}-\left(10+\dfrac{10}{r^2}\right)\sin\theta=-\dfrac{\partial\psi}{\partial r}=-10\sin\theta-\dfrac{10}{r^2}\sin\theta.\quad \dfrac{df}{d\theta}=0.\ f=\text{C}.$ 0

$\therefore \phi=10\left(r+\dfrac{1}{r}\right)\cos\theta\ \text{ or }\ \underline{\phi(x,y)=10x+\dfrac{10x}{x^2+y^2},}$

where we let $r\cos\theta = x$ and $r^2 = x^2+y^2$.

c) Along the x-axis, $v = -\frac{\partial\psi}{\partial x} = 0$ (let $y = 0$ in part (a)) and

$$u = \frac{\partial\psi}{\partial y} = 10 - \frac{10}{x^2 + y^2} + \frac{20y^2}{(x^2 + y^2)^2} = 10 - \frac{10}{x^2} \text{ with } y = 0.$$

Euler's Eq: $\rho u \frac{\partial u}{\partial x} = -\frac{\partial p}{\partial x}$. $\therefore \rho\left(10 - \frac{10}{x^2}\right)\left(\frac{20}{x^3}\right) = -\frac{\partial p}{\partial x}$.

$$\therefore p = \int \rho\left(\frac{200}{x^5} - \frac{200}{x^3}\right)dx = \rho\left[-\frac{50}{x^4} + \frac{100}{x^2}\right] + C. \quad C = 50\,000.$$

$$= 1000\left[\frac{100}{x^2} - \frac{50}{x^4}\right] + 50\,000 \text{ Pa.} \quad \text{(Could have used Bernoulli!)}$$

d) Let $u = 0$: $0 = 10 - \frac{10}{x^2}$. $\therefore x = \pm 1$. $\therefore$ Stag pts: (1,0), (-1,0)

8.68 a) $\frac{\partial^2\phi}{\partial x^2} + \frac{\partial^2\phi}{\partial y^2} = \frac{\partial}{\partial x}\left[10 + \frac{10x}{x^2 + y^2}\right] + \frac{\partial}{\partial y}\left[\frac{10y}{x^2 + y^2}\right] = \frac{(x^2 + y^2)10 - 10x(2x}{(x^2 + y^2)^2}$

$$+ \frac{(x^2 + y^2)10 - 10y(2y)}{(x^2 + y^2)^2} = \frac{10x^2 + 10y^2 - 20x^2 + 10x^2 + 10y^2 - 20y^2}{(x^2 + y^2)^2} = 0.$$

b) Polar coord: $\phi = 10r\cos\theta + 5\ell n\, r^2$. (See Eq. 8.5.14.)

$$\frac{\partial\phi}{\partial r} = 10\cos\theta + \frac{10r}{r^2} = \frac{1}{r}\frac{\partial\psi}{\partial\theta}. \quad \therefore \psi = 10r\sin\theta + 10\theta + f(r)$$

$$\frac{1}{r}\frac{\partial\phi}{\partial\theta} = -10\sin\theta = -\frac{\partial\psi}{\partial r} = -10\sin\theta - \frac{df}{dr}. \quad \therefore f = \not{C}^{\,0}. \quad \therefore \psi = 10r\sin\theta + 10\theta.$$

$$\therefore \psi(x,y) = 10y + 10\tan^{-1}\frac{y}{x}.$$

c) $v = \frac{\partial\phi}{\partial y} = \frac{10y}{x^2 + y^2}$. Along x-axis ($y = 0$) $v = 0$.

$u = \frac{\partial\phi}{\partial x} = 10 + \frac{10x}{x^2 + y^2}$. Along x-axis $u = 10 + \frac{10}{x}$.

Bernoulli: $\frac{V^2}{2} + \frac{p}{\rho} + \not{g}z = \frac{V_\infty^2}{2} + \frac{p_\infty}{\rho} + \not{g}z_\infty$ (assume $z = z_\infty$)

$$\frac{(10 + 10/x)^2}{2} + \frac{p}{\rho} = \frac{10^2}{2} + \frac{100\,000}{\rho}. \quad \therefore p = 100 - 50\left(\frac{2}{x} + \frac{1}{x^2}\right) \text{ kPa.}$$

d) $u = 0$: $0 = 10 + \frac{10}{x}$. $\therefore x = -1$. $\therefore$ Stag pt: (-1,0)

e) $a_y = v\frac{\partial v}{\partial y} + u\frac{\partial v}{\partial x} = 0$ on x – axis. $a_x = u\frac{\partial u}{\partial x} + v\frac{\partial u}{\partial y} = \left(10+\frac{10}{x}\right)\left(-\frac{10}{x^2}\right)$.

$\therefore a_x(-2,0) = (10-5)\left(-\frac{10}{4}\right) = \underline{-12.5 \text{ m/s}^2}$.

8.69 $u(x,y) = y - 5y^2 = \frac{\partial \psi}{\partial y}$. $\therefore \psi = \frac{y^2}{2} - \frac{5y^3}{3} + C$. $\therefore \underline{\psi = \frac{1}{6}(3y^2 - 10y^3)}$.

$q = \int_0^{.2} u\,dy = \int_0^{.2} (y - 5y^2)dy = \frac{0.2^2}{2} - 5\frac{0.2^3}{3} = \underline{6.667\times10^{-3} \text{ m}^2\text{/s}}$.

$\psi_2 - \psi_1 = \frac{1}{6}(3\times0.2^2 - 10\times0.2^3) - 0 = \underline{6.667\times10^{-3} \text{ m}^2\text{/s}}$.

$\omega = -\frac{\partial u}{\partial y} = -1 + 10y \neq 0$. $\therefore \phi$ doesn't exist.

8.70 $\psi = 10y + \frac{\pi/2}{2\pi}\theta = 10r\sin\theta + \frac{1}{4}\theta$.

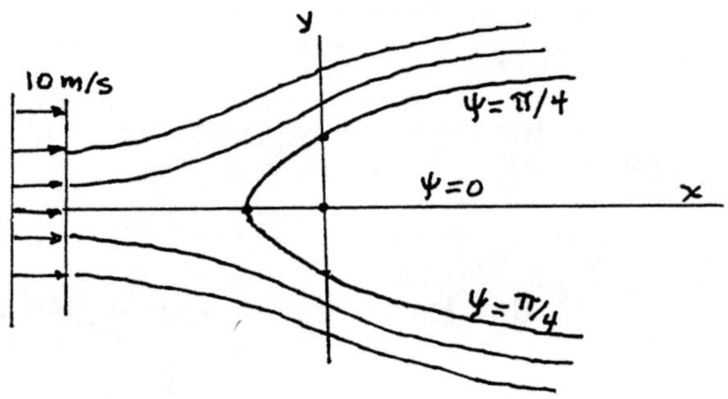

a) $v_r = \frac{1}{r}\frac{\partial \psi}{\partial \theta} = 10\cos\theta + \frac{1}{4r} = 0$.

At $\theta = \pi$, $\frac{1}{4r_s} = 10$. $\therefore r_s = \frac{1}{40}$ m.

Stag. pt: $\underline{(-2.5 \text{ cm}, 0)}$.

b) At $\theta = \pi$, $r = \frac{1}{40}$, $\psi_s = 10\times\frac{1}{40}\times 0 + \frac{\pi}{4} = 10r\sin\frac{\pi}{2} + \frac{\pi/2}{4}$.

$\therefore r = y_{\text{inter}} = \frac{\pi/8}{10} = \frac{\pi}{80}$ m or $\underline{3.93 \text{ cm}}$.

c) $q = U\times H = \Delta\psi$. $\frac{\pi}{4} = 10\,H$. $\therefore H = \frac{\pi}{40}$. Thickness = $2H = \frac{\pi}{20}$ m or $\underline{15.7 \text{ cm}}$.

d) $v_r(.04,\pi) = 10\cos\pi + \frac{1}{4\times.04} = -3.75$. $\therefore \underline{u = 3.75 \text{ m/s}}$.

8.70E $\psi = 30y + \frac{5\pi}{2\pi}\theta = 30r\sin\theta + \frac{5}{2}\theta$.

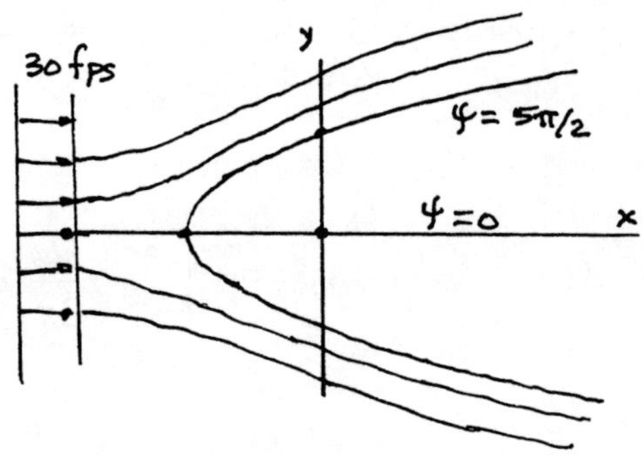

a) $v_r = \frac{1}{r}\frac{\partial \psi}{\partial \theta} = 30\cos\theta + \frac{5}{2r} = 0$.

At $\theta = \pi$, $\frac{5}{2r_s} = 30$. $\therefore r_s = 0.0833'$.

Stag. pt: $\underline{(-1'', 0)}$.

b) At $\theta = \pi,\ r = .0833,\quad \psi_s = \frac{5\pi}{2} = 30r\sin\frac{\pi}{2} + \frac{5}{2}\frac{\pi}{2}.$

$\therefore r = y_{\text{inter}} = \underline{.0119\text{ ft.}}$

c) $q = U \times H = \Delta\psi.\ \therefore 30H = \frac{5\pi}{2}.\ \therefore H = \frac{5\pi}{60}.$ Thickness $= 2H = \frac{5\pi}{30}$ ft or $\underline{1.257''}$.

d) $v_r(1,\pi) = 30\cos\pi + \frac{5}{2} = -30 + 2.5 = -27.5.\quad \therefore \underline{u = 27.5\text{ fps.}}$

8.71 $$\phi = \frac{\pi}{2\pi}\ell n\left[(x+1)^2 + y\right]^{1/2} - \frac{\pi}{2\pi}\ell n\left[(x-1)^2 + y^2\right]^{1/2} + 10x = \frac{1}{4}\ell n\left[(x+1)^2 + y^2\right]$$
$$-\frac{1}{4}\ell n\left[(x-1)^2 + y^2\right] + 10x.$$

$$u = \left.\frac{\partial\phi}{\partial x}\right|_{y=0} = \frac{\frac{1}{4}[2(x+1)]}{(x+1)^2} - \frac{\frac{1}{4}[2(x-1)]}{(x-1)^2} + 10 = \frac{1}{2(x+1)} - \frac{1}{2(x-1)} + 10.\quad v = 0 \text{ if } y = 0.$$

At the stagnation point, $u = 0$. $\therefore \frac{1}{2(x+1)} - \frac{1}{2(x-1)} + 10 = 0.\ \therefore \frac{2}{x^2 - 1} = 20.$

$\therefore x^2 = 1.1.\quad \therefore x = \pm 1.049$ m. $\quad\therefore$ oval length $= 2 \times 1.049 = \underline{2.098\text{ m}}$.

All the flow from the source goes to the sink, i.e., $\pi\ \text{m}^2/\text{s}$, or $\frac{\pi}{2}\ \text{m}^2/\text{s}$ for $y > 0$.

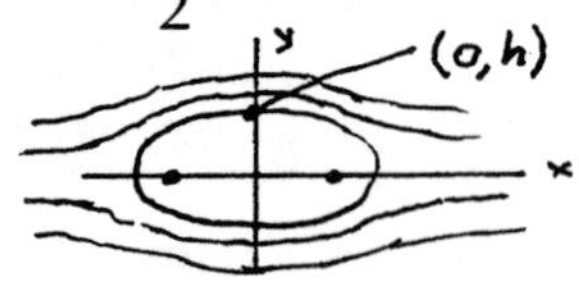

$$u(y) = \left.\frac{\partial\phi}{\partial x}\right|_{x=0} = \frac{\frac{1}{4}(2)}{1+y^2} - \frac{\frac{1}{4}(-2)}{1+y^2} + 10 = \frac{1}{1+y^2} + 10.$$

$$q = \int_0^h \left(\frac{1}{1+y^2} + 10\right)dy = \frac{\pi}{2}.\quad \therefore \tan^{-1}h + 10h = \frac{\pi}{2}.\quad \therefore h = 0.143\text{ m.}$$

Thickness $= 2h = \underline{0.286\text{ m}}$.

The minimum pressure occurs on the oval surface at $(0,h)$.

There $u = \frac{1}{1+.143^2} + 10 = 10.98\text{ m/s}.$

Bernoulli: $\frac{V^2}{2} + \frac{p}{\rho} = \frac{V_\infty^2}{2} + \frac{p_\infty}{\rho}.\quad \frac{10.98^2}{2} + \frac{p}{1000} = \frac{10^2}{2} + \frac{10\,000}{1000}.$

$\therefore \underline{p_{\min} = -280\text{ Pa.}}$

8.72 $$\phi = \frac{2\pi}{2\pi}\ell n\left[(x+1)^2 + y^2\right]^{1/2} + \frac{-2\pi}{2\pi}\ell n\left[(x-1)^2 + y^2\right]^{1/2} + 2x = \frac{1}{2}\ell n\left[(x+1)^2 + y^2\right]$$
$$-\frac{1}{2}\ell n\left[(x-1)^2 + y^2\right] + 2x.$$

$$u = \frac{\partial\phi}{\partial x} = \frac{\frac{1}{2}2(x+1)}{(x+1)^2 + y^2} - \frac{\frac{1}{2}2(x-1)}{(x-1)^2 + y^3} + 2.\quad v = \frac{y}{(x-1)^2 + y^2} - \frac{y}{(x-1)^2 + y^2}$$

Along the x-axis ($y = 0$), $v = 0$ and $u = \frac{1}{x+1} - \frac{1}{x-1} + 2.$

Set $u = 0$: $\frac{1}{x-1} - \frac{1}{x+1} = 2$, or $x^2 = 2$. $\therefore x = \pm\sqrt{2}.$

Stag. pts.: $\underline{(\sqrt{2},0),\ (-\sqrt{2},0)}.$

$$u(4,0) = \frac{1}{-4+1} - \frac{1}{-4-1} + 2 = \underline{1.867\ \text{m/s}}. \qquad v(-4,0) = \underline{0}.$$

$$u(0,4) = \frac{1}{1+4^2} - \frac{-1}{1+4^2} + 2 = \underline{2.118\ \text{m/s}}. \qquad v(0,4) = \frac{4}{1+4^2} - \frac{4}{1+4^2} = \underline{0}.$$

8.73 $$\phi = \frac{2\pi}{2\pi}\ell n[x^2+(y-1)^2]^{1/2} + \frac{2\pi}{2\pi}\ell n[x^2+(y+1)^2]^{1/2}$$

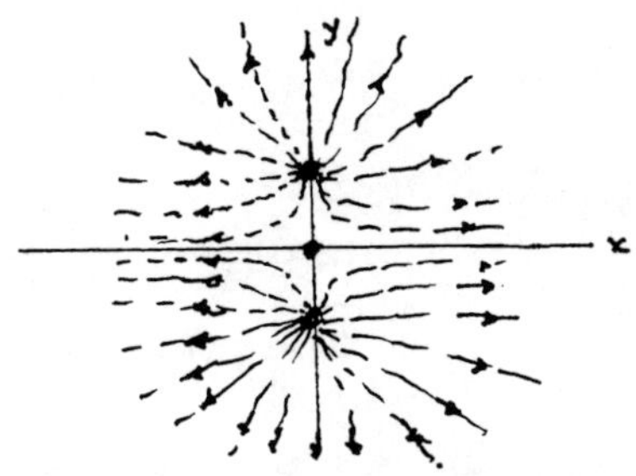

$$= \frac{1}{2}\ell n[x^2+(y-1)^2] + \frac{1}{2}\ell n[x^2+(y+1)^2].$$

$$u = \frac{\partial\phi}{\partial x} = \frac{x}{x^2+(y-1)^2} + \frac{x}{x^2+(y+1)^2}.$$

$$v = \frac{\partial\phi}{\partial y} = \frac{y-1}{x^2+(y-1)^2} + \frac{y+1}{x^2+(y-1)^2}.$$ <u>At (0,0)</u> $u = 0$ and $v = 0$.

At (1,1) $v = 0 + \frac{2}{2^2+1^2} = \frac{2}{5} = 0.4\ \text{m/s}.$ $u = \frac{1}{1^2} + \frac{1}{2^2+1^2} = 1.2\ \text{m/s}.$

$\therefore \underline{\vec{V} = 1.2\hat{i} + 0.4\hat{j}\ \text{m/s}}.$

8.73E $$\phi = \frac{20\pi}{2\pi}\ell n[(y-6)^2+x^2]^{1/2} + \frac{20\pi}{2\pi}\ell n[(y+6)^2+x^2]^{1/2}$$ (see figure above)

$$= 5\ell n[(y-6)^2+x^2] + 5\ell n[(y+6)^2+x^2].$$

$$v = \frac{\partial\phi}{\partial y} = \frac{10(y-6)}{(y-6)^2+x^2} + \frac{10(y+6)}{(y+6)^2+x^2}.$$

$$u = \frac{\partial\phi}{\partial x} = \frac{10x}{(y-6)^2+x^2} + \frac{10x}{(y+6)^2+x^2}.$$ <u>At (0,0)</u> $u = 0$, $v = 0$.

At (3,3) $v = \frac{-30}{3^2+3^2} + \frac{90}{9^2+3^2} = -0.667\ \text{fps}.$ $u = \frac{30}{3^2+3^2} + \frac{30}{9^2+3^2} = 2\ \text{fps}.$

$\therefore \underline{\vec{V} = 2\hat{i} - 0.667\hat{j}\ \text{fps}}.$

8.74 $$\phi = \frac{2\pi}{2\pi}\ell n[(y-1)^2+x^2]^{1/2} + \frac{2\pi}{2\pi}\ell n[(y+1)^2+x^2]^{1/2} + U_\infty x.$$

$$= \frac{1}{2}\ell n[(y-1)^2+x^2] + \frac{1}{2}\ell n[(y+1)^2+x^2] + U_\infty x.$$

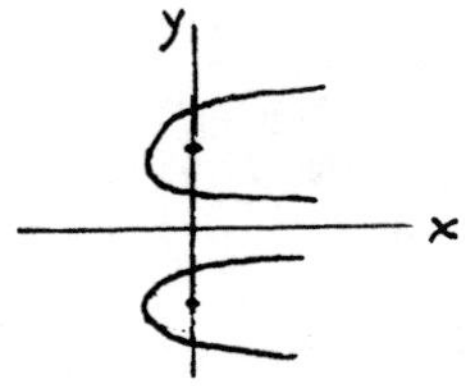

a) Stag. pts. may occur on x-axis, $y = 0$.

$$u = \left.\frac{\partial\phi}{\partial x}\right|_{y=0} = \frac{x}{1+x^2} + \frac{x}{1+x^2} + 10.$$

$\therefore x^2 + .2x + 1 = 0.$ $\therefore$ no stagnation points exist on the x-axis. (They do exist away from the x-axis.)

Along the y-axis: $u(y) = 10.$ $q = \int_0^h u dy = \frac{1}{2}(2\pi) = \pi \text{ m}^2/\text{s}.$

$\therefore \pi = \int_0^h 10 dy = 10\, h.$ $\therefore h = \underline{0.314 \text{ m}}.$

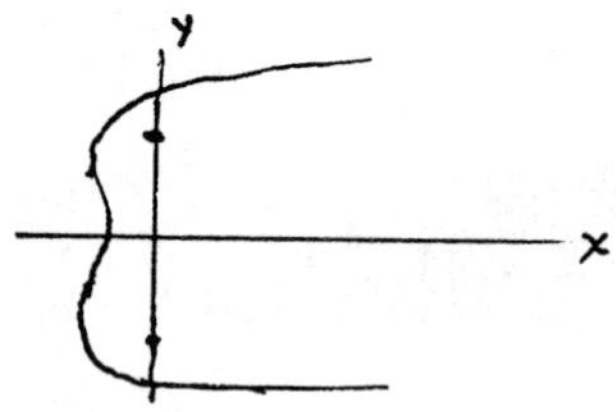

b) $u = \frac{2x}{1+x^2} + 1.$ $\therefore x^2 + 2x + 1 = 0.$ $\therefore x = -1 \text{ m}.$

Stag. pt.: $\underline{(-1,0)}$

Along the y-axis: $u = 1.0.$ $\therefore \pi = 1 \times h.$ $\therefore \underline{h = 3.14 \text{ m}}.$

c) $u = \frac{2x}{1+x^2} + 0.2.$ $\therefore x^2 + 10x + 1 = 0.$ $\therefore x = -9.90, -0.10 \text{ m}.$

Stag. pts.: $\underline{(-9.9,0)}$, $\underline{(-0.1,0)}$.

Along the y-axis: $u = 0.2.$ $\therefore \pi = 0.2\, h.$ $\therefore \underline{h = 15.71 \text{ m}}.$

8.75 $\phi = \frac{60}{r}\cos\theta + 8r\cos\theta.$

a) $v_r = \frac{\partial\phi}{\partial r} = -\frac{60}{r^2}\cos\theta + 8\cos\theta = \left(8 - \frac{60}{r^2}\right)\cos\theta.$

At the cylinder surface $v_r = 0$ for all θ. Hence,

$$\frac{60}{r_c^2} = 8. \quad \therefore \underline{r_c = 2.739 \text{ m}}$$

b) Bernoulli: $\Delta p = \rho\frac{U_\infty^2}{2} = 1000\frac{8^2}{2} = 32\,000$ Pa or $\underline{32 \text{ kPa}}$

c) $v_\theta = \frac{1}{r}\frac{\partial\phi}{\partial\theta} = -\frac{60}{r^2}\sin\theta - 8\sin\theta.$

At $r = r_c$, $v_\theta = -8\sin\theta - 8\sin\theta = \underline{-16\sin\theta}$

d) $\Delta p = \rho\frac{v_{90^\circ}^2}{2} = 1000\frac{16^2}{2} = 128\,000$ Pa or $\underline{128 \text{ kPa}}$

8.76 $\psi = \frac{4\pi}{2\pi}\theta + \frac{20\pi}{2\pi}\ell n\, r = 2\theta + 10\ell n\, r$

At $(x, y) = (0,1)$, $(r,\theta) = (1, \pi/2)$.

$$v_r(1,\pi/2) = \frac{1}{r}\frac{\partial\psi}{\partial\theta} = \frac{1}{1}(2) = 2.$$

$$v_\theta(1,\pi/2) = -\frac{\partial\psi}{\partial r} = -\frac{10}{1} = -10.$$

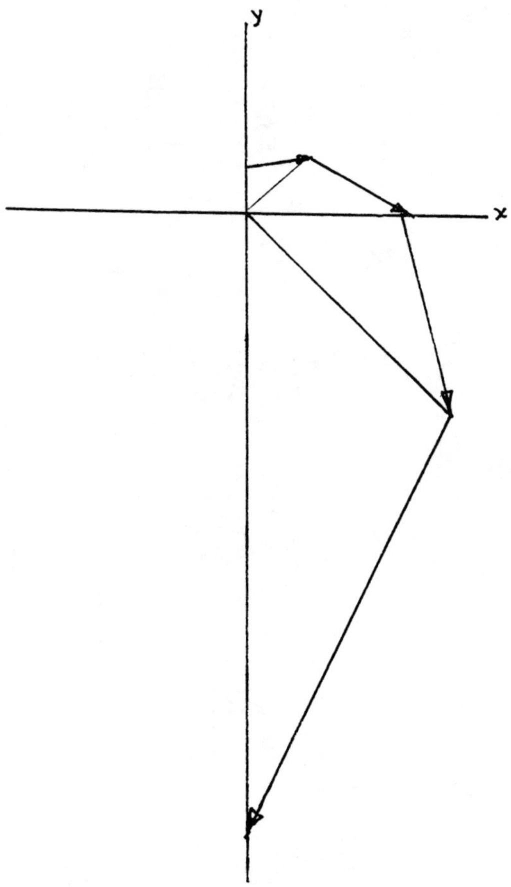

$$v_r(1.7,\pi/4) = \frac{2}{1.7} = 1.18$$

$$v_\theta(1.7,\pi/4) = \frac{-10}{1.7} = -5.88$$

$$v_r(3.2,0) = \frac{2}{3.2} = 0.625$$

$$v_\theta(3.2,0) = \frac{-10}{3.2} = 3.125$$

$$v_r(6,-\pi/4) = \frac{2}{6} = 0.333$$

$$v_\theta(6,-\pi/4) = \frac{-10}{6} = -1.67$$

etc.

Note: We scaled the radius at each 45° increment to find r.

b) $v_r = \frac{2}{r}$ and $v_\theta = -\frac{10}{r}$.

From Table 5.1 (use the l.h.s. of momentum)

$$a_r = \frac{Dv_r}{Dt} + \frac{v_\theta^2}{r} = v_r\frac{\partial v_r}{\partial r} - \frac{v_\theta^2}{r}$$

$$= \frac{2}{r}\left(-\frac{2}{r^2}\right) - \frac{100}{r^3} = -\frac{104}{1}$$

$$= -104 \text{ m/s}^2$$

$$a_\theta = \frac{Dv_\theta}{Dt} + \frac{v_r v_\theta}{r} = v_r\frac{\partial v_\theta}{\partial r} + \frac{v_r v_\theta}{r}$$

$$= \frac{2}{r}\left(\frac{10}{r^2}\right) + \frac{2(-10)}{r^3} = 0.$$

$$\therefore \vec{a}(0,1) = \underline{(-104,0) \text{ m/s}^2}$$

c) $v_r(14.14,\pi/4) = 2/14.14 = 0.1414, \quad v_\theta(14.14,\pi/4) = -\frac{10}{14.14} = -0.707$

$v_r(0.1,\pi/2) = 2/0.1 = 20, \quad v_\theta(0.1,\pi/2) = -10/0.1 = -100.$

Bernoulli: $\frac{20\,000}{1.2} + \frac{0.1414^2 + 0.707^2}{2} = \frac{p}{1.2} + \frac{20^2 + 100^2}{2}. \quad \therefore \underline{p = 13\,760 \text{ Pa}}$

We used $\rho_{air} = 1.2 \text{ kg/m}^3$ at standard conditions.

8.77 Along the y-axis $v_r = 0$ and $v_\theta = -10 - \frac{40}{r^2}$.

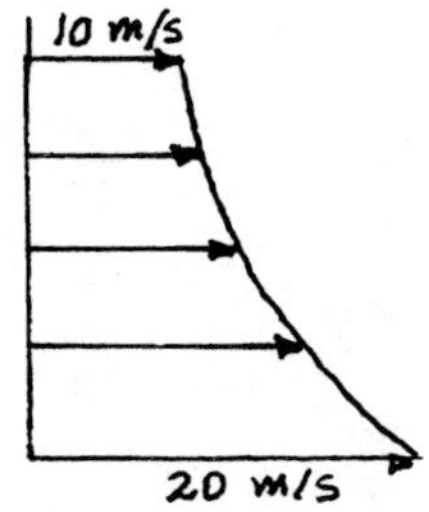

We have set $\theta = \frac{\pi}{2}$ in Eq. 8.5.27. $r_c = \sqrt{\frac{40}{10}} = 2.$

b) $v_r = 10\cos\theta - \frac{40}{r^2}\cos\theta.$ $(-4,3) \Rightarrow (5,126.9°).$

$v_\theta = -10\sin\theta - \frac{40}{r^2}\sin\theta.$ $\therefore \underline{v_r = -6.96 \text{ m/s}},\ \underline{v_\theta = -9.28 \text{ m/s}}.$

c) Use Eq. 8.5.28: $p = p_0 - 2\rho U_\infty^2 \sin^2\alpha$

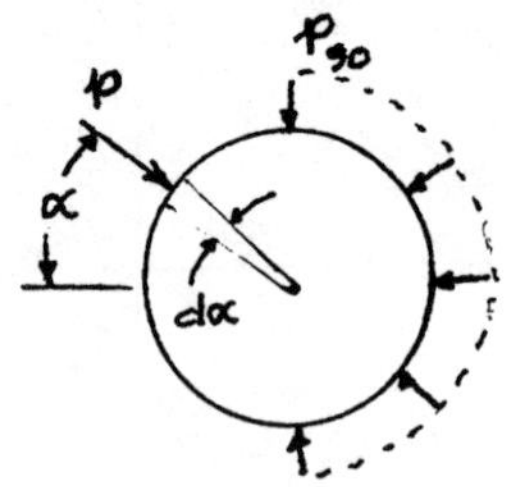

$$\text{Drag} = \int_{-\pi/2}^{\pi/2} p\cos\alpha\, r_c d\alpha L - p_{90} \times 2r_c L. \quad p_{90} = p_0 - 2\rho U_\infty^2.$$

$$= 2\int_0^{\pi/2} (p_0 - 2\rho U_\infty^2 \sin^2\alpha)\cos\alpha\, r_c L d\alpha - p_{90} \times 2r_c L$$

$$= 2r_c L\left[p_0 \sin\alpha - 2\rho U_\infty^2 \frac{\sin^3\alpha}{3}\right]_0^{\pi/2} - \left[p_0 - 2\rho U_\infty^2\right]2r_c L = \frac{8}{3} r_c L \rho U_\infty^2.$$

$$C_D = \frac{\text{Drag}}{\frac{1}{2}\rho U_\infty^2 A} = \frac{(8/3) r_c L \rho U_\infty^2}{\frac{1}{2}\rho U_\infty^2 2r_c L} = \frac{8}{3} = \underline{2.667}.$$

8.78 $v_r = U_\infty \cos\theta - \frac{\mu}{r^2}\cos\theta.$ Let $U_\infty = 4,\ \mu = r_c^2 U_\infty = 1^2 \times 4 = 4.$

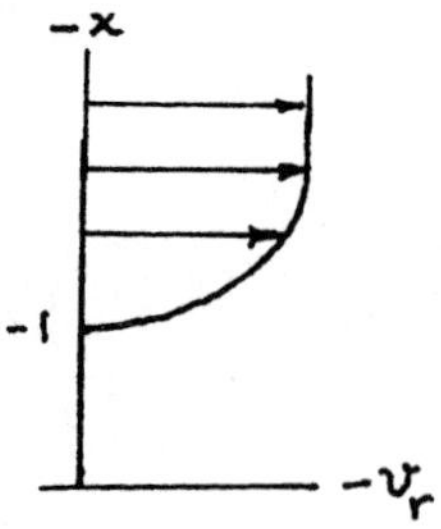

For $\theta = \pi,\ v_r = -4 + \frac{4}{r^2}.$

b) $v_\theta = -\frac{\partial\psi}{\partial r} = -U_\infty \sin\theta - \frac{\mu\sin\theta}{r_c^2} = \left(-4 - \frac{4}{1^2}\right)\sin\theta = \underline{-8\sin\theta}.$

c) $p_c = p_\infty + \rho\frac{V_\infty^2}{2} - \rho\frac{v_\theta^2}{2} = 50\,000 + 1000 \times \frac{4^2}{2} - 1000\frac{8^2\sin^2\theta}{2}.$

$\therefore p_c = \underline{58 - 32\sin^2\theta \text{ kPa.}}$

d) $\text{Drag} = 2\int_0^{\pi/2}(58 - 32\sin^2\alpha)\cos\alpha \times 1 \times 1 d\alpha - 26 \times 2 \times 1$

$= 2\left[58 - 32\left(\frac{1}{3}\right)\right] - 52 = \underline{42.7 \text{ kN}}.$ See figure in 8.77c

8.79 On the cylinder $v_\theta = -2U_\infty \sin\theta - \frac{\Gamma}{2\pi r_c} = -20\sin\theta - \frac{100}{\pi \times 2},\ r_c = 2.$

If $v_\theta = 0,\ \sin\theta = -\frac{50}{20\pi} = -0.7958.$ $\therefore \theta = 232.7°,\ 307.3°.$

Stag. pts.: $\underline{(2,232.7°)}$, $\underline{(2,307.3°)}$.

Max. pressure occurs on the cylinder at a stagnation pt.:

$$\therefore\ p_{max} = 0 - 1.22\frac{10^2}{2}\left[-2\times.7958 + \frac{200}{2\pi\times 2\times 10}\right]^2 + 1.22\frac{10^2}{2} = \underline{61\ \text{Pa}}.$$

Min. pressure occurs at the top of the cylinder where $\theta = 90°$:

$$p_{min} = 0 + 1.22\times\frac{10^2}{2} - 1.22\frac{10^2}{2}\left[2\times 1 + \frac{200}{2\pi\times 2\times 10}\right]^2 = \underline{-158\ \text{Pa}}.$$

8.80 $v_\theta = -2\times 20\sin\theta - \dfrac{\Gamma}{2\pi\times.4}$. For one stag. pt.: $v_\theta = 0$ at $\theta = 270°$:

$$0 = -2\times 20\sin 270° - \frac{\Gamma}{2\pi\times.4}.\quad \therefore\ \Gamma = 2\times 20\times 2\pi\times.4 = 100.5\ \text{m}^2/\text{s}.$$

$$\Gamma = 2\pi r_c^2\omega.\quad \therefore\ \omega = \frac{\Gamma}{2\pi r_c^2} = \frac{100.5}{2\pi\times.4^2} = \underline{100\ \text{rad/s}}.$$ (See Example 8.12.)

Min. pressure occurs where $|v_\theta|$ is max, i.e., $\theta = \pi/2$. There

$$v_\theta = -2\times 20\times 1 - \frac{100.5}{2\pi\times.4} = 80\ \text{m/s}.$$

$$\therefore\ p_{min} = p_\infty + \frac{V_\infty^2}{2}\rho - \frac{v_\theta^2}{2}\rho = 0 + \frac{20^2}{2}\times 1.22 - \frac{80^2}{2}\times 1.22 = \underline{-3660\ \text{Pa}}.$$

8.81 $\Gamma = 2\pi r_c^2\omega = 2\pi\times.6^2\times 120\times 2\pi/60 = 28.42\ \text{m}^2/\text{s}$. $\mu = r_c^2 U = .6^2\times 3 = 1.08\ \text{m}^3/\text{s}$.

$\therefore\ v_\theta = -2\times 3\sin\theta - \dfrac{28.42}{2\pi\times.6}$. $\therefore\ \sin\theta = -1.256$. Impossible. $\therefore$ Stag. pt. is off the cylinder at $\theta = 270°$, but $r > r_c$. From Eq. 8.5.29,

$$v_\theta = -\frac{\partial\psi}{\partial r} = -U_\infty\sin\theta - \frac{\mu}{r^2}\sin\theta - \frac{\Gamma}{2\pi r} = -3(-1) - \frac{1.08}{r^2}(-1) - \frac{28.42}{2\pi r} = 0.$$

$$\therefore\ 3 + \frac{1.08}{r^2} = \frac{4.523}{r}.\quad \therefore\ r^2 - 1.508r + 0.36 = 0.\quad \therefore\ r = 1.21\ \text{m}.$$

Stag. pt.: $\underline{(1.21, 270°)}$. $(v_\theta)_{90°} = -2\times 3 - \dfrac{28.42}{2\pi\times.6} = -13.54\ \text{m/s}$.

Min. pressure occurs at $\theta = 90°$, at $r = r_c$: $p_{min} = \left(\dfrac{3^2}{2} - \dfrac{13.54^2}{2}\right)1.22 = \underline{-106\ \text{Pa}}.$

Max. pressure occurs at $\theta = 270°$, at $r = r_c$: $p_{max} = \left(\dfrac{3^2}{2} - \dfrac{1.54^2}{2}\right)1.22 = \underline{-4.04\ \text{Pa}}.$

8.82 At 10 000 m, $\rho = .338\times 1.22 = .412\ \text{kg/m}^3$.

Lift $= \rho U_\infty \Gamma L = .412\times 100\times 1500\times 20 = \underline{1.24\times 10^6\ \text{N}}.$

8.82E At 15,000 ft, $\rho = .0015\ \text{slug/ft}^3$.

Lift $= \rho U_\infty \Gamma L = .0015\times 350\times 15{,}000\times 60 = \underline{472{,}000\ \text{lb}}.$

8.83 Place four sources as shown. Then, with $q = 2\pi$ for each:

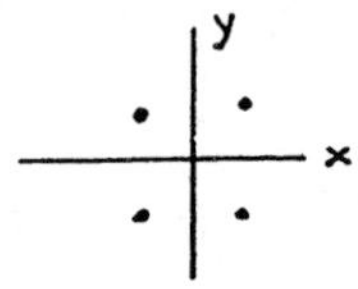

$$u(x,y) = \frac{x-2}{(x-2)^2+(y-2)^2} + \frac{x+2}{(x+2)^2+(y-2)^2} + \frac{x-2}{(x-2)^2+(y+2)^2} + \frac{x+2}{(x+2)^2+(y+2)^2}$$

$$v(x,y) = \frac{y-2}{(x-2)^2+(y-2)^2} + \frac{y+2}{(x-2)^2+(y+2)^2} + \frac{y-2}{(x+2)^2+(y-2)^2} + \frac{y+2}{(x+2)^2+(y+2)^2}$$

8.84 Place four sources with $q = 0.2\ \text{m}^2/\text{s}$, as shown.

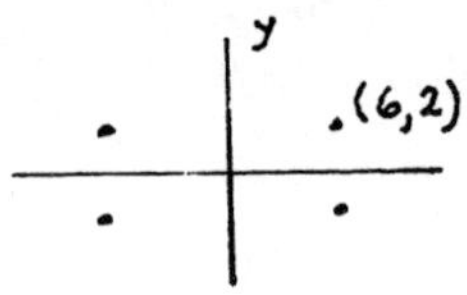

$$u(x,y) = -.0318\left[\frac{x-6}{(x-6)^2+(y-2)^2} + \frac{x-6}{(x-6)^2+(y+2)^2} + \frac{x+6}{(x+6)^2+(y-2)^2} + \frac{x+6}{(x+6)^2+(y+2)^2}\right]$$

$$v(x,y) = -.0318\left[\frac{y-2}{(x-6)^2+(y-2)^2} + \frac{y-2}{(x+6)^2+(y-2)^2} + \frac{y+2}{(x-6)^2+(y+2)^2} + \frac{y+2}{(x+6)^2+(y+2)^2}\right]$$

$$\frac{q}{2\pi} = \frac{-.2}{2\pi} = -.0318.$$

At (4,3) $u(4,3) = -.0318\left[\frac{-2}{4+1} + \frac{-2}{4+25} + \frac{10}{100+1} + \frac{10}{100+25}\right] = \underline{0.00922\ \text{m}/\text{s}}.$

$$v(4,3) = -.0318\left[\frac{1}{4+1} + \frac{1}{100+1} + \frac{5}{4+25} + \frac{5}{100+25}\right] = \underline{-0.01343\ \text{m}/\text{s}}.$$

8.85 $\text{Re}_{\text{crit}} = \frac{U_\infty x_T}{\nu}$. $6\times10^5 = \frac{100 x_T}{\nu}$.

a) $\nu = 1.47\times10^{-5}\ \text{m}^2/\text{s}$. $\therefore x_T = 6\times10^5 \times 1.47\times10^{-5}/100 = 0.0882$ m or $\underline{8.82\ \text{cm}}$

b) $\nu = \frac{\mu}{\rho} = \frac{1.67\times10^{-5}}{1.22\times.669} = 2.05\times10^{-5}$. $\therefore x_T = 6\times10^5\times2.05\times10^{-5}/100 = 0.123$

or $\underline{12.3\ \text{cm}}$.

c) $\nu = \frac{\mu}{\rho} = \frac{1.46\times10^{-5}}{1.22\times.338} = 3.54\times10^{-5}$. $\therefore x_T = 6\times10^5\times3.54\times10^{-5}/100 = 0.212$ m

or $\underline{21.2\ \text{cm}}$.

Note: $\mu = \mu(T)$. Find T in Table B.2. Find μ in Table B.3.

8.85E $\text{Re}_{\text{crit}} = \frac{U_\infty x_T}{\nu}$. $\therefore x_T = 6 \times 10^5 \nu / 300 = 2000\nu$.

a) $\nu = 1.56 \times 10^{-4}$ ft^2 / sec. $\therefore x_T = 2000 \times 1.56 \times 10^{-4} = 0.312'$ or $\underline{3.74''}$

b) $\nu = \frac{\mu}{\rho} = 2.1 \times 10^{-4}$ ft^2 / sec. $\therefore x_T = 2000 \times 2.1 \times 10^{-4} = 0.42'$ or $\underline{5.04''}$

c) $\nu = 3.47 \times 10^{-4}$ ft^2 / sec. $\therefore x_T = 2000 \times 3.47 \times 10^{-4} = 0.694'$ or $\underline{8.33''}$

8.86 a) Use $\text{Re}_{\text{crit}} = 3 \times 10^5 = 10x_T / 1.51 \times 10^{-5}$. $\therefore x_T = \underline{0.453 \text{ m}}$.
b) Use $\text{Re}_{\text{crit}} = 10^6 = 10x_T / 1.51 \times 10^{-5}$. $\therefore x_T = \underline{1.51 \text{ m}}$.
c) Use $\text{Re}_{\text{crit}} = 3 \times 10^5 = 10x_T / 1.51 \times 10^{-5}$. $\therefore x_T = \underline{0.453 \text{ m}}$.
d) Use $\text{Re}_{\text{crit}} = 3 \times 10^5 = 10x_T / 1.51 \times 10^{-5}$. $\therefore x_T = \underline{0.453 \text{ m}}$.
Note: A rough plate, high free-stream disturbances, or a vibrated smooth plate all experience transition at the lower Re_{crit}.

8.87 a) Use $\text{Re}_{\text{crit}} = 3 \times 10^5 = 10x_T / 10^{-6}$. $\therefore x_T = 0.03$ m or $\underline{3 \text{ cm}}$.
b) Use $\text{Re}_{\text{crit}} = 10^6 = 10x_T / 10^{-6}$. $\therefore x_T = 0.1$ m or $\underline{10 \text{ cm}}$.
c) Use $\text{Re}_{\text{crit}} = 3 \times 10^5 = 10x_T / 10^{-6}$. $\therefore x_T = 0.03$ m or $\underline{3 \text{ cm}}$.
d) Use $\text{Re}_{\text{crit}} = 3 \times 10^5 = 10x_T / 10^{-6}$. $\therefore x_T = 0.03$ m or $\underline{3 \text{ cm}}$.

8.88 $\text{Re}_{\text{crit}} = 6 \times 10^5 = \frac{U_\infty \times 2}{\nu}$. For a wind tunnel: $6 \times 10^5 = \frac{U_\infty \times 2}{1.5 \times 10^{-5}}$.
$\therefore \underline{U_\infty = 4.5 \text{ m / s}}$.

For a water channel: $6 \times 10^5 = \frac{U_\infty \times 2}{10^{-6}}$. $\therefore \underline{U_\infty = 0.3 \text{ m / s}}$.

8.89 The x-coordinate is measured along the cylinder surface as shown in Fig. 8.19. The pressure distribution (see solution 8.77) on the surface is
$p = p_0 - 2\rho U_\infty^2 \sin^2 \alpha$ where $r_c\alpha = x$ (α is zero at the stagnation point). Then

$$p(x) = 20\,000 - 2 \times 1000 \times 10^2 \sin^2(x/2)$$
$$= \underline{20 - 200 \sin^2(x/2) \text{ kPa}}$$

The velocity $U(x)$ at the edge of the b.l. is $U(x)$ on the cylinder wall:

$$v_\theta(r = 2) = -10 \sin\theta - 10 \sin\theta = -20 \sin(\pi - \alpha) = 20 \sin\alpha$$
$$\therefore U(x) = \underline{20 \sin(x/2)}$$

8.90 $U(x) = v_\theta$ at $r_c = 1$. $v_\theta = 8 \sin\alpha$. $\therefore U(x) = \underline{8 \sin x}$ since $x = \alpha r_c$.
$p(x) = 58 - 32 \sin^2\alpha = \underline{58 - 32 \sin^2 x \text{ kPa}}$

8.91 The height h above the plate is $h(x)=mx+.4$. $.1=m\times 2+.4$ $\therefore m=-.15$

$\therefore h(x)=0.4-.15x$. Continuity: $6\times .4=U(x)h$. $\therefore U(x)=\dfrac{2.4}{0.4-.15x}$ or

$U(x)=\dfrac{16}{2.67-x}$.

Euler's Eqn: $\rho u\dfrac{\partial u}{\partial x}=-\dfrac{\partial p}{\partial x}$. $\therefore \dfrac{dp}{dx}=\rho\dfrac{16}{2.67-x}\dfrac{16}{(2.67-x)^2}$

$$=\frac{256}{(2.67-x)^3}.$$

8.92 $\tau_0=-\delta\dfrac{dp}{dx}+U(x)\dfrac{d}{dx}\int_0^\delta \rho u dy-\dfrac{d}{dx}\int_0^\delta \rho u^2 dy$

$$=-\delta\frac{dp}{dx}+\frac{d}{dx}\int_0^\delta \rho uUdy-\frac{dU}{dx}\left(\int_0^\delta \rho udy\right)-\frac{d}{dx}\int_0^\delta \rho u^2 dy$$

where we have used $g\dfrac{df}{dx}=\dfrac{dfg}{dx}-f\dfrac{dg}{dx}$. $\left(\text{Here } U=g,\ f=\int_0^\delta \rho udy.\right)$

$$\therefore \tau_0=-\delta\frac{dp}{dx}+\frac{d}{dx}\int_0^\delta \rho u(U-u)dy-\rho\frac{dU}{dx}\int_0^\delta udy. \qquad (\rho=\text{const.})$$

8.93 $\dfrac{dp}{dx}=-\dfrac{\rho}{2}\dfrac{d}{dx}U^2=-\rho U\dfrac{dU}{dx}=-\rho\dfrac{dU}{dx}\left(\dfrac{1}{\delta}\int_0^\delta Udy\right)$ where $U=\dfrac{1}{\delta}\int_0^\delta Udy$.

$$\therefore \tau_0=-\delta\left[-\rho\frac{dU}{dx}\left(\frac{1}{\delta}\int_0^\delta Udy\right)\right]+\rho\frac{d}{dx}(\theta U^2)-\rho\frac{dU}{dx}\int_0^\delta udy$$

$$=\rho\frac{d}{dx}(\theta U^2)+\rho\frac{dU}{dx}\int_0^\delta (U-u)dy=\rho\frac{d}{dx}(\theta U^2)+\rho\frac{dU}{dx}U\delta_d.$$

8.94 If $dp/dx=0$ then $\dfrac{dU}{dx}=0$ and $\tau_0=\rho\dfrac{d}{dx}\int_0^\delta u(U_\infty-u)dy$.

$$\tau_0=\rho\frac{d}{dx}\int_0^\delta U_\infty^2\sin\frac{\pi y}{2\delta}\left(1-\sin\frac{\pi y}{2\delta}\right)dy=\rho U_\infty^2\frac{d}{dx}\left[-\frac{2\delta}{\pi}\cos\frac{\pi y}{2\delta}-\frac{y}{2}\right]_0^\delta=\rho U_\infty^2\frac{d}{dx}\left(\frac{2\delta}{\pi}-\frac{\delta}{2}\right)$$

$$\tau_0=\mu\left.\frac{\partial u}{\partial y}\right|_{y=0}=\mu U_\infty\frac{\pi}{2\delta}\cos 0.$$

$\therefore \mu U_\infty\dfrac{\pi}{2\delta}=.137\rho U_\infty^2\dfrac{d\delta}{dx}$. $\therefore \delta d\delta=11.5\dfrac{\nu}{U_\infty}dx$. $\therefore \delta=4.79\sqrt{\dfrac{\nu x}{U_\infty}}$.

b) $\tau_0 = \mu U_\infty \frac{\pi}{2}\frac{1}{4.79}\sqrt{\frac{U_\infty}{\nu x}} = \underline{0.328\mu U_\infty\sqrt{\frac{U_\infty}{\nu x}}}.$

c) $\frac{\partial u}{\partial x} = U_\infty \frac{\partial}{\partial x}\sin\left(\frac{\pi y}{2\times 4.79}\sqrt{\frac{U_\infty}{\nu x}}\right) = U_\infty \frac{\partial}{\partial x}\sin\left(\frac{a}{\sqrt{x}}\right) = U_\infty\left(-\frac{ax^{-3/2}}{2}\right)\cos\frac{a}{\sqrt{x}} = -\frac{\partial v}{\partial y}.$

$$\therefore v = \int_0^\delta U_\infty \frac{.164y}{x^{3/2}}\sqrt{\frac{U_\infty}{\nu}}\cos\left(.328y\sqrt{\frac{U_\infty}{\nu x}}\right)dy = \underline{\int_0^\delta .0316U_\infty\sqrt{\frac{U_\infty}{\nu}}\cos\left[\left(.189\sqrt{\frac{U_\infty}{\nu}}\right)y\right]dy}.$$

8.95 $u = U_\infty \frac{y}{\delta}.\qquad \tau_0 = \rho\frac{d}{dx}\int_0^\delta U_\infty^2\frac{y}{\delta}\left(1-\frac{y}{\delta}\right)dy$

$$= \rho\frac{d}{dx}U_\infty^2\left(\frac{\delta}{2}-\frac{\delta}{3}\right) = \frac{1}{6}\rho U_\infty^2\frac{d\delta}{dx}.$$

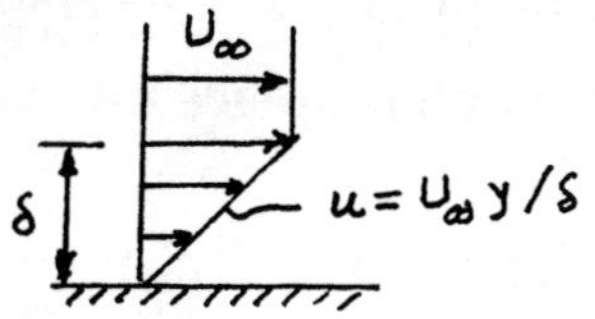

$\tau_0 = \mu\frac{\partial u}{\partial y} = \mu\frac{U_\infty}{\delta}.\qquad \therefore \mu\frac{U_\infty}{\delta} = \frac{1}{6}\rho U_\infty^2\frac{d\delta}{dx}.\qquad \therefore \delta d\delta = 6\frac{\nu}{U_\infty}dx$

$\therefore \delta^2 = 12\frac{\nu}{U_\infty}x.\quad \delta(x) = 3.46\sqrt{\frac{\nu x}{U_\infty}}.\quad \tau_0 = 0.289\mu U_\infty\sqrt{\frac{U_\infty}{\nu x}}.$

%error in $\delta(x) = \frac{5-3.46}{5}\times 100 = \underline{30.8\%\text{ low.}}$

%error in $\tau_0(x) = \frac{.332-.289}{.332}\times 100 = \underline{13\%\text{ low}}.$

8.96 $\tau_0 = \frac{d}{dx}\left\{\int_0^{\delta/6}\rho 3U_\infty^2\frac{y}{\delta}\left(1-3\frac{y}{\delta}\right)dy + \int_{\delta/6}^{\delta/2}\rho U_\infty^2\left(\frac{y}{\delta}+\frac{1}{3}\right)\left(1-\frac{y}{\delta}-\frac{1}{3}\right)dy\right.$

$$\left.+\int_{\delta/2}^{\delta}\rho U_\infty^2\left(\frac{y}{3\delta}+\frac{2}{3}\right)\left(1-\frac{y}{3\delta}-\frac{2}{3}\right)dy\right\}$$

$$= \frac{d}{dx}\rho U_\infty^2(0.1358\delta) = \mu\frac{3U_\infty}{\delta}.\qquad \therefore \delta\frac{d\delta}{dx} = 22.08\mu/\rho U_\infty.$$

Thus, $\delta(x) = 6.65\sqrt{\nu x/U_\infty},\ \tau_0(x) = 0.1358\rho U_\infty^2\left(\frac{6.65}{2}\sqrt{\frac{\nu}{U_\infty x}}\right) = \underline{0.451\rho U_\infty^2\,\text{Re}_x^{-1/2}}.$

%error for $\delta = \frac{6.65-5}{5}\times 100 = \underline{33\%}.\qquad$ %error for $\tau_0 = \frac{0.451-0.332}{0.332}\times 100 = \underline{36\%}$

8.97 Continuity from entrance to x: $U_0 H = 2\int_0^\delta u(y)dy + U(x)(H-2\delta).$

Write $U(x)\delta = U(x)\int_0^\delta dy = \int_0^\delta U(x)dy.$ Then, continuity provides

$$U_0H = 2\int_0^{\delta}(u-U)dy + UH$$

$$= UH - 2\int_0^{\delta}(U-u)dy$$

$$= UH - 2U\delta_d. \qquad \therefore U(x) = \frac{U_0H}{H-2\delta_d}.$$

If we were to move the walls out a distance $\delta_d(x)$, then $U(x)$ would be constant since $\left[(H-2\delta_d)+2\delta_d\right]$ would be constant; then $U(x)=U_0$. For a square wind tunnel, displace one wall outward $4\delta_d$ for $dp/dx=0$.

8.98 The given velocity profile is that used in Example 8.13. There we found $\delta = 5.48\sqrt{\nu x/U_\infty} = 5.48\sqrt{10^{-6}x/10} = 0.00173\sqrt{x} = 0.00173\sqrt{3} = 0.003$ m.
Assume the streamline is outside the b.l. Continuity is then

$$10\times 0.02 = \int_0^{.003} 10\left(\frac{2y}{.003} - \frac{y^2}{.003^2}\right)dy + (h-.003)10$$

$$= 0.02 + 10h - 0.03. \qquad \therefore h = 0.021 \text{ m or } \underline{2.1 \text{ cm}}$$

$$\delta_d = \frac{1}{10}\int_0^{.003}\left(10 - \frac{20y}{.003} + \frac{10y^2}{.003^2}\right)dy = \frac{1}{10}[.03-.03+.01] = \underline{0.001 \text{ m}}$$

$h-2 = 2.1-2 = 0.1$ cm or 0.001 m.
The streamline moves away from the wall a distance δ_d.

8.99 From Problem 8.97 we found that we should displace the one wall outward $4\delta_d$.
From the definition of δ_d:

$$h(x) = 4\delta_d = \frac{4}{10}\int_0^{\delta}\left(10 - \frac{20y}{\delta} + \frac{10y^2}{\delta^2}\right)dy = 4\left(\delta - \delta + \frac{\delta}{3}\right) = \frac{4}{3}\delta$$

$$= \frac{4}{3}\left(5.48\sqrt{\frac{1.86\times10^{-5}x/10}{160/(.287\times303)}}\right) = \underline{0.00735\sqrt{x}} \text{ m}$$

We used $\delta(x)$ found in Example 8.13, $\rho = p/RT$, and $\nu = \mu/\rho$.

8.100 a) $u = U_\infty\left(\frac{3}{2}\frac{y}{\delta} - \frac{1}{2}\frac{y^3}{\delta^3}\right)$. $\delta_d = \frac{1}{U_\infty}\int_0^{\delta} U_\infty\left(1 - \frac{3}{2}\frac{y}{\delta} + \frac{1}{2}\frac{y^3}{\delta^3}\right)dy = \delta - \frac{3}{4}\delta + \frac{1}{8}\delta = .375\delta.$

From Eq. 8.6.16, $\delta_d = .375\times 4.65\sqrt{\frac{\nu x}{U_\infty}} = \underline{1.74\sqrt{\frac{\nu x}{U_\infty}}}$. %error = $\underline{1.2\%}$.

$$\theta = \frac{1}{U_\infty^2}\int_0^\delta U_\infty^2\left(\frac{3}{2}\frac{y}{\delta} - \frac{1}{2}\frac{y^3}{\delta^3}\right)\left(1 - \frac{3}{2}\frac{y}{\delta} + \frac{1}{2}\frac{y^3}{\delta^3}\right)dy = 0.139\delta.$$

$$\therefore \theta = .139 \times 4.65\sqrt{\frac{\nu x}{U_\infty}} = \underline{0.648\sqrt{\frac{\nu x}{U_\infty}}}. \quad \%\text{error} = \frac{.648 - .644}{.644} \times 100 = \underline{0.62\%}$$

b) $u = U_\infty\left(2\frac{y}{\delta} - \frac{y^2}{\delta^2}\right)$. See Example 8.13. $\delta_d = \int_0^\delta \left(1 - \frac{2y}{\delta} + \frac{y^2}{\delta^2}\right)dy = \delta - \delta + \frac{\delta}{3} = \delta/3.$

$$\therefore \delta_d = \frac{5.48}{3}\sqrt{\frac{\nu x}{U_\infty}} = \underline{1.83\sqrt{\frac{\nu x}{U_\infty}}}. \quad \%\text{error} = \frac{1.83 - 1.72}{1.72} \times 100 = \underline{6.4\%}.$$

$$\theta = \int_0^\delta \left(2\frac{y}{\delta} - \frac{y^2}{\delta^2}\right)\left(1 - 2\frac{y}{\delta} + \frac{y^2}{\delta^2}\right)dy = \delta - \frac{1}{3}\delta - \frac{4}{3}\delta + \frac{2}{4}\delta + \frac{2}{4}\delta - \frac{1}{5}\delta = .1333\delta.$$

$$\therefore \theta = .1333 \times 5.48\sqrt{\frac{\nu x}{U_\infty}} = \underline{0.731\sqrt{\frac{\nu x}{U_\infty}}}. \quad \%\text{error} = \frac{.731 - .644}{.644} \times 100 = \underline{13.5\%}.$$

c) $\delta_d = \int_0^\delta \left(1 - \sin\frac{\pi y}{2\delta}\right)dy = \delta - \frac{2\delta}{\pi} = 0.363\delta.$ See Problem 8.94. $\delta = 4.79\sqrt{\frac{\nu x}{U_\infty}}.$

$$\therefore \delta_d = 0.363 \times 4.79\sqrt{\frac{\nu x}{U_\infty}} = \underline{1.74\sqrt{\frac{\nu x}{U_\infty}}}. \quad \%\text{error} = \frac{1.74 - 1.72}{1.72} \times 100 = \underline{1.2\%}$$

$$\theta = \int_0^\delta \sin\frac{\pi y}{2\delta}\left(1 - \sin\frac{\pi y}{2\delta}\right)dy = \left[-\frac{2\delta}{\pi}\cos\frac{\pi y}{2\delta} - \frac{y}{2} + \text{sin term}\right]_0^\delta = -\frac{\delta}{2} + \frac{2\delta}{\pi} = 0.137\delta.$$

(sin term → 0)

$$\therefore \theta = .137 \times 4.79\sqrt{\frac{\nu x}{U_\infty}} = \underline{0.654\sqrt{\frac{\nu x}{U_\infty}}}. \quad \%\text{error} = \frac{.654 - .644}{.644} \times 100 = \underline{1.6\%}.$$

8.101 a) $\delta = 5.48\sqrt{\frac{\nu x}{U_\infty}} = 5.48\left(\frac{1.5\times10^{-5}\times 6}{4}\right)^{1/2} = \underline{0.0260\text{ m}}.$

b) $\tau_0 = .365\rho U_\infty^2\sqrt{\frac{\nu}{xU_\infty}} = .365\times1.22\times4^2\left(\frac{1.5\times10^{-5}}{6\times4}\right)^{1/2} = \underline{0.00563\text{ Pa}}.$

c) Drag $= \int_0^6 .365\rho U_\infty^2\sqrt{\frac{\nu}{U_\infty}}x^{1/2}5dx = \frac{.365}{1/2}\times1.22\times4^2\sqrt{\frac{1.5\times10^{-5}}{4}}\times5\times\sqrt{6} = \underline{0.338\text{ N}}.$

d)

$$\frac{\partial u}{\partial x} = U_\infty\left[-\frac{2y}{\delta^2} + \frac{2y^2}{\delta^3}\right]\frac{d\delta}{dx} = 4\left[-\frac{2}{.0184^2}y + \frac{2}{.0184^3}y^2\right]\frac{5.48}{2}\sqrt{\frac{1.5\times10^{-5}}{4\times3}} = (-72.4y + 3930y^2).$$

$$\delta_{x=3} = 5.48\sqrt{\frac{1.5\times10^{-5}\times3}{4}} = .0184\text{ m}.$$

$$v = -\int_0^{\delta} -\frac{\partial u}{\partial x} dy = \left(\frac{72.4}{2}\times.0184^2 - \frac{3930}{3}\times.0184^3\right) = \underline{0.00410 \text{ m/s}}.$$

8.101E a) $\delta = 4.65\sqrt{\frac{\nu x}{U_\infty}} = 4.65\left(\frac{1.6\times10^{-4}\times20}{12}\right)^{1/2} = \underline{0.0759 \text{ ft}}.$

b) $\tau_0 = .323\rho U_\infty^2\sqrt{\frac{\nu}{xU_\infty}} = .323\times.0024\times12^2\left(\frac{1.6\times10^{-4}}{20\times12}\right)^{1/2} = \underline{9.11\times10^{-5} \text{ psf}}.$

c) Drag =

$$\frac{1}{2}\rho U_\infty^2\times20\times15\times1.29\sqrt{\frac{\nu}{LU_\infty}} = \frac{1}{2}\times.0024\times12^2\times300\times1.29\left(\frac{1.6\times10^{-4}}{20\times12}\right)^{1/2} = \underline{0.0546 \text{ lb}}.$$

d) $\delta_{x=10} = 4.65\sqrt{\frac{1.6\times10^{-4}\times10}{12}} = 0.0416 \text{ ft}.\quad \frac{\partial u}{\partial x} = U_\infty\left[-\frac{3y}{2\delta^2}+\frac{3y^3}{2\delta^4}\right]\frac{d\delta}{dx}.$

$$\therefore \frac{\partial u}{\partial x} = 12\left[-\frac{3y}{.2\times.0416^2}+\frac{3y^3}{.2\times.0416^4}\right]\frac{4.65}{2}\sqrt{\frac{1.6\times10^{-4}}{10\times12}} = -27.9y+16140y^3.$$

$$\therefore v = \int_0^{\delta} -\frac{\partial u}{\partial x}dy = \frac{27.9}{2}\times.0416^2 - \frac{16140}{4}\times.0416^4 = \underline{0.0121 \text{ fps}}.$$

8.102 a) $\delta = 4.65\sqrt{\frac{\nu x}{U_\infty}} = 4.65\left(\frac{1.5\times10^{-5}\times6}{4}\right)^{1/2} = \underline{0.0221 \text{ m}}.$

b) $\tau_0 = 0.323\rho U_\infty^2\sqrt{\frac{\nu}{xU_\infty}} = .323\times1.22\times4^2\left(\frac{1.5\times10^{-5}}{6\times4}\right)^{1/2} = \underline{0.00498 \text{ Pa}}.$

c) Drag = $\frac{1}{2}\rho U_\infty^2 Lw\times1.29\sqrt{\frac{\nu}{LU_\infty}} = \frac{1}{2}\times1.22\times4^2\times6\times5\times1.29\left(\frac{1.5\times10^{-5}}{6\times4}\right)^{1/2} = \underline{0.299 \text{ N}}.$

d)

$$\frac{\partial u}{\partial x} = U_\infty\left[-\frac{3y}{2\delta^2}+\frac{3}{2}\frac{y^3}{\delta^4}\right]\frac{d\delta}{dx} = 4\left[-\frac{3\times4}{2\times4.65^2\times1.5\times10^{-5}\times3}y+\frac{3}{2}\frac{4^2y^3}{4.65^4\times(1.5\times10^{-5}\times3)^2}\right]\frac{d\delta}{dx}.$$

$$\therefore \frac{\partial u}{\partial x} = 4\left[-6166y+2.53\times10^7y^3\right]\frac{4.65}{2}\sqrt{\frac{1.5\times10^{-5}}{4}}\frac{1}{\sqrt{3}} = -64.1y+2.63\times10^5y^3.$$

$$\therefore v = \int_0^{\delta} -\frac{\partial u}{\partial x}dy = \frac{64.1}{2}\times.0156^2 - \frac{2.63\times10^5}{4}\times.0156^4 = \underline{0.00391 \text{ m/s}},$$

where $\delta_{x=3} = 4.65\sqrt{\frac{1.5\times10^{-5}\times3}{4}} = .01560 \text{ m}.$

8.103 a) $\delta = 5\sqrt{\dfrac{\nu x}{U_\infty}} = 5\left(\dfrac{1.5\times10^{-5}\times 2}{10}\right)^{1/2} = \underline{0.00866\text{ m}}$. Use $\tau_0 = .332\rho U_\infty^2\sqrt{\dfrac{\nu}{xU_\infty}}$.

$$\text{Drag} = \int_0^L \tau_0 w dx = .332\times1.22\times10^2\sqrt{\frac{1.5\times10^{-5}}{10}}\frac{\sqrt{2}}{1/2}\times4 = \underline{0.561\text{ N}}.$$

b) $\delta = .38\times2\left(\dfrac{1.5\times10^{-5}}{10\times2}\right)^{.2} = \underline{0.0453\text{ m}}$.

$$\text{Drag} = \frac{1}{2}\rho U_\infty^2 Lw\times.074\left(\frac{\nu}{U_\infty L}\right)^{.2} = \frac{1}{2}\times1.22\times10^2\times2\times4\times.074\left(\frac{1.5\times10^{-5}}{10\times2}\right)^{.2} = \underline{2.15\text{ N}}.$$

8.104 a) $\delta = .38\times6\left(\dfrac{1.5\times10^{-5}}{20\times6}\right)^{.2} = \underline{0.0949\text{ m}}$. $\tau_0 = \dfrac{1}{2}\times1.22\times20^2\times.059\left(\dfrac{1.5\times10^{-5}}{20\times6}\right)^{.2}$
$= \underline{.6\text{ Pa}}$.

b) $\delta = .38\times6\left(\dfrac{10^{-6}}{20\times6}\right)^{.2} = \underline{0.0552\text{ m}}$. $\tau_0 = \dfrac{1}{2}\times1000\times20^2\times.059\left(\dfrac{10^{-6}}{20\times6}\right)^{.2}$
$= \underline{286\text{ Pa}}$.

8.105 $u(y=\delta) = U_\infty$. $\dfrac{\partial u}{\partial y} = \dfrac{1}{7}U_\infty y^{-6/7}\delta^{-1/7}$. $\left.\dfrac{\partial u}{\partial y}\right|_{y=\delta} = \dfrac{1}{7}U_\infty/\delta$.

$\left.\dfrac{\partial u}{\partial y}\right|_{y=\delta}$ should be zero. Thus, this condition is not satisfied.

$\tau_0 = \mu\left.\dfrac{\partial u}{\partial y}\right|_{y=0} = \mu\dfrac{1}{7}U_\infty\dfrac{1}{0}\delta^{-1/7} = \infty$. Thus, this is unacceptable and

$\dfrac{\partial u}{\partial y}$ at, and near, the wall is not valid.

$$u = U_\infty\left(\frac{3}{2}\frac{y}{\delta} - \frac{1}{2}\frac{y^3}{\delta^3}\right).$$

$$\bar{u} = U_\infty\left(\frac{y}{\delta}\right)^{1/7}.$$

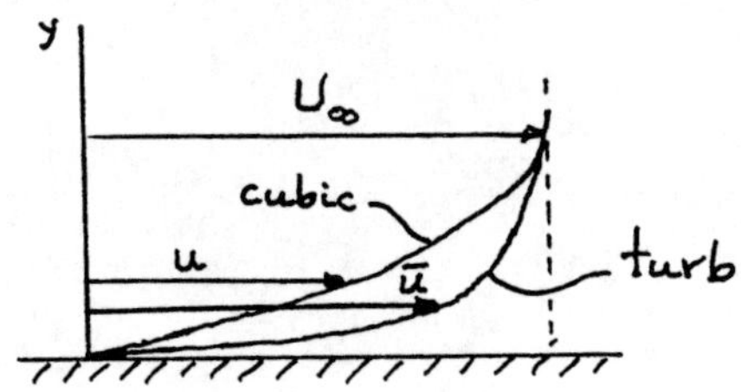

8.106 a) $\text{Re}_{\text{crit}} = 3\times10^5 = \dfrac{6x_T}{1.5\times10^{-5}}$. $\therefore x_T = .75\text{ m}$. Use Eq. 8.6.31:

$$c_f = .074\left(\frac{1.5\times10^{-5}}{6\times4}\right)^{.2} - 1060\left(\frac{1.5\times10^{-5}}{6\times4}\right) = 0.00359.$$

$$\therefore \text{Drag} = \frac{1}{2}\times1.22\times6^2\times(4\times5)\times.00359 = \underline{1.58\text{ N}}.$$

b) Drag $= \frac{1}{2} \times 1.22 \times 6^2 \times (4 \times 5)\left[.074\left(\frac{1.5 \times 10^{-5}}{6 \times 4}\right)^{.2} - 1700\left(\frac{1.5 \times 10^{-5}}{6 \times 4}\right)\right] = \underline{1.40 \text{ N}}.$

c) Drag $= \frac{1}{2} \times 1.22 \times 6^2 \times (4 \times 5)\left[.074\left(\frac{1.5 \times 10^{-5}}{6 \times 4}\right)^{.2} - 2080\left(\frac{1.5 \times 10^{-5}}{6 \times 4}\right)\right] = \underline{1.30 \text{ N}}.$

8.106E a) Drag =

$$\frac{1}{2}\times .0024 \times 20^2 \times (12 \times 15)\left[.074\left(\frac{1.58 \times 10^{-4}}{20 \times 12}\right)^{.2} - 1060\left(\frac{1.58 \times 10^{-4}}{20 \times 12}\right)\right] = \underline{0.31 \text{ lb}}.$$

b) Drag =

$$\frac{1}{2}\times .0024 \times 20^2 \times (12 \times 15)\left[.074\left(\frac{1.58 \times 10^{-4}}{20 \times 12}\right)^{.2} - 1700\left(\frac{1.58 \times 10^{-4}}{20 \times 12}\right)\right] = \underline{0.27 \text{ lb}}.$$

c) Drag =

$$\frac{1}{2}\times .0024 \times 20^2 \times (12 \times 15)\left[.074\left(\frac{1.58 \times 10^{-4}}{20 \times 12}\right)^{.2} - 2080\left(\frac{1.58 \times 10^{-4}}{20 \times 12}\right)\right] = \underline{0.25 \text{ lb}}.$$

8.107 a) Drag $= \frac{1}{2} \times 1000 \times 1.2^2 \times (1 \times 2)\left[.074\left(\frac{10^{-6}}{1.2 \times 1}\right)^{.2} - 1060\left(\frac{10^{-6}}{1.2 \times 1}\right)\right] = \underline{5.21 \text{ N}}.$

b) Drag $= \frac{1}{2} \times 1000 \times 1.2^2 \times (1 \times 2)\left[.074\left(\frac{10^{-6}}{1.2 \times 1}\right)^{.2} - 1700\left(\frac{10^{-6}}{1.2 \times 1}\right)\right] = \underline{4.44 \text{ N}}.$

c) Drag $= \frac{1}{2} \times 1000 \times 1.2^2 \times (1 \times 2)\left[.074\left(\frac{10^{-6}}{1.2 \times 1}\right)^{.2} - 2080\left(\frac{10^{-6}}{1.2 \times 1}\right)\right] = \underline{3.99 \text{ N}}.$

8.108 $U_\infty = 60\frac{1000}{3600} = 16.67 \text{ m/s}. \quad \delta = .38 \times 100\,000\left(\frac{1.5 \times 10^{-5}}{16.67 \times 10^5}\right)^{.2} = \underline{235 \text{ m}}.$

$$\tau_0 = \frac{1}{2}\rho U_\infty^2 c_f = \frac{1}{2} \times 1.22 \times 16.67^2 \times \left[.059\left(\frac{1.5 \times 10^{-5}}{16.67 \times 10^5}\right)^{.2}\right] = \underline{0.0618 \text{ Pa}}.$$

b) $\tau_0 = \frac{1}{2}\rho U_\infty^2 c_f = \frac{1}{2} \times 1.22 \times 16.67^2 \frac{.455}{\left[\ell n\left(.06\frac{16.67 \times 10^5}{1.5 \times 10^{-5}}\right)\right]^2} = \underline{0.151 \text{ Pa}}.$

$$\therefore u_\tau = \sqrt{\frac{.151}{1.22}} = .351 \text{ m/s}. \quad \therefore \frac{16.67}{.351} = 2.44 \ell n \frac{.351\delta}{1.5 \times 10^{-5}} + 7.4. \quad \therefore \underline{\delta = 585 \text{ m}}.$$

Both of the above are in error, however, part (b) is more accurate.
(Re = 1.1×10^{11})

8.109 a) $5 = \frac{u_\tau \delta_v}{\nu}$ (See Fig. 8.24 b). $\therefore \delta_v = \frac{5 \times 1.5 \times 10^{-5}}{.351} = \underline{2.14 \times 10^{-4}\ \text{m}}$.

b) $\delta_d = \frac{1}{U_\infty}\int_0^\delta (U_\infty - \bar{u})dy = \frac{\mu_\tau}{U_\infty}\int_{\delta_v}^{.15\delta}\left(2.5 - 2.44\ell n\frac{y}{\delta}\right)dy + \frac{u_\tau}{U_\infty}\int_{.15\delta}^{\delta} -3.74\ell n\frac{y}{\delta}dy$

$= \frac{u_\tau}{U_\infty}\left[2.5(.15\delta - \delta_v) - 2.44\left(y\ell n\frac{y}{\delta} - y\right)\Big|_{\delta_v}^{.15\delta = 87.8} - 3.74\left(y\ell n\frac{y}{\delta} - y\right)\Big|_{87.8}^{\delta = 585}\right]$ (neglect δ_v)

$= \frac{.351}{16.67}[219 + 620 - .008 + 2188 - 951] = \underline{43.7\ \text{m}}$. (neglect .008)

Note: We cannot use zero as a lower limit since the ln-profile does not go to the wall. Hence, we use δ_v; the lower limit provides a negligible contribution to the integral.

8.110 a) Use Eq. 8.6.40: $c_f = \frac{.455}{\left[\ell n\left(.06\frac{100 \times 6}{1.5 \times 10^{-5}}\right)\right]^2} = \underline{.00211}$.

b) $\tau_0 = \frac{1}{2}\rho U_\infty^2 c_f = \frac{1}{2} \times 1.22 \times 100^2 \times .00211 = \underline{12.9\ \text{Pa}}$. $u_\tau = \sqrt{\frac{12.9}{1.22}} = 3.25\ \text{m/s}$.

c) $\delta_v = \frac{5\nu}{u_\tau} = 5 \times 1.5 \times 10^{-5} / 3.25 = \underline{2.31 \times 10^{-5}\ \text{m}}$.

d) $\frac{100}{3.25} = 2.44\ell n\frac{3.25\delta}{1.5 \times 10^{-5}} + 7.4$. $\therefore \underline{\delta = 0.0666\ \text{m}}$.

8.110E a) Use Eq. 8.6.40: $c_f = \frac{.455}{\left[\ell n\left(.06\frac{300 \times 20}{1.58 \times 10^{-4}}\right)\right]^2} = \underline{0.00212}$.

b) $\tau_0 = \frac{1}{2}\rho U_\infty^2 c_f = \frac{1}{2} \times .0024 \times 300^2 \times .00212 = \underline{0.229\ \text{psf}}$. $u_\tau = \sqrt{\frac{.229}{.0024}} = 9.77\ \text{fps}$.

c) $\delta_v = \frac{5\nu}{u_\tau} = 5 \times 1.58 \times 10^{-4} / 9.77 = \underline{8.09 \times 10^{-5}\ \text{ft}}$.

d) $\frac{300}{9.77} = 2.44\ell n\frac{9.77\delta}{1.58 \times 10^{-4}} + 7.4$. $\therefore \underline{\delta = 0.228\ \text{ft}}$.

8.111 a) $\tau_0 = \frac{1}{2}\rho U_\infty^2 c_f = \frac{1}{2} \times 1000 \times 10^2 \frac{.455}{\left[\ell n\left(.06\frac{10 \times 3}{10^{-6}}\right)\right]^2} = 110\ \text{Pa}$.

$\therefore u_\tau = \sqrt{\frac{110}{1000}} = .332\ \text{m/s}$. $\therefore \delta_v = \frac{5\nu}{u_\tau} = \frac{5 \times 10^{-6}}{.332} = \underline{1.51 \times 10^{-5}\ \text{m}}$.

b) $\bar{u} = 5u_\tau = 5 \times .332 = \underline{1.66\ \text{m/s}}$.

c) $y = .15\delta$. — Do part (d) first! $\therefore y = .15 \times .0333 = \underline{0.005 \text{ m}}$.

d) $\frac{10}{.332} = 2.44 \ell n \frac{.332\delta}{10^{-6}} + 7.4.$ $\therefore \underline{\delta = 0.0333 \text{ m}}$.

8.112 Assume flat plates with $dp/dx = 0$. $C_f = \frac{.523}{\left[\ell n\left(.06 \frac{10 \times 100}{10^{-6}}\right)\right]^2} = .00163.$

$$\therefore \text{Drag} = 2 \times \frac{1}{2} \times 1000 \times 10^2 \times 10 \times 100 \times .00163 = \underline{163\ 000 \text{ N}}.$$

To find δ_{max} we need u_τ.

$$\tau_0 = \frac{1}{2} \times 1000 \times 10^2 \frac{.455}{\left[\ell n\left(.06 \frac{10 \times 100}{10^{-6}}\right)\right]^2} = 70.9 \text{ Pa}. \quad \therefore u_\tau = \sqrt{\frac{70.9}{1000}} = 0.266 \text{ m / s}.$$

$$\frac{10}{.266} = 2.44 \ell n \frac{.266\delta}{10^{-6}} + 7.4. \quad \therefore \underline{\delta_{max} = 0.89 \text{ m}}.$$

8.113 Assume a flat plate of width πD.

$$\therefore \text{Drag} = \frac{1}{2}\rho U_\infty^2 wLC_f = \frac{1}{2} \times 1.11 \times 20^2 \times \pi \times 30 \times 300 \frac{.523}{\left[\ell n\left(.06 \frac{20 \times 300}{1.6 \times 10^{-5}}\right)\right]^2} = 11\ 500 \text{ N}.$$

$$\therefore \dot{W} = \text{Drag} \times U_\infty = 11\ 500 \times 20 = 230\ 000 \text{ W or } \underline{307 \text{ Hp}}.$$

8.113E Assume a flat plate of width πD.

$$\therefore \text{Drag} = \frac{1}{2}\rho U_\infty^2 wLC_f = \frac{1}{2} \times .002 \times 60^2 \times \pi \times 100 \times 1000 \frac{.523}{\left[\ell n\left(.06 \frac{60 \times 1000}{1.7 \times 10^{-4}}\right)\right]^2} = 2080 \text{ lb}.$$

$$\therefore \dot{W} = \text{Drag} \times U_\infty = 2080 \times 60 = 125{,}000 \frac{\text{ft-lb}}{\text{sec}} \text{ or } \underline{227 \text{ Hp}}.$$

8.114 $u = \frac{\partial\psi}{\partial y}, \quad \frac{\partial u}{\partial x} = \frac{\partial^2\psi}{\partial x \partial y}, \quad v = -\frac{\partial\psi}{\partial x}, \quad \frac{\partial u}{\partial y} = \frac{\partial^2\psi}{\partial y^2}, \quad \frac{\partial^2 u}{\partial y^2} = \frac{\partial^3\psi}{\partial y^3}.$

Substitute into Eq. 8.6.45 (with $dp/dx = 0$):

$$\frac{\partial\psi}{\partial y}\frac{\partial^2\psi}{\partial x \partial y} - \frac{\partial\psi}{\partial x}\frac{\partial^2\psi}{\partial y^2} = \nu \frac{\partial^3\psi}{\partial y^3}.$$

8.115 We also have

$$\frac{\partial\psi}{\partial x} = \frac{\partial\psi}{\partial\phi}\frac{\partial\phi}{\partial x} + \frac{\partial\psi}{\partial\eta}\frac{\partial\eta}{\partial x}, \quad \frac{\partial^2\psi}{\partial x \partial y} = \frac{\partial(\partial\psi / \partial y)}{\partial\phi}\frac{\partial\phi}{\partial x} + \frac{\partial(\partial\psi / \partial y)}{\partial\eta}\frac{\partial\eta}{\partial x}$$

Recognizing that $\partial\phi / \partial x = 1$, $\partial\phi / \partial y = 0$, $\partial\eta / \partial x = -\frac{y}{2}\sqrt{U_\infty / \nu x^3}$, and

$\partial\eta / \partial y = \sqrt{U_\infty / \nu x}$,

$$\frac{\partial\psi}{\partial y} = \sqrt{\frac{U_\infty}{\nu\phi}}\frac{\partial\psi}{\partial\eta}, \quad \frac{\partial\psi}{\partial x} = \frac{\partial\psi}{\partial\phi} - \frac{y}{2}\sqrt{\frac{U_\infty}{\nu x^3}}\frac{\partial\psi}{\partial\eta}$$

$$\frac{\partial^2\psi}{\partial x\partial y} = \sqrt{\frac{U_\infty}{\nu\phi}}\frac{\partial^2\psi}{\partial\phi\partial\eta} - \frac{1}{2}\sqrt{\frac{U_\infty}{\nu\phi^3}}\frac{\partial\psi}{\partial\eta} + \sqrt{\frac{U_\infty}{\nu\phi}}\frac{\partial^2\psi}{\partial\eta^2}\left(-\frac{y}{2}\sqrt{\frac{U_\infty}{\nu\phi^3}}\right)$$

$$\frac{\partial^2\psi}{\partial y^2} = \sqrt{\frac{U_\infty}{\nu\phi}}\frac{\partial^2\psi}{\partial\eta^2}\left(\sqrt{\frac{U_\infty}{\nu\phi}}\right), \quad \frac{\partial^3\psi}{\partial y^3} = \frac{U_\infty}{\nu\phi}\frac{\partial^3\psi}{\partial\eta^3}\sqrt{\frac{U_\infty}{\nu\phi}}$$

Equation 8.6.47 then becomes, using $\sqrt{U_\infty / \nu\phi} = \eta / y$,

$$\frac{\eta}{y}\frac{\partial\psi}{\partial\eta}\left(\frac{\eta}{y}\frac{\partial^2\psi}{\partial\phi\partial\eta} - \frac{\eta}{2yx}\frac{\partial\psi}{\partial\eta} - \frac{\eta^2}{2yx}\frac{\partial^2\psi}{\partial\eta^2}\right) - \left(\frac{\partial\psi}{\partial\phi} - \frac{\eta}{2x}\frac{\partial\psi}{\partial\eta}\right)\left(\frac{\eta^2}{y^2}\right)\frac{\partial^2\psi}{\partial\eta^2}$$
$$= \nu\frac{U_\infty}{\nu x}\frac{\eta}{y}\frac{\partial^3\psi}{\partial\eta^3}$$

Multiply by y^2 / η^2 and Eq. 8.6.49 results:

$$-\frac{1}{2\phi}\left(\frac{\partial\psi}{\partial\eta}\right)^2 + \frac{\partial^2\psi}{\partial\phi\partial\eta}\frac{\partial\psi}{\partial\eta} - \frac{\partial\psi}{\partial\phi}\frac{\partial^2\psi}{\partial\eta^2} = \nu\frac{\partial^3\psi}{\partial\eta^3}\sqrt{\frac{U_\infty}{\nu\phi}}$$

8.116 $u = \frac{\partial\psi}{\partial y} = \sqrt{U_\infty \nu x}\frac{dF}{d\eta}\frac{\partial\eta}{\partial y} = \sqrt{U_\infty \nu x}\; F'(\eta)\sqrt{\frac{U_\infty}{\nu x}} = \underline{U_\infty F'(\eta)}.$

We used Eq. 8.6.50 and Eqs. 8.6.48.

$$v = -\frac{\partial\psi}{\partial x} = -\frac{\partial}{\partial x}\left(\sqrt{U_\infty \nu x}\; F\right) = -\frac{1}{2}\sqrt{\frac{U_\infty \nu}{x}}F - \sqrt{U_\infty \nu x}\frac{\partial F}{\partial\eta}\frac{\partial\eta}{\partial x}$$
$$= -\frac{1}{2}\sqrt{\frac{U_\infty \nu}{x}}F - \sqrt{U_\infty \nu x}\, F'\, y\sqrt{\frac{U_\infty}{\nu}}\left(-\frac{1}{2}x^{-3/2}\right)$$
$$= -\frac{1}{2}\sqrt{\frac{U_\infty \nu}{x}}F + \frac{y}{2}\sqrt{\frac{U_\infty}{\nu x}}\sqrt{\frac{U_\infty \nu}{x}}F' = \underline{\frac{1}{2}\sqrt{\frac{U_\infty \nu}{x}}(\eta F' - F)}.$$

8.117 The results are shown in Table 8.5.

8.118 a) $\tau_0 = 0.332 \times 1.22 \times 5^2\sqrt{\frac{1.5 \times 10^{-5}}{2 \times 5}} = \underline{0.0124 \text{ Pa}}.$

b) $\delta = 5\sqrt{\frac{1.5 \times 10^{-5} \times 2}{5}} = \underline{0.0122 \text{ m}}.$

c) $v_{max} = \sqrt{\frac{\nu U_\infty}{x}}\left[\frac{1}{2}(\eta F' - F)\right]_{max} = \sqrt{\frac{1.5 \times 10^{-5} \times 5}{2}} \times .8605 = \underline{0.00527 \text{ m / s}}.$

d) $Q = \int_0^\delta u dy = \int_0^\delta U_\infty \frac{dF}{d\eta} dy = \int_0^\delta U_\infty \frac{dF}{d\eta} d\eta \sqrt{\frac{\nu x}{U_\infty}}$

$= U_\infty \sqrt{\frac{\nu x}{U_\infty}} [F(\delta) - F(0)] = 5\sqrt{\frac{1.5 \times 10^{-5} \times 2}{5}} \times 3.28 = \underline{0.04 \text{ m}^2 / \text{s} / \text{m}}$

(with $F(0) = 0$)

8.118E a) $\tau_0 = .332 \times .0024 \times 15^2 \sqrt{\frac{1.6 \times 10^{-4}}{6 \times 15}} = \underline{2.39 \times 10^{-4} \text{ psf.}}$

b) $\delta = 5\sqrt{\frac{1.6 \times 10^{-4} \times 6}{15}} = 0.04 \text{ ft.}$

c) $v_{\max} = \sqrt{\frac{\nu U_\infty}{x}} \left[\frac{1}{2}(\eta F' - F)\right]_{\max} = \sqrt{\frac{1.6 \times 10^{-4} \times 15}{6}} \times .8605 = \underline{0.0172 \text{ fps.}}$

d) $Q = \int_0^\delta u dy = U_\infty \sqrt{\frac{\nu x}{U_\infty}} F(\delta) = 15\sqrt{\frac{1.6 \times 10^{-4} \times 6}{15}} \times 3.28 = \underline{0.394 \text{ ft}^2 / \text{sec} / \text{ft.}}$

8.119 At $x = 2$m, Re = $5 \times 2/10^{-6} = 10^7$. $\therefore$ Assume turbulent from the leading edge.

a) $\tau_0 = \frac{1}{2}\rho U_\infty^2 \frac{0.455}{(\ell n\ 0.06 \text{Re}_x)^2}$

$= \frac{1}{2} \times 1000 \times 5^2 \frac{0.455}{(\ell n\ 0.06 \times 10^7)^2} = \underline{32.1 \text{ Pa}}$

b) $u_\tau = \sqrt{\tau_0 / \rho} = \sqrt{32.1 / 1000} = 0.1792 \text{ m} / \text{s}$

$\frac{5}{0.1792} = 2.44 \ell n \frac{0.1792\delta}{10^{-6}} + 7.4. \quad \therefore \delta = 0.0248 \text{ m or } \underline{24.8 \text{ mm}}$

c) Use the 1/7 the power-law equation:

$Q = \int_0^{0.0248} 5(y / 0.0248)^{1/7} dy = \underline{0.109 \text{ m}^3 / \text{s} / \text{m}}$

8.120 From Table 8.5 we would select $\eta = 6$:

$\delta = 6\sqrt{\frac{\nu x}{U_\infty}} = 6\sqrt{\frac{1.5 \times 10^{-5} \times 2}{5}} = \underline{0.0147 \text{ m}}$

8.121 From Table 8.5 we interpolate for $F' = 0.5$ to be

$\eta = \frac{0.5 - 0.3298}{0.6298 - 0.3298}(2 - 1) + 1 = 1.57$

$= y\sqrt{\frac{5}{1.5 \times 10^{-5} \times 2}}. \quad \therefore y = 0.00385 \text{ m or } \underline{3.85 \text{ mm}}$

$$\tau = \mu\left(\frac{\partial u}{\partial y} + \frac{\partial v}{\partial x}\right) = F''\rho U_\infty^2 \sqrt{\frac{\nu}{xU_\infty}}$$

(neglect $\frac{\partial v}{\partial x}$)

$$= 0.291 \times 1.2 \times 5^2 \sqrt{\frac{1.5 \times 10^{-5}}{2 \times 5}} = \underline{0.0107 \text{ Pa}}$$

8.122

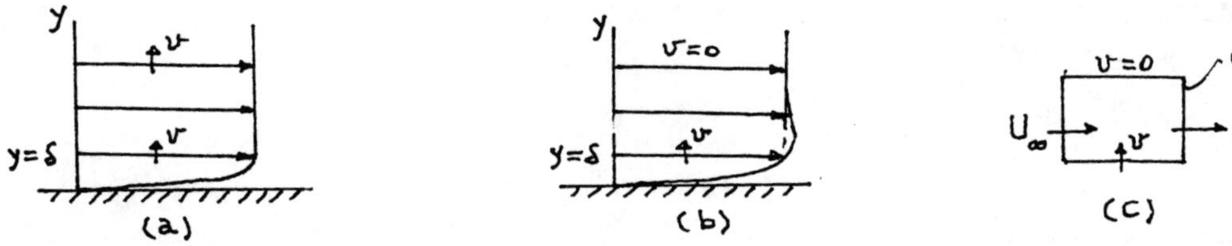

If $v = 0$ at $y = 10\delta$ and $v > 0$ at $y = \delta$, then $\partial v / \partial y < 0$ and continuity demands that $\partial u / \partial x > 0$. The u component, for $y > \delta$ must then be greater than U_∞, as shown in (b); there should be a slight "overshoot". Also, consider the control volume of (c) where the lower boundary is just above $y = \delta$. If $v = 0$ at large y, say $y = 10\delta$, then continuity demands that u out the right area be greater than U_∞: an "overshoot". It is not reasonable to assume that v = const as in (a); reality would demand a profile such as that sketched in (b). The overshoot would be quite small and is neglected in boundary layer theory.

8.123 $u = U_\infty\left(\frac{3}{2}\frac{y}{\delta} - \frac{1}{2}\frac{y^3}{\delta^3}\right)$

Blasius profile: see Table 8.5

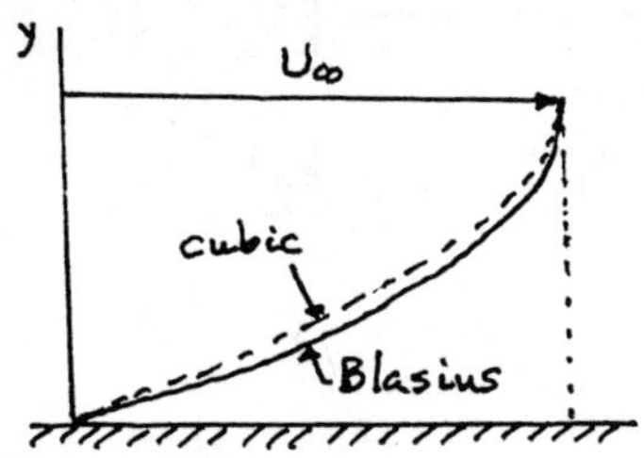

8.124

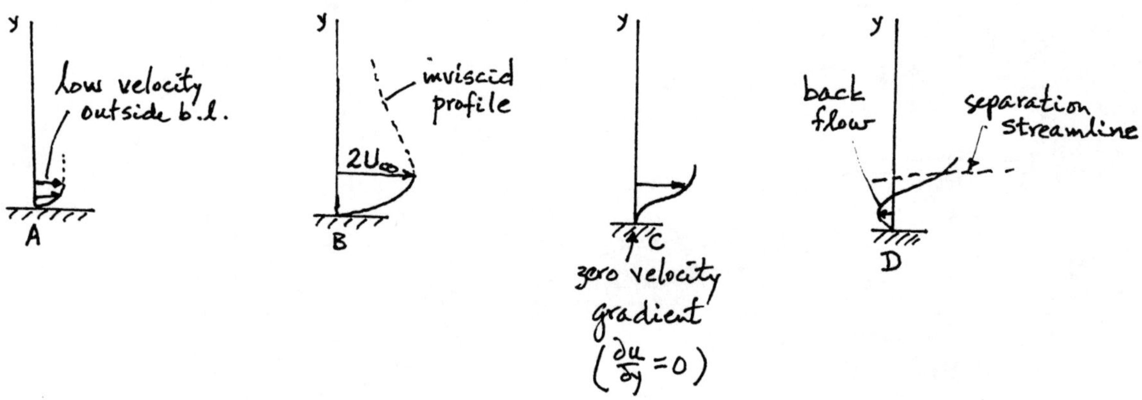

8.125 A: $\dfrac{\partial p}{\partial x} < 0.$ (favorable)

B: $\dfrac{\partial p}{\partial x} \cong 0.$

C: $\dfrac{\partial p}{\partial x} > 0.$ (unfavorable)

D: $\dfrac{\partial p}{\partial x} > 0.$

E: $\dfrac{\partial p}{\partial x} < 0.$

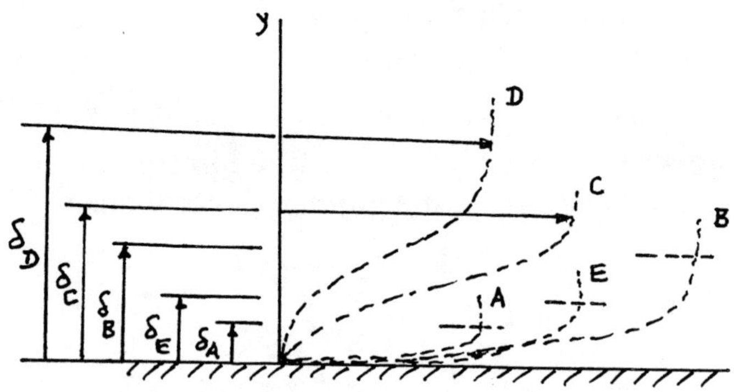

NINE

Compressible Flow

9.1 $c_p = 0.24 \dfrac{\text{Btu}}{\text{lbm-}^\circ\text{ R}}\ 778 \dfrac{\text{ft-lb}}{\text{Btu}}\ 32.2 \dfrac{\text{lbm}}{\text{slug}} = \underline{6012\ \text{ft-lb/slug-}^\circ\text{ R}}.$

$$c_v = c_p - R = 6012 - 1716 = \underline{4296\ \frac{\text{ft-lb}}{\text{slug-}^\circ\text{ R}}} = 4296\ \frac{\text{ft-lb}}{\text{slug-}^\circ\text{ R}}\frac{1}{778}\frac{\text{Btu}}{\text{ft-lb}}\frac{1}{32.2}\frac{\text{slug}}{\text{lbm}}$$

$= \underline{0.171\ \text{Btu/lbm-}^\circ\text{ R}}.$

9.2 $c_p = c_v + R. \qquad c_p = kc_v. \qquad \therefore c_p = \dfrac{c_p}{k} + R$ or $c_p\left(1 - \dfrac{1}{k}\right) = R.$

$\therefore \underline{c_p = Rk/(k-1)}.$

9.3 If $\Delta s = 0$, Eq. 9.1.9 can be written as

$$c_p \ell n \frac{T_2}{T_1} = R \ell n \frac{p_2}{p_1} \quad \text{or} \quad \ell n\left(\frac{T_2}{T_1}\right)^{c_p} = \ell n\left(\frac{p_2}{p_1}\right)^R$$

It follows that, using $c_p = c_v + R$ and $c_p / c_v = k$,

$$\frac{T_2}{T_1} = \left(\frac{p_2}{p_1}\right)^{R/c_p} = \left(\frac{p_2}{p_1}\right)^{1-\frac{1}{k}}.$$

Using Eq. 9.1.7,

$$\frac{T_2}{T_1} = \frac{p_2\rho_1}{\rho_2 p_1} = \left(\frac{p_2}{p_1}\right)^{1-\frac{1}{k}} \quad \text{or} \quad \frac{\rho_1}{\rho_2} = \left(\frac{p_2}{p_1}\right)^{-1/k}.$$

Finally, this can be written as

$$\frac{p_2}{p_1} = \left(\frac{\rho_2}{\rho_1}\right)^k.$$

9.4 Substitute Eq. 4.4.18 into Eq. 4.4.17 and neglect potential energy change:

$$\frac{\dot{Q} - \dot{W}_S}{\dot{m}} = \frac{V_2^2 - V_1^2}{2} + \frac{p_2}{\rho_2} - \frac{p_1}{\rho_1} + \tilde{u}_2 - \tilde{u}_1.$$

Enthalpy is defined in Thermodynamics as $h = \tilde{u} + pv = \tilde{u} + p/\rho$. Therefore,

$$\frac{\dot{Q} - \dot{W}_S}{\dot{m}} = \frac{V_2^2 - V_1^2}{2} + h_2 - h_1.$$

Assume the fluid is an ideal gas with constant specific heat so that $\Delta h = c_p \Delta T$. Then

$$\frac{\dot{Q} - \dot{W}_S}{\dot{m}} = \frac{V_2^2 - V_1^2}{2} + c_p(T_2 - T_1).$$

Next, let $c_p = c_v + R$ and $k = c_p / c_v$ so that $c_p / R = k(k-1)$. Then, with the ideal gas law $T = p / R\rho$, the first law takes the form

$$\frac{\dot{Q} - \dot{W}_S}{\dot{m}} = \frac{V_2^2 - V_1^2}{2} + \frac{k}{k-1}\left(\frac{p_2}{\rho_2} - \frac{p_1}{\rho_1}\right).$$

9.5 Differentiate $p\rho^{-k} = c$ using $d(xy) = ydx + xdy$:

$$\rho^{-k} dp - pk\rho^{-k-1} d\rho = 0.$$

Rewrite:

$$\frac{dp}{d\rho} = k\frac{p}{\rho}.$$

9.6 The speed of sound is given by

$$c = \sqrt{dp / d\rho}.$$

For an isothermal process $TR = p / \rho = K$, where K is a constant. This can be differentiated:

$$dp = Kd\rho = RTd\rho.$$

Hence, the speed of sound is

$$c = \sqrt{RT}.$$

9.7 Eq. 9.1.4 with $\dot{Q} = \dot{W}_S = 0$ is: $\dfrac{V^2}{2} + c_p T = \text{cons't}.$

$$\frac{V^2}{2} + c_p T = \frac{(V + \Delta V)^2}{2} + c_p(T + \Delta T) = \frac{V^2 + 2V\Delta V + (\Delta V)^2}{2} + c_p T + c_p \Delta T.$$

neglect

$$\therefore 0 = \frac{2V\Delta V}{2} + \frac{(\Delta V)^2}{2} + c_p \Delta T. \quad \therefore -V\Delta V = c_p \Delta T = \Delta h.$$

The velocity of a small wave is $V = c$. $\therefore \underline{\Delta h = -c\Delta V}.$

9.8 For water

$$\rho\frac{dp}{d\rho} = 2110 \times 10^6 \text{ Pa}$$

Since $\rho = 1000 \text{ kg / m}^3$, we see that

$$c = \sqrt{dp / d\rho}$$

$$= \sqrt{2110 \times 10^6 / 1000} = \underline{1453 \text{ m / s}}$$

9.9 For water $c = \sqrt{\frac{\Delta p}{\Delta \rho}} \cong \sqrt{\frac{dp}{d\rho}} = \sqrt{\frac{2110 \times 10^6}{1000}} = 1453 \text{ m / s.}$

$L = \text{velocity} \times \text{time} = 1453 \times 0.6 = \underline{872 \text{ m}}.$

9.10 Since $c = 1450$ m/s for the small wave, the time increment is

$$\Delta t = \frac{d}{c} = \frac{10}{1450} = \underline{0.0069 \text{ seconds}}$$

9.11 a) $\text{M} = \frac{V}{c} = \frac{200}{\sqrt{1.4 \times 287 \times 288}} = \underline{0.588}.$

b) $\text{M} = 200 / \sqrt{1.4 \times 287 \times 256} = \underline{0.624}.$

c) $\text{M} = 200 / \sqrt{1.4 \times 287 \times 223} = \underline{0.668}.$

d) $\text{M} = 200 / \sqrt{1.4 \times 287 \times 217} = \underline{0.677}.$

e) $\text{M} = 200 / \sqrt{1.4 \times 287 \times 238} = \underline{0.647}.$

9.11E a) $\text{M} = \frac{600}{\sqrt{1.4 \times 1716 \times 519}} = \underline{0.537}.$

b) $\text{M} = \frac{600}{\sqrt{1.4 \times 1716 \times 466}} = \underline{0.567}.$

c) $\text{M} = \frac{600}{\sqrt{1.4 \times 1716 \times 412}} = \underline{0.603}.$

d) $\text{M} = \frac{600}{\sqrt{1.4 \times 1716 \times 392}} = \underline{0.618}.$

e) $\text{M} = \frac{600}{\sqrt{1.4 \times 1716 \times 409}} = \underline{0.605}.$

9.12 $c = \sqrt{kRT} = \sqrt{1.4 \times 287 \times 263} = 256 \text{ m / s.} \quad \therefore d = ct = 256 \times 1.21 = \underline{309 \text{ m}}.$

9.13 a) Assume $T = 20°\text{C}$:

$c = \sqrt{kRT} = \sqrt{1.4 \times 287 \times 293} = 343 \text{ m / s.}$

$d = c\Delta t = 343 \times 2 = \underline{686 \text{ m}}$

b) Assume $T = 70°\text{F}$:

$c = \sqrt{kRT} = \sqrt{1.4 \times 1716 \times 530} = 1130 \text{ fps.}$

$d = c\Delta t = 1130 \times 2 = \underline{2260 \text{ ft}}.$

For every second that passes, the lightning flashed about 1000 ft. away. Count 5 seconds and it is one mile away.

9.14 $c=\sqrt{1.4\times 287\times 263}=256\text{ m/s}.\quad \sin\alpha=\dfrac{1}{M}=\dfrac{c}{V}.$

$\alpha=\sin^{-1}\dfrac{256}{1000}=14.83^\circ.\qquad \therefore L=\underline{3776\text{ m}}.$

$\Delta t=\dfrac{3776}{1000}=\underline{3.776\text{ s}}.$

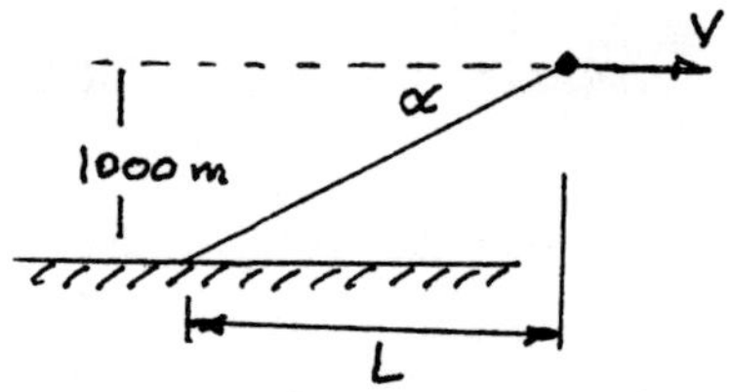

9.15 Use Eq. 9.2.13:

a) $\dfrac{c}{V}=\sin\alpha$ or $V=\dfrac{\sqrt{1.4\times 287\times 288}}{\sin 22^\circ}=\underline{908\text{ m/s}}$

b) $\dfrac{c}{V}=\sin\alpha$ or $V=\dfrac{\sqrt{1.4\times 1716\times 519}}{\sin 22^\circ}=\underline{2980\text{ fps}}$

9.16 Eq. 9.2.4: $\Delta V=-\dfrac{\Delta p}{\rho c}=-\dfrac{\Delta p}{\rho\sqrt{kRT}}=-\dfrac{15}{1.22\sqrt{1.4\times 287\times 288}}=\underline{-0.0361\text{ m/s}}.$

Energy Eq: $\dfrac{V^2}{2}+c_pT=\dfrac{(V+\Delta V)^2}{2}+c_p(T+\Delta T).\quad \therefore 0=V\Delta V+c_p\Delta T.$

$\therefore \Delta T=-c\Delta V/c_p=-\sqrt{1.4\times 287\times 288}(-.0361)/1003=\underline{0.0122^\circ\text{C}}.$

9.16E Eq. 9.2.4: $\Delta V=-\dfrac{\Delta p}{\rho c}=-\dfrac{\Delta p}{\rho\sqrt{kRT}}=-\dfrac{0.3}{.00237\sqrt{1.4\times 1716\times 519}}=\underline{-0.113\text{ fps}}.$

Energy Eq: $\dfrac{V^2}{2}+c_pT=\dfrac{(V=\Delta V)^2}{2}+c_p(T+\Delta T).\quad \therefore 0=V\Delta V+c_p\Delta T.$

$\therefore \Delta T=-c\Delta V/c_p=-\sqrt{1.4\times 1716\times 519}(-.113)/(0.24\times 778\times 32.2)=\underline{0.021^\circ\text{F}}.$

Note: $c_p=.24\dfrac{\text{Btu}}{\text{lbm-}^\circ\text{F}}\times 778\dfrac{\text{ft-lb}}{\text{Btu}}\times 32.2\dfrac{\text{lbm}}{\text{slug}}=.24\times 778\times 32.2\dfrac{\text{ft-lb}}{\text{slug-}^\circ\text{F}}.$

Then $\dfrac{\text{ft}^2/\text{sec}^2}{\text{ft-lb}/(\text{slug-}^\circ\text{F})}=\dfrac{\text{ft}^2-\text{lb-sec}^2-^\circ\text{F}}{\text{sec}^2-\text{ft-lb-ft}}={}^\circ\text{F}.$ (units can be a pain!)

9.17 a) $\rho AV=\rho AV+\rho AdV+AVd\rho+Ad\rho dV+\rho VdA+\rho dAdV+Vd\rho dA+d\rho dAdV$

Keep only the first order terms (the higher order terms — those with more than one differential quantity — will be negligible):

$0=\rho AdV+AVd\rho+\rho VdA$

Divide by ρAV:

$$\frac{dV}{V}+\frac{d\rho}{\rho}+\frac{dA}{A}=0$$

b) Expand the r.h.s. of Eq. 9.3.5 (keep only first order terms):

$$\frac{V^2}{2}+\frac{k}{k-1}\frac{p}{\rho}=\frac{V^2+2VdV}{2}+\frac{k}{k-1}\frac{p+dp}{\rho+d\rho}.$$

Hence,

$$\begin{aligned}0&=\frac{2VdV}{2}+\frac{k}{k-1}\left(\frac{p+dp}{\rho+d\rho}-\frac{p}{\rho}\right)\\&=VdV+\frac{k}{k-1}\left(\frac{\rho p+\rho dp-p\rho-pd\rho}{\rho^2+\rho d\rho}\right)\\&=VdV+\frac{k}{k-1}\left(\frac{\rho dp-pd\rho}{\rho^2}\right)\end{aligned}$$

where we neglected $\rho d\rho$ compared to ρ^2. For an isentropic process Eq. 9.2.8 gives $\rho dp = kpd\rho$, so the above becomes

$$\begin{aligned}0&=VdV+\frac{k}{k-1}\frac{kpd\rho-pd\rho}{\rho^2}\\&=VdV+\frac{k}{k-1}\frac{(k-1)pd\rho}{\rho^2}\\&=VdV+k\frac{p}{\rho^2}d\rho\end{aligned}$$

But $d\rho/\rho=-dV/V-dA/A$ so that the above equation is

$$0=VdV+k\frac{p}{\rho}\left(-\frac{dV}{V}-\frac{dA}{A}\right)$$

which can be written as

$$\frac{dA}{A}=\left(\frac{V^2\rho}{kp}-1\right)\frac{dV}{V}.$$

b) Since $c^2=kp/\rho$, this is put in the form

$$\frac{dA}{A}=\left(\frac{V^2}{c^2}-1\right)\frac{dV}{V}$$

Introducing M=V/c, we have

$$\frac{dA}{A}=\left(\mathrm{M}^2-1\right)\frac{dV}{V}$$

c) Substituting in $V=\mathrm{M}c$, $c^2=kRT$, and $R/c_p=(k-1)/k$, we find

$$\begin{aligned}\frac{T_0}{T}&=\frac{V^2}{2c_pT}+1=\frac{\mathrm{M}^2c^2}{2c_pT}+1=\frac{\mathrm{M}^2kRT}{2c_pT}+1\\&=\frac{\mathrm{M}^2k(k-1)}{2k}+1=1+\frac{k-1}{2}\mathrm{M}^2.\end{aligned}$$

d) $$\dot{m} = p\sqrt{\frac{k}{TR}}AM = \frac{p_0\left(1+\frac{k-1}{2}M^2\right)^{k/(1-k)}}{\sqrt{T_0}\left(1+\frac{k-1}{2}M^2\right)^{-1/2}}\sqrt{\frac{k}{R}}AM$$

$$= p_0\sqrt{\frac{k}{RT_0}}MA\left(1+\frac{k-1}{2}M^2\right)^{\frac{k+1}{2(1-k)}}$$

At the critical area A^*, $M^* = 1$. Hence,

$$\dot{m} = p_0\sqrt{\frac{k}{RT_0}}A^*\left(\frac{k+1}{2}\right)^{\frac{k+1}{2(1-k)}}.$$

e) Since $\dot{m}$ is constant throughout the nozzle, we can equate Eq. 9.3.17 to Eq. 9.3.18:

$$p_0\sqrt{\frac{k}{RT_0}}MA\left(1+\frac{k-1}{2}M^2\right)^{\frac{k+1}{2(1-k)}} = p_0\sqrt{\frac{k}{RT_0}}A^*\left(\frac{k+1}{2}\right)^{\frac{k+1}{2(1-k)}}$$

or

$$\frac{A}{A^*} = \frac{1}{M}\left[\frac{2+(k-1)M^2}{k+1}\right]^{\frac{k+1}{2(k-1)}}$$

9.18 a) $p_s = p_{atm} + 10 = 69.9 + 10 = 79.9$ kPa abs. $p_1 = 69.9$ kPa abs.

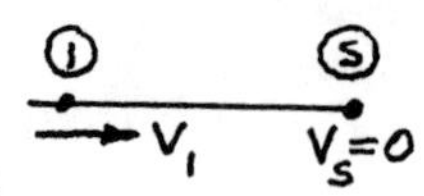

From ① →Ⓢ: $\frac{V_1^2}{2} + \frac{p_1}{\rho_1} = \frac{p_s}{\rho_s}$. $\rho_s = \rho_1\left(\frac{p_s}{p_1}\right)^{1/k} = .906\left(\frac{79.9}{69.9}\right)^{1/1.4} = 0.997 \text{ kg/m}^3$.

$\therefore \frac{V_1^2}{2} + \frac{69\,900}{.906} = \frac{79\,900}{.997}$. $\therefore V_1 = 77.3$ m/s.

b) $p_s = 26.4 + 10 = 36.4$ kPa abs. $p_1 = 26.4$ kPa abs.

From ① →Ⓢ: $\frac{V_1^2}{2} + \frac{p_1}{\rho_1} = \frac{p_s}{\rho_s}$. $\rho_s = \rho_1\left(\frac{p_s}{p_1}\right)^{1/k} = 0.412\left(\frac{36.4}{26.4}\right)^{1/1.4} = 0.518 \text{ kg/m}^3$.

$\frac{V_1^2}{2} + \frac{26\,400}{.412} = \frac{36\,400}{.518}$. $\therefore V_1 = 111$ m/s.

9.19 a) $\frac{V_1^2}{2} + \frac{p_1}{\rho_1} = \frac{p_s}{\rho_s}$. $\rho_s = \rho_1\left(\frac{p_s}{p_1}\right)^{1/k} = 1.22\left(\frac{105}{101}\right)^{1/1.4} = 1.254 \text{ kg/m}^3$.

$\frac{V_1^2}{2} + \frac{1.4}{.4}\frac{101\,000}{1.22} = \frac{105\,000}{1.254}\frac{1.4}{.4}$. $\therefore V_1 = 81.3$ m/s

b) $\frac{V_1^2}{2} = \frac{4000}{1.22}$. $\therefore V_1 = 81.0$ m/s. % error $= \frac{81.3 - 81}{81.3} \times 100 = 0.42\%$.

9.20 Is $p_r < .5283 p_0$? $0.5283 \times 200 = 105.7$ kPa.

a) $p_r < .5283 p_0$. $\therefore$ choked flow. $\therefore \text{M}_e = 1$. $\therefore V_e^2 = kRT_e$. $p_e = 105.7$ kPa.

$$1000 \times 298 = \frac{1.4 \times 287 T_e}{2} + 1000 T_e. \quad \therefore T_e = 248.1 \text{ K}, \quad V_e = 315.8 \text{ m / s}.$$

$$\rho_e = \frac{105.7}{.287 \times 248.1} = 1.484 \text{ kg / m}^3. \quad \therefore \dot{m} = 1.484 \times \pi \times .01^2 \times 315.8 = \underline{0.1473 \text{ kg / s}}.$$

b) $p_r > .5283 p_0$. $\therefore \text{M}_e < 1$. $1000 \times 298 = \frac{V_e^2}{2} + \frac{1.4}{.4}\frac{130\ 000}{\rho_e}$. $\frac{130}{200} = \left(\frac{\rho_e}{2.338}\right)^{1.4}$

$$\rho_0 = \frac{200}{.287 \times 298} = 2.338. \quad \therefore \rho_e = 1.7187 \text{ kg / m}^3. \quad \therefore V_e = 257.9 \text{ m / s}.$$

$$\therefore \dot{m} = 1.7187 \times \pi \times .01^2 \times 257.9 = \underline{0.1393 \text{ kg / s}}.$$

9.20E Is $p_r < .5283 p_0$? $0.5283 \times 30 = 15.85$ psia.

a) $p_r < 15.85$. $\therefore$ choked flow and $\text{M}_e = 1$, $p_e = 15.85$ psia. $V_e^2 = kRT$.

$$0.24 \times 530 = \frac{1.4 \times 1716 \times T_e}{2(778 \times 32.2)} + 0.24\ T_e. \quad \therefore T_e = 441.7^\circ \text{R}, \quad V_e = 1030 \text{ fps}.$$

$$\rho_e = \frac{15.85 \times 144}{1716 \times 441.7} = 0.003011 \text{ slug / ft}^3. \quad \therefore \dot{m} = .003011 \times \pi \left(\frac{.5}{12}\right)^2 \times 1030 = \underline{0.01692 \text{ slug / sec}}.$$

b) $p_r > 15.85$. $\therefore \text{M}_e < 1$, and $p_e = 20$ psia. $\rho_0 = \frac{30 \times 144}{1716 \times 530} = .00475 \text{ slug / ft}^3$.

$$\therefore \rho_e = .00475\left(\frac{20}{30}\right)^{1/1.4} = .003556 \text{ slug / ft}^3. \quad 0.24 \times 530(778 \times 32.2) = \frac{V_e^2}{2} + \frac{1.4}{.4}\frac{20 \times 144}{.003556}.$$

$$\therefore V_e = 838.9 \text{ fps}. \quad \therefore \dot{m} = .003556 \times \pi \left(\frac{.5}{12}\right)^2 \times 838.9 = \underline{0.01627 \text{ slug / sec}}.$$

(Note: $c_p = 0.24 \text{ Btu / lbm-}^\circ\text{R} = 0.24 \times 778 \frac{\text{ft-lb}}{\text{lbm-}^\circ\text{R}} = 0.24 \times 778 \times 32.2 \frac{\text{ft-lb}}{\text{slug-}^\circ\text{R}}$.)

9.21 a) $p_r < .5283\, p_0$. $\therefore \text{M}_e = 1$. $\therefore p_e = .5283 \times 200 = 105.7$ kPa. $T_e = .8333 \times 298 = 248.3$ K.

$$\rho_e = \frac{105.7}{.287 \times 248.3} = 1.483 \text{ kg / m}^3. \quad V_e = \sqrt{1.4 \times 287 \times 248.3} = 315.9 \text{ m / s}.$$

$$\therefore \dot{m} = 1.483 \times \pi \times .01^2 \times 315.9 = \underline{0.1472 \text{ kg / s}}.$$

b) $p_r > .5283\, p_0$. $\therefore p_e = 130$ kPa, $\frac{p_e}{p_0} = 0.65$. $\therefore \text{M}_e = .81$, $T_e = .884 T_0$

$$\rho_e = \frac{130}{.287 \times 263.4} = 1.719 \text{ kg / m}^3, \quad V_e = .81\sqrt{1.4 \times 287 \times 263.4} = 263.5 \text{ m / s}.$$

$$\therefore \dot{m} = 1.719 \times \pi \times .01^2 \times 263.5 = \underline{0.1423 \text{ kg / s}}.$$

9.21E a) $p_r < .5283\, p_0$. $\therefore \mathrm{M}_e = 1$. $\therefore p_e = .5283 \times 30 = 15.85$ psia.

$T_e = .8333 \times 530 = 441.6°\mathrm{R}$.

$$\therefore \rho_e = \frac{15.85 \times 144}{1716 \times 441.6} = .003012\ \frac{\text{slug}}{\text{ft}^3}.\quad V_e = \sqrt{1.4 \times 1716 \times 441.6} = 1030 \text{ fps.}$$

$$\dot{m} = .003012 \times \pi\left(\frac{.5}{12}\right)^2 \times 1030 = \underline{0.01692 \text{ slug / sec.}}$$

b) $p_r > .5283\, p_0$. $\therefore p_e = 20$ psia. $\dfrac{p_e}{p_0} = \dfrac{20}{30} = .6667$. $\therefore \mathrm{M}_e = .785$. $T_e = 0.890 T_0$.

$$\therefore \rho_0 = \frac{20 \times 144}{1716 \times 472} = .00356.\quad V_e = .785\sqrt{1.4 \times 1716 \times 472} = 836 \text{ fps.}$$

$$\therefore \dot{m} = .00356\pi\left(\frac{.5}{12}\right)^2 \times 836 = \underline{0.01664 \text{ slug / sec.}}$$

9.22 $p_e = .5283 \times 400 = \underline{211.3 \text{ kPa abs}}$. $T_e = .8333 \times 303 = 252.5$ K.

$$V_e = \sqrt{1.4 \times 287 \times 252.5} = 318.5 \text{ m / s.}\quad \therefore \dot{m} = \frac{211.3}{.287 \times 252.5}\pi \times .05^2 \times 318.5 = \underline{7.29 \text{ kg / s.}}$$

9.23 $p_e = .5283\, p_0 = 101$ kPa. $\therefore p_0 = \underline{191.2 \text{ kPa abs}}$. $T_e = .8333 \times 283 = 235.8$ K.

$$V_e = \sqrt{1.4 \times 287 \times 235.8} = 307.8 \text{ m / s.}\quad \therefore \dot{m} = \frac{101}{.287 \times 235.8}\pi \times .03^2 \times 307.8 = \underline{1.30 \text{ kg / s.}}$$

$p_0 = 2 \times 191.2 = 382.4$ kPa abs. $p_e = .5283\, p_0 = 202.0$ kPa abs. $T_e = 235.8$ K.

$$V_e = 307.8 \text{ m / s since } \mathrm{M}_e = 1.\quad \therefore \dot{m} = \frac{202}{.287 \times 235.8}\pi \times .03^2 \times 307.8 = \underline{2.60 \text{ kg / s.}}$$

9.23E $p_e = .5283\, p_0 = 14.7$ psia. $\therefore p_0 = \underline{27.83 \text{ psia}}$. $T_e = .8333 \times 500 = 416.6°\mathrm{R}$.

$$V_e = \sqrt{1.4 \times 1716 \times 416.6} = 1000 \text{ fps.}\quad \therefore \dot{m} = \frac{14.7 \times 144}{1716 \times 416.6}\pi\left(\frac{1.25}{12}\right)^2 \times 1000 = \underline{0.101 \text{ slug / sec.}}$$

$p_0 = 2 \times 27.83$. $p_e = .5283\, p_0 = 29.4$ psia, $T_e = 416.6°\mathrm{R}$, $V_e = 1000$ fps.

$\therefore \underline{\dot{m} = 0.202 \text{ slug / sec.}}$

9.24 Treat the pipeline as a reservoir. Then, $p_e = .5283\, p_0 = 264.5$ kPa abs.

$\mathrm{M}_e = 1$ and $V_e = \sqrt{1.4 \times 287(.8333 \times 283)} = 307.8$ m / s.

$$\dot{m} = \frac{264.5}{.287 \times (.8333 \times 283)} \times 30 \times 10^{-4} \times 307.8 = 3.61 \text{ kg / s.}$$

$$\Delta V\!\!\!/ = \frac{\dot{m}\Delta t}{\rho} = \frac{3.61 \times 6 \times 60}{264.5 / (.287 \times .8333 \times 283)} = \underline{333 \text{ m}^3}.$$

9.25 $5193 \times 300 = \dfrac{1.667 \times 2077\ T_e}{2} + 5193\ T_e.\quad \therefore T_e = 225\ \text{K}.\quad \therefore p_e = 200\left(\dfrac{225}{300}\right)^{\frac{1.667}{.667}}$
$= \underline{97.45\ \text{kPa}}.$

Next, $T_t = 225$ K, $p_t = 97.45$ kPa; $\therefore V_t = \sqrt{1.667 \times 2077 \times 225} = 882.6$ m / s.

$$\rho_t = \frac{97.45}{2.077 \times 225} = 0.2085\ \text{kg / m}^3.\quad .2085 \times \pi \times .03^2 \times 882.6 = \rho_e \pi \times .075^2 V_e$$

$$5193 \times 300 = \frac{V_e^2}{2} + \frac{1.667}{.667}\frac{p_e}{\rho_e}.\quad p_e = 200\left(\frac{\rho_e}{200 / 2.077 \times 300}\right)^{1.667} = 1330\rho_e^{\ 1.667}\ \text{kPa}.$$

$$= \frac{V_e^2}{2} + 3324 \times 10^3 \times 9.54 V_e^{-.667}.$$

or $3.116 \times 10^6 = V_e^2 + 63\ 420 \times 10^3 V_e^{-.667}$. Trial - and - error: $V_e = 91.8$ m / s.

$\therefore \rho_e = 0.3203$ kg / m^3 and $p_e = \underline{199.35\ \text{kPa abs.}}$

9.26 $\rho_1 = \dfrac{p_1}{RT_1} = \dfrac{300 + 100}{.287 \times 293} = 4.757\ \text{kg / m}^3.\quad \rho_2 = 4.757\left(\dfrac{340}{400}\right)^{1/1.4} = 4.236\ \text{kg / m}^3.$

$V_1 \times 4.757 \times 10^2 = V_2 \times 4.236 \times 5^2.\quad \therefore V_2 = 4.492\ V_1.$

$$\frac{V_1^2}{2} + \frac{k}{k-1}\frac{p_1}{\rho_1} = \frac{V_2^2}{2} + \frac{k}{k-1}\frac{p_2}{\rho_2}.\quad \frac{V_1^2}{2} + \frac{1.4}{.4}\frac{400\ 000}{4.757} = \frac{4.492^2 V_1^2}{2} + \frac{1.4}{.4}\frac{340\ 000}{4.236}.$$

$$\therefore V_1 = 37.35\ \frac{\text{m}}{\text{s}}.$$

$\therefore \dot{m} = \rho_1 A_1 V_1 = 4.757 \times \pi \times .05^2 \times 37.35 = \underline{1.395\ \text{kg / s}}.$

9.26E $\rho_1 = \dfrac{p_1}{RT_1} = \dfrac{(45 + 14.7)144}{1716 \times 520} = 0.009634\ \dfrac{\text{slug}}{\text{ft}^3}.$

$$\rho_2 = .009634\left(\frac{50.7}{59.7}\right)^{1/1.4} = .008573\ \text{slug / ft}^3.$$

$V_1 \times .009634 \times 4^2 = V_2 \times .008573 \times 2^2.\quad \therefore V_2 = 4.495\ V_1.$

$$\frac{V_1^2}{2} + \frac{1.4}{.4}\frac{59.7 \times 144}{.009634} = \frac{4.495^2 V_1^2}{2} + \frac{1.4}{.4}\frac{50.7 \times 144}{.008573}.\quad \therefore V_1 = 121.9\ \text{fps}.$$

$\therefore \dot{m} = .009634\pi \times (2 / 12)^2 \times 121.9 = \underline{0.1025\ \text{slug / sec.}}$

9.27 Energy $0 \to 2$: $1000 \times 303 = \dfrac{V_2^2}{2} + 1000\, T_2$. $V_2 = 3\sqrt{kRT_2}$

$\therefore 303\ 000 = \dfrac{9 \times 1.4 \times 287\, T_2}{2} + 1000\, T_2$. $\therefore T_2 = 107.9$ K.

$\therefore p_2 = 200\left(\dfrac{107.9}{303}\right)^{\frac{1.4}{.4}} = 5.390$ kPa. $\rho_2 = \dfrac{5.39}{.287 \times 107.9} = 0.1740$ kg / m^3.

Energy $0 \to 1$: $1000 \times 303 = \dfrac{V_1^2}{2} + 1000\, \dfrac{V_1^2}{1.4 \times 287}$. $\therefore V_1 = 318.4$ m / s, $T_1 = 252.3$ K.

$p_1 = 200\left(\dfrac{252.3}{303}\right)^{\frac{1.4}{.4}} = 105.4$ kPa. $\rho_1 = \dfrac{105.4}{.287 \times 252.3} = 1.455$ kg / m^3.

Continuity: $1.455\pi \times .05^2 \times 318.4 = .174\pi \dfrac{d_2^2}{4} \times 3\sqrt{1.4 \times 287 \times 107.9}$. $\therefore \underline{d_2 = 0.2065 \text{ m.}}$

9.28 $V_t^2 = kRT_t$. $1000 \times 293 = \dfrac{1.4 \times 287\, T_t}{2} + 1000\, T_t$. $\therefore T_t = 244.0$ K. $V_t = 313.1$ m / s.

$\therefore p_t = 500\left(\dfrac{244}{293}\right)^{\frac{1.4}{.4}} = 263.5$ kPa abs. $\therefore \rho_t = \dfrac{263.5}{.287 \times 244} = 3.763$ kg / m^3.

$1000 \times 293 = \dfrac{V_e^2}{2} + \dfrac{1.4}{.4}\dfrac{p_e}{\rho_e}$. $3.763 \times \pi \times .025^2 \times 313.1 = \rho_e \pi \times .075^2 V_e$. $\dfrac{p_e}{\rho_e^{1.4}} = \dfrac{263\ 500}{3.763^{1.4}}$

$\therefore 293\ 000 = \dfrac{V_e^2}{2} + 1.014 \times 10^6 V_e^{-.4}$. Trial - and - error: $V_e = 22.2$ m / s, 659 m / s.

$\therefore \rho_e = 5.897,\ 0.1987$ kg / m^3. $\therefore p_e = \underline{494.2 \text{ kPa}}$, $\underline{4.29 \text{ kPa abs}}$.

9.29 $\dfrac{A_e}{A^*} = 9$. $\therefore \dfrac{p_e}{p_0} = 0.997$ from Table C.1. $\therefore p_e = 500 \times .997 = \underline{498.5 \text{ kPa}}$.

and $\dfrac{p_e}{p_0} = 0.00855$ from Table C.1. $\therefore p_e = \underline{4.28 \text{ kPa abs}}$.

9.30 $\mathrm{M_t} = 1$. $\therefore p_t = .5283 \times 800 = 422.6$ kPa, $T_t = .8333 \times 300 = 250$ K.

$\therefore \rho_t = 5.890$ kg / m^3.

$\dot{m} = 6 = 5.89 \dfrac{\pi d_t^2}{4}\sqrt{1.4 \times 287 \times 250}$. $\therefore \underline{d_t = 0.0640 \text{ m}}$.

$\dfrac{p}{p_0} = \dfrac{100}{800} = .125$. $\therefore \mathrm{M_e} = 2.014$, $\dfrac{A}{A^*} = 1.708$. $\therefore A_e = 1.708\, A_t$.

$\pi d_e^2 / 4 = (\pi \times .064^2 / 4)1.708$. $\therefore \underline{d_e = 0.0836 \text{ m}}$.

$T_e = .552\, T_0 = .552 \times 300 = 165.6$ K. $\therefore V_e = 2.014\sqrt{1.4 \times 287 \times 165.6} = \underline{520 \text{ m / s}}$.

9.30E $M_t = 1. \quad \therefore p_t = .5283 \times 120 = 63.4 \text{ psia}, \ T_t = .8333 \times 520 = 433.3^\circ \text{R}.$

$$\therefore \rho_t = .01228 \ \frac{\text{slug}}{\text{ft}^3}.$$

$$\dot{m} = 1 = .01228 \frac{\pi d_t^2}{4} \sqrt{1.4 \times 1716 \times 433.3}. \quad \therefore \underline{d_t = 0.319 \text{ ft.}}$$

$$\frac{p}{p_0} = \frac{15}{120} = .125. \quad \therefore M_e = 2.014, \ T_e = .552 \times 520 = 287^\circ \text{R}, \ V_e = 2.014\sqrt{1.4 \times 1716 \times 287}$$

$$= \underline{684 \text{ fps}}.$$

$$\frac{A}{A^*} = 1.708. \quad \therefore \frac{\pi d_e^2}{4} = 1.708 \frac{\pi \times .319^2}{4}. \quad \therefore \underline{d_e = 0.417 \text{ ft}}$$

9.31 $M_e = 4. \quad \frac{A}{A^*} = 10.72, \ p_e = .006586 \times 2000 = 13.17 \text{ kPa}, \ T_e = .2381 \times 293 = 69.76 \text{ K}.$

For $\frac{A}{A^*} = 10.72, \ M_e = .0584. \quad \therefore p_e = .9976\, p_0 = .9976 \times 2000 = \underline{1995.2 \text{ kPa abs}}.$

9.32 Let $M_t = 1$. Neglect viscous effects. $M_1 = \frac{150}{\sqrt{1.4 \times 287 \times 303}} = 0.430.$

$$\therefore \frac{A}{A^*} = 1.5007. \quad \therefore A_t = \frac{A_1}{1.5007} = \frac{\pi \times .05^2}{1.5007} = \frac{\pi d_t^2}{4}. \quad \therefore d_t = 0.0816 \text{ m or } \underline{8.16 \text{ cm}}.$$

9.33 $p_e = .5283 \times 400 = 211.3 \text{ kPa abs}. \ T_{es} = .8333 \times 303 = 252.5.$

$$.96 = \frac{303 - T_e}{303 - 252.5}. \quad \therefore T_e = 254.5 \text{ K}. \quad \therefore V_e = \sqrt{1.4 \times 287 \times 254.5} = 319.8 \text{ m/s}.$$

$$\therefore \dot{m} = \frac{211.3}{.287 \times 254.5} \pi \times .05^2 \times 319.8 = \underline{7.27 \text{ kg/s}}.$$

9.34 Isentropic flow. Since $k = 1.4$ for nitrogen, the isentropic table may be used.

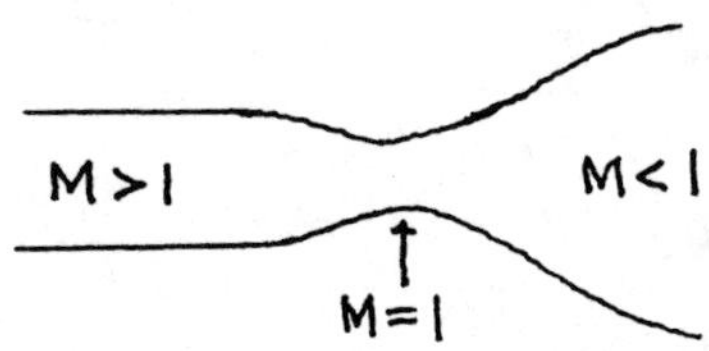

$M = 3: \quad \frac{A}{A^*} = 4.235.$

$$V_i = 3\sqrt{1.4 \times 297 \times 373} = 1181 \text{ m/s}. \quad \rho_i = \frac{100}{.297 \times 373} = .9027 \text{ kg/m}^3.$$

$$\therefore A_i = \frac{\dot{m}}{\rho_i V_i} = \frac{10}{.9027 \times 1181} = 0.00938 \text{ m}^2. \quad \therefore A_t = \frac{.00938}{4.235} = \underline{0.00221 \text{ m}^2}.$$

At $M = 3$, $T = .3571\, T_0$, $p = .02722\, p_0$.

$$\therefore T_0 = T_e = \frac{373}{.3571} = 1044 \text{ K or } \underline{772^\circ\text{C}}. \quad p_0 = \frac{100}{.02722} = p_e = \underline{3670 \text{ kPa}}.$$

9.34E Isentropic flow. Since $k = 1.4$ for nitrogen, the isentropic table may be used. (See figure for 9.34)

$\mathrm{M} = 3$: $\dfrac{A}{A^*} = 4.235.$

$V_i = 3\sqrt{1.4 \times 1776 \times 660} = 3840 \text{ fps}.$ $\rho_i = \dfrac{15 \times 144}{1776 \times 660} = .001843 \dfrac{\text{slug}}{\text{ft}^3}.$

$A_i = \dfrac{.2}{.001843 \times 3840} = .0283 \text{ ft}^2.$ $\therefore A_t = \dfrac{.0283}{4.235} = \underline{0.00667 \text{ ft}^2}.$

At $\mathrm{M} = 3$, $T = .3571\, T_0$, $p = .02722\, p_0$.

$\therefore T_0 = T_e = \dfrac{660}{.3571} = 1848^\circ\text{R}$ or $\underline{1388^\circ\text{F}}$. $p_0 = p_e = \dfrac{15}{.02722} = \underline{551 \text{ psia}}.$

9.35 Assume $p_e = 101$ kPa. Then $\rho_e = \dfrac{101}{.189 \times 1273} = .4198 \text{ kg/m}^3.$

Momentum: $F = \dot{m}V = \rho A V^2.$ $\dfrac{80\,000 \times 9.81}{6} = .4198\pi \times .25^2 V^2.$ $\therefore \underline{V = 1260 \text{ m/s}}.$

9.36 $F = \dot{m}V = \rho A V^2.$ $\rho = \dfrac{101}{.287 \times 873} = .403 \text{ kg/m}^3.$ (Assume gases are air.)

$100 \times 9.81 = .403 \times 200 \times 10^{-4} V^2.$ $\therefore V = \underline{349 \text{ m/s}}.$

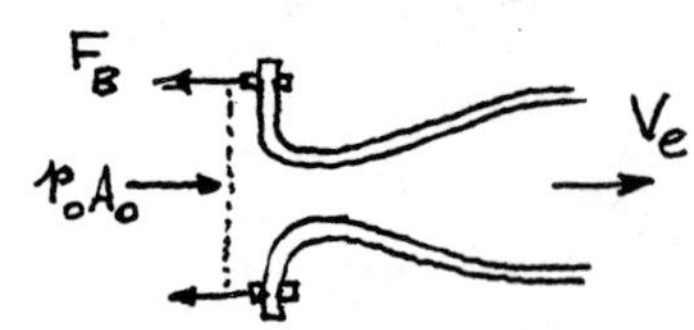

9.37 $\mathrm{M_t} = 1.$ $\dfrac{A_e}{A^*} = 4;$ $\therefore \mathrm{M_e} = 2.94,$ $p_e = .02980\, p_0.$

$T_e = .3665\, T_0 = .3665 \times 300 = 109.95$ K, $p_e = 100 = .0298\, p_0.$ $\therefore p_0 = 3356$ kPa abs.

$\therefore V_e = 2.94\sqrt{1.4 \times 287 \times 109.95} = 618 \text{ m/s}.$

$\therefore F_B = \dfrac{-100}{.287 \times 109.95}\pi \times .05^2 \times 618^2 + 3\,356\,000\pi \times .2^2 = \underline{412\,000 \text{ N}}.$

9.38 Assume an isentropic flow; Eq. 9.3.13 provides

$$\frac{1.03p}{p} = \left(1 + \frac{k-1}{2}\mathrm{M}^2\right)^{\frac{1}{k-1}}.$$

Using $k = 1.4$ this gives $\mathrm{M}^2 = 0.0424$ or $\mathrm{M} = 0.206.$

For standard conditions $V = \mathrm{M}c = 0.206\sqrt{1.4 \times 287 \times 288} = \underline{70 \text{ m/s}}.$

9.39 a) $.9850 \times 1000 = \rho_2 V_2.$ $80\,000 - p_2 = .985 \times 1000(V_2 - 1000)$

$$\frac{V_2^2 - 1000^2}{2} + \frac{1.4}{.4}\left(\frac{p_2}{\rho_2} - 287 \times 283\right) = 0. \quad \left(\rho_1 = \frac{80}{.287 \times 283} = .9850 \text{ kg/m}^3.\right)$$

$$\frac{V_2^2}{2} - \frac{1000^2}{2} + \frac{1.4}{.4}\frac{V_2}{985}(-985V_2 + 1\,065\,000) - 284\,300 = 0$$

$\therefore 3V_2^2 - 3784V_2 + 784\,300 = 0.$ $\therefore V_2 = 261$ m / s. $\rho_2 = \underline{3.774 \text{ kg / m}^3}.$

$p_2 = \underline{808 \text{ kPa}}.$

$$\text{M}_1 = \frac{1000}{\sqrt{1.4 \times 287 \times 283}} = \underline{2.966}. \quad T_2 = \frac{808}{.287 \times 3.774} = 746 \text{ K or } \underline{473^\circ\text{C}}.$$

$$\text{M}_2 = \frac{261}{\sqrt{1.4 \times 287 \times 746}} = \underline{0.477}.$$

b) $\text{M}_1 = 1000 / \sqrt{1.4 \times 287 \times 283} = \underline{2.97}.$ $\therefore \text{M}_2 = \underline{0.477}.$ $p_2 = 10.12\,p_1 = \underline{809.6 \text{ kPa}}.$

$$T_2 = 2.644 \times 283 = 748 \text{ K or } \underline{475^\circ\text{C}}. \quad \therefore \rho_2 = \frac{809.6}{.287 \times 748} = \underline{3.771 \text{ kg / m}^3}.$$

9.39E a) $\rho_1 = \dfrac{12 \times 144}{1716 \times 500} = .002014 \dfrac{\text{slug}}{\text{ft}^3}.$ $.002014 \times 3000 = \rho_2 V_2.$

$$12 \times 144 - p_2 = .002014 \times 3000(V_2 - 3000). \quad \frac{V_2^2 - 3000^2}{2} + \frac{1.4}{.4}\left(\frac{p_2}{\rho_2} - 1716 \times 500\right) = 0.$$

$$V_2^2 - 3000^2 + 7\left(\frac{V_2}{6.042}\right)(19{,}854 - 6.042V_2) - 6.006 \times 10^6 = 0.$$

$$\therefore 6V_2^2 - 23{,}000V_2 + 15 \times 10^6 = 0. \quad \therefore V_2 = 833 \text{ fps.} \quad \rho_2 = \underline{0.00725 \frac{\text{slug}}{\text{ft}^3}}.$$

$$p_2 = \underline{102.9 \text{ psia}}. \quad \text{M}_1 = \frac{3000}{\sqrt{1.4 \times 1716 \times 500}} = \underline{2.74}. \quad T_2 = \frac{102.9 \times 144}{1716 \times .00725} = 1191^\circ\text{R or } \underline{731^\circ\text{F}}.$$

$$\text{M}_2 = \frac{833}{\sqrt{1.4 \times 1716 \times 1191}} = \underline{0.492}.$$

b) $\text{M}_1 = 3000 / \sqrt{1.4 \times 1716 \times 500} = \underline{2.74}.$ $\therefore \text{M}_2 = \underline{0.493}$ $p_2 = 8.592 \times 12 = \underline{103.1 \text{ psia}}.$

$$T_2 = 2.386 \times 500 = 1193^\circ\text{R or } \underline{733^\circ\text{F}}. \quad \therefore \rho_2 = \frac{103.1 \times 144}{1716 \times 1193} = \underline{0.00725 \text{ slug / ft}^3}.$$

9.40 $$\frac{\rho_2}{\rho_1} = \frac{p_2}{p_1}\frac{T_1}{T_2} = \frac{2k\text{M}_1^2 - k + 1}{k+1}\,\frac{(k+1)^2\text{M}_1^2}{\left[1 + \frac{k-1}{2}\text{M}_1^2\right]\left[4k\text{M}_1^2 - 2k + 2\right]} = \frac{(k+1)\text{M}_1^2}{2 + (k-1)\text{M}_1^2}.$$

$\text{M}_1^2 = \dfrac{k+1}{2k}\dfrac{p_2}{p_1} + \dfrac{k-1}{2k}.$ (This is Eq. 9.4.12). Substitute into above:

$$\frac{\rho_2}{\rho_1} = \frac{(k+1)\left[(k+1)\dfrac{p_2}{p_1}+(k-1)\right]}{4k+(k-1)\left\{(k+1)\dfrac{p_2}{p_1}+(k-1)\right\}} = \frac{(k+1)\left[(k+1)\dfrac{p_2}{p_1}+k-1\right]}{(k+1)^2+(k-1)(k+1)\dfrac{p_2}{p_1}}.$$

$$= \frac{k-1+(k+1)p_2/p_1}{k+1+(k-1)p_2/p_1}.$$

For a strong schock in which $\frac{p_2}{p_1} >> 1,\ \underline{\frac{\rho_2}{\rho_1} = \frac{k+1}{k-1}}.$

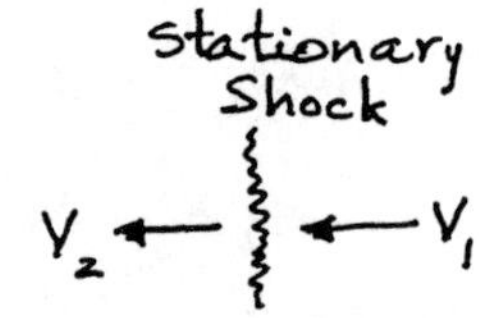

9.41 Assume standard conditions: $T_1 = 15°C,\ \rho_1 = 101$ kPa.
$\therefore V_1 = 2\sqrt{1.4\times 287\times 288} = 680$ m / s.
$M_1 = 2.\ \ \therefore M_2 = .5774.\ \ T_2 = 1.688\times 288 = 486$ K. $\ p_2 = 4.5\times 101 = \underline{454\text{ kPa}}.$
$\therefore V_2 = .5774\sqrt{1.4\times 287\times 486} = 255$ m / s. $\ \therefore V_{\text{induced}} = V_1 - V_2 = 680 - 255 = \underline{425\text{ m / s}}.$
The high pressure and high induced velocity cause extreme damage.

9.42 If $M_2 = .5$, then $M_1 = 2.645.\ \ \therefore V_1 = 2.645\sqrt{1.4\times 287\times 293} = \underline{908\text{ m / s}}.$
$p_2 = 8.00\times 200 = \underline{1600\text{ kPa abs}}.\ \ \rho_2 = \dfrac{1600}{.287\times(2.285\times 293)} = \underline{8.33\text{ kg / m}^3}.$

9.42E If $M_2 = .5$, then $M_1 = 2.645.\ \ \therefore V_1 = 2.645\sqrt{1.4\times 1716\times 520} = \underline{1118\text{ fps}}.$
$p_2 = 8.00\times 30 = \underline{240\text{ psia}}.\ \ \rho_2 = \dfrac{240\times 144}{1716\times(2.285\times 520)} = \underline{0.01695\text{ slug / ft}^3}.$

9.43 $p_1 = .2615\times 101 = 26.4$ kPa. $T_1 = 223.3$ K. $M_1 = 1000/\sqrt{1.4\times 287\times 223.3} = 3.34.$
$\therefore M_2 = .4578.\ \ p_2 = 12.85\times 26.4 = 339$ kPa. $T_2 = 3.101\times 223.3 = 692.5$ K.
For isentropic flow from ② → ⓪: For $M = .458,\ p = .866\,p_0$ and
$T = .960\ T_0.\ \ \therefore p_0 = 339/.866 = \underline{391\text{ kPa abs}}.\ \ T_0 = 692.5/.960 = 721$ K or $\underline{448°C}.$

9.44 After the schock $M_2 = .4752,\ p_2 = 10.33\times 800 = 8264$ kPa abs.
For isentropic flow from ② → ⓪: For $M = .475,\ p = .857\,p_0.$
$\therefore p_0 = 8264/.857 = \underline{9640\text{ kPa abs}}.$

9.45 $\dfrac{A}{A^*} = 4.\ \ \therefore M_e = .147.\ \ p_e = .985\,p_0\ \ \therefore p_0 = 101/.985 = \underline{102.5\text{ kPa abs}}.$
$M_t = 1.\ \ p_t = .5283\times 102.5 = 54.15$ kPa. $\ T_t = .8333\times 298 = 248.3$ K.

$\therefore \rho_t = \dfrac{54.15}{.287 \times 248.3} = .7599\ \mathrm{kg/m^3}$. $V_t = \sqrt{1.4 \times 287 \times 248.3} = 315.9\ \mathrm{m/s}$.

$\therefore \dot{m} = .7599 \times \pi \times .025^2 \times 315.9 = \underline{0.471\ \mathrm{kg/s}}$. If throat area is reduced, M_t remains at 1, $\rho_t = .7599\ \mathrm{kg/m^3}$ and $\dot{m} = .7599 \times \pi \times .02^2 \times 315.9 = \underline{0.302\ \mathrm{kg/s}}$.

9.46 $p_e = 101\ \mathrm{kPa} = p_2$. $\dfrac{A}{A^*} = 4$. $\therefore M_1 = 2.94$, and $p_2 / p_1 = 9.918$.

$\therefore p_1 = 101 / 9.918 = 10.18\ \mathrm{kPa}$. At $M_1 = 2.94$, $p / p_0 = .0298$.

$\therefore p_0 = 10.18 / .0298 = \underline{342\ \mathrm{kPa\ abs}}$.

$M_t = 1$, $p_t = .5283 \times 342 = \underline{181\ \mathrm{kPa\ abs}}$. $T_t = .8333 \times 293 = 244.1\ \mathrm{K}$.

$\therefore V_t = \sqrt{1.4 \times 287 \times 244.1} = \underline{313\ \mathrm{m/s}}$.

$M_1 = 2.94$, $p_1 = \underline{10.18\ \mathrm{kPa\ abs}}$. $T_1 = .3665 \times 293 = 107.4\ \mathrm{K}$.

$\therefore V_1 = 2.94\sqrt{1.4 \times 287 \times 107.4} = \underline{611\ \mathrm{m/s}}$.

$M_2 = .4788$, $p_e = \underline{101\ \mathrm{kPa}}$. $T_e = T_2 = 2.609 \times 107.4 = 280.2\ \mathrm{K}$.

$\therefore V_2 = .4788\sqrt{1.4 \times 287 \times 280.2} = \underline{161\ \mathrm{m/s}}$.

9.46E $p_e = 14.7\ \mathrm{psia} = p_2$. $\dfrac{A}{A^*} = 4$. $\therefore M_1 = 2.94$, and $p_2 / p_1 = 9.918$.

$\therefore p_1 = 14.7 / 9.918 = 1.482\ \mathrm{psia}$. At $M_1 = 2.94$, $p / p_0 = .0298$.

$\therefore p_0 = 1.482 / .0298 = \underline{49.7\ \mathrm{psia}}$.

$M_t = 1$, $p_t = .5283 \times 49.7 = \underline{26.3\ \mathrm{psia}}$. $T_t = .8333 \times 520 = 433.3^\circ\mathrm{R}$.

$\therefore V_t = \sqrt{1.4 \times 1716 \times 433.3} = \underline{1020\ \mathrm{fps}}$.

$M_1 = 2.94$, $p_1 = \underline{1.482\ \mathrm{psia}}$. $T_1 = .3665 \times 520 = 190.6^\circ\mathrm{R}$.

$\therefore V_1 = 2.94\sqrt{1.4 \times 1716 \times 190.6} = \underline{1989\ \mathrm{fps}}$.

$M_2 = .4788$, $p_e = \underline{14.7\ \mathrm{psia}}$. $T_e = T_2 = 2.609 \times 190.6 = 497.3^\circ\mathrm{R}$.

$\therefore V_2 = .4788\sqrt{1.4 \times 1716 \times 497.3} = \underline{523\ \mathrm{fps}}$.

9.47 $M_t = 1$. $p_t = .5283 \times 500 = 264\ \mathrm{kPa}$. $T_t = .8333 \times 298 = 248.3\ \mathrm{K}$.

$\dfrac{A_1}{A^*} = \dfrac{8^2}{5^2} = 2.56$. $\therefore M_1 = 2.47$, $p_1 = .0613 \times 500 = 30.65$.

$T_1 = .451 \times 298 = 134.4\ \mathrm{K}$. $\therefore V_1 = 2.47\sqrt{1.4 \times 287 \times 134.4} = \underline{574\ \mathrm{m/s}}$.

$M_2 = .516$, $p_2 = 6.95 \times 30.65 = 213\ \mathrm{kPa}$. $T_2 = 2.108 \times 134.4 = 283.3\ \mathrm{K}$.

After the shock it's isentropic flow. At $M = .516$, $\dfrac{A}{A^*} = 1.314$.

$$p_{02} = .511 \times 500 = 255.5 \text{ kPa.} \quad A^* = \frac{\pi \times .04^2}{1.314} = .003825 \text{ m}^2.$$

$$\frac{A_e}{A^*} = \frac{\pi \times .05^2}{.003825} = 2.05. \quad \therefore p_e = .940 \times 255.5 = \underline{240 \text{ kPa abs.}} = p_r. \quad \text{M}_e = .298.$$

$$T_e = 283.3\left(\frac{213}{240}\right)^{.2857} = 273.8 \text{ K.} \quad \therefore V_e = .298\sqrt{1.4 \times 287 \times 273.8} = \underline{99 \text{ m/s.}}$$

9.48 $$p_t = .546\, p_0 = .546 \times 1200 = 655 \text{ kPa.} \quad T_t = 673\left(\frac{655}{1200}\right)^{.3/1.3} = 585 \text{ K.}$$

$$\therefore \rho_t = \frac{655}{.462 \times 585} = 2.42 \text{ kg/m}^3. \quad V_t = \sqrt{1.3 \times 462 \times 585} = 593 \text{ m/s.} \quad (\text{M}_t = 1.)$$

$$\dot{m} = \rho_t A_t V_t. \quad \therefore 4 = 2.42 \times \frac{\pi \times d_t^2}{4} \times 593. \quad \therefore d_t = 0.060 \text{ m or } \underline{6 \text{ cm}}.$$

$$T_e = 673\left(\frac{101}{1200}\right)^{.3/1.3} = 380.2 \text{ K} \quad \therefore \rho_e = \frac{101}{.462 \times 380.2} = .575 \text{ kg/m}^3.$$

$$\frac{V_e^2}{2} + 1872 \times 380.2 = 1872 \times 673. \quad (\text{Energy from } ⓪ \rightarrow \text{e.}) \quad (c_p = 1872 \text{ J/kg} \cdot \text{K})$$

$$\therefore V_e = 1050 \text{ m/s.} \quad \therefore 4 = .575\frac{\pi d_e^2}{4} \times 1050. \quad \therefore d_e = 0.092 \text{ m or } \underline{9.2 \text{ cm}}.$$

9.49 $$\text{M}_e = 1. \quad p_e = .546\, p_0 = .546 \times 1000 = 546 \text{ kPa.}$$

$$T_e = 623\left(\frac{546}{1000}\right)^{\frac{.3}{1.3}} = 542 \text{ K.} \quad \therefore \rho_e = \frac{546}{.462 \times 542} = 2.18 \frac{\text{kg}}{\text{m}^3}.$$

$$V_e = \sqrt{1.3 \times 462 \times 542} = 571 \text{ m/s.} \quad 15 = 2.18\frac{\pi d_e^2}{4} \times 571. \quad \therefore d_e = 0.124 \text{ m or } \underline{12.4 \text{ cm}}.$$

9.49E $$\text{M}_e = 1. \quad p_e = .546 \times 150 = 81.9 \text{ psia.} \quad T_e = 1160\left(\frac{81.9}{150}\right)^{\frac{.3}{1.3}} = 1009^\circ\text{R.}$$

$$\therefore \rho_e = \frac{81.9 \times 144}{2762 \times 1009} = .00423 \frac{\text{slug}}{\text{ft}^3}. \quad V_e = \sqrt{1.3 \times 2760 \times 1009} = 1903 \text{ fps.}$$

$$.25 = .00423\frac{\pi d_e^2}{4} \times 1903. \quad \therefore d_e = 0.199 \text{ ft. or } \underline{2.39''}.$$

9.50 $$\text{M}_t = 1. \quad p_t = .546 \times 1200 = 655 \text{ kPa.} \quad T_t = 673\left(\frac{655}{1200}\right)^{.3/1.3} = 585 \text{ K.}$$

$$\therefore V_t = \sqrt{1.3 \times 462 \times 585} = 593 \text{ m/s.} \quad \rho_t = \frac{655}{.462 \times 585} = 2.42 \text{ kg/m}^3.$$

$\therefore \dot{m} = 2.42 \times \pi \times .0075^2 \times 593 = \underline{0.254 \text{ kg / s per nozzle}}$

$T_e = 673\left(\dfrac{120}{1200}\right)^{.3/1.3} = \underline{396 \text{ K}}.$

9.51 $\quad M_1 = \dfrac{800}{\sqrt{1.4 \times 287 \times 303}} = 2.29.$

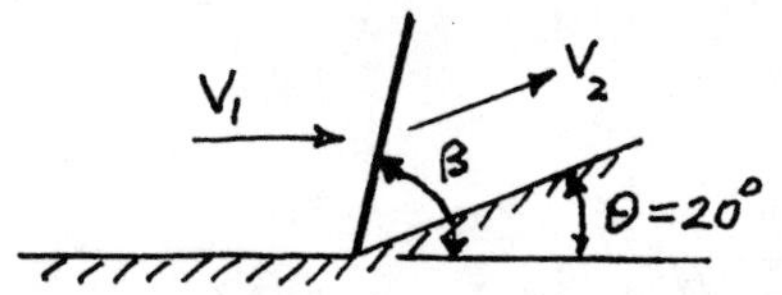

From Fig. 9.15, $\beta = 46^\circ,\ 79^\circ.$

a) $\beta = 46^\circ. \quad \therefore M_{1n} = 2.29 \sin 46^\circ = 1.65. \quad \therefore M_{2n} = .654 = M_2 \sin(46^\circ - 20^\circ).$

$\therefore \underline{M_2 = 1.49}.$

$p_2 = 3.01 \times 40 = \underline{120.4 \text{ kPa abs}}. \quad T_2 = 1.423 \times 303 = 431 \text{ K}.$

$V_2 = \sqrt{1.4 \times 287 \times 431} \times 1.49 = \underline{620 \text{ m / s}}.$

b) $\beta = 79^\circ. \quad \therefore M_{1n} = 2.29 \sin 79^\circ = 2.25. \quad \therefore M_{2n} = .541 = M_2 \sin(79^\circ - 20^\circ).$

$\therefore \underline{M_2 = 0.631}.$

$p_2 = 5.74 \times 40 = \underline{230 \text{ kPa abs}}. \quad T_2 = 1.90 \times 303 = 576 \text{ K}.$

c) 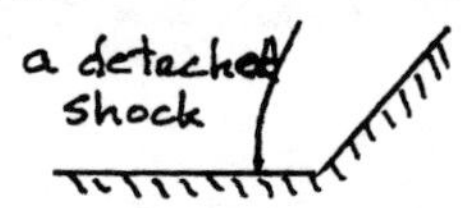

$V_2 = \sqrt{1.4 \times 287 \times 576} \times .631 = \underline{303 \text{ m / s}}.$

9.52 $\quad \beta_1 = 40^\circ. \quad \therefore \theta = 10^\circ.$ (See Fig. E9.11.)

$M_{1n} = 2 \sin 40^\circ = 1.29. \quad \therefore M_{2n} = .791 = M_2 \sin(40^\circ - 10^\circ). \quad \therefore M_2 = 1.58.$

If $\theta_2 = 10^\circ$ then, with $M = 1.58$, $\beta_2 = 51^\circ$. $1.58 \sin 51^\circ = M_{2n}. \quad \therefore M_{2n} = 1.23.$

$\therefore M_{3n} = .824 = M_3 \sin(51^\circ - 10^\circ). \quad \therefore \underline{M_3 = 1.26}. \quad \beta = \beta_2 - 10 = 51 - 10 = \underline{41^\circ}.$

9.53 $\quad M_{1n} = 3.5 \sin 35^\circ = 2.01. \quad \therefore M_{2n} = .576. \quad T_2 = 1.696 \times 303 = 514 \text{ K}.$

$M_2 = \dfrac{.576}{\sin(35^\circ - 20^\circ)} = 2.26. \quad \theta_1 = 20^\circ = \theta_2. \quad \therefore \beta_2 = 47^\circ.$ (See Fig. E9.11.)

$M_{2n} = 2.26 \sin 47^\circ = 1.65. \quad \therefore M_{3n} = .654 = M_3 \sin(47^\circ - 20^\circ). \quad \therefore M_3 = 1.44.$

$T_3 = 1.423 \times 514 = 731 \text{ K}. \quad V_3 = M_3\sqrt{kRT_3} = 1.44\sqrt{1.4 \times 287 \times 731} = \underline{780 \text{ m / s}}.$

9.53E $\quad M_{1n} = 3.5 \sin 35^\circ = 2.01. \quad \therefore M_{2n} = .576. \quad T_2 = 1.696 \times 490 = 831^\circ\text{R}.$

$M_2 = \dfrac{.576}{\sin(35^\circ - 20^\circ)} = 2.26. \quad \theta_1 = 20^\circ = \theta_2. \quad \therefore \beta_2 = 47^\circ.$

$M_{2n} = 2.26 \sin 47^\circ = 1.65. \quad \therefore M_{3n} = .654 = M_3 \sin(47^\circ - 20^\circ). \quad \therefore M_3 = 1.44.$

$T_3 = 1.423 \times 831 = 1180^\circ\text{R}. \quad V_3 = M_3\sqrt{kRT_3} = 1.44\sqrt{1.4 \times 1716 \times 1180} = \underline{2420 \text{ fps}}.$

9.54 $M_1 = 3,\ \theta = 10^\circ.\ \therefore \beta_1 = 28^\circ.\ M_{1n} = 3\sin 28^\circ = 1.41.\ \therefore M_{2n} = .736.$

$\therefore p_2 = 2.153 \times 40 = 86.1$ kPa.

$M_2 = \dfrac{.736}{\sin(28^\circ - 10^\circ)} = 2.38.\ \therefore p_3 = 6.442 \times 86.1 = \underline{555\text{ kPa}}.$

$(p_3)_{normal} = 10.33 \times 40 = \underline{413\text{ kPa}}.$

9.55 At $M_1 = 3,\ \theta_1 = 49.8^\circ,\ \mu_1 = 19.47^\circ.$ (See fig. E9.12.)

$\theta_1 + \theta_2 = 49.8 + 25 = 74.8^\circ.\ \therefore M_2 = 4.78.$

From isentropic flow table: $p_2 = p_1 \dfrac{p_0}{p_1}\dfrac{p_2}{p_0} = 20 \times \dfrac{1}{.02722} \times .002452 = \underline{1.80\text{ kPa}}.$

$T_2 = T_1 \dfrac{T_0}{T_1}\dfrac{T_2}{T_0} = 253 \times \dfrac{1}{.3571} \times .1795 = 127\text{K}$ or $\underline{-146^\circ\text{C}}.\ \mu_2 = 12.08^\circ.$

$V_2 = 4.78\sqrt{1.4 \times 287 \times 127} = \underline{1080\text{ m/s}}.\ \alpha = 90 + 25 - 70.53 - 12.08 = \underline{32.4^\circ}.$

9.56 $\theta_1 = 26.4^\circ.$ For $M = 4,\ \theta = 65.8^\circ.$ (See fig. E.9.12.)

$\therefore \theta = 65.8 - 26.4 = \underline{39.4^\circ}.$

$T_2 = T_1 \dfrac{T_0}{T_1}\dfrac{T_2}{T_0} = 273 \dfrac{1}{.5556} \times .2381 = 117\text{ K}.\ \therefore V_2 = 4\sqrt{1.4 \times 287 \times 117} = \underline{867\text{ m/s}}.$

$T_2 = \underline{-156^\circ\text{C}}.$

9.56E $\theta = 26.4^\circ.$ For $M = 4,\ \theta = 65.8^\circ.\ \therefore \theta = 65.8 - 26.4 = \underline{39.4^\circ}.$

$T_2 = T_1 \dfrac{T_0}{T_1}\dfrac{T_2}{T_0} = 490 \dfrac{1}{.5556} \times .2381 = 210^\circ\text{R}$ or $\underline{-250^\circ\text{F}}.$

$V_2 = 4\sqrt{1.4 \times 1716 \times 210} = \underline{2840\text{ fps}}.$

9.57 a) $\theta_1 = 39.1^\circ.\ \theta_2 = 39.1 + 5 = 44.1^\circ.\ \therefore M_u = 2.72.\ p_{2u} = 20 \dfrac{1}{.0585} \times .04165$

$= \underline{14.24\text{ kPa}}.$

For $\theta = 5^\circ$ and $M = 2.5,\ \beta = 27^\circ.\ M_{1n} = 2.5\sin 27^\circ = 1.13.\ \therefore M_{2n} = .889.$

$\therefore p_{2\ell} = 1.32 \times 20 = \underline{26.4\text{ kPa}}.$

$M_\ell = M_2 = \dfrac{.889}{\sin(27^\circ - 5^\circ)} = 2.37.$

b) $M = 2.72,\ \theta = 5^\circ. \quad \therefore \beta = 25^\circ.\ \ M_{1n} = 2.72 \sin 25^\circ = 1.15,\ M_{2n} = .875.$

$$M_{2u} = \frac{.875}{\sin(25^\circ - 5^\circ)} = \underline{2.56}.$$

For $M = 2.37,\ \theta = 36.0^\circ.$ For $\theta = 36 + 5 = 41^\circ,\ M_{2\ell} = \underline{2.58}.$

c) Force on plate $= (26.4 - 14.24) \times 1000 \times A = F.$

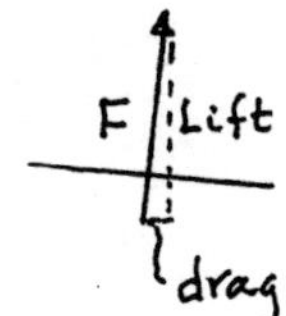

$$C_L = \frac{F \cos 5^\circ}{\frac{1}{2}\rho_1 V_1^2 A} = \frac{12.2 \times .996 \times 1000A}{\frac{1}{2} \times 1.4 \times 2.5^2 \times 20\ 000A} = \underline{0.139}.$$

d) $$C_D = \frac{F \sin 5^\circ}{\frac{1}{2}\rho_1 V_1^2 A} = \frac{12.2 \times 1000A \times .0872}{\frac{1}{2} \times 1.4 \times 2.5^2 \times 20\ 000A} = \underline{0.0122}.$$

9.58 $\beta = 19^\circ.\ \ M_{1n} = 4 \sin 19^\circ = 1.30. \quad \therefore p_2 = 1.805 \times 20 = 36.1 \text{ kPa}.\ \ M_{2n} = .786.$

$$M_2 = \frac{.786}{\sin(19^\circ - 5^\circ)} = 3.25. \quad \theta_1 = 54.36. \quad \theta_2 = 59.36. \quad \therefore M_3 = 3.55.$$

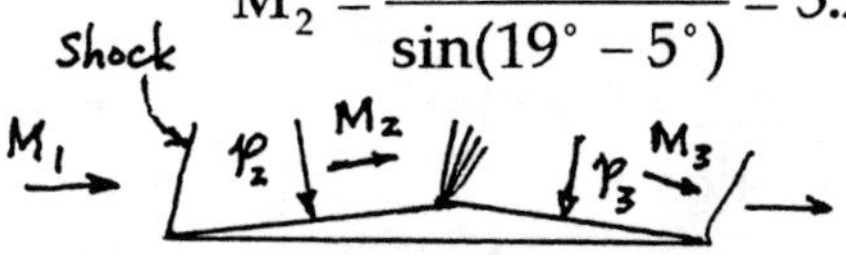

$$p_3 = p_2 \frac{p_0}{p_2}\frac{p_3}{p_0} = 36.1 \times \frac{1}{.0188} \times .0122 = 23.4 \text{ kPa}.$$

$$C_D = \frac{\left(36.1\frac{A}{2} - 23.4\frac{A}{2}\right)\sin 5^\circ}{\frac{1}{2}\rho V_1^2 A} = \frac{6.35 \times .0872}{\frac{1}{2} \times 1.4 \times 4^2 \times 20} = \underline{0.0025}.$$

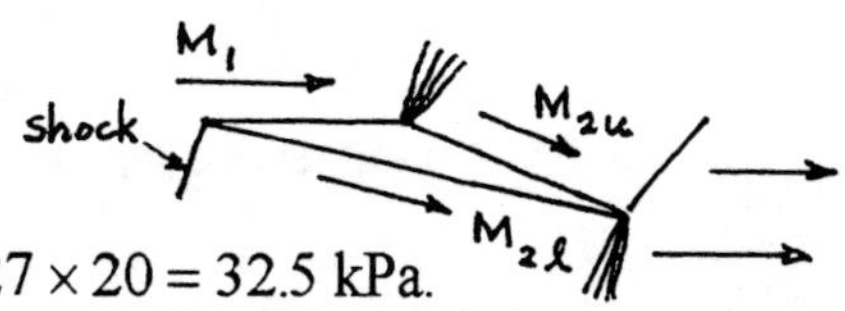

9.59 If $\theta = 5^\circ$ with $M_1 = 4$, then Fig. 9.15 $\rightarrow \beta = 18^\circ$.

$M_{1n} = 4 \sin 18^\circ = 1.24. \quad \therefore M_{2n} = .818. \quad \therefore p_{2\ell} = 1.627 \times 20 = 32.5 \text{ kPa}.$

$$\therefore M_{2\ell} = \frac{.818}{\sin(18^\circ - 5^\circ)} = 3.64.$$

At $M_1 = 4,\ \theta_1 = 65.8^\circ$. At 75.8° $M_{2u} = 4.88.$ $p_{2u} = p_1 \frac{p_0}{p}\frac{p_2}{p_0} = 20\frac{.002177}{.006586}$

$= 6.61$ kPa.

$$C_L = \frac{\text{Lift}}{\frac{1}{2}\rho V_1^2 A} = \frac{32.5A\cos 5^\circ - 20 \times A/2 - 6.61 \times A/2 \times \cos 10^\circ}{\frac{1}{2} \times 1.4 \times 4^2 \times 20A} = \underline{0.0854}.$$

$$C_D = \frac{\text{Drag}}{\frac{1}{2}\rho V_1^2 A} = \frac{32.5A \sin 5^\circ - 6.61 \times A/2 \times \sin 10^\circ}{\frac{1}{2} \times 1.4 \times 4^2 \times 20A} = \underline{0.010}.$$

TEN

Flow in Open Channels

10.1

Type	Closed Conduit	Open Channel
a) Steady, uniform	Pumping between two reservoirs in constant diameter pipe	Terminal flow in a very long prismatic channel
b) Unsteady, nonuniform	Water hammer	Flood wave in river, or hydraulic bore
c) Steady, nonuniform	Constant discharge in diffuser	Backwater conditions behind a dam
d) Unsteady, uniform	Gradual deceleration of flow in a pipe of constant diameter	"Practically impossible situation" (Chow); however, the kinematic wave concept assumes this type (Henderson)

10.2 $\alpha = \cos^{-1}(1 - 2 \times 0.3) = 1.159 \text{ rad or } 66.4°$,

$$\frac{Q^2 B}{gA^3} = \frac{Q^2 d \sin\alpha}{g\left[d^2/4\ (\alpha - \sin\alpha\cos\alpha)\right]^3}$$

$$= \frac{4^2 \sin(1.159)}{9.81 \times (d^5/64) \times 1.159 - \sin 1.159 \cos 1.159)^3} = 1.$$

Eq. reduces to $192.1 = d^5$,

$\therefore d = (192.1)^{1/5} = \underline{2.86\text{ m}}$.

10.3 a) Steady, nonuniform

b) Unsteady, nonuniform

c) Unsteady, nonuniform

d) steady, nonuniform

10.4

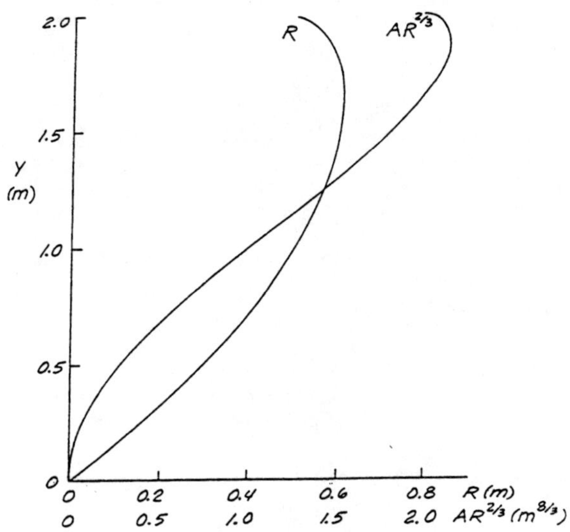
2.0
1.5
1.0
0.5
0
Y
(m)
R
$AR^{2/3}$
0 0.2 0.4 0.6 0.8 R (m)
0 0.5 1.0 1.5 2.0 $AR^{2/3}$ ($m^{8/3}$)

a)

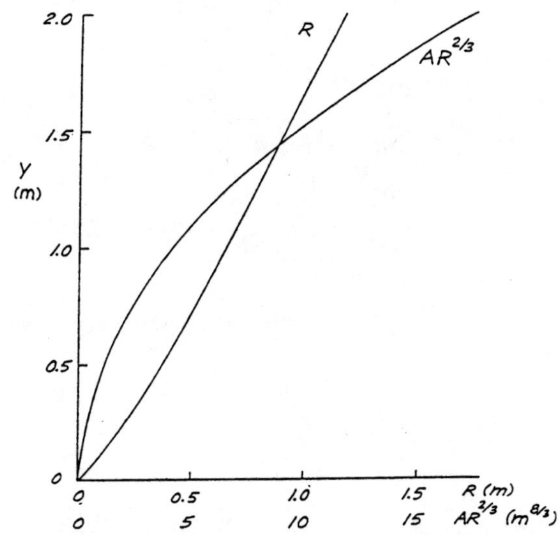
2.0
1.5
1.0
0.5
0
Y
(m)
R
$AR^{2/3}$
0 0.5 1.0 1.5 R (m)
0 5 10 15 $AR^{2/3}$ ($m^{8/3}$)

b)

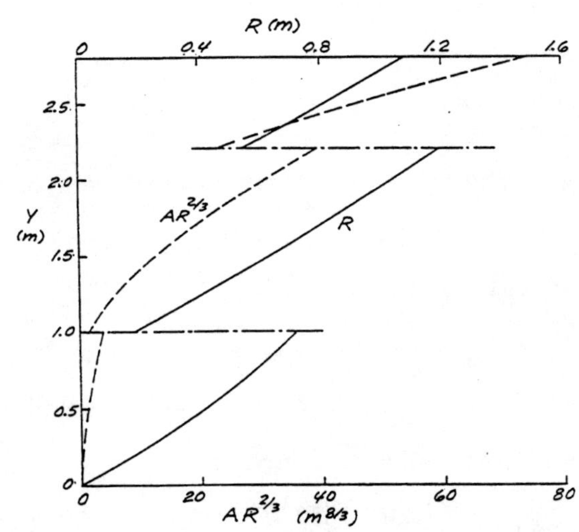
R (m)
0 0.4 0.8 1.2 1.6
2.5
2.0
1.5
1.0
0.5
0
Y
(m)
$AR^{2/3}$
R
0 20 40 60 80
$AR^{2/3}$ ($m^{8/3}$)

c)

10.5 $A = by + my^2$; solve for b: $b = A/y - my$; then

$P = b + 2y\sqrt{1+m^2} = A/y - my + 2y\sqrt{1+m^2}$.

Set $dP/dm = 0$, solve for m:

$$\frac{dP}{dm} = -y + \frac{2my}{\sqrt{1+m^2}} = 0,$$

$$2m = \sqrt{1+m^2},$$

$$3m^2 = 1,$$

$$\therefore \underline{m = \sqrt{3}/3}.$$

Set $dP/dy = 0$, noting that $dA/dy = 0$ likewise:

$$\frac{dP}{dy} = \frac{1}{y}\frac{dA}{dy} - \frac{A}{y^2} - m + 2\sqrt{1+m^2} = 0$$

$$\therefore A = -my^2 + 2y^2\sqrt{1+m^2}$$

$$= y^2\left(-\frac{\sqrt{3}}{3} + 2\sqrt{1+\frac{1}{3}}\right)$$

$$= y^2\left(-\frac{\sqrt{3}}{3} + 4\frac{\sqrt{3}}{3}\right),$$

$$\therefore \underline{A = \sqrt{3}y^2}.$$

One can also show that $b = \frac{2}{3}\sqrt{3}y$ and $P = 2\sqrt{3}y = 3b$.

10.6 Use Chezy-Manning equation in the form $Q = \frac{c_1}{n}\frac{A^{5/3}}{P^{2/3}}\sqrt{S_0}$.

Substitute in appropriate expressions for A and P, and solve for y_0 by trial and error.

a) $$35 = \frac{\left[4.5y_0 + \frac{1}{2}y_0^2(2.5+3.5)\right]^{5/3}}{0.015\left[4.5 + y_0(\sqrt{1+2.5^2} + \sqrt{1+3.5^2})\right]^{2/3}}\sqrt{0.00035}.$$

Reduces to $28.06 = \frac{(4.5y_0 + 3y_0^2)^{5/3}}{(4.5+6.33y_0)^{2/3}}$,

$\therefore$Solving, $y_0 = \underline{2.15\text{ m}}$, $A = 4.5 \times 2.15 + 3 \times 2.15^2 = \underline{23.5\text{ m}^2}$,

$P = 4.5 + 6.33 \times 2.15 = \underline{18.1\text{ m}}$.

b) $$425 = \frac{\left[2.1^2/4\,(\alpha - \sin\alpha\cos\alpha)\right]^{5/3}}{0.012(2.1\alpha)^{2/3}}\sqrt{0.001}.$$

Reduces to $2.248 = \dfrac{(\alpha - \sin\alpha\cos\alpha)^{5/3}}{\alpha^{2/3}}$,

$\therefore$ Solving, $\alpha = \underline{1.825\text{ rad}}$, $y_0 = \dfrac{2.1}{2}(1 - \cos 1.825) = \underline{1.31\text{ m}}$,

$A = \dfrac{2.1^2}{4}(1.825 - \sin 1.825\cos 1.825) = \underline{2.28\text{ m}^2}$,
$P = 2.1 \times 1.825 = \underline{3.83\text{ m}}$.

c) $$120 = \frac{1.49\left[6^2/4\,(\alpha - \sin\alpha\cos\alpha)\right]^{5/3}}{0.012(6\alpha)^{2/3}}\sqrt{0.001}.$$

Reduces to $2.593 = \dfrac{(\alpha - \sin\alpha\cos\alpha)^{5/3}}{\alpha^{2/3}}$,

$\therefore$ Solving, $\alpha = \underline{1.965\text{ rad}}$, $y_0 = \dfrac{6}{2}(1 - \cos 1.965) = \underline{4.15\text{ ft}}$,

$A = \dfrac{36}{4}(1.965 - \sin 1.965\cos 1.965) = \underline{20.9\text{ ft}^2}$,
$P = 6 \times 1.965 = \underline{11.8\text{ ft}}$.

d) From Pbm. 10.5, $m = \sqrt{3}/3$, $A = \sqrt{3}\,y_0^2$, $P = 2\sqrt{3}y_0$.
Substitute into Chezy-Manning equation:

$$\frac{(\sqrt{3}\,y_0^2)^{5/3}}{(2\sqrt{3}\,y_0)^{2/3}} = \frac{Qn}{\sqrt{s_0}} = \frac{15 \times 0.011}{\sqrt{0.0013}} = 4.576.$$

Reduces to $y_0^{8/3} = 4.195$,
$\therefore y_0 = 4.195^{3/8} = \underline{1.71\text{ m}}$, $A = \sqrt{3} \times 1.71^2 = \underline{5.08\text{ m}^2}$,
$P = 2\sqrt{3} \times 1.71 = \underline{5.93\text{ m}}$.

e) $$1200 = \frac{1.49(25y_0)^{5/3}}{0.02(25 + 2y_0)^{2/3}}\sqrt{0.0004}.$$

Reduces to $3.768 = \dfrac{y_0^{5/3}}{(25 + 2y_0)^{2/3}}$,

$\therefore$ Solving, $y_0 = \underline{10.2\text{ ft}}$, $A = 25 \times 10.2 = \underline{255\text{ ft}^2}$,
$P = 25 + 2 \times 10.2 = \underline{45.4\text{ ft}}$.

10.7 $A = Q/V = 4/1.4 = 2.857 \text{ m}^2$

$$V = \frac{1}{n}\left(\frac{A}{P}\right)^{2/3}\sqrt{S_0}; \text{ solve for } P.$$

$$P = A\left(\frac{\sqrt{S_0}}{Vn}\right)^{3/2} = 2.857\left(\frac{\sqrt{0.001}}{1.4\times 0.015}\right)^{3/2} = 5.280 \text{ m},$$

$$\therefore P = b + 2y_0\sqrt{1+m^2};\ 5.280 = b + 2y_0\sqrt{1+1.75^2};$$

$b = 5.280 - 4.031y_0$. Substitute into area function:

$$A = by_0 + my_0^2;\ 2.857 = (5.280 - 4.031y_0) + 1.75y_0^2.$$

Relation reduces to $y_0^2 - 2.315y_0 + 1.253 = 0$,

$$\therefore y_0 = \frac{2.315}{2} \pm \frac{1}{2}\sqrt{2.315^2 - 4\times 1.253} = 1.452 \text{ or } 0.863 \text{ m}.$$

Use lower value of y_0, since $y_0 = 1.452$ results in a negative b:
$\therefore y_0 = \underline{0.86 \text{ m}},\ b = 5.280 - 4.031\times 0.863 = \underline{1.80 \text{ m}}.$

10.8 $b = 0,\ m_1 = 10,\ m_2 = 0$

$$A = by + \frac{1}{2}y^2(m_1 + m_2) = \frac{1}{2}m_1y^2 = \frac{10}{2}y^2 = 5y^2$$

$$P = b + y\left(\sqrt{1+m_1^2} + \sqrt{1+m_2^2}\right) = y\left(\sqrt{1+m_1^2} + 1\right) = y\left(\sqrt{101} + 1\right) = 11.05y$$

$$Q = AR^{2/3}\frac{\sqrt{S_0}}{n} = 5y^2\left(\frac{5y^2}{11.05y}\right)^{2/3}\frac{\sqrt{0.0005}}{0.013} = 5.068y^{8/3}$$

a) $y = 0.01 \text{ m},\ Q = 5.068(0.01)^{8/3} = \underline{0.012 \text{ m}^3}$

b) $Q = 0.075 \text{ m}^3/\text{s},\ y = \left(\frac{0.075}{5.068}\right)^{3/8} = \underline{0.207 \text{ m}}.$

10.9

$$Q=\frac{1}{n}\frac{A^{5/3}}{P^{2/3}}\sqrt{S_0}=\frac{1}{n}\frac{d^2}{4}\frac{(\alpha+\sin\alpha\cos\alpha)^{5/3}}{(\alpha d)^{2/3}}\sqrt{S_0}\,, \qquad \alpha=\frac{\pi}{2}$$

$$\therefore Q=\frac{1}{n}\frac{d^2}{4}\left(\frac{\pi}{2}\right)d^{-2/3}\sqrt{S_0}=\frac{\pi}{8n}d^{4/3}\sqrt{S_0}$$

a) $Q=1.75\text{ m}^3/\text{s},\ S_0=0.0003,\ n=0.014$

$$\therefore d=\left(\frac{8nQ}{\pi\sqrt{S_0}}\right)^{3/4}=\left(\frac{8\times0.014\times1.75}{\pi\times\sqrt{0.0003}}\right)^{3/4}=\underline{2.62\text{ m}}$$

b) $S_0=0.00005,\ d=1.30$

$$Q=\frac{\pi}{8\times0.014}\times1.3^{4/3}\times\sqrt{0.00005}=\underline{0.281\text{ m}^3/\text{s}}$$

c) $d=0.75\text{ m},\ Q=0.45\text{ m}^3/\text{s}$

$$\therefore S_0=\left(\frac{8nQ}{\pi d^{4/3}}\right)^2=\left(\frac{8\times0.014\times0.45}{\pi\times0.75^{4/3}}\right)^2=\underline{0.000554}$$

10.10 $q=\sqrt{2gy^2(E-y)},\quad E=\text{constant},$

$$\frac{dq}{dy}=\frac{1}{2}\frac{4gy(E-y)-2gy^2}{\sqrt{2gy^2(E-y)}}=\frac{2gy(E-y)-gy^2}{q}.$$

Setting $dq/dy=0$ and noting that $q\neq0,$ so that the numerator is zero, one can solve for y at the condition $q=q\text{max}$. (Note: $y=0$ is a trivial solution)

$$gy[2(E-y)-y]=0, \qquad 2E-3y=0, \qquad \therefore\ y=\frac{2}{3}E=y_c.$$

10.11 $E=y+\dfrac{q^2}{2gy^2},$

$$\frac{E}{y_c}=\frac{y}{y_c}+\frac{q^2}{2gy_cy^2}=\frac{y}{y_c}+\frac{q^2}{2gy_c^3(y/y_c)^2}.$$

But $q/gy_c^3=1,\qquad \therefore \underline{\dfrac{E}{y_c}=\dfrac{y}{y_c}+\dfrac{1}{2}\dfrac{1}{(y/y_c)^2}},$

$$\frac{d}{dy}\left(\frac{E}{y_c}\right)=1-\frac{1}{(y/y_c)^3}=0$$

$\therefore \underline{y/y_c=1}$ and $E/y_c=1+1/2=\underline{3/2}$

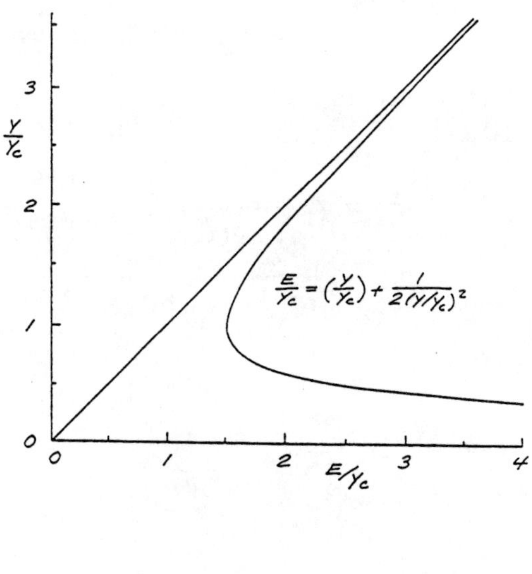

10.12 $q = V_1 y_1 = 3 \times 2.5 = 7.5 \text{ m}^2/\text{s}$,

$$E_1 = y_1 + \frac{q_2}{2gy_1^2} = 2.5 + \frac{3^2}{2 \times 9.81} = 2.96 \text{ m},$$

$$y_c = \sqrt[3]{q^2/g} = \sqrt{7.5^2/9.81} = 1.79 \text{ m}, \ E_c = \frac{3}{2}y_c = \frac{3}{2} \times 1.79 = 2.68 \text{ m}.$$

a) $E_2 = E_1 - h = 2.96 - 0.2 = 2.76$ m, $E_2 > E_c$. $\therefore y_2$ is subcritical.

$$\therefore E_2 = y_2 + \frac{q^2}{2gy_2^2}, \text{ or } 2.76 = y_2 + \frac{7.5^2}{2 \times 9.81 y_2^2} = y_2 + \frac{2.87}{y_2^2}$$

$\therefore$ Solving, $y_2 = 2.13$ m, and $y_2 + h - y_1 = 2.13 + 0.2 - 2.5 = \underline{-0.17 \text{ m}}$.

b) $y_1 + (1 + K_c)\frac{q^2}{2gy_1^2} = y_2 + (1 + K_c)\frac{q^2}{2gy_2^2} + h$; $K_c = 0.1$, $h = 0.15$ m.

$$\therefore 2.5 + (1 + 0.1)\frac{3^2}{2 \times 9.81} = y_2 + (1 + 0.1)\frac{7.5^2}{2 \times 9.81 y_2^2} + 0.15.$$

Equation reduces to $2.855 = y_2 + \frac{3.154}{y_2^2}$. $\therefore$ Solving, $\underline{y_2 = 2.20 \text{ m}}$,

$y_2 + h - y_1 = 2.2 + 0.15 - 2.5 = \underline{-0.15 \text{ m}}$.

c) Set $E_2 = E_c = 2.68$ m and $\therefore$ maximum h is
$h_{max} = E_1 - E_2 = E_1 - E_c = 2.96 - 2.68 = \underline{0.28 \text{ m}}$.

d) $y_1 + (1 + K_e)\frac{q^2}{2gy_1^2} = y_2 + (1 - K_e)\frac{q^2}{2gy_2^2} + h$; $K_e = 0.2$, $h = -0.2$ m.

$$\therefore 2.5 + (1 - 0.2)\frac{3^2}{2 \times 9.81} = y_2 + (1 - 0.2)\frac{7.5^2}{2 \times 9.81 y_2^2} - 0.2.$$

Equation reduces to $3.067 = y_2 + \frac{2.294}{y_2^2}$. $\therefore$ Solving, $\underline{y_2 = 2.77 \text{ m}}$,

$y_2 + h - y_1 = 2.77 - 0.2 - 2.5 = \underline{+0.07 \text{ m}}$.

10.12E $q = V_1 y_1 = 10 \times 8 = 80 \text{ ft}^2/\text{sec}$,

$$E_1 = y_1 + \frac{q^2}{2gy_1^2} = 8 + \frac{10^2}{2 \times 32.2} = 9.55 \text{ ft},$$

$$y_c = \sqrt[3]{q^2/g} = \sqrt[3]{80^2/32.2} = 5.83 \text{ ft}, \ E_c = \frac{3}{2} \times 5.83 = 8.74 \text{ ft}.$$

a) $E_2 = E_1 - h = 9.55 - 8/12 = 8.88$ ft, $E_2 > E_c$. $\therefore y_2$ is subcritical.

$\therefore E_2 = y_2 + \dfrac{q^2}{2gy_2^2}$, or $8.88 = y_2 + \dfrac{80^2}{2 \times 32.2 y_2^2} = y_2 + \dfrac{99.38}{y_2^2}$.

$\therefore$ Solving, $\underline{y_2 = 6.60 \text{ ft}}$, $y + h - y_1 = 6.6 + 0.67 - 8 = \underline{-0.73 \text{ ft}}$.

b) $y_1 + (1+K_c)\dfrac{q^2}{2gy_1^2} = y_2 + (1+K_c)\dfrac{q^2}{2gy_2^2} + h$; $K_c = 0.1$, $h = \dfrac{6}{12}$ ft

$\therefore 8 + (1+0.1)\dfrac{10^2}{2 \times 32.2} = y_2 + (1+0.1)\dfrac{80^2}{2 \times 32.2 y_2^2} + 0.5$.

Equation reduces to $9.208 = y_2 + \dfrac{109.3}{y_2^2}$. $\therefore$ Solving, $\underline{y_2 = 6.94 \text{ ft}}$,

$y_2 + h - y_1 = 6.94 + 0.5 - 8 = \underline{-0.56 \text{ ft}}$.

c) Set $E_2 = E_c = 8.74$ ft and $\therefore$ maximum h is

$h_{max} = E_1 - E_2 = E_1 - E_c = 9.55 - 8.74 = \underline{0.81 \text{ ft}}$.

d) $y_1 + (1+K_e)\dfrac{q^2}{2gy_1^2} = y_2 + (1-K_e)\dfrac{q^2}{2gy_2^2} + h$; $K_e = 0.2$, $h = -\dfrac{8}{12}$ ft.

$\therefore 8 + (1-0.2)\dfrac{10^2}{2 \times 32.2} = y_2 + (1-0.2)\dfrac{80^2}{2 \times 32.2 y_2^2} - 0.667$. Equation

reduces to $9.909 = y_2 + \dfrac{79.50}{y_2^2}$. $\therefore$ Solving, $\underline{y_2 = 8.91 \text{ ft}}$,

$y_2 + h - y_1 = 8.91 - 0.67 - 8 = \underline{+0.24 \text{ ft}}$

10.13 $q_1 = \dfrac{Q}{b_1} = \dfrac{25}{5} = 5 \text{ m}^2/\text{s}$, $V_1 = \dfrac{q_1}{y_1} = \dfrac{5}{2} = 2.5 \text{ m/s}$,

$y_{c_1} = \sqrt[3]{q_1^2/g} = \sqrt[3]{5^2/9.81} = 1.37$ m, $E_1 = y_1 + \dfrac{q_1^2}{2gy_1^2} = 2 + \dfrac{2.5^2}{2 \times 9.81} = 2.32$ m.

a) $q_2 = \dfrac{Q}{b_2} = \dfrac{25}{5.5} = 4.54 \text{ m}^2/\text{s}$, $E_{c_2} = \dfrac{3}{2}\sqrt[3]{q_2^2/g} = \dfrac{3}{2}\sqrt[3]{\dfrac{4.54^2}{9.81}} = 1.92$ m.

Set $E_1 = E_2$ and solve for y_2, noting that y_2 is subcritical, since $E_{c_2} < E_2$:

$2.32 = y_2 + \dfrac{4.54^2}{2 \times 9.81 y_2^2} = y_2 + \dfrac{1.051}{y_2^2}$. $\therefore$ Solving, $y_2 = \underline{2.07 \text{ m}}$

b) $q_2 = \frac{Q}{b_2} = \frac{25}{4.75} = 5.263 \text{ m}^2/\text{s}, \ \ K_c = 0.1.$

$$y_1 + (1+K_c)\frac{q_1^2}{2gy_1^2} = y_2 + (1-K_c)\frac{q_2^2}{2gy_2^2},$$

$\therefore 2 + (1+0.1)\frac{5^2}{2\times 9.81\times 2^2} = y_2 + (1+0.1)\frac{5.263^2}{2\times 9.81 y_2^2}$. Equation reduces to $2.35 = y_2 + \frac{1.553}{y_2^2}$. $\therefore$ Solving (assuming subcritical conditions at location 2), $\underline{y_2 = 1.94 \text{ m.}}$

10.14 $q_1 = V_1 y_1 = 3\times 3 = 9 \text{ m}^2/\text{s}, \ \ Q = b_1 q_1 = 3\times 9 = 27 \text{ m}^3/\text{s},$

$E_1 = y_1 + \frac{q_1^2}{2gy_1^2} = 3 + \frac{3^2}{2\times 9.81} = 3.46 \text{ m}, \ \ y_{c_1} = \sqrt[3]{9^2/9.81} = 2.02 \text{ m},$

$E_{c_1} = \frac{3}{2}\times 2.02 = 3.03$ m. Without change in width at loc. 2,

$E_2 = E_1 - h = 3.46 - 0.7 = 2.76$ m. Since $E_2 < E_{c_1}$, width must change to prevent choking. $\therefore$ Set $E_{c_2} = 2.76$ m.

$\therefore y_{c_2} = \frac{2}{3}\times 2.76 = 1.84 \text{ m}, \ \ q_2 = \sqrt{g y_{c2}^3} = \sqrt{9.81\times 1.84^3} = 7.81 \text{ m}^2/\text{s}.$

$\therefore b_2 = Q/q_2 = 27/7.81 = \underline{3.46 \text{ m}}.$

10.14E $q_1 = V_1 y_1 = 10\times 10 = 100 \text{ ft}^2/\text{sec}, \ \ Q = b_1 q_1 = 10\times 100 = 1000 \text{ ft}^3/\text{sec},$

$E_1 = y_1 + \frac{q_1^2}{2gy_1^2} = 10 + \frac{10^2}{2\times 32.2} = 11.55 \text{ ft}, \ \ y_{c_1} = \sqrt[3]{100^2/32.2} = 6.76 \text{ ft},$

$E_{c_1} = \frac{3}{2}\times 6.76 = 10.14$ ft. Without change in width at loc. 2,

$E_2 = E_1 - h = 11.55 - 2.3 = 9.25$ ft. Since $E_2 < E_{c_1}$, width must change to prevent choking. $\therefore$ Set $E_{c_2} = 9.25$ ft.

$\therefore y_{c_2} = \frac{2}{3}\times 9.25 = 6.17 \text{ ft}, \ \ q_2 = \sqrt{g(y_{c_2})^3} = \sqrt{32.2\times 6.17^3} = 86.9 \text{ ft}^2/\text{sec}.$

$\therefore b_2 = Q/q_2 = \frac{1000}{86.9} = \underline{11.51 \text{ ft}}.$

10.15 $q = Q/b_1 = 4.8/2 = 2.4\ \text{m}^2/\text{s},\ y_{c_1} = \sqrt[3]{2.4^2/9.81} = 0.84\ \text{m}.$

a) Desire $y_2 = y_1 + h = 1.22 + 0.1 = 1.32$ m. Write energy eqn. from upstream loc. (1) into transition loc. (2):

$$y_1 + \frac{q_1^2}{2gy_1^2} + h = y_2 + \frac{q_2^2}{2gy_2^2},$$ and solve for the unknown q_2.

$$\therefore 1.22 + \frac{2.4^2}{2 \times 9.81 \times 1.22^2} + 0.1 = 1.52 = 1.32 + \frac{q_2^2}{2 \times 9.81 \times 1.32^2}.$$

Reducing, $q_2^2 = 6.84$, $\therefore q_2 = \sqrt{6.84} = 2.62\ \text{m}^2/\text{s}$.

$\therefore b_2 = Q/q_2 = 4.8/2.62 = \underline{1.84\ \text{m}}.$

b) Let $E_{\bar{2}} = E_{c_{\bar{2}}} = E_1 + h = 1.52$ m. Then,

$$y_{c_{\bar{2}}} = \frac{2}{3}E_{c_{\bar{2}}} = \frac{2}{3} \times 1.52 = 1.01\ \text{m},\ q_{\bar{2}} = \sqrt{1.01^3 \times 9.81} = 3.195\ \text{m},$$

and $\therefore b_{\bar{2}} = Q/q_{\bar{2}} = 4.8/3.195 = \underline{1.50\ \text{m}}.$

10.15E $q = Q/b_1 = 170/6.5 = 26.15\ \text{ft}^2/\text{sec},\ y_c = \sqrt[3]{26.15^2/32.2} = 4.33\ \text{ft}.$

a) Desire $y_2 = y_1 + h = 4 + \frac{4}{12} = 4.33$ ft. Write energy eqn. from upstream loc. (1) into transition loc. (2):

$$y_1 + \frac{q_1^2}{2gy_1^2} + h = y_2 + \frac{q_2^2}{2gy_2^2},$$ and solve for the unknown q_2.

$$\therefore 4 + \frac{26.15^2}{2 \times 32.2 \times 4^2} + 0.333 = 4.997 = 4.33 + \frac{q_2^2}{2 \times 32.2 \times 4.33^2}.$$

Reducing, $q_2^2 = 80.13$, $\therefore q_2 = \sqrt{80.13} = 28.31\ \text{ft}^2/\text{sec}$.

$\therefore b_2 = Q/q_2 = 170/28.31 = \underline{6.00\ \text{ft}}.$

b) Let $E_{\bar{2}} = E_{c_{\bar{2}}} = E_1 + h = 4.997$ ft. Then,

$y_{c_{\bar{2}}} = 2/3E_{c_{\bar{2}}} = 2/3 \times 4.997 = 3.33\ \text{ft},\ q_{\bar{2}} = \sqrt{3.33^3 \times 32.2} = 34.48\ \text{ft}^2/\text{sec},$

and $\therefore b_{\bar{2}} = Q/q_{\bar{2}} = 170/34.48 = \underline{4.93\ \text{ft}}.$

10.16

$$E_1 = y_1 + \frac{q^2}{2gy_1^2} = 2.15 + \frac{5.5^2}{2 \times 9.81 \times 2.15^2} = 2.484 \text{ m}$$

$$\text{Fr}_1 = \frac{q}{\sqrt{gy_1^3}} = \frac{5.5}{\sqrt{9.81 \times 2.15^3}} = 0.557$$

$$y_c = \sqrt[3]{\frac{q^2}{g}} = \sqrt[3]{\frac{5.5^2}{9.81}} = 1.456 \approx 1.46 \text{ m}$$

$$E_c = \frac{3}{2}y_c = \frac{3}{2} \times 1.456 = 2.183 \approx 2.18 \text{ m}$$

a) The maximum height of the raised bottom at location 2 will be one for which the energy is a minimum:

$E_1 = E_c + h$, $2.484 = 2.183 + h$, $\therefore h = 2.484 - 2.183 =$ 0.30 m

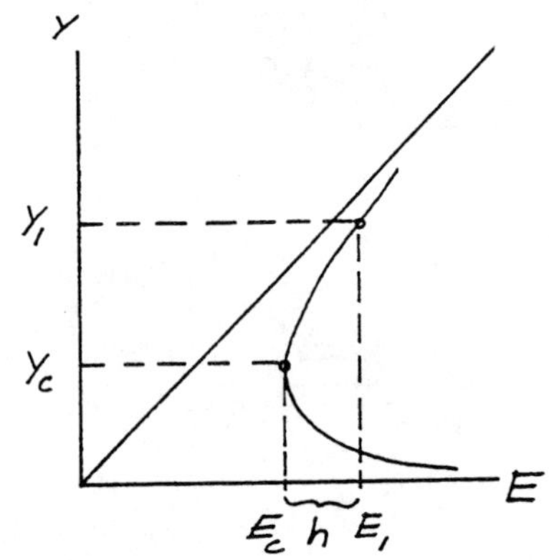

b)

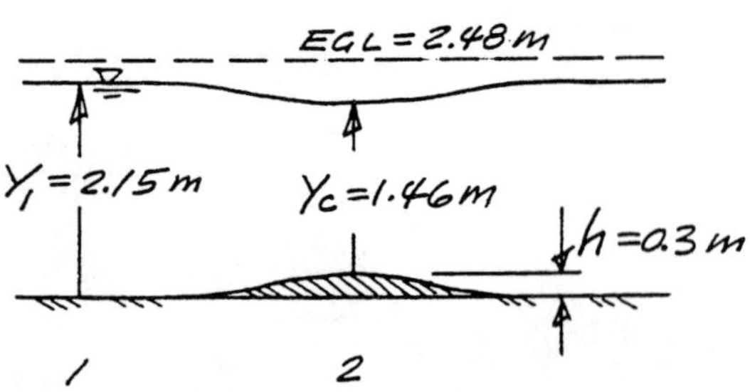

c)

d) Since $\text{Fr}_1 < 1$, if $h > 0.30$ m, subcritical nonuniform flow will occur upstream of the transition.

10.17 Refer to Fig. 10.8. Since a steep channel exists downstream of the entrance, critical conditions occur at the entrance. The two eqns. to be satisfied are

$$y_1 - z_2 = y_c + \frac{A_c}{2B_c} \text{ and } Q = \sqrt{gA_c^3 / B_c}.$$

The first eqn. is solved independently for y_c, and then y_c is substituted into the second eqn. to find Q. Note: A_c and B_c refer to critical conditions.

a) Rectangular channel: $A_c = by_c = 4y_c$, $B_c = b = 4$.

$$y_1 - z_2 = y_c + \frac{by_c}{2b} = \frac{3}{2}y_c, \quad \therefore y_c = \frac{2}{3}(y_1 - z_2) = \frac{2}{3} \times 2.5 = 1.67 \text{ m},$$

$$\therefore Q = \sqrt{gb^3y_c^3 / b} = \sqrt{9.81 \times 4^2 \times 1.67^3} = \underline{27.0 \text{ m}^3/\text{s}}.$$

b) Trapezoidal channel: $A_c = by_c + my_c^2$, $B_c = b + 2my_c$.

$$y_1 - z_2 = y_c + \frac{by_c + my_c^2}{2(b + 2my_c)}; \; 2.5 = y_c + \frac{3y_c + 2.5y_c^2}{2(3 + 2 \times 2.5y_c)};$$

reduces to $12.5y_c^2 - 16y_c - 15 = 0$

$$\therefore y_c = \frac{1}{2 \times 12.5}\left(16 \pm \sqrt{16^2 + 4 \times 12.5 \times 15}\right) = 1.91 \text{ m} \quad \text{(use positive root).}$$

$$A_c = 3 \times 1.91 + 2.5 \times 1.91^2 = 14.85 \text{ m}^2, \; B_c = 3 + 5 \times 1.91 = 12.55 \text{ m}.$$

$$\therefore Q = \sqrt{9.81 \times 14.85^3 / 12.55} = \underline{50.6 \text{ m}^3 / \text{s}}.$$

c) Circular channel: $A_c = \frac{d^2}{4}(\alpha - \sin\alpha\cos\alpha)$, $B_c = d\sin\alpha$,

$\alpha = \cos^{-1}(1 - 2y_c / d)$.

$$y_1 - z_2 = y_c + \frac{d^2 / 4(\alpha - \sin\alpha\cos\alpha)}{2d\sin\alpha};$$

$$2.5 = y_c + \frac{3.5^2 / 4(\alpha - \sin\alpha\cos\alpha)}{2 \times 3.5\sin\alpha}; \text{ reduces to}$$

$$2.5 = 0.4375(\alpha - \sin\alpha\cos\alpha) / \sin\alpha, \; \alpha = \cos^{-1}(1 - .5714y_c).$$

$\therefore$ Solving, $y_c = 1.79$ m ($\alpha = 1.594$ rad),

$$A_c = 3.5^2 / 4(1.594 - \sin 1.594 \cos 1.594) = 4.95 \text{ m}^2,$$

$$B_c = 2 \times 3.5 \times \sin 1.594 = 3.50 \text{ m}.$$

$$\therefore Q = \sqrt{9.81 \times 4.95^3 / 3.50} = \underline{18.4 \text{ m}^3 / \text{s}}.$$

10.18 At the canal entrance, the condition of critical flow is $\text{Fr}^2 = \frac{Q^2B}{gA^3} = 1$, to be solved for the unknown width b.

a) Rectangular channel: $B = b$, $A = by$, $y = y_c$.

$$\therefore \frac{Q^2b}{gb^3y^3} = \frac{Q^2}{gb^2y^3} = 1. \quad \therefore b = \frac{Q}{\sqrt{gy^3}} = \frac{18}{\sqrt{9.81 \times 1}} = \underline{5.75 \text{ m}}.$$

b) Trapezoidal channel: $B = b + 2my = b + 2 \times 3 \times 1 = b + 6,$

$A = by + my^2 = b \times 1 + 3 \times 1^2 = b + 3.$

$\therefore \frac{Q^2 B}{gA^3} = \frac{18^2(b+6)}{9.81(b+3)^3} = 1, \text{ or } \frac{(b+3)^3}{(b+6)} = 33.0.$

$\therefore$ Solving, $\underline{b = 3.89 \text{ m}}$.

10.18E At the canal entrance, the condition of critical flow is $\text{Fr}^2 = \frac{Q^2 B}{gA^3} = 1$, to be solved for the unknown width b.

a) Rectangular channel: $B = b, \ A = by, \ y = y_c.$

$\therefore \frac{Q^2 B}{gA^3} = \frac{Q^2}{gb^2 y^3} = 1. \quad \therefore b = \frac{Q}{\sqrt{gy^3}} = \frac{635}{\sqrt{32.2 \times 3^3}} = \underline{21.5 \text{ ft}}.$

b) Trapezoidal channel: $B = b + 2my = b + 2 \times 3 \times 3 = b + 18,$

$A = by + my^2 = b \times 3 + 3 \times 3^2 = 3b + 27.$

$\therefore \frac{Q^2 B}{gA^3} = \frac{635^2(b+18)}{32.2(3b+27)^3} = 1, \text{ or } \frac{(b+9)^3}{(b+18)} = 463.8.$

$\therefore$ Solving, $\underline{b = 16.1 \text{ ft}}$.

10.19 a) Write energy eqn. from reservoir (section R) to section C, where critical conditions exist:

$y_R + z_R = E_c + z_c, \ E_c = y_R + z_R - z_c = 103.6 - 102 = 1.6 \text{ m}.$

$\therefore y_c = \frac{2}{3} E_c = \frac{2}{3} \times 1.6 = 1.067 \text{ m}.$

$\therefore q = \sqrt{g y_c^3} = \sqrt{9.81 \times 1.067^3} = \underline{3.45 \text{ m}^2/\text{s}}.$

b) Energy eqn. from reservoir to section A:

$103.6 = y_A + \frac{3.45^2}{2 \times 9.81 y_A^2} + 100, \text{ or } 3.6 = y_A + \frac{0.6067}{y_A^2}.$

$\therefore$ Solving, $\underline{y_A = 3.55 \text{ m}}$.

Energy eqn. from reservoir to section B:

$103.6 = y_B + \frac{3.45^2}{2 \times 9.81 y_B^2} + 101, \text{ or } 2.6 = y_B + \frac{0.6067}{y_B^2}.$

$\therefore$ Solving, $\underline{y_B = 2.50 \text{ m}}$.

10.20 Objective is to determine the zero of the function $f(y) = \dfrac{Q^2 B(y)}{g[A(y)]^3} - 1$, in which eqns. for $B(y)$ and $A(y)$ representing either a circular or trapezoidal section area substituted. The false position algorithm is explained in Example 10.10. The solutions are:

(a) $y_c = 1.00$ m, (b) $y_c = 1.50$ ft, (c) $y_c = 1.81$ m, (d) $y_c = 3.48$ ft.

10.21 Assume $y_c > 1.5$ m; then $A = (y_c - 1.5) \times 10 + \frac{3}{2} y_c^2$ and $B = 10 + 3y_c$. The criterion for critical flow is $\dfrac{A^3}{B} = \dfrac{Q^2}{g}$, or

$$\frac{\left[(y_c - 1.5) \times 10 + \frac{3}{2} y_c^2\right]^3}{10 + 3y_c} = \frac{16.5^2}{9.81} = 27.75$$

A trial-and-error solution yields $\underline{y_c \approx 1.78 \text{ m}}$.

10.22 In the lower section ($y \le 1$ m), the area is

$$A = \int_0^y b\,d\eta = \int_0^y 3\sqrt{\eta}\,d\eta = 2y^{3/2}. \text{ Assume } y < 1 \text{ m; then } \mathrm{Fr}^2 = \frac{Q^2 B}{gA^3} = \frac{Q^2 3\sqrt{y}}{g8y^{9/2}} = \frac{3Q^2}{8gy^4}$$

If $y_c = 1$ m, then $\dfrac{3Q^2}{8 \times 9.81 \times 1^4} = 1$, or $Q = \sqrt{8 \times 9.81 / 3} = 5.1 \text{ m}^3/\text{s}$

$\therefore$ when $Q > 5.1 \text{ m}^3/\text{s}$, critical depth will be > 1 m. Cross-sectional area including upper region (*y*>*1* m) is $A = 2(1)^{3/2} + 23(y_c - 1) = 23y_c - 21$, and $B = 23$ m.

a) $\dfrac{Q^2 B}{gA^3} = \dfrac{55^2 \times 23}{9.81(23y_c - 21)^3} = 1$, $(23y_c - 21)^3 = 7092$,

$\therefore y_c = \dfrac{(7092)^{1/3} + 21}{23} = \underline{1.75 \text{ m}}$.

b) $\dfrac{Q^2 B}{gA^3} = \dfrac{3 \times 55^2}{8 \times 9.81 y_c^4} = 1$, $y_c^4 = 0.468$, $\therefore \underline{y_c = 0.83 \text{ m}}$.

10.23 $x=\sqrt{10\eta},\ A=\int_0^y 2\times d\eta=\int_0^y 2\sqrt{10\eta}=\frac{4}{3}\sqrt{10}y^{3/2}$, or

$A=4.216y^{3/2}$.

$$E=y+\frac{Q^2}{2gA^2}=y+\frac{25^2}{2\times 9.81\times 4.216^2 y^3}=y+\frac{1.792}{y^3}.$$

Determine depth alternate to $y = 2.0$ m:

$$\therefore 2+\frac{1.792}{2^3}=2.224=y_{\text{alt}}+\frac{1.792}{y_{\text{alt}}^3},$$

$\therefore$ Solving, $y_{\text{alt}} = \underline{1.21\text{ m}}$.

10.24 Determine the zero of the function $f(y)=E-y-\frac{Q^2}{g[A(y)]^3}$, in which E is known.

An interval-halving algorithm is listed in statements 1890 - 2010, Figure E.1.

10.25

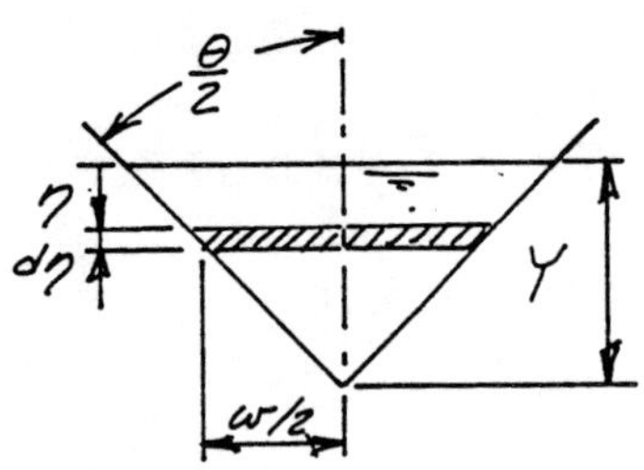

$$v=\sqrt{2g\eta},\quad \tan\frac{\theta}{2}=\frac{w/2}{Y-\eta},$$

$$Q=\int_0^Y vw\,d\eta$$

$$=\int_0^Y \sqrt{2g\eta}\,2\left(\tan\frac{\theta}{2}\right)(Y-\eta)d\eta$$

$$=2\sqrt{2g}\left(\tan\frac{\theta}{2}\right)\int_0^Y\left(Y\eta^{1/2}-\eta^{3/2}\right)d\eta,$$

$$Q=2\sqrt{2g}\left(\tan\frac{\theta}{2}\right)\left(\frac{2}{3}Y^{5/2}-\frac{2}{5}Y^{5/2}\right)=\frac{8}{15}\sqrt{2g}\left(\tan\frac{\theta}{2}\right)Y^{5/2}.$$

Multiply by the discharge coefficient c_d:

$$\therefore Q=c_d\frac{8}{15}\sqrt{2g}\left(\tan\frac{\theta}{2}\right)Y^{5/2}.$$

10.26 a)

$$q = \frac{22.5}{15} = 1.5 \text{ m}^2/\text{s}$$

$$y_c = \sqrt[3]{\frac{1.5^2}{9.81}} = 0.612 \text{ m}$$

$$E_c = \frac{3}{2} \times 0.612 = 0.919 \text{ m}$$

$$E_0 = 1.2 + \frac{1.5^2}{2 \times 9.81 \times 1.2^2} = 1.28 \text{ m} = E_3$$

$$E_2 = E_0 - h = 1.28 - 0.2 = 1.08 \text{ m} > E_3 \quad \therefore \text{no choking.}$$

Since losses are neglected throughout the transition, $y_1 = y_0$, and y_2 can be computed by writing the energy equation between locations 1 and 2:

$$1.\ 8 = y_2 + \frac{1.5^2}{2 \times 9.81 \times 1.2^2} + h = y_2 + \frac{0.115}{y_2^2} + 0.2, \quad \therefore \underline{y_2 \approx 0.95 \text{ m}}.$$

b)

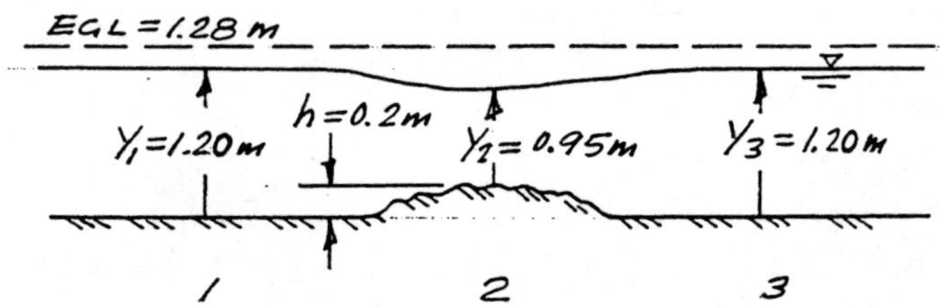

10.27

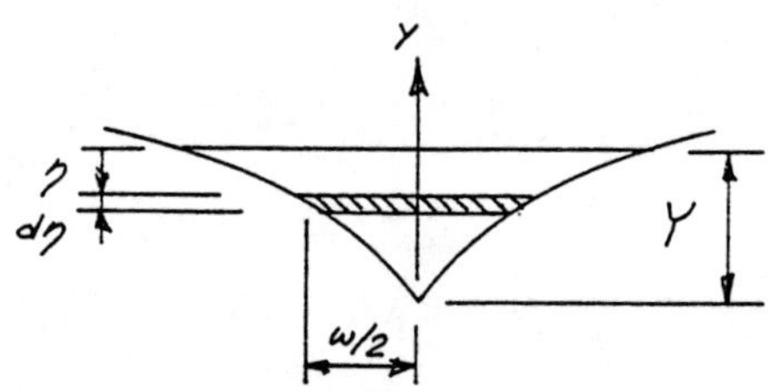

$$v = \sqrt{2g\eta}$$
$$w/2 = x$$
$$= Y^2$$
$$= (Y - \eta)^2$$

$$Q = \int_0^Y vw\,d\eta = \int_0^Y \sqrt{2g\eta}\,2(Y-\eta)^2 d\eta = 2\sqrt{2g}\int_0^Y \eta^{1/2}(Y^2 - 2Y\eta + \eta^2)d\eta$$

$$= 2\sqrt{2g}\int_0^Y \left(Y^2\eta^{1/2} - 2Y\eta^{3/2} + \eta^{5/2}\right)d\eta = 2\sqrt{2g}\left(\frac{2}{3}Y^{7/2} - \frac{4}{5}Y^{7/2} + \frac{2}{7}Y^{7/2}\right)$$

$$= 2\sqrt{2g}\left(\frac{70 - 84 + 30}{105}\right)Y^{7/2}, \text{ or } \therefore \underline{Q = \frac{32}{105}\sqrt{2g}Y^{7/2}}.$$

10.28E $B = 3$ ft, $H = 1.2$ ft

$$Q = 4BH^{1.522B^{0.026}} = 4 \times 3 \times 1.2^{1.522 \times 3^{0.026}} = 16 \text{ ft}^3/\text{sec}$$

H (ft)	0.5	0.8	0.9	1.0	1.1	1.2	1.3	1.4	1.5
Q (ft³/sec)	4.05	8.46	10.2	12.0	13.9	16.0	18.1	20.3	22.6

10.29 Let location 1 be upstream of the transition section, and location 2 be in the transition. For both a) and b), $b = 3.5$ m, and $h = 0.6$ m.

a) $y_1 = 1.5$ m, $Q = 4$ m³/s.

$$q = \frac{4}{3.5} = 1.143 \text{ m}^2/\text{s}, \qquad y_c = \sqrt[3]{\frac{1.143^2}{9.81}} = 0.511 \text{ m},$$

$$E_c = \frac{3}{2} \times 0.511 = 0.767 \text{ m}, \qquad E_1 = 1.5 + \frac{1.14^2}{2 \times 9.81 \times 1.5^2} = 1.529 \text{ m},$$

Check: $E_2 = E_1 - h = 1.529 - 0.6 = 0.929 > E_c$

$$\therefore 0.929 = y_2 + \frac{1.14^2}{2 \times 9{,}81 \times y_2^2} = y_2 + \frac{0.0662}{y_2^2} \Rightarrow \therefore \underline{y_2 \approx 0.835 \text{ m}} \text{ (T\&E)}$$

b) $y_1 = 1.0$ m, $Q = 4$ m³/s.

$$q = \frac{3}{3.5} = 0.857 \text{ m}^2/\text{s}, \qquad y_c = \sqrt[3]{\frac{0.857^2}{9.81}} = 0.422 \text{ m},$$

$$E_c = \frac{3}{2} \times 0.422 = 0.633 \text{ m}, \qquad E_1 = 1.0 + \frac{0.857^2}{2 \times 9.81 \times 1.0^2} = 1.037 \text{ m},$$

Check: $E_2 = E_1 - h = 1.037 - 0.6 = 0.437 < E_c$

$\therefore$ Choking occurs at location 2. Set $E_2 = E_c$, hence $E_1 = E_c + h$, or

$$y_1 + \frac{0.857^2}{2 \times 9.81 \times y_1^2} = 0.633 + 0.6, \text{ which reduces to}$$

$$y_1 + \frac{0.0374}{y_1^2} = 1.233 \Rightarrow \therefore \underline{y_1 \approx 1.21 \text{ m}} \text{ (T\&E)}$$

c) For situation b), the raised section acts as a broad-crested weir, since critical flow occurs over it.

10.30 $Q_1 = b\left(\frac{2}{3}\sqrt{\frac{2}{3}g}\right)(y_1 - h)^{3/2}, \; Q_2 = b\left(\frac{2}{3}\sqrt{\frac{2}{3}g}\right)(y_2 - h)^{3/2};$

given $Q_1, \; y_1, \; Q_2, \; y_2,$ solve for b and h.

$$\frac{y_1 - h}{y_2 - h} = \left(\frac{Q_1}{Q_2}\right)^{2/3}, \text{ or } h = \frac{y_1 - y_2(Q_1 / Q_2)^{2/3}}{1 - (Q_1 / Q_2)^{2/3}}.$$

$$\therefore h = \frac{1.05 - 1.75(0.15 / 30)^{2/3}}{1 - (0.15 / 30)^{2/3}} = \underline{1.03 \text{ m}},$$

$$\therefore b = \frac{Q_2}{\frac{2}{3}\left(\sqrt{\frac{2}{3}g}\right)(y_2 - h)^{3/2}} = \frac{30}{\frac{2}{3}\left(\sqrt{\frac{2}{3} \times 9.81}\right)(1.75 - 1.03)^{3/2}} = \underline{28.8 \text{ m}}.$$

10.30E (See Pbm. 10.30 for development.)

$$\therefore h = \frac{3.45 - 5.75(5 / 1000)^{2/3}}{1 - (5 / 1000)^{2/3}} = \underline{3.38 \text{ ft}},$$

$$\therefore b = \frac{1000}{\frac{2}{3}\left(\sqrt{\frac{2}{3} \times 32.2}\right)(5.75 - 3.38)^{3/2}} = \underline{88.7 \text{ ft}}.$$

10.31 Assume uniform, subcritical flow throughout the canal; then,

$$S_0 = \frac{\text{Elev. A - Elev. B}}{L}, \; Q = \frac{(by)^{5/3}}{n(b + 2y)^{2/3}}\sqrt{S_0}.$$

Write energy eqn. from reservoir to canal entrance:

$h = y + \frac{Q^2}{2gA^2} = y + \frac{Q^2}{2gb^2y^2}.$ Unknowns are Q and y.

Eliminate Q in energy eqn. and solve for y:

$$h = \frac{1}{2gb^2y^2}\left(\frac{b^{10/3}S_0}{n^2}\right)\frac{y^{10/3}}{(b + 2y)^{4/3}} = y + \left(\frac{b^{4/3}S_0}{2gn^2}\right)\frac{y^{4/3}}{(b + 2y)^{4/3}}.$$

Substitute given data into above relations, and compute $S_0, \; y,$ and Q:

$\therefore S_0 = (501.8 - 500.2) / 1500 = \underline{0.00107},$

$$2 = y + \left(\frac{2.5^{4/3} \times 0.00107}{2 \times 9.81 \times 0.014^2}\right)\frac{y^{4/3}}{(2.5 + 2y)^{4/3}} = y + \frac{0.944y^{4/3}}{(2.5 + 2y)^{4/3}},$$

$\therefore$ Solving, $y = 1.82$ m,

$$\therefore Q = \frac{(2.5 \times 1.82)^{5/3}\sqrt{0.00107}}{0.014(2.5 + 2 \times 1.82)^{2/3}} = \underline{8.70 \text{ m}^3 / \text{s.}}$$

Check for $y > y_c$:

$y_c = \sqrt[3]{q^2 / g} = \sqrt[3]{(8.70/2.5)^2 / 9.81} = \underline{1.07 \text{ m}} < 1.82$ m.

$\therefore$ Assumption of subcritical flow valid.

10.31E (See Pbm. 10.31 for development.)

$\therefore S_0 = (1646 - 1641) / 4920 = \underline{0.00102}$,

$$6.5 = y + \left(\frac{1.49^2 \times 8^{4/3} \times 0.00102}{2 \times 32.2 \times 0.014^2}\right)\frac{y^{4/3}}{(8+2y)^{4/3}} = y + \frac{2.868y^{4/3}}{(8+2y)^{4/3}},$$

$\therefore$ Solving, $y = \underline{5.93 \text{ ft}}$,

$$\therefore Q = \frac{1.49(8 \times 5.93)^{5/3}\sqrt{0.00102}}{0.014(8 + 2 \times 5.93)^{2/3}} = \underline{288 \text{ ft}^3 / \text{sec.}}$$

$y_c = \sqrt[3]{(288/8)^2 / 32.2} = \underline{3.42 \text{ ft}} < 5.93$ ft.

10.32 $M = b\left(\frac{y^2}{2} + \frac{q^2}{gy}\right)$

$$\frac{M}{by_c^2} = \frac{1}{2}\left(\frac{y}{y_c}\right)^2 + \frac{q^2}{gy_c^2 y}$$

$$= \frac{1}{2}\left(\frac{y}{y_c}\right)^2 + \frac{q^2}{gy_c^3 (y/y_c)}.$$

But $q^2 / gy_c^3 = 1$,

$$\therefore \underline{\frac{M}{by_c^2} = \frac{1}{2}\left(\frac{y}{y_c}\right)^2 + \frac{1}{(y/y_c)}}$$

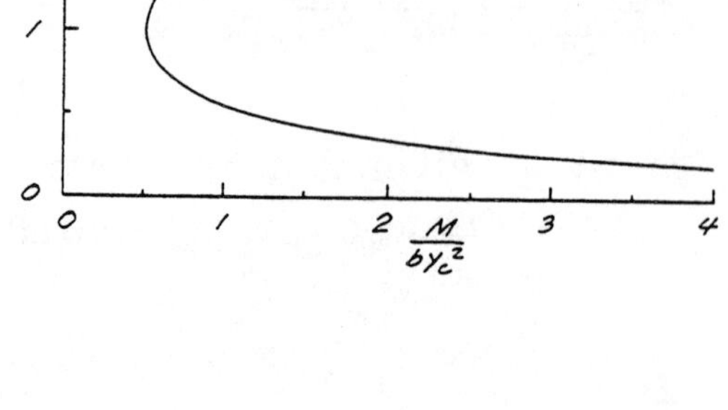

10.33 Determine the zero of the function $f(y) = A_2(y)\bar{y}(y) + \dfrac{Q^2}{gA_2(y)} - M_1 + \dfrac{F}{\gamma}$,

in which M_1 and F are known. An interval-halving algorithm is listed in statements 1890 - 2010, Figure E.1.

10.34 Conditions at location 1:

$$E_1 = y_1 + \frac{q_1^2}{2gy_1^2} = 1.8 + \frac{1.5^2}{2 \times 9.81 \times 1.8^2} = 1.835 \text{ m},$$

$$y_c = \sqrt[3]{q_1^2 / g} = \sqrt[3]{1.5^2 / 9.81} = 0.612 \text{ m}.$$

a) The smallest constriction at location 2 is one that establishes critical flow, i.e., where minimum energy exists. $\therefore E_2 = E_c$, or since $E_1 = E_2$,

$$y_{c_2} = \frac{2}{3}E_{c_2} = \frac{2}{3}E_1 = \frac{2}{3} \times 1.835 = 1.22 \text{ m}.$$

$$\therefore q_2 = \sqrt{g(y_{c_2})^3} = \sqrt{9.81 \times 1.22^3} = 4.22 \text{ m}^2/\text{s},$$

$$b_2 = q_1 b_1 / q_2 = 1.5 \times 6 / 4.22 = 2.13 \text{ m}.$$

Hence the maximum diameter is

$d = b_1 - b_2 = 6 - 2.13 = \underline{3.87 \text{ m}}$

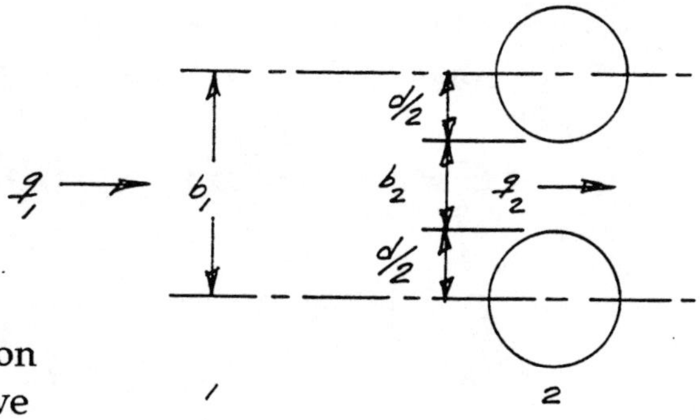

b) Momentum eqn. cannot be used since no information is given with regard to location downstream of cofferdam. But we do have available the drag relation, which will provide the required drag force.

Frontal area: $A = y_1 d = 1.8 \times 3.87 = 6.97 \text{ m}^2$,

Approach velocity: $V_1 = q_1 / y_1 = 1.5 / 1.8 = 0.833 \text{ m/s}$.

$\therefore F = C_D A \rho V_1^2 / 2 = 0.15 \times 6.97 \times 1000 \times 0.833^2 / 2 = \underline{363 \text{ N}}$,

a rather insignificant drag force!

10.34E (See Pbm. 10.34 for development and figure.)

$$E_1 = y_1 + \frac{q_1^2}{2gy_1^2} = 6 + \frac{16^2}{2 \times 32.2 \times 6^2} = 6.11 \text{ ft}, \quad y_c = \sqrt[3]{q_1^2 / g} = \sqrt[3]{16^2 / 32.2} = 1.99 \text{ ft}.$$

a) $y_{c_2} = \dfrac{2}{3}E_{c_2} = \dfrac{2}{3}E_1 = \dfrac{2}{3} \times 6.11 = 4.07$ ft,

$$q_2 = \sqrt{(gy_{c_2})^3} = \sqrt{32.2 \times 4.07^3} = 46.6 \text{ ft}^2/\text{sec},$$

$$b_2 = q_1 b_1 / q_2 = 16 \times 20 / 46.6 = 6.87 \text{ ft},$$

$d = b_1 - b_2 = 20 - 6.87 = \underline{13.1 \text{ ft}}.$

b) Frontal area: $A = y_1 d = 6 \times 13.1 = 78.6\ \text{ft}^2$,
Approach velocity: $V_1 = q_1 / y_1 = 16 / 6 = 2.67$ ft / sec.
$\therefore F = C_D A \rho V_1^2 / 2 = 0.15 \times 78.6 \times 1.94 \times 2.67^2 / 2 = \underline{81.5\ \text{lb.}}$

10.35

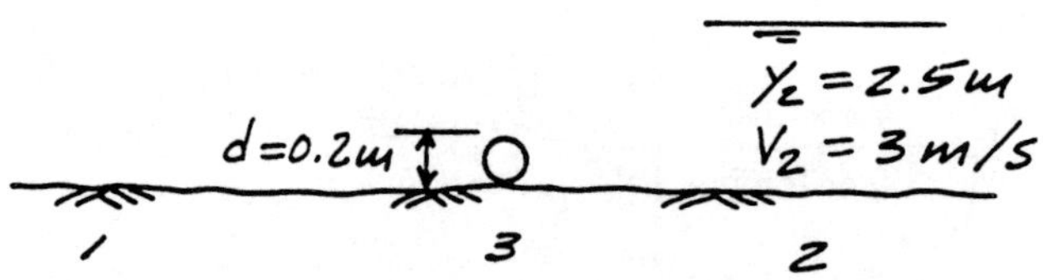

Check flow conditions at location 2:

$q = V_2 y_2 = 3 \times 2.5 = 7.5\ \text{m}^2/\text{s},\ y_c = \sqrt[3]{q^2/g} = \sqrt[3]{7.5/9.81} = 1.79$ m.

$\therefore$ Flow is subcritical, and provided pipeline does not "choke" flow at location 3, conditions will remain subcritical from locs. 1 to 2.

$E_3 = E_2 - d = 2.5 + 7.5^2 / (2 \times 9.81 \times 2.5^2) - 0.2 = 2.76$ m,

$E_c = \frac{3}{2} y_c = \frac{3}{2} \times 1.79 = 2.68$ m. $\therefore E_c < E_2 - d$, and no choking occurs.

a) Write momentum eqn. between locations 1 and 2:

$$\frac{y_1^2}{2} + \frac{q^2}{g y_1} - \frac{F}{\gamma b} = \frac{y_2^2}{2} + \frac{q^2}{g y_2}.\ \text{Now,}\ \frac{F}{\gamma b} = \frac{1}{\gamma b}\left(C_D A_p \frac{V_1^2}{2}\right) = \frac{C_D b q^2}{2 g y_1^2}.$$

Substitute this along with known data into momentum eqn:

$$\frac{y_1^2}{2} + \frac{7.5^2}{9.81 y_1} - \frac{0.3 \times 0.2 \times 7.5^2}{2 \times 9.81 y_1^2} = \frac{2.5^2}{2} + \frac{7.5^2}{9.81 \times 2.5},$$

or $\frac{y_1^2}{2} + \frac{5.734}{y_1} - \frac{0.172}{y_1^2} = 5.419.$ $\therefore$ Solving, $y_1 = \underline{2.52\ \text{m}}$,

$\therefore V_1 = q / y_1 = 7.5 / 2.52 = \underline{2.98\ \text{m / s}}.$

b) $F = C_D b d \rho V_1^2 / 2 = 0.3 \times 100 \times 0.2 \times 1000 \times 2.98^2 / 2 = \underline{2.66 \times 10^4\ \text{N.}}$

10.36

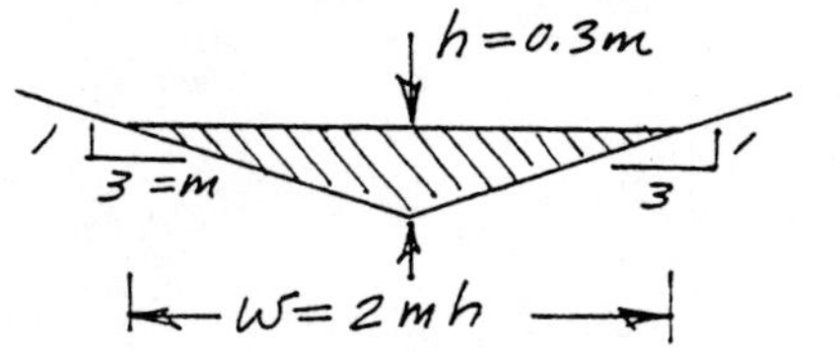

$$A_{\text{sill}} = \frac{1}{2}wh = \frac{1}{2} \times 2 \times 3 \times 0.3 = 0.27 \text{ m}^2$$

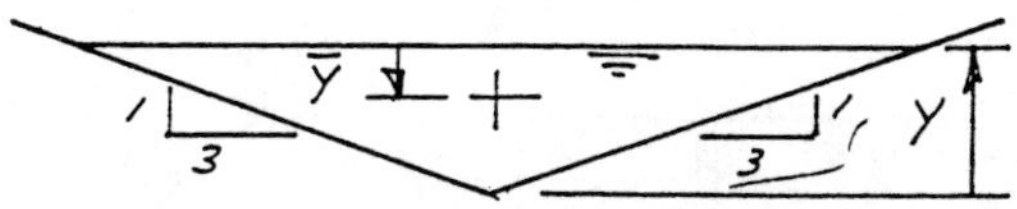

$$\bar{y} = \frac{1}{3}y$$

$$A = my^2 = 3y^2$$

$$M_1 = A_1\bar{y}_1 + \frac{Q^2}{gA_1} = 3 \times 0.5^2 \times \frac{1}{3} \times 0.5 + \frac{Q^2}{9.81 \times 3 \times 0.5^2} = 0.125 + \frac{Q^2}{7.358},$$

$$M_2 = A_2\bar{y}_2 + \frac{Q^2}{gA_2} = 3 \times 1.8^2 \times \frac{1}{3} \times 1.8 + \frac{Q^2}{9.81 \times 3 \times 1.8^2} = 5.832 + \frac{Q^2}{95.35},$$

$$\frac{F}{\gamma} = \frac{C_D A_{\text{sill}} Q^2}{2gA_1^2} = \frac{0.4 \times 0.27Q^2}{2 \times 9.81 \times (3 \times 0.5^2)^2} = \frac{Q^2}{102.2}.$$

Substitute into momentum eqn: $M_1 = M_2 + F/\gamma$.

$$\therefore 0.125 + \frac{Q^2}{7.358} = 5.832 + \frac{Q^2}{95.35} + \frac{Q^2}{102.2},$$

$$Q^2 = (5.832 - 0.125)\left(\frac{1}{7.58} - \frac{1}{95.35} - \frac{1}{102.2}\right)^{-1} = 49.37,$$

$$\therefore \underline{Q = 7.03 \text{ m}^3/\text{s}}.$$

10.37

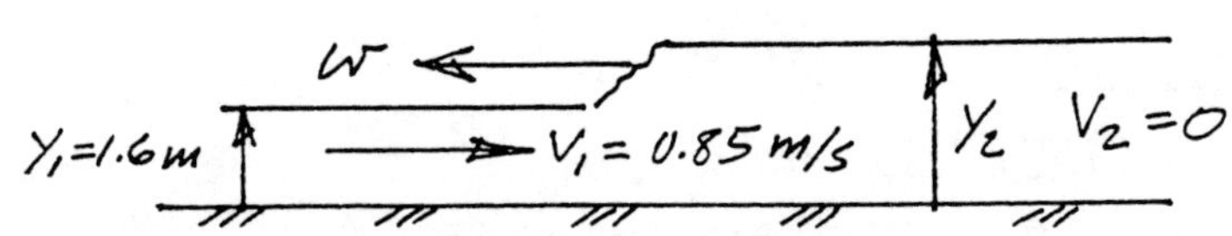

Continuity eqn: $\dfrac{y_2}{y_1} = \dfrac{V_1 + w}{w}$.

Momentum eqn: $\dfrac{y_2}{y_1} = \dfrac{1}{2}\left[\sqrt{1 + 8\dfrac{(V_1 + w)^2}{gy_1}} - 1\right]$.

Eliminate y_2 / y_1, substitute in known data, and solve for w:

$$\frac{V_1 + w}{w} = \frac{1}{2}\left[\sqrt{1 + 8\frac{(V_1 + w)^2}{gy_1}} - 1\right],$$

$$\frac{0.85 + w}{w} = \frac{1}{2}\left[\sqrt{1 + 8\frac{(0.85 + w)^2}{9.81 \times 1.6}} - 1\right],$$

$$\frac{0.85}{w}+1=\frac{1}{2}\left[\sqrt{1+0.51(0.85+w)^2}-1\right],$$

$$3=\sqrt{1+0.51(0.85+w)^2}-\frac{1.70}{w}.$$

$\therefore$Solving, $w = \underline{3.8 \text{ m/s}}$, and $y_2 = y_1\left(\frac{V_1+w}{w}\right) = 1.6\left(\frac{0.85+3.8}{3.8}\right) = \underline{1.96 \text{ m}}$.

10.38

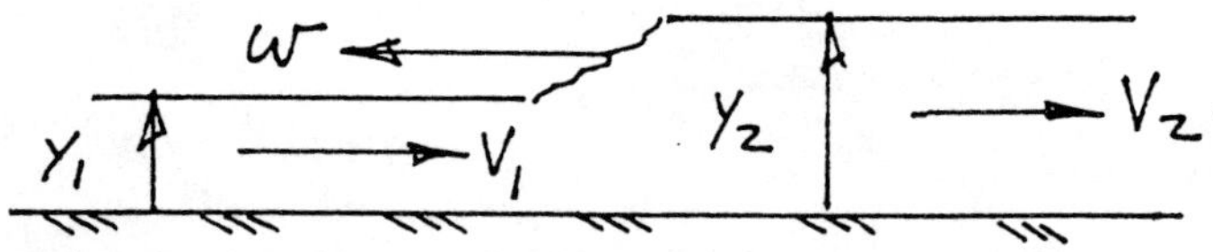

Given condition: $V_2 y_2 = 0.4 V_1 y_1$,

Continuity eqn: $y_1(V_1+w) = y_2(V_2+w)$,

Momentum eqn: $\frac{y_2}{y_1}=\frac{1}{2}\left[\sqrt{1+8\frac{(V_1+w)^2}{gy_1}}-1\right].$

Unknowns are V_2, y_2 and w. Eliminate V_2 and w, and solve for y_2:

$V_2 y_2 = 0.4\times 1\times 1.5 = 0.6,$

$1.5(1+w) = y_2V_2 + y_2 w = 0.6 + y_2 w$, or

$w = 0.9/(y_2-1.5).$

$$\therefore \frac{y_2}{1.5}=\frac{1}{2}\left[\sqrt{1+8\left(1+\frac{0.9}{y_2-1.5}\right)^2/(9.81\times 1.5)}-1\right], \text{ or}$$

$$1-\sqrt{1+0.5437\left(1+\frac{0.9}{y_2-1.5}\right)^2}+1.333y_2=0.$$

$\therefore$Solving, $y_2 = \underline{1.77 \text{ m}}$,

$\therefore V_2 = 0.6/y_2 = 0.6/1.77 = \underline{0.339 \text{ m/s}}$,

$\therefore w = 0.9/(y_2-1.5) = 0.9/(1.77-1.5) = \underline{3.33 \text{ m/s}}$.

10.38E (See Pbm. 10.38 for development and figure.)

$V_2 y_2 = 0.4\times 3\times 5 = 6,$

$5(3+w) = y_2V_2 + y_2 w = 0.6 + y_2 w$, or

$w = 9/(y_2-5).$

$$\therefore \frac{y_2}{5}=\frac{1}{2}\left[\sqrt{1+8\left(3+\frac{9}{y_2-5}\right)^2/(32.2\times 5)}-1\right], \text{ or}$$

$$1-\sqrt{1+0.0497\left(3+\frac{9}{y_2-5}\right)^2}+0.4y_2=0.$$

$\therefore$ Solving, $y_2 = \underline{5.80 \text{ ft}}$,

$\therefore V_2 = 0.6 / y_2 = 0.6 / 5.80 = \underline{1.03 \text{ ft / sec}}$,

$\therefore w = 9 / (y_2 - 5) = 9 / (5.80 - 5) = \underline{11.2 \text{ ft / sec}}$.

10.39 Since a hydraulic jump is to occur between locs. 1 and 2, apply momentum eqn. to find y_2. Then use energy eqn. to determine h, with critical conditions on sill, loc. 3.

$q = Q / b = 20 / 7.5 = 2.667 \text{ m}^2 / \text{s}$,

$\text{Fr}_1 = q / \sqrt{g y_1^3} = 2.667 / \sqrt{9.81 \times 0.5^3} = 2.408 > 1.$

a) $\therefore y_2 = \frac{y_1}{2}\left(\sqrt{1+8\text{Fr}_1^2}-1\right) = \frac{0.5}{2}\left(\sqrt{1+8\times 2.408^2}-1\right) = \underline{1.47 \text{ m}}$,

$\therefore y_3 = y_c = \sqrt[3]{q^2/g} = \sqrt[3]{2.667^2/9.81} = \underline{0.898 \text{ m}}$.

b) $h = E_2 - E_3 = E_2 - E_c = y_2 + \frac{q^2}{2g y_2^2} - \frac{3}{2} y_c$,

$$\therefore h = 1.47 + \frac{2.667^2}{2\times 9.81\times 1.47^2} - \frac{3}{2}\times 0.898 = \underline{0.29 \text{ m}}.$$

c) Downstream of the sill the depth y_4 is the d epth alternate to y_2:

$E_2 = 1.638 = y_4 + \frac{2.667^2}{2\times 9.81\times y_4^2}$, $\quad\therefore$ Solving, $y_4 = 0.59$ m.

$$F = \gamma\left(M_2 - M_4\right) = \gamma b\left(\frac{y_2^2}{2}+\frac{q^2}{g y_2}-\frac{y_4^2}{2}-\frac{q^2}{g y_4}\right)$$

$$= 9810\times 7.5\times\left(\frac{1.47^2}{2}+\frac{2.667^2}{9.81\times 1.47}-\frac{0.59^2}{2}-\frac{2.667^2}{9.81\times 0.59}\right) = \underline{1.26\times 10^4 \text{ N}}$$

d)

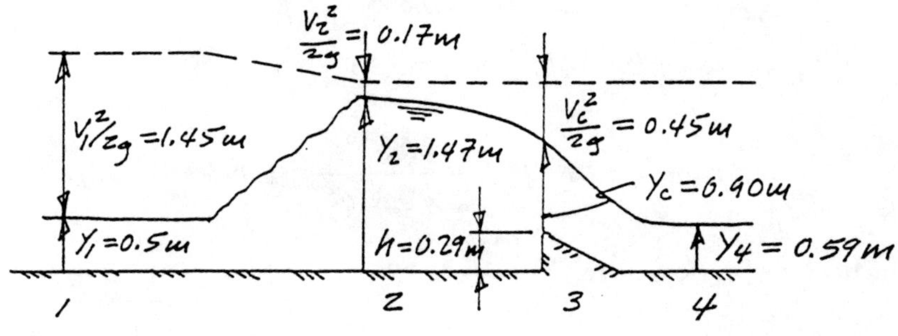

e) $\text{Fr}_1 = 2.4$, hence from Table 10.2 the jump is characterized as weak.

10.40 a) To compute the discharge, write the energy equation between locations 1 and 2, and solve for q:

$$q = \sqrt{\frac{2g(y_2 - y_1)}{\left(\frac{1}{y_1^2} - \frac{1}{y_2^2}\right)}} = \sqrt{\frac{2 \times 9.81(0.10 - 2.5)}{\left(\frac{1}{2.5^2} - \frac{1}{0.10^2}\right)}} = 0.687 \text{ m}^2/\text{s}$$

$$\therefore Q = bq = 5 \times 0.687 = \underline{3.43 \text{ m}^3/\text{s}}$$

b) To compute depth downstream of jump, first compute the Froude number at location 2:

$$\text{Fr}_2 = \frac{q}{\sqrt{gy_2^3}} = \frac{0.687}{\sqrt{9.81 \times 0.10^3}} = 6.934$$

$$\therefore y_3 = \frac{y_2}{2}\left(\sqrt{1 + 8Fr_2^2} - 1\right) = \frac{0.10}{2}\left(\sqrt{1 + 8 \times 6.934^2} - 1\right) = \underline{0.932 \text{ m}}$$

c) To calculate power lost, first compute the head loss in the jump:

$$h_j = \frac{(y_3 - y_2)^3}{4y_3y_2} = \frac{(0.932 - 0.10)^3}{4 \times 0.932 \times 0.10} = 1.544 \text{ m}$$

$$\therefore \dot{W} = \gamma Qh_j = 9810 \times 3.43 \times 1.544 = \underline{5.20 \times 10^4 \text{ watt, or } 52.0 \text{ kW}}$$

10.41 $q = Q/b = \dfrac{5.55}{3} = 1.85 \text{ m}^2/s, \ \text{Fr}_3 = q/\sqrt{gy_3^3},$

a) $y_3 = \sqrt[3]{\dfrac{1}{g}(q/\text{Fr}_3)^2} = \sqrt[3]{\dfrac{1}{9.81}\left(\dfrac{1.85}{0.75}\right)^2} = \underline{0.853 \text{ m}}.$

$\left(y_c = \sqrt[3]{q^2/g} = \sqrt[3]{1.85^3/9.81} = 0.704 \text{ m.}\right)$

$$E_3 = y_3 + \frac{q^2}{2gy_3^2} = 0.853 + \frac{1.85^2}{2 \times 9.81 \times 0.853^2} = 1.093 \text{ m},$$

$$E_2 = E_3 + h_3, \text{ but } h_3 = \frac{1}{2}\sqrt{2} = 0.707.$$

$$\therefore y_2 + \frac{q^2}{2gy_2^2} = E_3 + h_3, \text{ or } y_2 + \frac{1.85^2}{2 \times 9.81y_2^2} = 1.093 + 0.707,$$

Equation reduces to $y_2 + \dfrac{0.1744}{y_2^2} = 1.80 \qquad \therefore \ \underline{y_2 = 1.74 \text{ m.}}$

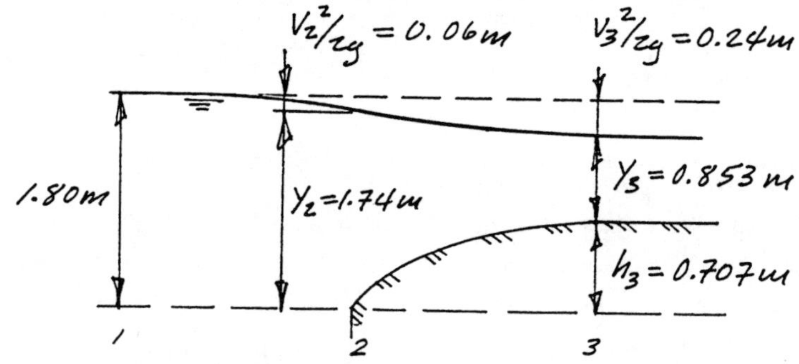

At loc. 1, the water surface elev. relative to the datum is $E_3 + h_3 = \underline{1.80\text{ m}}$.

b) $F = \gamma(M_2 - M_3) = \gamma b\left(\dfrac{y_2^2}{2} + \dfrac{q^2}{gy_2} - \dfrac{y_3^2}{2} - \dfrac{q^2}{gy_3}\right)$,

$$= 9810 \times 3\left(\frac{1.74^2}{2} + \frac{1.85^2}{9.81 \times 1.74} - \frac{0.853^2}{2} - \frac{1.85^2}{9.81 \times 0.853}\right),$$

$\therefore F = \underline{2.77 \times 10^4\text{ N}}.$

10.42 Use Eqn. 10.5.16 with $F = 0$:

$$M_1 = \frac{y_1^2}{6}(2my_1 + 3b) + \frac{Q^2}{g(by_1 + my_1^2)}$$

$$= \frac{1.1^2}{6}(2 \times 3 \times 1.1 + 3 \times 5) + \frac{60^2}{9.81(5 \times 1.1 + 3 \times 1.1^2)} = 44.55\text{ m}^3.$$

$\therefore M_2 = M_1$, or

$$\frac{y_2^2}{6}(2my_2 + 3b) + \frac{Q^2}{g(by_2 + my_2^2)} = 44.55,$$

$$\frac{y_2^2}{6}(2 \times 3y_2 + 3 \times 5) + \frac{60^2}{9.81(5y_2 + 3y_2^2)} = 44.55,$$

$$y_2^3 + 2.5y_2^2 + \frac{367}{5y_2 + 3y_2^2} = 44.55.$$ $\therefore$ Solving, $y_2 = \underline{2.55\text{ m}}$.

Energy loss across jump is h_j:

$$h_j = E_1 - E_2 = y_1 + \frac{Q^2}{2gA_1^2} - y_2 - \frac{Q^2}{2gA_2^2},$$

$A_1 = 5 \times 1.1 + 3 \times 1.1^2 = 9.13\text{ m}^2$, $A_2 = 5 \times 2.55 + 3 \times 2.55^2 = 32.26\text{ m}^2$,

$$\therefore h_j = 1.1 + \frac{60^2}{2 \times 9.81 \times 9.13^2} - 2.55 + \frac{60^2}{2 \times 9.81 \times 32.26^2} = 0.575\text{ m}.$$

$\therefore$ Power dissipated is $\dot{W}_j$: $\dot{W}_j = \gamma Q h_j = 9810 \times 60 \times 0.575 = \underline{3.38 \times 10^5\text{ W}}.$

10.43 a) Compute the height of the step:

$$q_1 = V_1 \times y_1 = 8 \times 0.5 = 4 \text{ m}^2/\text{s}$$

$F = \gamma(M_1 = M_2)$, or

$$C_D hb\rho \frac{V_1^2}{2} = \gamma b\left(\frac{y_1^2}{2} + \frac{q^2}{gy_1} - \frac{y_2^2}{2} - \frac{q^2}{gy_2}\right)$$

$$\therefore h = \frac{9.81\left(0.5^2/2 + 4^2/(9.81 \times 0.5) - 2^2/2 - 4^2/(9.81 \times 2)\right)}{1.2 \times 8^2/2}$$

$= 0.146 \approx \underline{0.15 \text{ m}}$

b) Compute the downstream depth if step were not present:

$$\text{Fr}_1 = \frac{q}{\sqrt{gy_1^3}} = \frac{4}{\sqrt{9.81 \times 0.5^3}} = 3.613$$

$$\therefore y_2 = \frac{y_1}{2}\left(\sqrt{1+8\text{Fr}_1^2} - 1\right) = \frac{0.5}{2}\left(\sqrt{1+8 \times 3.613^2} - 1\right) = \underline{2.317 \text{ m, or } 2.32 \text{ m}}$$

Note that the width b was not used in the problem.

10.44 First find Q, then compute y_1:

$V_2 = Fr_2\sqrt{gy_2} = 0.4\sqrt{9.81 \times 1.5} = 1.534$ m / s,

$\therefore Q = V_2 A_2 = 1.534 \times 5 \times 1.5 = \underline{11.51 \text{ m}^3 / \text{s}}.$

$q = Q/b = 11.51/5 = 2.30$ m^2 / s,

$\left(y_c = \sqrt[3]{q^2/g} = \sqrt[3]{2.30^2/9.81} = 0.814 \text{ m}\right)$

To compute y_2, use momentum eqn, $M_1 - F/\gamma = M_2$, or

$$\left(\frac{y_1^2}{2} + \frac{q^2}{gy_1}\right) - \frac{C_D A\rho V_1^2/2}{b\gamma} = \left(\frac{y_2^2}{2} + \frac{q^2}{gy_2}\right).$$

Substitute in known data, plus the relation $V_1 = q/y_1$:

$$\frac{y_1^2}{2} + \frac{2.30^2}{9.81y_1} - \frac{0.35 \times 5 \times 0.17 \times 1000 \times 2.30^2}{5 \times 9810y_1^2} = \frac{1.5^2}{2} + \frac{2.30^2}{9.81 \times 1.5^2},$$

which reduces to

$$y_1^2 + \frac{1.078}{y_1} - \frac{0.0321}{y_1^2} = 2.97.$$

$\therefore$ Solving, $y_1 = \underline{0.346 \text{ m}}.$

10.45 Assume steep channel slope, with critical flow at the entrance. Then, from the minimum energy concept,

$y_c = \frac{2}{3}(103.6 - 101.5) = \underline{1.40 \text{ m}},$

a) $\therefore Q = b\sqrt{g y_c^3} = 8\sqrt{9.81 \times 1.40^3} = \underline{41.5 \text{ m}^3/\text{s}}.$

b) Compute y_0 using Chezy-Manning eqn:

$$\frac{A^{5/3}}{P^{2/3}} = \frac{Qn}{c_1\sqrt{S_0}}, \quad \frac{(8y_0)}{(8+2y_0)^{2/3}} = \frac{41.5 \times 0.014}{1 \times \sqrt{0.004}} = 9.186,$$

$\therefore$ Solving, $y_0 = \underline{1.21 \text{ m}}.$

Since $y_c > y_0$, slope is <u>steep</u>, and assumption of critical flow at entrance is valid.

c) A possible jump is located at B. Note that it's location must be determined using numerical analysis.

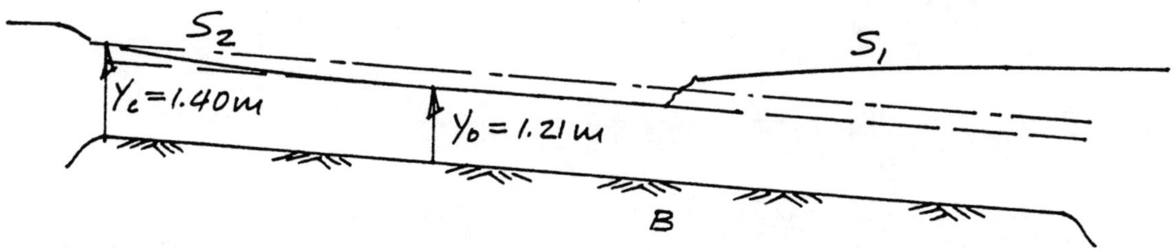

10.46 $q = Q/b = 0.35/1.8 = \underline{0.194 \text{ m}^2/\text{s}},$

$y_c = \sqrt[3]{q^2/g} = \sqrt[3]{0.194^2/9.81} = \underline{0.157 \text{ m}}.$

Use Chezy-Manning eqn. to compute y_0:

$$\frac{Qn}{c_1\sqrt{S_0}} = \frac{0.35 \times 0.012}{1 \times \sqrt{0.0163}} = 0.0329 = \frac{(1.8y_0)^{5/3}}{(1.8+2y_0)^{2/3}},$$

$\therefore$ Solving, $y_0 = \underline{0.094 \text{ m}}.$

$$E_c + h = \frac{3}{2}y_c + h = \frac{3}{2} \times 0.157 + 0.1 = 0.335 \text{ m},$$

$$E_0 = y_0 + \frac{q^2}{2gy_0^2} = 0.094 + \frac{0.194^2}{2 \times 9.81 \times 0.094^2} = 0.311 \text{ m}.$$

Since $E_0 < E_c + h$, normal conditions cannot exist at loc. 1, and choking will occur at loc. 2. Compute alternate depths at locs. 1 and 3: $E_1 = E_3 = E_c + h$.

$$y + \frac{0.194^2}{2 \times 9.81 y^2} = y + \frac{0.00192}{y^2} = 0.335,$$

$\therefore$ Solving, $y_1 = \underline{0.316 \text{ m}}$, $y_3 = \underline{0.088 \text{ m}}$. (Note: loc. 2 is a critical control, with subcritical flow upstream and supercritical flow downstream.) A jump is located upstream of loc. 1. Find the depth conjugate to y_0:

$$\text{Fr}_0^3 = q^2 / g y_0^3 = 0.194^2 / (9.81 \times 0.094^3) = 4.62,$$

$$\therefore y_{cj} = \frac{y_0}{2}\left(\sqrt{1 + 8\text{Fr}_0^2} - 1\right) = \frac{0.094}{2}\left(\sqrt{1 + 8 \times 4.62} - 1\right) = \underline{0.243 \text{ m}}$$

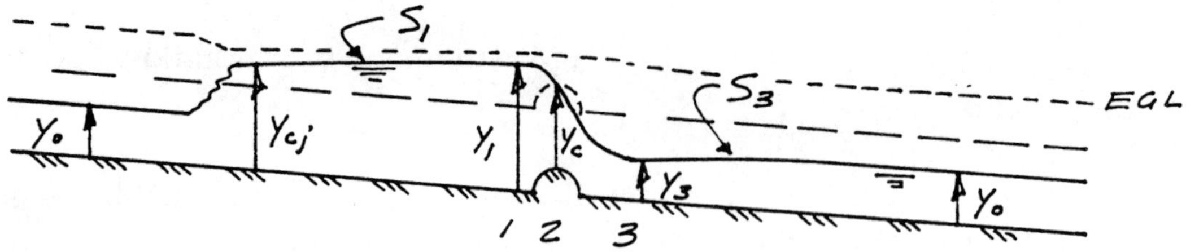

10.46E (See Pbm 10.46 for development and figure.)

$$q = Q / b = 12.5 / 6 = \underline{2.08 \text{ ft}^2 / \text{s}},$$

$$y_c = \sqrt[3]{q^2 / g} = \sqrt[3]{2.08^2 / 32.2} = \underline{0.51 \text{ ft}}.$$

Use Chezy-Manning eqn. to compute y_0:

$$\frac{Qn}{c_1\sqrt{S_0}} = \frac{12.5 \times 0.012}{1.49 \times \sqrt{0.0163}} = 0.7885 = \frac{(6y_0)^{5/3}}{(6 + 2y_0)^{2/3}},$$

$\therefore$ Solving, $y_0 = \underline{0.31 \text{ ft}}$

$$E_c + h = \frac{3}{2} y_c + h = \frac{3}{2} \times 0.51 + 0.33 = 1.10 \text{ ft},$$

$$E_0 = y_0 + \frac{q^2}{2g y_0^2} = 0.31 + \frac{2.08^2}{2 \times 32.2 \times 0.31^2} = 1.01 \text{ ft}.$$

$$y + \frac{2.08^2}{2 \times 32.2 y^2} = y + \frac{0.0672}{y^2} = 1.10,$$

$\therefore$ Solving, $y_1 = \underline{1.04 \text{ ft}}$, $y_3 = \underline{0.29 \text{ ft}}$.

$$\text{Fr}_0^3 = q^2 / g y_0^3 = 2.08^2 / (32.2 \times 0.31^3) = 4.51,$$

$$\therefore y_{cj} = \frac{y_0}{2}\left(\sqrt{1 + 8\text{Fr}_0^2} - 1\right) = \frac{0.31}{2}\left(\sqrt{1 + 8 \times 4.51} - 1\right) = \underline{0.79 \text{ ft}}$$

10.47 From Pbm. 10.46, $q = \underline{0.194\ \text{m}^2/\text{s}}$,

$y_c = \underline{0.157\ \text{m}}$, and $E_c + h = \underline{0.335\ \text{m}}$.

Use Chezy-Manning eqn. to compute y_0:

$$\frac{Qn}{c_1\sqrt{S_0}} = \frac{0.35 \times 0.012}{1 \times \sqrt{0.0013}} = 0.1165 = \frac{(1.8y_0)^{5/3}}{(1.8 + 2y_0)^{2/3}},$$

$\therefore$ Solving, $y_0 = \underline{0.21\ \text{m}}$.

$$E_0 = y_0 + \frac{q^2}{2gy_0^2} = 0.21 + \frac{0.194^2}{2 \times 9.81 \times 0.21^2} = 0.254\ \text{m}.$$

Since $E_0 < E_c + h$, normal conditions cannot exist at loc. 1, and choking will occur at loc. 2. Compute alternate depths at locs. 1 and 3: $E_1 = E_3 = E_c + h$.

$$y + \frac{0.194^2}{2 \times 9.81y^2} = y + \frac{0.00192}{y^2} = 0.335,$$

$\therefore$ Solving, $y_1 = \underline{0.316\ \text{m}}$, $y_3 = \underline{0.088\ \text{m}}$. (Note: loc. 2 is a critical control, with subcritical flow upstream & supercritical flow downstream.) A jump is located downstream of location 3. Find depth conjugate to y_0:

$Fr_0^2 = q^2 / gy_0^3 = 0.194^2 / (9.81 \times 0.21^3) = 0.414,$

$$\therefore y_{cj} = \frac{y_0}{2}\left(\sqrt{1 + 8Fr_0^2} - 1\right) = \frac{0.21}{2}\left(\sqrt{1 + 8 \times 0.414} - 1\right) = \underline{0.113\ \text{m}}.$$

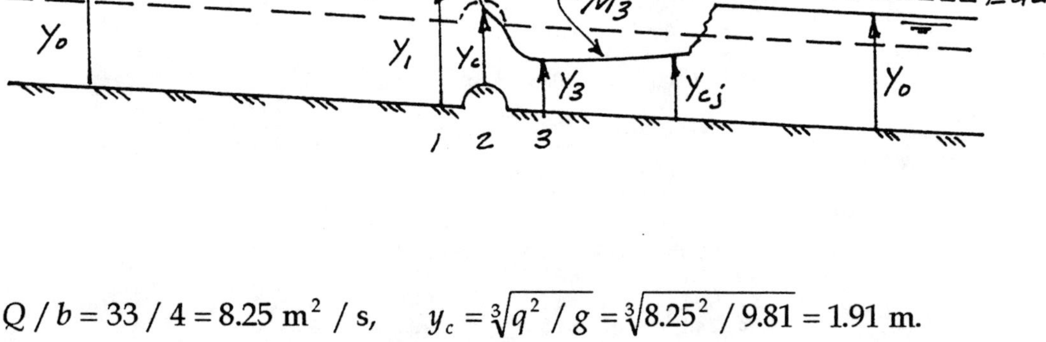

10.48 $q = Q/b = 33/4 = 8.25\ \text{m}^2/\text{s}$, $\quad y_c = \sqrt[3]{q^2/g} = \sqrt[3]{8.25^2/9.81} = 1.91\ \text{m}.$

Use Chezy-Manning eqn. to compute y_0:

$$\frac{Qn}{c_1\sqrt{S_0}} = \frac{33 \times 0.012}{1 \times \sqrt{0.00087}} = 13.43 = \frac{(4y_0)^{5/3}}{(4 + 2y_0)^{2/3}}, \qquad \therefore \text{Solving, } y_0 = \underline{2.98\ \text{m}}.$$

Since $y_0 > y_c$, mild slope conditions prevail. With $y_{\text{entrance}} < y_c$, an M_3 profile exists downstream of the entrance. For free outfall conditions, $y_{\text{exit}} \cong y_c$, and an M_2 profile exists upstream of the exit. The M_3 and M_2 profiles are separated by a hydraulic jump located approximately 260 m downstream of the entrance (determined by numerical analysis).

10.49 $q_1 = Q / b_1 = 8.5 / 3 = 2.83 \text{ m}^2 / \text{s}$,

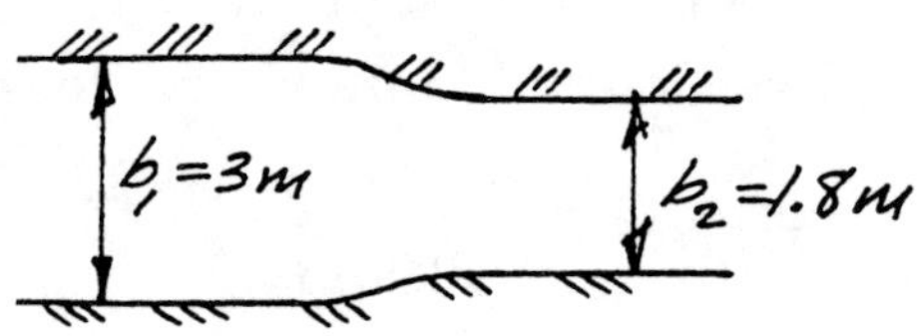

$y_{c_1} = \sqrt[3]{q_1^2 / g} = \sqrt[3]{2.83^2 / 9.81} = \underline{0.935 \text{ m}}$,

$q_2 = Q / b_2 = 8.5 / 1.8 = 4.72 \text{ m}^2 / \text{s}$,

$y_{c_2} = \sqrt[3]{q_2^2 / g} = \sqrt[3]{4.72^2 / 9.81} = 1.314 \text{ m}$,

$E_{c_2} = \frac{3}{2} y_{c_2} = \frac{3}{2} \times 1.314 = \underline{1.97 \text{ m}}$,

$$E_0 = y_0 + \frac{q_1^2}{2gy_0^2} = 1.54 + \frac{2.83^2}{2 \times 9.81 \times 1.54^2} = \underline{1.71 \text{ m}}.$$

a) Since $E_0 < E_{c_2}$, choking occurs at loc. 2, and the depth at loc. 1 is greater than y_0. To compute y_1, set $E_1 = E_{c_2}$:

$$y_1 + \frac{q_1^2}{2gy_1^2} = y_1 + \frac{2.83^2}{2 \times 9.81 y_1^2} = y_1 + \frac{0.4082}{y_1^2} = 1.97,$$

$\therefore$ Solving, $y_1 = \underline{1.85 \text{ m}}$.

b) Since $y_0 > y_{c_1}$, a mild slope condition exists, and an M_1 curve occurs upstream of the constriction.

10.50 Output is from program listed in Fig.E.1. A hydraulic jump occurs at $x \cong 150$ m.

Q =	20	N =	.014	SO =	.001
L =	300	B =	0	M1 =	3.5
M2 =	2.5	Y1 =	.5	Y2 =	2.5
H =	10	UNITS =	SI		

YO =	1.809	EO =	2.020
YC =	1.554	EC =	1.942

X	Y	E	YCJ	
0.000	0.500	36.744	3.769	M_3
33.077	0.605	17.470	3.304	
67.953	0.711	9.586	2.951	
104.241	0.816	5.921	2.670	
141.482	0.922	4.062	2.439	
179.082	1.027	3.063	2.242	
216.212	1.132	2.510	2.071	
251.650	1.238	2.203	1.920	
283.462	1.343	2.039	1.786	
308.287	1.449	1.963	1.664	
300.000	2.500	2.558	0.892	M_1
222.975	2.431	2.496	0.925	
144.184	2.362	2.435	0.961	
63.055	2.293	2.375	0.998	
-21.258	2.224	2.316	1.038	

10.50E Output is from program listed in Fig.E.1. A hydraulic jump occurs at $x \cong 530$ ft.

```
    Q =    700            N =    .014        SO =    .001
    L =    1000           B =    0           M1 =    3.5
   M2 =    2.5           Y1 =    1.65        Y2 =    8
    H =    10         UNITS =    EN

   YO =    5.909         EO =    6.603

   YC =    5.080         EC =    6.349
```

X	Y	E	YCJ
0.000	1.650	115.710	12.240
108.330	1.993	55.582	10.747
222.462	2.336	30.731	9.610
341.138	2.679	19.095	8.703
462.869	3.022	13.161	7.952
585.715	3.365	9.960	7.314
706.973	3.708	8.181	6.760
822.660	4.051	7.191	6.271
926.481	4.394	6.662	5.834
1007.476	4.737	6.416	5.439

M_3

X	Y	E	YCJ
1000.000	8.000	8.206	2.998
763.549	7.791	8.020	3.104
521.190	7.582	7.838	3.215
271.049	7.373	7.659	3.331
10.353	7.164	7.485	3.454
-265.238	6.955	7.316	3.582

M_1

10.51

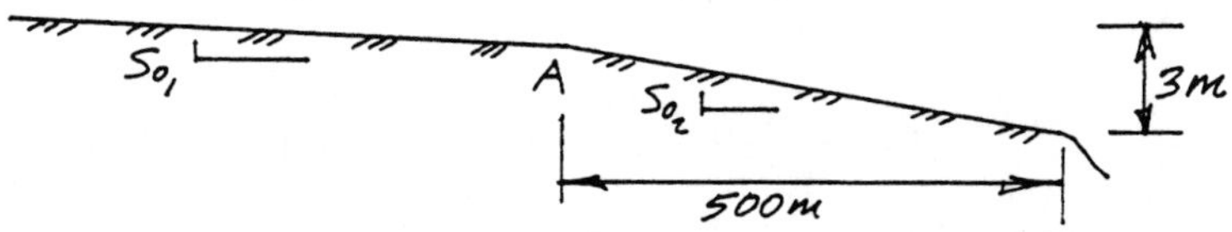

First, compute y_{0_1}, y_{0_2} and y_c. This can be done by trial & error, or by use of the program in Fig. E.1 (see Pbm. 10.57). The results are:
$y_{0_1} = \underline{1.82 \text{ m}}$, $y_{0_2} = \underline{0.89 \text{ m}}$, $y_c = \underline{1.19 \text{ m}}$.
Since $y_{0_1} > y_c$ and $y_{0_2} < y_c$, the slope upstream of loc. A is mild, and downstream of A the slope is steep. Hence, loc. A acts as a critical control, provided no backwater from the reservoir influences the depth at A. (This is verified by the varied flow calculations shown in Pbm. 10.57.) Upstream of loc. A there is an M_2 curve, with critical flow at A, followed by an S_2 curve terminating in a hydraulic jump. An S_1 curve exists behind the jump.

10.52 $q = Q/b = 15/4 = 3.75 \text{ m}^2/\text{s}$,
$y_c = \sqrt[3]{q^2/g} = \sqrt{3.75^2/9.81} = \underline{1.13 \text{ m}}$.

$y_{0_1} = 0.93m,\ y_c > y_{0_1},\ \ \therefore$ upstream reach has a <u>steep</u> slope.
$y_{0_2} = 1.42m,\ y_c < y_{0_2},\ \ \therefore$ downstream reach has a <u>mild</u> slope.

Compute depth conjugate to y_{0_1}:

$$\mathrm{Fr}_{01}^2 = q^2/(gy_{01}^3) = 3.75^2/(9.81 \times 0.93^3) = 1.782$$

$$\therefore y_{cj} = \frac{y_{0_1}}{2}\left[\sqrt{1+8\mathrm{Fr}_{01}^2} - 1\right] = \frac{0.93}{2}\left[\sqrt{1+8\times 1.782} - 1\right] = 1.35 \text{ m}$$

$\therefore$ Hydraulic jump occurs upstream of transition, followed by an S_1 curve up to the transition.

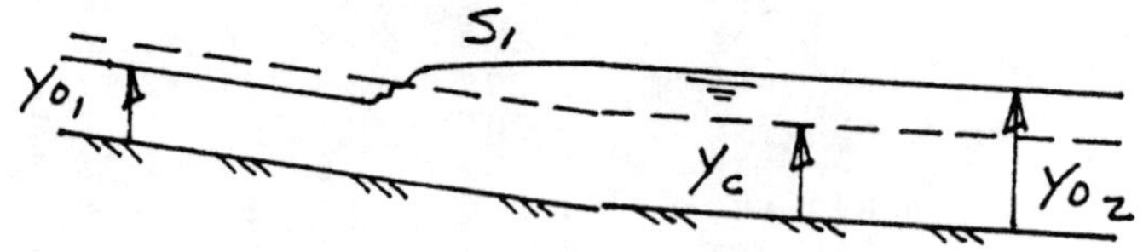

10.53 a) $q = \dfrac{Q}{b} = \dfrac{5}{3} = 1.67 \text{ m}^3/\text{s}$, and $y_c = \sqrt[3]{\dfrac{q^2}{g}} = \sqrt[3]{\dfrac{1.67^2}{9.81}} = 0.657 \text{ m}$

Since $y_0 > y_c$, the channel upstream of A is steep. Compute the depth conjugate to y_0:

$$\mathrm{Fr}_0^2 = \frac{q^2}{gy_0^3} = \frac{1.67^2}{9.81 \times 0.4^3} = 4.43$$

$$y_{cj} = \frac{y_0}{2}\left(\sqrt{1+8\mathrm{Fr}_0^1} - 1\right) = \frac{0.4}{2}\left(\sqrt{1+8\times 4.43} - 1\right) = 1.01 \text{ m}$$

The depth at the outfall is $1.6 + y_c = 2.26$ m. There is an H_2 profile upstream of the weir up to location A; upstream of A an S_1 profile exists. Between location A and the outfall the water depth is always greater than y_c, therefore the jump will be upstream of A, where the depth on the S_1 curve is equal to y_{cj}.

b)

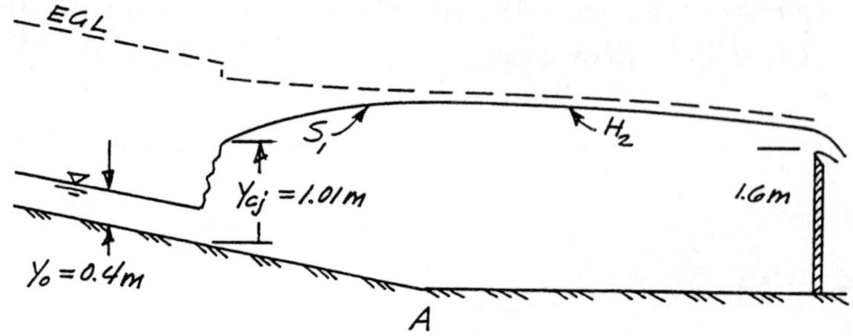

10.54 a) $y_c = \sqrt[3]{\dfrac{q^2}{g}} = \sqrt[3]{\dfrac{5.75^2}{9.81}} = 1.50\text{ m}$. Since $y_{01} < y_c$ and $y_{02} > y_c$, there will be a hydraulic jump between A and C.

b) Either an M_3 curve exists downstream of B, with the depth equal to y_{01} upstream be, or an S_1 curve exists upstream of B, with the depth equal to y_{02} downstream of B.

c) Compute the depth conjugate to y_{01}:

$$\text{Fr}_1^2 = \frac{5.75^2}{9..81 \times 1.0^3} = 3.37\ ;\ \ y_{cj} = \frac{1.0}{2}\left(\sqrt{1+8\times 3.37} - 1\right) = 2.14\text{ m}$$

Since $y_{cj} > y_{02}$, the hydraulic jump must be downstream of B. Hence, an M_1 curve exists downstream of B, increasing from y_{01} at B to y_{cj}, followed by a jump to y_{02}.

10.55 a) $y_c = \sqrt[3]{\dfrac{q^2}{g}} = \sqrt[3]{\dfrac{4^2}{9.81}} = 1.18\text{ m}$. Since $y_{01} > y_c$ and $y_{02} < y_c$, an M_2 curve is located upstream of C and an S_2 curve is downstream of C. Note that the depth at C is critical.

b) Extend the downstream channel slope from C to B, with an abrupt drop of height h immediately upstream of B. Write the energy equation from B to C to determine the magnitude of h:

$$h = y_{01} + \frac{q^2}{2gy_{01}^2} - \frac{3}{2}y_c = 1.6 + \frac{4^2}{2\times 9.81\times 1.6^2} - \frac{3}{2}\times 1.18 \approx 0.15\text{ m}$$

c) Just as in part a), an S_2 profile exists downstream of location B.

10.56 Compute y_c using $Fr^2 = 1$:

$$\frac{Q^2B}{gA^3} = \frac{Q^2 d\sin\alpha}{g\left[d^2/4(\alpha - \sin\alpha\cos\alpha)\right]^3} = \frac{2.5^2\times 2.5\sin\alpha}{9.81\left[2.5^2/4(\alpha-\sin\alpha\cos\alpha)\right]^3}$$

$$= 0.4175\frac{\sin\alpha}{(\alpha - \sin\alpha\cos\alpha)^3} = 1,$$

$\therefore$ Solving, $\alpha = 1.115$ rad, and $\therefore\ y_c = \dfrac{d}{2}(1-\cos\alpha) = \dfrac{2.5}{2}(1-\cos 1.115) = \underline{0.70\text{ m}}$.

Compute y_0 using

$$\frac{Qn}{c_1\sqrt{S_0}}=\frac{\left[d^2/4(\alpha-\sin\alpha\cos\alpha)\right]^{5/3}}{(\alpha d)^{2/3}}, \quad \frac{2.5\times 0.015}{1\times\sqrt{0.001}}=\frac{\left[2.5^2/4(\alpha-\sin\alpha\cos\alpha)\right]^{5/3}}{(\alpha\times 2.5)^{2/3}},$$

which reduces to $1.039=\dfrac{(\alpha-\sin\alpha\cos\alpha)^{5/3}}{\alpha^{2/3}}$.

$\therefore$ Solving, $\alpha = 1.361$ rad, and $\therefore y_0=\dfrac{d}{2}(1-\cos\alpha)=\dfrac{2.5}{2}(1-\cos 1.361)=\underline{0.99\text{ m}}$.

Since $y_0 > y_c$, a mild slope condition exists. The water surface consists of an M_3 curve beginning at the inlet, followed by a hydraulic jump to an M_2 curve, which terminates at critical depth at the outlet. The numerically-predicted water surface and energy grade line are provided in the following table.

x (m)	y (m)	E (m)	x (m)	y (m)	E (m)
0	0.40	1.64	406	0.90	1.03
7	0.43	1.44	442	0.87	1.01
14	0.46	1.29	465	0.84	1.00
20*	0.49*	1.18*	480	0.82	0.98
20*	0.97*	1.07*	490	0.79	0.97
230	0.96	1.065	500	0.70	0.95
350	0.93	1.05	*hydraulic jump		

10.56E See Pbm. 10.56 for development. The numerical results are given as follows:

$y_c = \underline{2.31\text{ ft}}$, $y_0 = \underline{3.28\text{ ft}}$,

x (ft)	y (ft)	E (ft)	x (ft)	y (ft)	E (ft)
0	1.30	5.57	1285	2.99	3.40
23	1.40	4.84	1405	2.89	3.34
46	1.50	4.32	1482	2.795	3.29
69*	1.61*	3.92*	1566	2.60	3.20
69*	3.21*	3.55*	1586	2.505	3.17
697	3.18	3.53	1600	2.31	3.14
1087	3.085	3.46	*hydraulic jump		

10.57 Output is from program listed in Fig. E.1. Profile upstream of location A:

```
   Q =    17.5           N =    .012          SO =    .0003
   L =    1000           B =    3             M1 =    1.8
  M2 =    1.8           Y1 =    10            Y2 =    1.188
   H =    5          UNITS =    SI

  YO =    1.823         EO =    1.942

  YC =    1.188         EC =    1.607

      X            Y           E          YCJ

 1000.000        1.188       1.607       1.187
  977.914        1.315       1.628       1.067
  890.044        1.442       1.682       0.961
  673.051        1.569       1.756       0.866
  126.537        1.696       1.844       0.780
  -73.463        1.823       1.942       0.704
```

Profile downstream of location A:

```
   Q =    17.5           N =    .012          SO =    .005
   L =    500            B =    3             M1 =    1.8
  M2 =    1.8           Y1 =    1.188         Y2 =    3
   H =    5          UNITS =    SI

  YO =    0.892         EO =    1.818

  YC =    1.188         EC =    1.607

      X            Y           E          YCJ
    0.000        1.188       1.607       1.188
    1.890        1.128       1.613       1.248
    9.117        1.069       1.632       1.313
   26.224        1.010       1.669       1.382
   67.658        0.951       1.729       1.455
  167.658        0.892       1.818       1.533
  267.658        0.892       1.818       1.533
  367.658        0.892       1.818       1.533
  467.658        0.892       1.818       1.533
  567.658        0.892       1.818       1.533

  500.000        3.000       3.025       0.288
  429.387        2.638       2.675       0.374
  360.188        2.275       2.335       0.492
  293.987        1.913       2.015       0.655
  235.220        1.550       1.744       0.879
```

A hydraulic jump is located ~230 m downstream of location A.

10.58 a)

$$q = \frac{30}{5} = 6 \text{ m}^2/\text{s}$$

$$y_c = \sqrt[3]{\frac{6^2}{9.81}} = 1.54 \text{ m}$$

$$E_1 = 2 + \frac{6^2}{2 \times 9.81 \times 2^2} = 2.459 \text{ m}$$

$$E_2 = 1.8 + \frac{6^2}{2 \times 9.81 \times 1.8^2} = 2.366 \text{ m}$$

$$y_m = 1.9 \text{ m}, \ A_m = 5 \times 1.9 = 9.5 \text{ m}^2, \ P_m = 5 + 2 \times 1.9 = 8.8 \text{ m}, \ R_m = 9.5/8.8 = 1.08 \text{ m}$$

$$S(y_m) = \frac{Q^2 n^2}{A_m^2 R_m^{4/3}} = \left(\frac{30 \times 0.012}{9.5}\right)^2 \frac{1}{1.08^{4/3}} = 0.0036$$

$$\therefore \Delta x = \frac{E_2 - E_1}{S_0 - S(y_m)1} = \frac{2..366 - 2.459}{-0.001 - 0.0036} = \underline{20.2 \text{ m}}$$

b) An A_2 profile is contained within the reach (which has an adverse slope with $y_2 < y_1$).

10.59 a) Use Eq. 10.4.27 to compute Q:

$$Q = 0.58 \times \frac{8}{15}\sqrt{2 \times 9.81} \tan 60^\circ \times 0.37^{5/2} = \underline{0.198 \text{ m}^3/\text{s}}$$

b) Compute q, y_c, and y_0:

$$q = \frac{0.198}{1.0} = 1.98 \text{ m}^2/\text{s}, \ y_c = \sqrt[3]{\frac{0.198^2}{9.81}} = 0.16 \text{ m},$$

$$\frac{Qn}{\sqrt{S_0}} = \frac{0.198 \times 0.017}{\sqrt{0.0002}} = 0.238 = \frac{y_0^{5/3}}{(1+2y_0)^{2/3}}, \text{ and by trial and error } y_0 \approx 0.57 \text{ m}$$

Since $y_0 > y_c$, the slope is mild. The depth upstream of the weir is $0.5 + \text{Y} = 0.87$ m. Hence, an M_1 curve is situated upstream of the weir.

10.60 Evaluate Q using $\Delta x = (E_2 - E_1)/\left[S_0 - S(y_m)\right]$.

$$E_1 = y_1 + \frac{Q^2}{2gA_1^2} = 1.05 + \frac{Q^2}{2 \times 9.81 \times (2.5 \times 1.05)^2} = 1.05 + \frac{Q^2}{135.2},$$

$$E_2 = y_2 + \frac{Q^2}{2gA_2^2} = 1.2 + \frac{Q^2}{2 \times 9.81 \times (2.5 \times 1.2)^2} = 1.2 + \frac{Q^2}{176.6},$$

$$y_m = \frac{1}{2}(y_1 + y_2) = \frac{1}{2}(1.05 + 1.2) = 1.125 \text{ m},$$

$$S(y_m) = \left(\frac{Qn}{c_1}\right)^2 \frac{(b + 2y_m)^{4/3}}{(by_m)^{10/3}}$$

$$= \frac{Q^2 \times 0.013^2 \times (2.5 + 2 \times 1.125)^{4/3}}{1^2 \times (2.5 \times 1.125)^{10/3}}$$

$$= 4.296 \times 10^{-5} Q^2.$$

$$\therefore \Delta x\left[S_0 - S(y_m)\right] = E_2 - E_1,$$

$$50(0.005 - 4.296 \times 10^{-5} Q^2) = 1.2 + \frac{Q^2}{176.6} - 1.05 - \frac{Q^2}{135.2},$$

which reduces to $4.141 \times 10^{-4} Q^2 = 0.10$.

$\therefore$ Solving, $Q = \underline{15.5 \text{ m}^3/\text{s}}$.

$$y_c = \sqrt[3]{\left(\frac{Q}{b}\right)^2 / g} = \sqrt[3]{\left(\frac{15.5}{2.5}\right)^2 / 9.81} = 1.58 \text{ m},$$

$\therefore$ The profile is an $\underline{S_3 \text{ curve}}$.

10.61 Consider location 1, immediately upstream of transition:

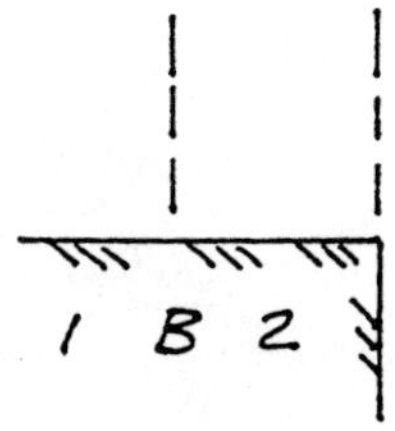

$$q_1 = Q/b_1 = 5.5/3 = 1.83 \text{ m}^2/\text{s},$$

$$y_{c_1} = \sqrt[3]{q_1^2/g} = \sqrt[3]{1.83^2/9.81} = \underline{0.70 \text{ m}}.$$

$\therefore y_{c_1} > y_0$, and steep slope condition prevails. Compute specific energy at normal depth:

$$E_0 = y_0 + \frac{q_1^2}{2gy_0^2} = 0.5 + \frac{1.83^2}{2 \times 9.81 \times 0.5^2} = \underline{1.18 \text{ m}}.$$

At location 2, within the constriction:

$$q_2 = Q / b_2 = 5.5 / 1.5 = 3.67 \text{ m}^2 / \text{s},$$
$$y_{c_2} = \sqrt[3]{q_2^2 / g} = \sqrt[3]{3.67^2 / 9.81} = \underline{1.11 \text{ m}},$$
$$E_{c_2} = \frac{3}{2} y_{c_2} = \frac{3}{2} \times 1.11 = \underline{1.67 \text{ m}}.$$

Since $E_{c_2} > E_0$, uniform flow cannot exist at loc. 1, hence choking occurs at loc. 2. Compute y_1 by setting $E_1 = E_{c_2}$:

$$y_1 + \frac{q_1^2}{2gy_1^2} = y_1 + \frac{1.83^2}{2 \times 9.81 y_1^2} = y_1 + \frac{0.71}{y_1^2} = 1.67,$$

$\therefore$ Solving, $y_1 = \underline{1.60 \text{ m}}$.

Since y_1 is subcritical, there will be a hydraulic jump some distance upstream of the transition, followed b an S_1 curve. The depth behind the jump is that depth conjugate to y_0:

$$\text{Fr}_0^2 = q_1^2 / g y_0^3 = 1.83^2 / (9.81 \times 0.5^3) = 2.73$$
$$\therefore y_{cj} = \frac{y_0}{2}\left(\sqrt{1 + 8\text{Fr}_0^2} - 1\right) = \frac{0.5}{2}\left(\sqrt{1 + 8 \times 2.73} - 1\right)$$
$$= \underline{0.94 \text{ m}}.$$

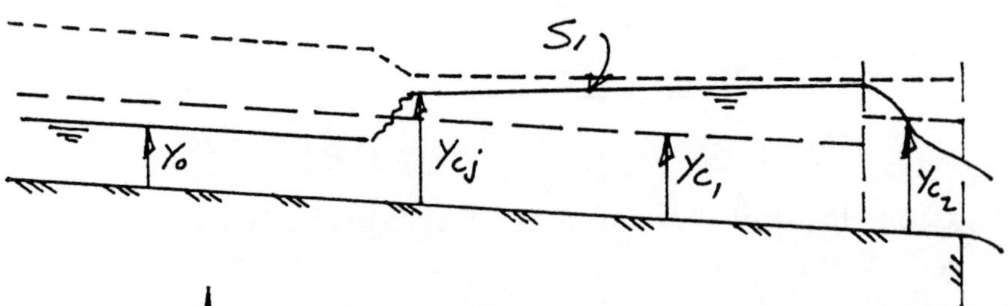

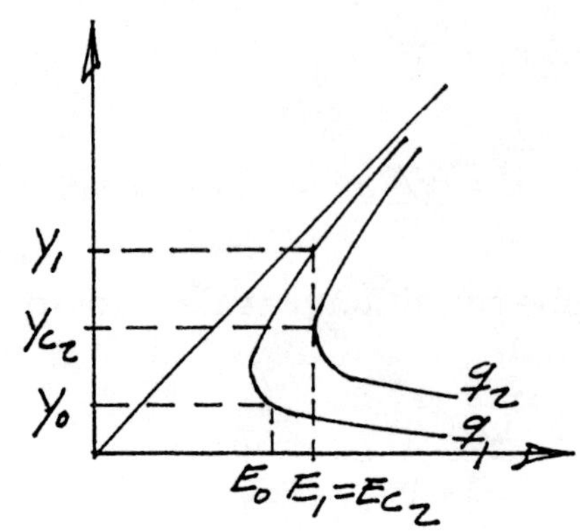

10.61E See Pbm. 10.61 for development and figures.

$$q_1 = Q / b_1 = 200 / 10 = 20 \text{ ft}^2 / \text{sec},$$

$$y_{c_1} = \sqrt[3]{q_1^2 / g} = \sqrt[3]{20^2 / 32.2} = \underline{3.67 \text{ ft}},$$

$$E_0 = y_0 + \frac{q_1^2}{2gy_0^2} = 1.65 + \frac{20^2}{2 \times 32.2 \times 1.65^2} = \underline{3.93 \text{ ft}}.$$

$$q_2 = Q / b_2 = 200 / 5 = 40 \text{ ft}^2 / \text{sec},$$

$$y_{c_2} = \sqrt[3]{q_2^2 / g} = \sqrt[3]{40^2 / 32.2} = \underline{3.67 \text{ ft}},$$

$$E_{c_2} = \frac{3}{2} y_{c_2} = \frac{3}{2} \times 3.67 = \underline{5.505 \text{ ft}}.$$

$$y_1 + \frac{q_1^2}{2gy_1^2} = y_1 + \frac{20^2}{2 \times 32.2 y_1^2} = y_1 + \frac{6.211}{y_1^2} = 5.505,$$

$\therefore$ Solving, $y_1 = \underline{5.28 \text{ ft}}$.

$$\text{Fr}_0^2 = q_1^2 / gy_0^3 = 20^2 / (32.2 \times 1.65^3) = 2.76,$$

$$\therefore y_{cj} = \frac{y_0}{2}\left(\sqrt{1 + 8\text{Fr}_0^2} - 1\right) = \frac{1.65}{2}\left(\sqrt{1 + 8 \times 2.76} - 1\right) = \underline{3.14 \text{ ft}}.$$

10.62 Write energy eqn. across the gate to compute the discharge:

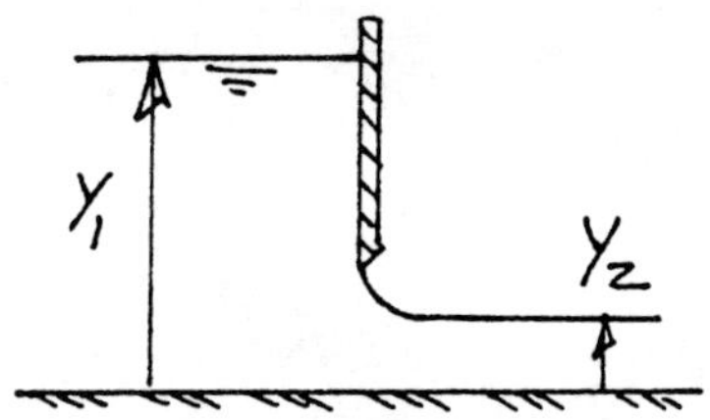

$$E_1 = E_2 \text{ , or } y_1 + \frac{q^2}{2gy_1^2} = y_2 + \frac{q^2}{2gy_2^2}.$$

$$\therefore q = \sqrt{2g(y_1 - y_2)\left(\frac{1}{y_2^2} - \frac{1}{y_1^2}\right)^{-1}}$$

$$= \sqrt{2 \times 9.81 \times (1.85 - 0.35)\left(\frac{1}{0.35^2} - \frac{1}{1.85^2}\right)^{-1}} = \underline{1.93 \text{ m}^2 / \text{s}},$$

$$\therefore Q = bq = 4 \times 1.93 = \underline{7.72 \text{ m}^3 / \text{s}}.$$

$$y_c = \sqrt[3]{q^2 / g} = \sqrt{1.93^2 / 9.81} = \underline{0.72 \text{ m}}.$$

Compute y_0 using Chezy-Manning relation:

$$\frac{Qn}{c_1\sqrt{S_0}} = \frac{7.72 \times 0.014}{1 \times \sqrt{0.0008}} = 3.821 = \frac{(4y_0)^{5/3}}{(4 + 2y_0)^{2/3}},$$

$\therefore$ Solving, $y_0 = \underline{1.17 \text{ m}}$. Since $y_0 > y_c$, mild channel conditions exist.

Upstream of the gate there will be an M_1 profile, with an M_3 downstream of the gate, terminating in a hydraulic jump to normal flow conditions.

Compute the depth upstream of the jump, conjugate to y_0:

$\mathrm{Fr}_0^2 = q^2 / gy_0^3 = 1.93^2 / (9.81 \times 1.17^3) = 0.237$,

$\therefore y_{cj} = \frac{y_0}{2}\left(\sqrt{1+8\mathrm{Fr}_0^2} - 1\right) = \frac{1.17}{2}\left(\sqrt{1+8\times 0.237} - 1\right) = \underline{0.41\ \mathrm{m}}.$

Use step method to compute water surface and energy grade line:

$$E = y + \frac{q^2}{2gy^2} = y + \frac{1.93^2}{2\times 9.81 y^2} = y + \frac{0.190}{y^2}$$

$$S(y) = \frac{Q^2 n^2 (4+2y)^{4/3}}{(4y)^{10/3}} = \frac{7.72^2 \times 0.014^2 (4+2y)^{4/3}}{4^{10/3} y^{10/3}}$$

$$= 1.15 \times 10^{-4} (4+2y)^{4/3} y^{-10/3}$$

M_1 curve upstream of gate:

y (m)	E (m)	y_m (m)	$S(y_m)$	Δx (m)	x (m)
1.85	1.906				0
		1.775	2.514×10^{-4}	-255	
1.7	1.766				-255
		1.6	3.336×10^{-4}	-390	
1.5	1.584				-645
		1.4	4.824×10^{-4}	-542	
1.3	1.412				-1187
		1.25	6.627×10^{-4}	-583	
1.2	1.332				-1770

M_3 curve downstream of gate:

y (m)	E (m)	y_m (m)	$S(y_m)$	Δx (m)	x (m)
0.35	1.901				0
		0.36	0.02741	5	
0.37	1.758				5
		0.38	0.02315	5	
0.39	1.639				10
		0.40	0.01973	5	
0.41	1.540				15

Hydraulic jump occurs ~5 m from the gate.

10.63

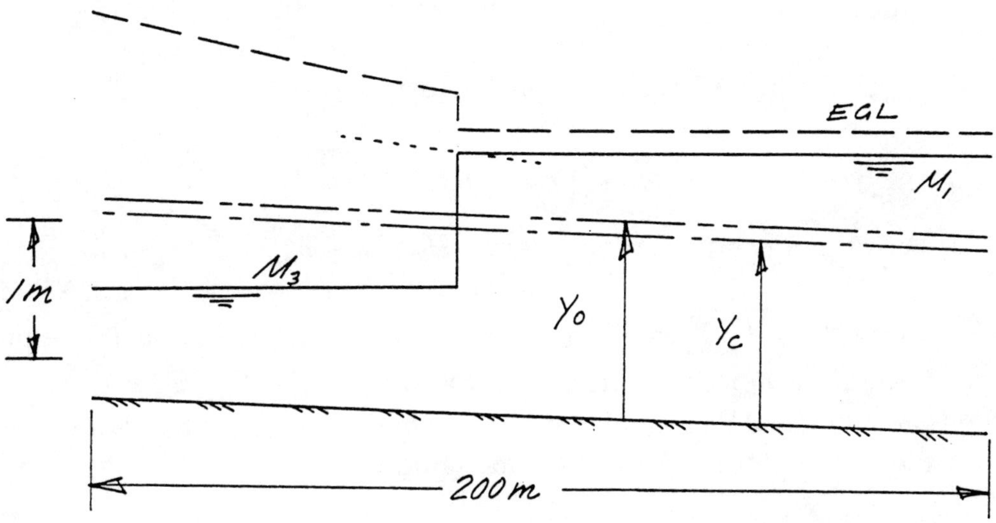

10.64, 10.64E

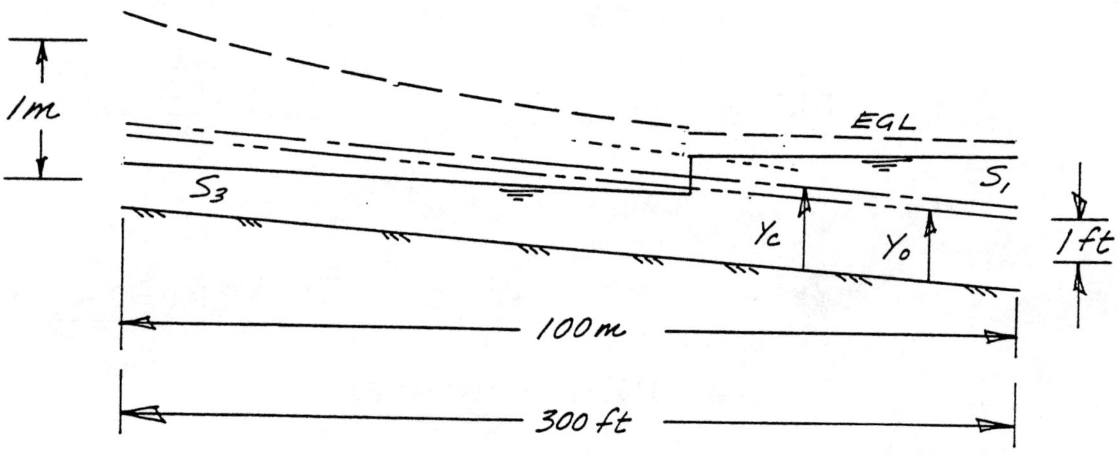

11.15 $W = \frac{fL}{D}\frac{V^2}{2g} \propto \frac{fV^2}{D}$, and since W is constant for all pipes, $V \propto \sqrt{\frac{D}{fL}}$.

$$V_1 \propto \sqrt{\frac{450}{0.012\times 2000}} = 4.33, \quad V_2 \propto \sqrt{\frac{150}{0.02\times 650}} = 3.40, \quad V_3 \propto \sqrt{\frac{300}{0.015\times 1650}} = 3.48$$

Hence, pipe 1 has the largest velocity.

11.16 a) Write energy eqn. across system: $H_P = a_0 - a_1Q_1^2 = \overline{R}_1Q_1^2 + \left[\sum_{i=2}^{4}\frac{1}{\sqrt{\overline{R}_i}}\right]^{-2} Q_1^2$,

with $Q_i = \sqrt{H/\overline{R}_i}$, $i = 2, 3, 4$, and $H = H_P - \overline{R}_1Q_1^2$, in which H = piezometric head at the junction. The unknowns are Q_1, Q_2, Q_3, Q_4 and H. The system can be treated as a parallel piping system because the hydraulic grade line at the exit of pipes 2, 3 and 4 are the same.

b) As opposed to a branching system, a parallel system requires no trial-and-error solution.

c) Substitute known data into energy eqn., and solve for Q_1:

$$a_0 = \left\{a_1 + \overline{R}_1 + \left[\sum\frac{1}{\sqrt{\overline{R}_i}}\right]^{-2}\right\}Q_1^2,$$

$$\therefore 45 = \left\{10^4 + 34\,650 + \left[\frac{1}{\sqrt{82\,500}} + \frac{1}{\sqrt{127\,900}} + \frac{1}{\sqrt{115\,500}}\right]^{-2}\right\}Q_1^2 = 56\,410Q_1^2,$$

$$\therefore Q_1 = \sqrt{45/56\,410} = \underline{0.0282 \text{ m}^3/\text{s}}.$$

$$\therefore H = 45 - (10^4 + 34\,650)\times 0.0282^2 = \underline{9.49 \text{ m}},$$

$$Q_2 = \sqrt{9.49/82\,500} = \underline{0.0107 \text{ m}^3/\text{s}}, \qquad Q_3 = \sqrt{9.49/127\,900} = \underline{0.0086 \text{ m}^3/\text{s}},$$

$$Q_4 = \sqrt{9.49/115\,500} = \underline{0.0091 \text{ m}^3/\text{s}}.$$

d) By increasing the diameter of pipe 1 by approximately 50%,

$\overline{R}_1 = 34\,650/1.5^5 = 4563$, $Q_1 \cong 0.041 \text{ m}^3/\text{s}$, and $H \cong 20.5$ m.

a) Find depth conjugate to y_{0_2}:

$\mathrm{Fr}_{02}^2 = q^2 / g y_{02}^3 = 4.20^2 / (9.81 \times 1.65^3) = 0.400,$

$\therefore y_{cj} = \dfrac{y_{0_2}}{2}\left(\sqrt{1+8\mathrm{Fr}_{02}^2} - 1\right) = \dfrac{1.65}{2}\left(\sqrt{1+8\times 0.4} - 1\right) = \underline{0.87\ \mathrm{m}}.$

b) Step method:

$$E = y + \frac{q^2}{2gy^2} = y + \frac{4.20^2}{2\times 9.81 y^2} = y + \frac{0.900}{y^2},$$

$$S(y) = \frac{Q^2 n^2 (b+2y)^{4/3}}{(by)^{10/3}} = \frac{15.38^2 \times 0.017^2 (3.66+2y)^{4/3}}{(3.66)^{10/3} y^{10/3}}$$
$$= 9.052\times 10^{-4}(3.66+2y)^{4/3} y^{-10/3}.$$

y (m)	E (m)	y_m (m)	$S(y_m)$	Δx (m)	x (m)
0.6	3.100				0
		0.65	0.03216	18.8	
0.7	2.537				18.8
		0.75	0.02104	17.6	
0.8	2.206				36.4
		0.835	0.01536	11.2	
0.87	2.059				$\underline{47.6} \cong L_j$

c) An M_3 curve exists downstream of the change in channel slope, terminating in a hydraulic jump, approximately 50 m from the slope change.

10.67

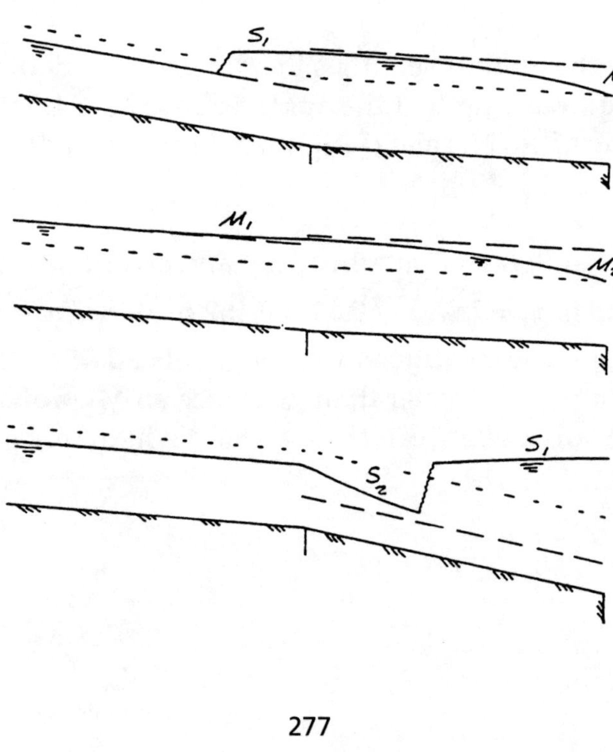

10.68 This is a design analysis problem. Some extensive calculations are required.

a) Find the depth of flow immediately upstream of the jump (call it location 1):

$$V_2 = \frac{Q}{A_2} = \frac{19}{0.75 \times 20} = 1.267 \text{ m/s}, \quad q = \frac{19}{20} = 0.95 \text{ m}^2/\text{s},$$

$$\text{Fr}_2 = \frac{1.267}{\sqrt{9.81 \times 0.75}} = 0.467, \quad y_1 = \frac{0.75}{2}\left(\sqrt{1 + 8 \times 0.467^2} - 1\right) = 0.246 \text{ m}$$

Compute the loss across the jump, and subsequently the dissipated power:

$$h_j = E_1 - E_2 = 0.10 + \frac{0.95^2}{2 \times 9.81 \times 0.10^2} - 0.75 - \frac{0.95^2}{2 \times 9.81 \times 0.75^2} = 3.87 \text{ m}$$

$$\therefore \dot{W} = \gamma Q h_j = 9810 \times 19 \times 3.87 = 7.21 \times 10^5 \text{ watt, or } \underline{721 \text{ kW}}$$

b) The length of the apron is the distance from the toe to downstream of the jump. First, compute the distance from y_{toe} to y_1 using a single increment of length:

$$E_{toe} = 0.10 + \frac{0.95^2}{2 \times 9.81 \times 0.10^2} = 4.700 \text{ m}, \quad E_1 = 0.246 + \frac{0.95^2}{2 \times 9.81 \times 0.246^2} = 1.006 \text{ m},$$

$$y_m = \frac{1}{2}(0.10 + 0.246) = 0.173 \text{ m}, \quad A_m = 20 \times 0.173 = 3.46 \text{ m}^2,$$

$$P_m = 20 + 2 \times 0.173 = 20.35 \text{ m}, \quad S_m = \left(\frac{19 \times 0.014}{3.46}\right)^2 \frac{1}{(3.46/20.35)^{4/3}} = 0.06275,$$

$$\therefore x_1 - x_{toe} = \frac{E_1 - E_{toe}}{S_0 - S_m} = \frac{1.006 - 4.700}{0.0005 - 0.06275} = 59.3 \text{ m}$$

The length of the jump is six times the downstream depth, or $6 \times 0.75 = 4.5$ m. Hence the required length of the apron is L = 59.3 + 4.5 = 63.8 m, or approximately 65 m. Normal design would require additional length as a safety factor.

c) The normal depth is computed by trial and error to be $y_0 \approx 0.755$ m. The critical depth is $y_c = \sqrt[3]{0.95^2/9.81} = 0.452 \text{ m}$. Since $y_0 > y_c$, the channel exhibits mild slope conditions. The depth ahead of the dam is $h + y_c = 6 + 0.45 = 6.45$ m, and is greater than y_0. Hence an M_1 profile is upstream of the dam. On the apron, between the toe and the hydraulic jump, an M_3 curve exists.

10.69 a)

$$y_0 = \left(\frac{q^2 n^2}{S_0}\right)^{3/10} = \left(\frac{1.2^2 \times 0.017^2}{0.0005}\right)^{3/10} = 0.946 \text{ m},$$

$$y_c = \left(\frac{q^2}{g}\right)^{1/3} = \left(\frac{1.2^2}{9.81}\right)^{1/3} = 0.528 \text{ m}$$

Since $y_0 > y_c$, the channel possesses a mild slope. With $y_A = 0.65$ m and $y_B = 0.90$ m, we see that $y_c < y_A < y_0$, and $y_c < y_B < y_0$, so that an M_2 curve exists. Because an M_2 profile decreases with increasing x, location B is upstream of location A.

b) From Ex. 10.16 we find, for a wide rectangular channel, $J = 2.5$, $M = 3$, and N = 3.33. Using Eq. 10.7.13,

$$x = \frac{0.946}{0.005}\left[u - F(u,3.33) + \left(\frac{0.528}{0.946}\right)^3 \frac{2.5}{3.33} F(v,2.5)\right]$$

$$= 1892\left[u - F(u,3.33) - 0.1305F(v,2.5)\right],$$

$$\frac{N}{J} = \frac{3.33}{2.5} = 1.33$$

Compute x_A :

$$u_A = \frac{0.65}{0.946} = 0.687, \quad v_A = 0.687^{1.33} = 0.607, \text{ and from Table 10.4,}$$

$$F(0.687,3.33) \approx 0.742, \quad F(0.607,2.5) \approx 0.668, \text{ so that}$$

$$\therefore x_A = 1892(0.687 - 0.742 + 0.1305 \times 0.668) = 61 \text{ m}$$

Compute x_B:

$$u_B = \frac{0.90}{0.946} = 0.951, \quad v_B = 0.951^{1.33} = 0.935,$$

$$F(0.951,3.33) \approx 1.615, \quad F(0.935,2.5) \approx 1.488,$$

$$\therefore x_B = 1892(0.951 - 1.615 + 0.1305 \times 1.488) = -889 \text{ m}$$

Therefore the distance between A and B is $\underline{y_A - y_B = 61 - (-889) = 950 \text{ m}}$.

10.70 Use the varied flow function to compute the water surface profile.

$$y_0 = \left(\frac{q^2 n^2}{S_0}\right)^{3/10} = \left(\frac{1.5^2 \times 0.015^2}{0.0001}\right) = 1.627 \text{ m},$$

$$y_c = \left(\frac{q^2}{g}\right)^{1/3} = \left(\frac{1.5^2}{9.81}\right) = 0.612 \text{ m}$$

Therefore the estuary has a mild slope and an M_1 curve exists upstream of the outlet. For a wide rectangular channel (Ex. 10.16) $J = 2.5$, $N = 3.33$, $M = 3$, and $N/J = 1.33$. Use Eq. 10.7.13 to compute x:

$$x = \frac{1.627}{0.0001}\left[u - F(u,3.33) + \left(\frac{0.612}{1.627}\right)^3 \frac{2.5}{3.33} F(v,2.5)\right] = 16270\left[u - F(u,3.33) - 0.04F(v,2.5)\right]$$

The results of the calculations are as follows:

y (m)	u	$F(u,3.33)$	v	$F(v,2.5)$	x (m)	x - 26 900 (m)
7	4.30	0.015	6.96	0.038	69 800	0
6	3.67	0.021	5.67	0.053	59 700	-10 100
5	3.07	0.032	4.45	0.074	49 500	-20 300
4	2.46	0.052	3.31	0.117	39 500	-30 300
2	1.23	0.348	1.32	0.572	14 700	-55 100

10.71 Let subscript 1 denote the side channel, and subscript 2 the main channel.

a)

$$A_1 = 1 \times 150 = 150 \text{ m}^2, \quad P_1 = 1 + 150 = 151 \text{ m},$$

$$R_1 = \frac{150}{151} = 0.993 \text{ m}, \quad K_1 = \frac{A_1 R_1^{2/3}}{n_1} = \frac{150 \times 0.993^{2/3}}{0.03} = 4977,$$

$$A_2 = 5 \times 5 = 25 \text{ m}^2, \quad P_2 = 4 + 5 + 5 = 14 \text{ m},$$

$$R_2 = \frac{25}{14} = 1.786 \text{ m}, \quad K_2 = \frac{A_2 R_2^{2/3}}{n_2} = \frac{25 \times 1.786^{2/3}}{0.02} = 1840$$

Compute α using Eq. 10.7.9:

$$\alpha = \frac{(A_1 + A_2)^2}{(K_1 + K_2)^3}\left(\frac{K_1^3}{A_1^2} + \frac{K_2^3}{A_2^3}\right) = \frac{(150 + 25)^2}{(4977 + 1840)^3}\left(\frac{4977^3}{150^2} + \frac{1840^3}{25^2}\right) = 1.493 \approx \underline{1.49}$$

b) $Q = (K_1 + K_2)\sqrt{S} = (4977 + 1840)\sqrt{0.0005} = \underline{152 \text{ m}^3/\text{s}}$

10.72 This is a design analysis problem, and requires extensive calculations that would best be performed on a spreadsheet. Let location A be 400 m upstream of location B. Conditions at B are known. The energy equation between A and B is predicted using total energy $H = y + z + \alpha V^2 / 2g$:

$$H_A^{*} \approx H_B + h_L \approx H_B + \frac{L}{2}(S_A + S_B)$$

In the equation, H_B and S_B are known, and H_A and S_A are unknown. A trial and error solution is required. With the table of given data, location A is associated with $x = 0$, and location B with $x = 400$ m. The following parameters are computed at location B, where $y_B = 3.0$ m:

Location	A_B (m²)	P_B (m)	R_B (m)	K_B (Eq. 10.7.8)
Side channel	60	149	0.40	11
Main channel	273	97	2.83	18 210

Now α_B, V_B, H_B, and S_B can be computed:

$$\alpha_B = \frac{(60+273)^2}{(11+18210)^3}\left(\frac{11^3}{60^2} + \frac{18210^3}{273^2}\right) = 1.48 \qquad \text{(Eq. 10.7.9)}$$

$$V_B = \frac{Q}{A_B} = \frac{280}{60+273} = 0.84 \text{ m/s}$$

$$S_B = \frac{280^2}{(11+18210)^3} = 0.000236 \qquad \text{(Eq. 10.7.7)}$$

$$H_B = 3.0 + 15.1 + 1.48 \times \frac{0.84^2}{2 \times 9.81} = 18.153 \text{ m}$$

With $L = 400$ m, the energy equation between A and B is predicted:

$$H_A^{*} = 18.153 + \frac{400}{2}(S_A + 0.000236) = 18.20 + 200 S_A$$

The total energy at location A is $H_A = y_A + z_A + \alpha_A \frac{V_A^2}{2g} = y_A + 15.0 + \alpha_A \frac{V_A^2}{2g}$

The trial and error solution proceeds by assuming values of y_A, computing the corresponding α_A, V_A, S_A, H_A, and H_A^{*} until the latter two are in close agreement.

For y_A = 3.2 m, the hydraulic parameters at location A are

Location	A_A (m²)	P_A (m)	R_A (m)	K_A (Eq. 10.7.8)
Side channel	172	116	1.48	4470
Main channel	342	112	3.05	24 000

The corresponding values of H_A and H_A^* are

$$\alpha_A = \frac{(172+342)^2}{(4470+24000)^3}\left(\frac{4470^3}{172^2}+\frac{24000^3}{342^2}\right) = 1.39 \qquad \text{(Eq. 10.7.9)}$$

$$V_A = \frac{Q}{A_A} = \frac{280}{172+342} = 0.54 \text{ m/s}$$

$$S_A = \frac{280^2}{(4470+24000)^3} = 0.0000967 \qquad \text{(Eq. 10.7.7)}$$

$$H_A = 3.2 + 15.0 + 1.39 \times \frac{0.54^2}{2\times 9.81} = 18.22 \text{ m}$$

$$H_A^* = 18.200 + 200 \times 0.0000967 = 18.22 \text{ m}$$

Hence the depth at the upstream location is $\underline{y_A = 3.2 \text{ m.}}$

ELEVEN

Flows in Piping Systems

11.1 $\dfrac{V_1^2}{2g} = \dfrac{Q^2}{2gA_1^2} = \dfrac{0.1^2}{2 \times 9.81(\pi / 4 \times 0.2^2)^2} = 0.52 \text{ m},$

$\dfrac{V_2^2}{2g} = \dfrac{Q^2}{2gA_2^2} = \dfrac{0.1^2}{2 \times 9.81(\pi / 4 \times 0.25^2)^2} = 0.21 \text{ m}.$

Location	EGL (m)	HGL (m)	Loss (m)
A	10.00	10.00	$K_1V_1^2 / 2g = 0.5 \times 0.52 = 0.26$
entrance	9.74	9.22	$-H_P = -45.2^{(1)}$
d/s pump	54.94	54.42	$K_vV_1^2 / 2g = 2 \times 0.52 = 1.04$
d/s valve	53.90	53.38	$R_1Q^2 = 53.4 \times 0.1^2 = 0.53$
u/s bend 1	53.37	52.85	$K_eV_2^2 / 2g = 0.25 \times 0.21 = 0.05^{(2)}$
d/s bend 1	53.32	53.11	$R_2 \times \dfrac{10}{500} \times Q^2 = 904 \times \dfrac{10}{500} \times 0.1^2 = 0.18^{(3)}$
u/s bend 2	53.14	52.93	$K_eV_2^2 / 2g = 0.05$
d/s bend 2	53.09	52.88	$R_2 \times \dfrac{490}{500} \times Q^2 = 904 \times \dfrac{490}{500} \times 0.1^2 = 8.86$
u/s exit	44.23	44.02	$K_2V_2^2 / 2g = 1 \times 0.21 = 0.21$
B	44.02	44.02	

(1) The pump adds energy, hence the minus sign.
(2) Note change in kinetic energy term.
(3) Note the 10 m length between elbows.

According to Ex. 11.2, the EGL an HGL at loc. B should be:

$\left(\dfrac{p}{\gamma} + z\right)_B = \dfrac{200 \times 10^3}{9810 \times 0.85} + 20 = 44.0 \text{ m}.$

Some round-off error is present.

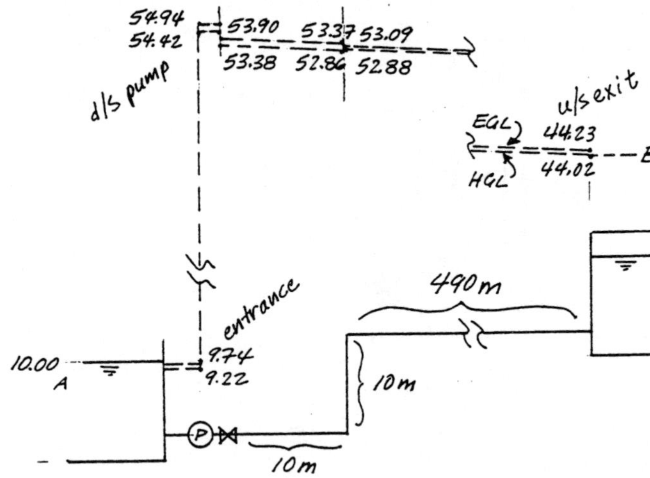

11.2 Substitute $V = \frac{Q}{A}$ and $H_P = \frac{\dot{W}_f}{\gamma Q}$ into the energy equation:

$$\frac{p_1}{\gamma} + \frac{Q^2}{2gA_1^2} + \frac{\dot{W}_f}{\gamma Q} = \frac{p_2}{\gamma} + \frac{Q^2}{2gA_2^2} + h_L$$

a) Note that the kinetic energy terms can be neglected:

$$A_1 = \frac{\pi}{4} \times 0.05^2 = 0.00196 \text{ m}^2, \quad A_2 = \frac{\pi}{4} \times 0.08^2 = 0.00503 \text{ m}^2,$$

$$Q = \frac{0.095}{60} = 0.00158 \text{ m}^3/\text{s}$$

$$\frac{350 \times 10^3}{9810} + \frac{0.00158^2}{2 \times 9.81 \times 0.00196^2} + \frac{\dot{W}_f}{9810 \times 0.00158} = \frac{760 \times 10^3}{9810} + \frac{0.00158^2}{2 \times 9.81 \times 0.00503^2} + 6.6$$

$$35.7 + 0.03 + 0.0645\dot{W}_f = 77.5 + 0.005 + 6.6$$

$$\therefore \dot{W}_f = 750 \text{ watt}$$

b) Neglect the kinetic energy terms:

$$Q = \frac{25}{7.48 \times 60} = 0.0557 \text{ ft}^3/\text{sec}$$

$$\frac{50 \times 114}{62.4} + 0 + \frac{\dot{W}_f}{62.4 \times 0.0557} = \frac{110 \times 144}{62.4} + 0 + 20$$

$$115.4 + 0.288\dot{W}_f = 253.8 + 20$$

$$\therefore \dot{W}_f = 550 \text{ ft-lb/sec, or 1 hp}$$

11.3 a) Use a numerical solution. The system demand curve is

$$H_P = \Delta z + \left(f\frac{L}{D} + \Sigma K \right) \frac{Q^2}{2gA^2}$$

$$= 32 + \left(0.02\frac{2440}{0.20} + 12.5 \right) \frac{Q^2}{2 \times 9.81 \times (0.7854 \times 0.20^2)^2}$$

$$= 32 + 13250Q^2$$

The pump data can be approximated by the quadratic equation (for example, using a least squares fit):

$$H_P = 54.5 + 38.6Q - 2330Q^2$$

Eliminating H_P from the two relations and solving yields $Q \approx 0.039$ m³/s and $H_P \approx 52.3$ m.

b) A numerical approximation to the efficiency curve is

$$\eta = 29.4Q - 349.2Q^2 + 1045Q^3$$

$$= 29.4 \times 0.039 - 349.2 \times 0.039^2 + 1045 \times 0.039^3$$

$$= 0.68$$

Hence the required pump power is

$$\dot{W}_P = \frac{\gamma Q H_P}{\eta} = \frac{0.82 \times 9810 \times 0.039 \times 52.3}{0.68} = 2.42 \times 10^4 \text{ watt, or } 24.2 \text{ kW}$$

11.4 Write energy eqn. between locs. 1 and 2:

$$H_P + z_1 = z_2 + \frac{f}{D}(L + L_e)\frac{Q^2}{2g\left(\frac{\pi}{4}D^2\right)^2}.$$

Solving for D_1, $D = \left[\frac{Q^2(L + L_e)f}{2g(\pi/4)^2(H_P - (z_2 - z_1))} \right]^{1/5}$.

Substitute the relations

$$f = 1.325\left[\ln\left(0.27\frac{e}{D} + 5.74\,\text{Re}^{-0.9} \right) \right]^{-2} \text{ and } \text{Re} = \frac{4Q}{\pi D \nu}, \quad Le = \frac{\Sigma k D}{f},$$

and solve for D by successive substitution.

a) Compute H_P from pump data using linear interpolation:

$$H_P = 453 - (453 - 423)\frac{(12{,}000 - 15{,}000)}{(12{,}000 - 16{,}000)} = 430.5 \text{ ft.}$$

$$Q = 15{,}000 \text{ gpm} \times \frac{1 \text{ ft}^3}{7.48 \text{ gal}} \times \frac{1 \text{ min}}{60 \text{ sec}} = 33.4 \text{ ft}^3/\text{sec},$$

$$\therefore D = \left[\frac{33.4^2(1500 + Le)f}{2 \times 32.2(\pi/4)^2(430.5 - 120)}\right]^{1/5} = 0.6184[(1500 + Le)f]^{1/5}.$$

D (ft)	e/D	$\text{Re} = \frac{3.016 \times 10^6}{D}$	f	$Le = \frac{2.50}{f}$ (ft)
1	0.003	3.02×10^6	0.026	96
1.30	0.0023	2.32×10^6	0.024	135
1.29	0.0023	2.34×10^6	0.024	134
1.29				

$\therefore$ $\underline{D = 1.29 \text{ ft.}}$

b) Compute H_P from pump data using linear interpolation:

$$H_P = 154 - (154 - 148)\frac{(0.8 - 1.1)}{(0.8 - 1.2)} = 149.5 \text{ m.}$$

$$\therefore D = \left[\frac{1.1^2(500 + Le)f}{2 \times 9.81(\pi/4)^2(149.5 - 40)}\right]^{1/5} = 0.2467[(500 + Le)f]^{1/5}.$$

D (m)	e/D	$\text{Re} = \frac{1.401 \times 10^6}{D}$	f	$Le = \frac{2.50}{f}$ (m)
0.3	0.0033	4.67×10^6	0.027	28
0.42	0.0024	3.34×10^6	0.025	42
0.42				

$\therefore$ $\underline{D = 420 \text{ mm.}}$

11.5 a) $z_A + H_P = z_B + \left(f\frac{L}{D} + \Sigma K\right)\frac{Q^2}{2gA^2},$

$$\frac{Q^2}{2gA^2} = \frac{0.4^2}{2 \times 9.81 \times \left(\frac{\pi}{4} \times 0.5^2\right)^2} = 0.2115 \text{ m,}$$

$$\text{Re} = \frac{4Q}{\pi D \nu} = \frac{4 \times 0.4}{\pi \times 0.5 \times 4.26 \times 10^{-7}} = 2.39 \times 10^6,$$

$$f = 1.325\left[\ln\left(0.27\frac{e}{D} + 5.74\,\text{Re}^{-0.9}\right)\right]^{-2}$$

$$= 1.325\left[\ln\left(0.27 \times \frac{1}{500} + 5.74(2.39 \times 10^{-6})^{-0.9}\right)\right]^{-2} = 0.0235.$$

$$\therefore H_P = z_B - z_A + \left(f\frac{L}{D} + \Sigma K\right)\frac{Q^2}{2gA^2}$$

$$= 220 - 100 + \left(\frac{0.0235 \times 5000}{0.5} + 2 + 2\right) \times 0.2115 = \underline{170.5\text{ m}};$$

$$\therefore \dot{W}_f = \gamma Q H_P = 0.81 \times 9810 \times 0.4 \times 170.5 = 542\ 000\text{ W}, \text{ or } \dot{W}_f = \underline{542\text{ kW}}.$$

b) Find the elevation at C when the pressure at that location is equal to the vapor pressure:

$$p_c = p_v - p_{\text{atm}} = 55.2 - 100 = 44.8\text{ kPa},$$

$$z_c = z_1 + H_P - \left(f\frac{L_{AC}}{D} + \Sigma K\right)\frac{Q^2}{2gA^2} - \frac{p_c}{\gamma} - \frac{Q^2}{2gA^2}$$

$$= 100 + 170.5 - \left(\frac{0.0235 \times 4000}{0.5} + 2\right) \times 0.2115 + \frac{44.8 \times 10^3}{0.81 \times 9810} - 0.2115$$

$$= \underline{235.7\text{ m}}$$

c)

Location	HGL (m)	Loss (m)
A	100.0	-170.5
d/s pump	270.5	$2 \times 0.2115 = 0.4$
d/s valve	270.1	$\frac{0.0235 \times 4000 \times 0.2115}{0.5} = 39.8$
C	230.3	$\frac{0.0235 \times 1000 \times 0.2115}{0.5} = 9.94$
u/s valve	220.4	$2 \times 0.2115 = 0.4$
B	220.0	

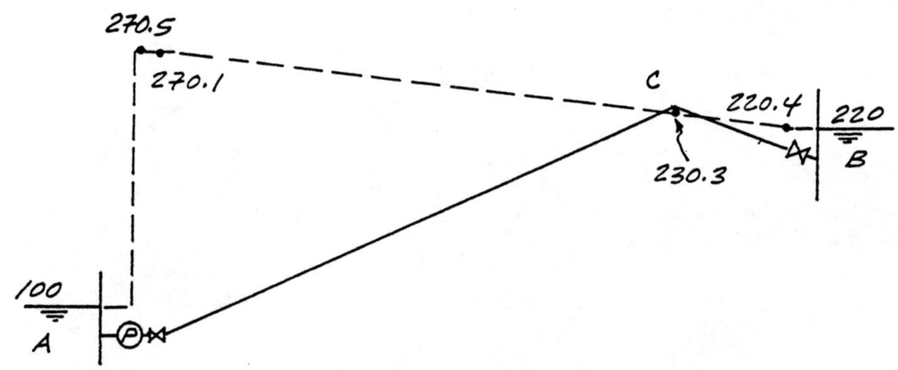

11.6 Energy eqn. from A to B:

$$\left(\frac{p}{\gamma}+z\right)_A-\left(\frac{p}{\gamma}+z\right)_B=(R_1+R_2+R_3)Q^{1.85}+\left(\frac{\Sigma K_1}{2gA_1^2}+\frac{\Sigma K_2}{2gA_2^2}\right)Q^2.$$

Use Hazen-Williams resistance formula for R:

$$R=\frac{10.59L}{C^{1.85}D^{4.87}}.$$

$$\therefore R_1=\frac{10.59\times 200}{100^{1.85}\times 0.2^{4.87}}=1071,$$

$$\frac{\Sigma K_1}{2gA_1^2}=\frac{2}{2\times 9.81\times(\pi/4)^2\times(0.2)^4}=103.3, \qquad R_2=\frac{10.59\times 150}{120^{1.85}\times 0.25^{4.87}}=193.4,$$

$$\frac{\Sigma K_2}{2gA_2^2}=\frac{3}{2\times 9.81\times(\pi/4)^2\times(0.3)^4}=63.5, \qquad R_3=\frac{10.59\times 300}{90^{1.85}\times 0.3^{4.87}}=271.1.$$

Substitute known data into energy eqn:

$$250-107=(1071+193+271)Q^{1.85}+(103.3+63.5)Q^2, \quad 143=1535Q^{1.85}+167Q^2,$$

$$F(Q)=1535Q^{1.85}+167Q^2-143, \qquad F'(Q)=2840Q^{0.85}+334Q$$

Iter.	$Q(\text{m}^3/\text{s})$	F	F'	$\Delta Q=-F/F'$
1	0.277*	12.60	1046	-0.01205
2	0.265	0.284	1007	-0.00028
3	0.265			

$$\therefore Q=\underline{0.265\ \text{m}^3/\text{s}}$$

$$^*Q\cong\left(\frac{143}{1535}\right)^{1/1.85} \quad \text{(1st iter.)}$$

11.6E See Pbm. 11.6 for development.

$$R_1=\frac{4.72\times 600}{100^{1.85}\times\left(\frac{8}{12}\right)^{4.87}}=4.071,$$

$$\frac{\Sigma K_1}{2gA_1^2}=\frac{2}{2\times 32.2\times(\pi/4)^2\times(8/12)^4}=0.255, \qquad R_2=\frac{4.72\times 300}{120^{1.85}\times\left(\frac{10}{12}\right)^{4.87}}=0.490,$$

$$\frac{\Sigma K_2}{2gA_2^2}=\frac{3}{2\times 32.2\times(\pi/4)^2\times(10/12)^4}=0.157, \qquad R_3=\frac{4.72\times 900}{90^{1.85}\times(1)^{4.87}}=1.030.$$

Substitute known data into energy eqn:

$$820 - 351 = (4.971 + 0.49 + 1.03)Q^{1.85} + (0.255 + 0.157)Q^2,$$
$$469 = 5.591Q^{1.85} + 0.412Q^2,$$

$F(Q) = 5.591Q^{1.85} + 0.412Q^2 - 469,\qquad F'(Q) = 10.343Q^{0.85} + 0.824Q$

Iter.	$Q(\text{ft}^3/\text{sec})$	F	F'	$\Delta Q = -F/F'$
1	10.96*	49.45	88.19	-0.561
2	10.40	1.161	84.27	-0.038
3	10.39	0.319	84.20	-0.0038
4	10.39			

$\therefore Q = \underline{10.39\ \text{ft}^3/\text{sec.}}$

$^{*}Q \cong \left(\dfrac{469}{5.591}\right)^{1/1.85}$ (1st iter.)

11.7 a) Let H_A = piezometric head at pipe inlet, and H_B = piezometric head at pipe outlet. Write energy eqn. across pipeline:

$$H_A + \frac{\dot{W}_{f_1}}{\gamma Q} + \frac{\dot{W}_{f_2}}{\gamma Q} + \frac{\dot{W}_{f_3}}{\gamma Q} = (\overline{R}_1 + \overline{R}_2 + \overline{R}_3)Q^2 + H_B,$$

$$H_A - H_B + \frac{1}{\gamma Q}\sum_{i=1}^{3}\dot{W}_{f_i} - Q^2\sum_{i=1}^{3}\overline{R}_i = 0.$$

But $H_A = H_B$, $\therefore$ solving for Q, $\underline{Q = [\Sigma\dot{W}_{f_i}/(\gamma\Sigma\overline{R}_i)]^{1/3}}$.

b) $\sum_{i=1}^{3}\dot{W}_{f_i} = 200\times10^3 + 200\times10^3 + 250\times10^3 = 650\times10^3,$

$\sum_{i=1}^{3}\overline{R}_i = 4\times10^4 + 3\times10^4 + 2\times10^5 = 2.7\times10^5,$

$\therefore Q = [650\times10^3/(8330\times2.7\times10^5)]^{1/3} = \underline{0.0663\ \text{m}^3/\text{s.}}$

11.8 Write the energy equation from the liquid surface (location 1) to the valve exit (location 2):

$$\frac{p_1}{\gamma} + z_1 + \frac{\dot{W}_f}{\gamma Q} = \left(\Sigma K + f\frac{L}{D} + 1\right)\frac{Q^2}{2gA^2} + z_2$$

a) Rearrange and substitute in known data:

$$\frac{P_1}{\gamma}+z_1-z_2+\frac{\dot{W}_f}{\gamma}\frac{1}{Q}-\left(\Sigma K+\frac{fL}{D}+1\right)\frac{Q^2}{2gA^2}=0$$

$$\frac{110\times10^3}{0.68\times9810}+24-18+\frac{1\times10^4}{0.68\times9810}\frac{1}{Q}$$

$$-\frac{(0.5+3\times0.26+2+0.015\times450/0.30+1)}{2\times9.81\times(0.7854\times0.30^2)^2}Q^2=0$$

$$22.4+\frac{1.50}{Q}-273.2Q^2=0$$

By trial and error, $\underline{Q\approx 0.32\ \text{m}^3/\text{s}}$.

b) Substitute in known data:

$$\frac{15\times144}{0.68\times62.4}+75-60+\frac{15\times550}{0.68\times62.4}\frac{1}{Q}$$

$$-\frac{(0.5+3\times0.26+2+0.015\times1500/0.667+1)}{2\times32.2\times(0.7854\times0.667^2)^2}Q^2=0$$

$$65.9+194.4\frac{1}{Q}-4.8Q^2=0$$

By trial and error, $\underline{Q\approx 4.7\ \text{ft}^3/\text{sec}}$.

11.9 Write energy eqn. between reservoirs:

$$H_P=\left(R_1+\frac{2}{2gA_1^2}+R_2+R_3+\frac{10}{2gA_3^2}\right)Q^2+50;$$

$$H_P=\frac{\dot{W}_P\eta}{\gamma Q}=\frac{1920\times10^3\times0.82}{9800\ Q}=\frac{160.7}{Q},$$

$$\frac{2}{2gA_1^2}=\frac{2}{2\times9.81\times(\pi/4)^2\times1.5^4}=0.03264,$$

$$\frac{10}{2gA_3^2}=\frac{10}{2\times9.81\times(\pi/4)^2\times1.2^4}=0.3985.$$

Compute the resistance coefficients assuming high Re-number flows:

$$R_1\cong1.07\frac{L_1}{gD_1^5}\left[\ln\left(0.27\frac{e_1}{D_1}\right)\right]^{-2}=1.07\frac{200}{9.81\times1.5^5}\left[\ln\left(0.27\times\frac{1}{1500}\right)\right]^{-2}=0.03864,$$

$$R_2=1.07\frac{300}{9.81\times1^5}\left[\ln\left(0.27\times\frac{1}{1000}\right)\right]^{-2}=0.4846,$$

$$R_3 = 1.07\frac{120}{9.81\times 1.2^5}\left[\ln\left(0.27\times\frac{1}{1200}\right)\right]^{-2} = 0.07455.$$

Substitute the data into the energy eqn:

$$\frac{160.7}{Q} = (0.03864 + 0.03264 + 0.4846 + 0.07455 + 0.3985)Q^2 + 50,$$

$$\frac{160}{Q} = 1.029Q^2 + 50$$

Solution using Newton's method:

$$F(Q) = 1.029Q^3 + 50Q - 160.7, \qquad F'(Q) = 3.087Q^2 + 50$$

Iter.	$Q(\mathrm{m}^3/\mathrm{s})$	F	F'	$\Delta Q = -F/F'$
1	2.0	-52.47	62.35	+0.842
2	2.842	5.020	74.93	-0.067
3	2.775	0.0389	73.77	-0.00053
4	2.774			

$\therefore Q \cong \underline{2.77\ \mathrm{m}^3/\mathrm{s}}.$

Recalculating the resistance coefficients based on the current value of Q yields $R_1 = 0.03916$, $R_2 = 0.4878$, $R_3 = 0.07522$. The new estimate of Q remains $2.77\ \mathrm{m}^3/\mathrm{s}$.

11.10 Initially work between locs. B and C:

$$Le = \frac{D\Sigma k}{f}, \quad \overline{R} = \frac{8f(L+Le)}{g\pi^2 D^5}$$

$$Le_1 = \frac{1.2\times 2}{0.015} = 160, \quad \overline{R}_1 = \frac{8\times 0.015\times 260}{9.81\times\pi^2\times 1.2^5} = 0.1295,$$

$$Le_2 = \frac{1\times 3}{0.02} = 150, \quad \overline{R}_2 = \frac{8\times 0.02\times 1150}{9.81\times\pi^2\times 1^5} = 1.900,$$

$$Le_3 = \frac{0.5\times 2}{0.018} = 56, \quad \overline{R}_3 = \frac{8\times 0.018\times 1556}{9.81\times\pi^2\times 0.5^5} = 74.05,$$

$$Le_4 = \frac{0.75\times 4}{0.021} = 143, \quad \overline{R}_4 = \frac{8\times 0.021\times 943}{9.81\times\pi^2\times 0.75^5} = 6.895.$$

$$\therefore W = \frac{Q^2}{\left(\sum_{i=2}^{4} \frac{1}{\sqrt{\bar{R}_i}}\right)^2} = \frac{3^2}{\left(\frac{1}{\sqrt{1.9}} + \frac{1}{\sqrt{74.05}} + \frac{1}{\sqrt{6.895}}\right)^2} = 6.022 \text{ m},$$

$$\therefore Q_2 = \sqrt{W / \bar{R}_2} = \sqrt{6.022 / 1.9} = \underline{1.780 \text{ m}^3 / \text{s}},$$

$$Q_3 = \sqrt{W / \bar{R}_3} = \sqrt{6.022 / 74.05} = \underline{0.285 \text{ m}^3 / \text{s}},$$

$$Q_4 = \sqrt{W / \bar{R}_4} = \sqrt{6.022 / 6.895} = \underline{0.935 \text{ m}^3 / \text{s}}.$$

Check continuity: $Q_2 + Q_3 + Q_4 = 1.78 + 0.285 + 0.935 = 3 = Q_1, \ \therefore$ ok.

To find $\dot{W}_P$, write energy eqn. from loc. A to B:

$$H_P = \bar{R}_1 Q_1^2 + \left(\frac{p}{\gamma} + z\right)_B = \bar{R}_1 Q_1^2 + W + 20,$$

$$\therefore H_P = 0.1295 \times 3^2 + 6.022 + 20 = \underline{27.2 \text{ m}}.$$

$$\therefore \dot{W}_P = \frac{\gamma Q_1 H_P}{\eta} = \frac{9810 \times 3 \times 27.2}{0.75} = 1.07 \times 10^6 \text{ W}, \quad \text{or } \dot{W}_P = \underline{1.07 \text{ MW}}.$$

11.11 a) Compute resistance coefficients:

$$Le_1 = D_1 \Sigma K / f_1 = 0.24 \times 2 / 0.02 = 24,$$

$$\bar{R}_1 = \frac{8 f_1 (L_1 + Le_1)}{g \pi^2 D_1^5} = \frac{8 \times 0.02 \times 54}{9.81 \times \pi^2 \times 0.24^5} = 112.1,$$

$$R_2 = \frac{8 f_2 L_2}{g \pi^2 D_2^5} = \frac{8 \times 0.015 \times 60}{9.81 \times \pi^2 \times 0.2^5} = 232.4,$$

$$R_3 = \frac{8 f_3 L_3}{g \pi^2 D_3^5} = \frac{8 \times 0.025 \times 90}{9.81 \times \pi^2 \times 0.16^5} = 1773.$$

Let H = hydraulic grade line at the junction. Write energy eqns. for each branch:

Line 1: $10 + H_P = \overline{R}_1 Q_1^2 + H$, or

$H = 10 + (20 - 30Q_1^2) - 112.1Q_1^2 = 30 - 142.1Q_1^2$.

Line 2: $H = R_2 Q_2^2 + 20$, or $Q_2 = \sqrt{(H - 20) / 232.4}$.

Line 3: $H = R_3 Q_3^2 + 18$, or $Q_3 = \sqrt{(H - 18) / 1773}$.

Continuity at junction: $\Sigma Q = Q_1 - Q_2 - Q_3 = 0$.
Tabulate solution (assume Q_1 for each iteration, then solve for H, Q_2, Q_3 and ΣQ):

Q_1 (m^3/s)	H (m)	Q_2 (m^3/s)	Q_3 (m^3/s)	ΣQ (m^3/s)
0.15	26.80	0.1711	0.0705	-0.0916
0.20	24.32	0.363	0.0597	+0.004

$\therefore Q_1 \cong \underline{0.20 \text{ m}^3/\text{s}}$, $Q_2 \cong \underline{0.14 \text{ m}^3/\text{s}}$, $Q_3 \cong \underline{0.06 \text{ m}^3/\text{s}}$.

b) $Le_1 = 10/12 \times 2/0.02 = 83.3$, $\overline{R}_1 = \dfrac{8 \times 0.02 \times 183.3}{32.2 \times \pi^2 \times (10/12)^5} = 0.2296$,

$R_2 = \dfrac{8 \times 0.015 \times 200}{32.2 \times \pi^2 \times (8/12)^5} = 0.5735$, $R_3 = \dfrac{8 \times 0.025 \times 300}{32.2 \times \pi^2 \times (6/12)^5} = 6.042$.

Line 1: $H = 20 + (55 - 0.1Q_1^2) - 0.2296Q_1^2$

$= 75 - 0.33Q_1^2$,

Line 2: $Q_2 = \sqrt{(H - 50)/0.5735}$,

Line 3: $Q_3 = \sqrt{(H - 45)/6.042}$.

Q_1 (ft^3/sec)	H (ft)	Q_2 (ft^3/sec)	Q_3 (ft^3/sec)	ΣQ (ft^3/sec)
6	63.12	4.783	1.732	-0.5148
6.5	61.06	4.391	1.630	+0.4788
6.26*	62.07	4.587	1.681	-0.008

*Estimated by linear interpolation between $Q = 6$ and $Q = 6.5$.

$\therefore Q_1 \cong \underline{6.26 \text{ ft}^3/\text{sec}}$, $Q_2 \cong \underline{4.59 \text{ ft}^3/\text{sec}}$, $Q_3 \cong \underline{1.68 \text{ ft}^3/\text{sec}}$.

11.12 Use SI data from Pbm. 11.11 (a).

$$\Sigma Q = Q_1 - Q_2 - Q_3 = 0, \text{ or } \qquad w(H) = \sqrt{\frac{30-H}{142.1}} - \sqrt{\frac{H-20}{232.4}} - \sqrt{\frac{H-18}{1773}}.$$

Use false position method of solution: $H_r = \dfrac{H_\ell w(H_u) - H_u w(H_\ell)}{w(H_u) - w(H_\ell)}$

Iter	H_ℓ	H_u	$w(H_\ell)$	$w(H_u)$	H_r	$w(H_r)$	Sign $w(H_\ell)w(H_r)$	ε
1	25	20	-0.02193	0.2317	24.57	-0.00562	+	—
2	24.57	20	-0.00562	0.2317	24.46	-0.00144	+	0.0045
3	24.46	20	-0.00144	0.2317	24.43	-0.00030	+	0.0012

$$\therefore Q_1 = \sqrt{(30-24.43)/142.1} = \underline{0.198 \text{ m}^3/\text{s}},$$

$$Q_2 = \sqrt{(24.43-20)/232.4} = \underline{0.138 \text{ m}^3/\text{s}},$$

$$Q_3 = \sqrt{(24.43-18)/1773} = \underline{0.060 \text{ m}^3/\text{s}}.$$

11.13 To compute a constant friction factor, use $f = 1.325\left[\ln\left(0.27\dfrac{e}{D}\right)\right]^{-2}$. Equivalent length and resistance coefficient are given by $Le = \Sigma kD/f$, and $\overline{R} = 8f(L+Le)/(g\pi^2 D^5)$. Finally the discharge in each pipe is $Q = \sqrt{W/\overline{R}}$, in which $W = (p/\gamma + z)_A - (p/\gamma + z)_B$.

Pipe 1: $f = 1.325\left[\ln\left(0.27 \times \dfrac{0.1}{1000}\right)\right]^{-2} = 0.012, \quad Le = 2 \times 1/0.012 = 167,$

$\overline{R} = 8 \times 0.012 \times 767/(9.81 \times \pi^2 \times 1^5) = 0.7605,$

$\therefore Q_1 = \sqrt{50/0.7605} = \underline{8.81 \text{ m}^3/\text{s}}.$

Pipe 2: $f = 1.325\left[\ln\left(0.27 \times \dfrac{0.15}{1200}\right)\right]^{-2} = 0.0125, \quad Le = 0$

$\overline{R} = 8 \times 0.0125 \times 1000/(9.81 \times \pi^2 \times 1.2^5) = 0.4151,$

$\therefore Q_2 = \sqrt{50/0.4151} = \underline{10.98 \text{ m}^3/\text{s}}.$

Pipe 3: $f = 1.325\left[\ln\left(0.27 \times \frac{0.2}{850}\right)\right]^{-2} = 0.0142, \quad Le = 4 \times 0.85 / 0.0142 = 239,$

$\overline{R} = 8 \times 0.0142 \times 789 / (9.81 \times \pi^2 \times 0.85^5) = 2.086,$

$\therefore Q_3 = \sqrt{50 / 2.086} = \underline{4.90\ \mathrm{m}^3/\mathrm{s}}.$

Pipe 4: $f = 1.325\left[\ln\left(0.27 \times \frac{0.1}{1000}\right)\right]^{-2} = 0.012, \quad Le = 1 \times 1 / 0.012 = 83,$

$\overline{R} = 8 \times 0.012 \times 883 / (9.81 \times \pi^2 \times 1^5) = 0.8755,$

$\therefore Q_4 = \sqrt{50 / 0.8755} = \underline{7.56\ \mathrm{m}^3/\mathrm{s}}.$

The total discharge through the piping is $Q = Q_1 + Q_2 + Q_3 + Q_4 = \underline{32.25\ \mathrm{m}^3/\mathrm{s}}.$

11.14 Continuity at junction: $Q_1 - (Q_2 + Q_3 + Q_4) = 0.$

Energy eqn. across pipe 1: $\frac{p_0}{\gamma} = \overline{R}_1 Q_1^2 + H$, or $Q_1 = \sqrt{(p_0/\gamma - H)/\overline{R}_1}$

in which H = piezometric head at junction.

For each branch (i= 2, 3, 4): $H = \overline{R}_i Q_1^2$, or $Q_i = \sqrt{H / \overline{R}_i}$.

Substitute into continuity eqn: $\left[\frac{p_0/\gamma - H}{\overline{R}_1}\right]^{1/2} = \sum_{i=2}^{4}\left[\frac{H}{\overline{R}_i}\right]^{1/2} = \sqrt{H}\sum_{i=2}^{4}\frac{1}{\sqrt{\overline{R}_i}}.$

$\therefore$ Solving for H,

$$H = \frac{p_0 \gamma}{1 + R_1\left[\sum \frac{1}{\sqrt{R_i}}\right]^2} = \frac{300 \times 10^3 / 9810}{1 + 1.6 \times 10^4\left[\frac{1}{\sqrt{5.3 \times 10^5}} + \frac{1}{\sqrt{1 \times 10^6}} + \frac{1}{\sqrt{1.8 \times 10^6}}\right]^2}$$

$$= \frac{30.58}{1.556} = \underline{26.46\ \mathrm{m}}.$$

$\therefore Q_1 = \sqrt{(30.58 - 26.46) / 1.6 \times 10^4} = \underline{0.01605\ \mathrm{m}^3/\mathrm{s}},$

$Q_2 = \sqrt{26.46 / 5.3 \times 10^5} = \underline{0.00707\ \mathrm{m}^3/\mathrm{s}}, \quad Q_3 = \sqrt{26.46 / 1 \times 10^6} = \underline{0.00514\ \mathrm{m}^3/\mathrm{s}},$

$Q_4 = \sqrt{26.46 / 1.8 \times 10^6} = \underline{0.00383\ \mathrm{m}^3/\mathrm{s}}.$

11.15 $W = \frac{fL}{D}\frac{V^2}{2g} \propto \frac{fV^2}{D}$, and since W is constant for all pipes, $V \propto \sqrt{\frac{D}{fL}}$.

$$V_1 \propto \sqrt{\frac{450}{0.012 \times 2000}} = 4.33, \quad V_2 \propto \sqrt{\frac{150}{0.02 \times 650}} = 3.40, \quad V_3 \propto \sqrt{\frac{300}{0.015 \times 1650}} = 3.48$$

Hence, pipe 1 has the largest velocity.

11.16 a) Write energy eqn. across system: $H_P = a_0 - a_1 Q_1^2 = \overline{R}_1 Q_1^2 + \left[\sum_{i=2}^{4} \frac{1}{\sqrt{\overline{R}_i}}\right]^{-2} Q_1^2$,

with $Q_i = \sqrt{H / \overline{R}_i}$, $i = 2, 3, 4$, and $H = H_P - \overline{R}_1 Q_1^2$, in which H = piezometric head at the junction. The unknowns are Q_1, Q_2, Q_3, Q_4 and H. The system can be treated as a parallel piping system because the hydraulic grade line at the exit of pipes 2, 3 and 4 are the same.

b) As opposed to a branching system, a parallel system requires no trial-and-error solution.

c) Substitute known data into energy eqn., and solve for Q_1:

$$a_0 = \left\{a_1 + \overline{R}_1 + \left[\sum \frac{1}{\sqrt{\overline{R}_i}}\right]^{-2}\right\} Q_1^2,$$

$$\therefore 45 = \left\{10^4 + 34\,650 + \left[\frac{1}{\sqrt{82\,500}} + \frac{1}{\sqrt{127\,900}} + \frac{1}{\sqrt{115\,500}}\right]^{-2}\right\} Q_1^2 = 56\,410 Q_1^2,$$

$$\therefore Q_1 = \sqrt{45 / 56\,410} = \underline{0.0282 \text{ m}^3/\text{s}}.$$

$$\therefore H = 45 - (10^4 + 34\,650) \times 0.0282^2 = \underline{9.49 \text{ m}},$$

$$Q_2 = \sqrt{9.49 / 82\,500} = \underline{0.0107 \text{ m}^3/\text{s}}, \qquad Q_3 = \sqrt{9.49 / 127\,900} = \underline{0.0086 \text{ m}^3/\text{s}},$$

$$Q_4 = \sqrt{9.49 / 115\,500} = \underline{0.0091 \text{ m}^3/\text{s}}.$$

d) By increasing the diameter of pipe 1 by approximately 50%,

$\overline{R}_1 = 34\,650 / 1.5^5 = 4563$, $Q_1 \cong 0.041 \text{ m}^3/\text{s}$, and $H \cong 20.5$ m.

11.17 Let H = piezometric head at the junction. Write energy eqn. for line 2 to find H: $H = z_2 + \overline{R}_2 Q_2^2 = 15 + 2000 \times 0.035^2 = 17.45$ m. The discharges Q_3 and Q_4 can now be computed directly:

$$Q_3 = \sqrt{\frac{H - z_3}{\overline{R}_3}} = \sqrt{\frac{17.45 - 12}{1500}} = 0.0603 \text{ m}^3 / \text{s},$$

$$Q_4 = \sqrt{\frac{H - z_4}{\overline{R}_4}} = \sqrt{\frac{17.45 - 10}{1000}} = 0.0863 \text{ m}^3 / \text{s}.$$

$$\therefore Q_1 = \sum_{i=2}^{4} Q_i = 0.035 + 0.0603 + 0.0863 = \underline{0.182 \text{ m}^3 / \text{s}},$$

and $H_P = H + \overline{R}_1 Q_1^2 - z_1 = 17.45 + 1400 \times 0.182^2 - 3 = \underline{60.82 \text{ m}}$.

11.18 For each pipe compute equivalent length and resistance coefficient:

Pipe 1: $Le = D\Sigma K / f = 0.05 \times 3 / 0.02 = 7.5,$

$$\overline{R}_1 = \frac{8f(L + Le)}{g\pi^2 D^5} = \frac{8 \times 0.02 \times 37.5}{9.81 \times \pi^2 \times 0.05^5} = 1.983 \times 10^5,$$

Pipe 2: $Le = 0.075 \times 5 / 0.025 = 15, \quad \overline{R}_2 = \dfrac{8 \times 0.025 \times 55}{9.81 \times \pi^2 \times 0.075^5} = 4.788 \times 10^4,$

Pipe 3: $Le = 0.06 \times 1 / 0.022 = 2.7, \quad \overline{R}_3 = \dfrac{8 \times 0.022 \times 62.7}{9.81 \times \pi^2 \times 0.06^5} = 1.466 \times 10^5.$

$$W = \left[\frac{Q}{\sum_{i=1}^{3} (\overline{R}_i)^{-1/2}} \right]^2 = \frac{(0.6/60)^2}{[(1.983 \times 10^5)^{-1/2} + (4.788 \times 10^4)^{-1/2} + (1.466 \times 10^5)^{-1/2}]^2} = 1.125 \text{ m}.$$

$$\therefore Q_1 = \sqrt{W / \overline{R}_1} = \sqrt{1.125 / 1.983 \times 10^5} = \underline{0.00238 \text{ m}^3 / \text{s}, \text{ or } 143 \text{ L/min}}$$

$$Q_2 = \sqrt{W / \overline{R}_2} = \sqrt{1.125 / 4.788 \times 10^4} = \underline{0.00485 \text{ m}^3 / \text{s}, \text{ or } 290 \text{ L/min}}$$

$$Q_3 = \sqrt{W / \overline{R}_3} = \sqrt{1.125 / 1.466 \times 10^5} = \underline{0.00277 \text{ m}^3 / \text{s}, \text{ or } 166 \text{ L/min}}$$

11.18E See Pbm. 11.19 for development.

Pipe 1: $Le = \frac{2}{12} \times 3/0.02 = 25, \quad \overline{R}_1 = \frac{8 \times 0.02 \times 115}{32.2 \times \pi^2 \times (2/12)^5} = 450.2,$

Pipe 2: $Le = \frac{3}{12} \times 5/0.025 = 50, \quad \overline{R}_2 = \frac{8 \times 0.025 \times 170}{32.2 \times \pi^2 \times (3/12)^5} = 109.6,$

Pipe 3: $Le = \frac{2.5}{12} \times 1/0.022 = 9.5, \quad \overline{R}_3 = \frac{8 \times 0.022 \times 189.5}{32.2 \times \pi^2 \times (2.5/12)^5} = 267.4.$

$$\therefore W = \frac{0.35^2}{[450.2^{-/2} + 109.6^{-1/2} + 267.4^{-1/2}]^2} = 2.949 \text{ ft},$$

$Q_1 = \sqrt{2.949/450.2} = \underline{0.081 \text{ ft}^3/\text{sec}},$

$Q_2 = \sqrt{2.949/109.6} = \underline{0.164 \text{ ft}^3/\text{sec}},$

$Q_3 = \sqrt{2.949/267.4} = \underline{0.105 \text{ ft}^3/\text{sec}}.$

11.19 Write energy eqn. across system, excluding pipe 3 with loc. A at lower reservoir, and loc. B at upper reservoir:

$z_A + H_P = \overline{R}_1 Q_1^2 + \overline{R}_2 Q_2^2 + z_B.$

But $\overline{R}_2 Q_2^2 = \overline{R}_3 Q_3^2$, so that $Q_3 = (\overline{R}_2/\overline{R}_3)^{1/2} Q_2$.

Write continuity eqn. at junction: $Q_1 = Q_2 + Q_3 = Q_2[1 + (\overline{R}_2/\overline{R}_3)^{1/2}]$, or

$Q_2 = Q_1[1 + (\overline{R}_2/\overline{R}_3)^{1/2}]^{-1}$. Substitute this expression along with

$H_P = 150 - 5Q_1^2$ into energy eqn. and solve for Q_1:

$z_A + 150 - 5Q_1^2 = \overline{R}_1 Q_1^2 + \overline{R}_2 Q_1^2 [1 + (\overline{R}_2/\overline{R}_3)^{1/2}]^{-2} + z_B,$

$$10 + 150 - 5Q_1^2 = \left\{ 400 + 1000 \left[1 + \left(\frac{1000}{1500} \right)^{1/2} \right]^{-2} \right\} Q_1^2 + 40,$$

which reduces to $120 = 708.1Q_1^2$,

$$\therefore Q_1 = \sqrt{120 / 708.1} = \underline{0.412 \text{ m}^3 / \text{s}}, \qquad Q_2 = 0.412\left[1 + \left(\frac{1000}{1500}\right)^{1/2}\right]^{-1} = \underline{0.227 \text{ m}^3 / \text{s}},$$

$$Q_3 = \left(\frac{1000}{1500}\right)^{1/2} \times 0.227 = \underline{0.185 \text{ m}^3 / \text{s}}.$$

$$H_P = 150 - 5 \times 0.412^2 = 149 \text{ m},$$

$$\therefore \dot{W}_P = \frac{\gamma Q H_P}{\eta} = \frac{9810 \times 0.412 \times 149}{0.75} = 8.03 \times 10^5 \text{ W}, \text{ or } \dot{W}_P = \underline{803 \text{ kW}}$$

11.20 Write energy eqn. from junction at A to suction side of pump at B:

$$H_{P_A} = \left(R_1 + \frac{\Sigma K}{2gA_1^2}\right)Q^2 + \frac{p_B}{\gamma} + z_B, \quad \text{where } H_{P_A} = \text{pump head and } p_B = \text{pressure.}$$

The only unknown in the relation is Q, since

$$H_{P_A} = \frac{\dot{W}_{P_A}\eta}{\gamma Q} = \frac{1 \times 10^6 \times 0.76}{0.81 \times 9810 \times Q} = \frac{95.6}{Q}.$$

Evaluate the constants:

$$R_1 = \frac{8 f_1 L_1}{g\pi^2 D_1^5} = \frac{8 \times 0.23 \times 5000}{9.81 \times \pi^2 \times 0.75^5} = 40.0, \quad \frac{\Sigma K}{2gA_1^2} = \frac{2}{2 \times 9.81 \times (\pi/4)^2 \times 0.75^4} = 0.52,$$

$$\frac{p_B}{\gamma} + z_B = \frac{150 \times 10^3}{0.81 \times 9810} + 27 = 45.9 \text{ m}.$$

Substituting into the energy eqn., $\dfrac{95.6}{Q} = 40.5Q^2 + 45.9$. $\therefore$ Solving using Newton's method, $Q = \underline{1.053 \text{ m}^3 / \text{s}}$, and $H_{P_A} = 95.6 / 1.053 = \underline{90.8 \text{ m}}$.

The second reach is now analyzed. Write energy eqn. from suction side of pump at B to downstream reservoir:

$$\frac{p_B}{\gamma} + z_B + H_{P_B} = \left(R_2 + \frac{\Sigma K}{2gA_2^2}\right)Q^2 + 50.$$

Evaluate the constants:

$$R_2 = \frac{8f_2L_2}{g\pi^2D_2^5} = \frac{8\times 0.023\times 7500}{9.81\times \pi^2\times 0.75^5} = 60.1, \quad \frac{\Sigma K}{2gA_1^2} = \frac{10}{2\times 9.81\times(\pi/4)^2\times 0.75^4} = 2.61,$$

$$\therefore 45.9 + H_{P_B} = (60.1+2.6)\times 1.053^2 + 50, \text{ or } H_{P_B} = \underline{73.6 \text{ m}}.$$

$$\therefore \dot{W}_{P_B} = \frac{\gamma Q H_{P_B}}{\eta} = \frac{0.81\times 9810\times 1.053\times 73.6}{0.76} = 8.10\times 10^5 \text{ W}, \text{ or } \dot{W}_{P_B} = \underline{810 \text{ kW}}.$$

11.21 Write energy equation from loc. A to loc. D:

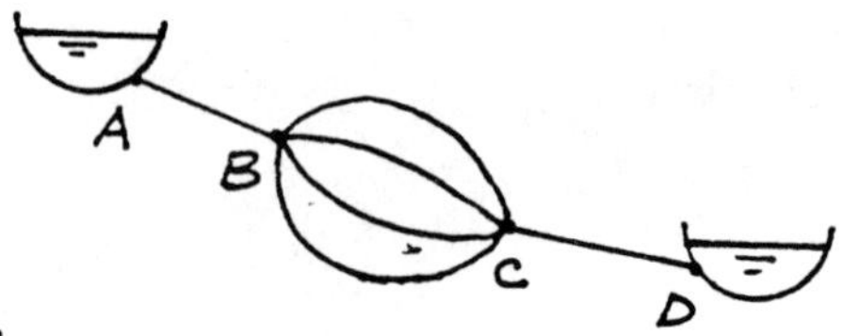

$$\left(\frac{p}{\gamma}+z\right)_A = W_{AB} + W_{BC} + W_{CD} + \left(\frac{p}{\gamma}+z\right)_D,$$

$$W_{AB} = \overline{R}_1 Q_1^2,\ W_{BC} = \left(\sum_{i=2}^{5}\frac{1}{\sqrt{\overline{R}_i}}\right)^{-2} Q_1^2,\ W_{CD} = \overline{R}_6 Q_1^2.$$

Since the friction factors are unknown, an iterative solution is required. For the first iteration assume flow in the fully rough zone, so that $f = 1.325\left[\ln 0.27\frac{e}{D}\right]^{-2}$.

Then the equivalent lengths and resistance coefficients are tabulated:

Pipe	$\frac{e}{D}$	f	$Le = \frac{D\Sigma K}{f}$	$\overline{R} = \frac{8f(L+Le)}{g\pi^2D^5}$
1	0.0005	0.0167	36	4.47×10^3
2	0.004	0.0284	0	4.81×10^7
3	0.004	0.0284	0	6.01×10^7
4	0.0033	0.0269	2	3.13×10^7
5	0.0025	0.0249	0	8.44×10^6
6	0.00057	0.0172	51	4.77×10^3

$$\left(\sum_{i=2}^{5}\frac{1}{\sqrt{\overline{R}_i}}\right)^{-2} = 1.578\times 10^6, \quad \therefore \text{energy eqn. becomes}$$

$$(4.47\times 10^3 + 1.578\times 10^6 + 4.77\times 10^3)Q_1^2 = 70-10,$$

$$\therefore Q_1^2 = 3.78\times 10^{-5}, \text{ and } Q_1 = 0.00615 \text{ m}^3/\text{s}.$$

$$W_{BC} = 1.578\times 10^6(3.78\times 10^{-5}) = 59.6 \text{ m},$$

$$Q_2 = \sqrt{W_{BC} / \overline{R}_2} = \sqrt{59.65 / 4.81 \times 10^7} = 0.00111 \text{ m}^3 / \text{s},$$
$$Q_3 = \sqrt{W_{BC} / \overline{R}_3} = \sqrt{59.65 / 6.01 \times 10^7} = 0.0010 \text{ m}^3 / \text{s},$$
$$Q_4 = \sqrt{W_{BC} / \overline{R}_4} = \sqrt{59.65 / 3.13 \times 10^7} = 0.00138 \text{ m}^3 / \text{s},$$
$$Q_5 = \sqrt{W_{BC} / \overline{R}_5} = \sqrt{59.65 / 8.44 \times 10^6} = 0.00266 \text{ m}^3 / \text{s}.$$

For the second iteration, assume water temperature of 20°C, so that $\nu = 1 \times 10^{-6}$ m²/s.

$$\text{Re} = \frac{QD}{A\nu} = \frac{4Q}{\pi D \nu} = 1.27 \times 10^6 \frac{Q}{D}, \text{ and } f = 1.325\left[\ln\left(0.27\frac{e}{D} + 5.74\,\text{Re}^{-0.9}\right)\right]^{-2},$$

and the updated resistance coefficients are tabulated:

Pipe	Re	f	Le	$\overline{R}$
1	3.90×10^4	0.0236	25	6.25×10^3
2	5.64×10^4	0.0306	0	5.18×10^7
3	5.08×10^4	0.0308	0	6.52×10^7
4	5.84×10^4	0.0292	2	3.40×10^7
5	8.45×10^4	0.0268	0	9.08×10^6
6	4.46×10^4	0.0233	38	6.31×10^3

$$\left(\sum_{i=2}^{5} \frac{1}{\sqrt{\overline{R}_i}}\right)^{-2} = 1.704 \times 10^6, \quad \therefore \text{energy eqn. becomes}$$

$$(6.25 \times 10^3 + 1.704 \times 10^6 + 6.31 \times 10^3)Q_1^2 = 60,$$

$$\therefore Q_1^2 = 3.494 \times 10^{-5}, \text{ and } Q_1 = 0.00591 \text{ m}^3/\text{s}.$$

$$W_{BC} = 1.704 \times 10^6 (3.494 \times 10^{-5}) = 59.54 \text{ m},$$

$$Q_2 = \sqrt{59.54 / 5.18 \times 10^7} = 0.00107 \text{ m}^3 / \text{s},$$
$$Q_3 = \sqrt{59.54 / 6.12 \times 10^7} = 0.00095 \text{ m}^3 / \text{s},$$
$$Q_4 = \sqrt{59.54 / 3.4 \times 10^7} = 0.00132 \text{ m}^3 / \text{s},$$
$$Q_5 = \sqrt{59.54 / 9.08 \times 10^6} = 0.00256 \text{ m}^3 / \text{s}.$$

Calculation of the friction factors with the current estimates of the discharges reveals that they change only in the third significant figure, hence new estimates of Q_i will not change significantly.

$\therefore Q_1 = Q_6 \cong \underline{5.91 \text{ L / s}}$, $Q_2 \cong \underline{1.07 \text{ L / s}}$,

$Q_3 \cong \underline{0.95 \text{ L / s}}$, $Q_4 \cong \underline{1.32 \text{ L / s}}$, $Q_5 \cong \underline{2.56 \text{ L / s}}$.

11.22 Compute resistance coefficients:

$$R_1 = \frac{10.59\ L}{C^{1.85}D^{4.87}} = \frac{10.59 \times 200}{130^{1.85} \times 0.5^{4.87}} = 7.6, \qquad R_2 = \frac{10.59 \times 600}{130^{1.85} \times 0.3^{4.87}} = 274.6,$$

$$R_3 = \frac{10.59 \times 1500}{130^{1.85} \times 0.3^{4.87}} = 686.5, \qquad R_4 = \frac{10.59 \times 1500}{130^{1.85} \times 0.4^{4.87}} = 169.1.$$

a) Let H_J = piezometric head at junction J. Assume $Q_1 = 2\ \text{m}^3/\text{s}$ (out of J).

Then $H_J = 30 + R_1 Q_1^2 = 30 + 7.6 \times 2^2 = 60.4$ m,

$Q_2 = \sqrt{(250 - 60.4)/274.6} = 0.831\ \text{m}^3/\text{s}$ (into J),

$Q_3 = \sqrt{(300 - 60.4)/686.5} = 0.591\ \text{m}^3/\text{s}$ (into J),

$Q_4 = \sqrt{(200 - 60.4)/169.1} = 0.909\ \text{m}^3/\text{s}$ (into J),

$\therefore \Delta Q = -2 + 0.831 + 0.591 + 0.909 = +0.331.$

Assume $Q_1 = 2.5\ \text{m}^3/\text{s}$. Then: $H_J = 30 + 7.6 \times 2.5^2 = 77.5$ m,

$Q_2 = \sqrt{(250 - 77.5)/274.6} = 0.793\ \text{m}^3/\text{s}$,

$Q_3 = \sqrt{(300 - 77.5)/686.5} = 0.569\ \text{m}^3/\text{s}$,

$Q_4 = \sqrt{(200 - 77.5)/169.1} = 0.851\ \text{m}^3/\text{s}$,

$\therefore \Delta Q = -2.5 + 0.793 + 0.569 + 0.851 = -0.287.$

Linear interpolation to find the next estimate of Q_1 yields

$$\frac{-0.287 - 0.331}{2.5 - 2} = \frac{-0.287}{2.5 - Q_1}, \qquad \therefore Q_1 = \underline{2.27\ \text{m}^3/\text{s}}.$$

Then: $H_J = 30 + 7.6 \times 2.27^2 = \underline{69.2\ \text{m}}$,

$Q_2 = \sqrt{(250 - 69.2)/274.6} = \underline{0.811\ \text{m}^3/\text{s}}$,

$Q_3 = \sqrt{(300 - 69.2)/686.5} = \underline{0.580\ \text{m}^3/\text{s}}$,

$Q_4 = \sqrt{(200 - 69.2)/169.1} = \underline{0.880\ \text{m}^3/\text{s}}$,

$\therefore \Delta Q = -2.27 + 0.811 + 0.580 + 0.880 = -0.001.$ $\therefore$ Ok.

b) Assume $Q_1 = 1\ \text{m}^3/\text{s}$ (into J). Then:

$H_J = 30 + (250 - 0.4Q_1 - 0.1Q_1^2) - R_1 Q_1^2 = 280 - 0.4 \times 1 - 7.7 \times 1^2 = 271.9$ m,

$Q_2 = \sqrt{(279.1 - 250) / 274.6} = 0.282 \text{ m}^3 / \text{s (out of J)},$
$Q_3 = \sqrt{(300 - 279.1) / 686.5} = 0.202 \text{ m}^3 / \text{s (into J)},$
$Q_4 = \sqrt{(279.1 - 200) / 169.1} = 0.652 \text{ m}^3 / \text{s (out of J)},$
$\therefore \Delta Q = 1 - 0.282 + 0.202 - 0.652 = +0.268.$

Assume $Q_1 = 0.5 \text{ m}^3 / \text{s}$ (into J). Then:
$H_J = 280 - 0.4 \times 0.5 - 7.7 \times 0.5^2 = 277.9 \text{ m},$
$Q_2 = \sqrt{(277.9 - 250) / 274.6} = 0.319 \text{ m}^3 / \text{s},$
$Q_3 = \sqrt{(300 - 277.9) / 686.5} = 0.179 \text{ m}^3 / \text{s},$
$Q_4 = \sqrt{(279.9 - 250) / 169.1} = 0.679 \text{ m}^3 / \text{s},$
$\therefore \Delta Q = 0.5 - 0.319 + 0.179 - 0.679 = -0.319.$

Use linear interpolation to find the new estimate of Q_1:

$$\frac{0.268 - (-0.319)}{1 - 0.5} = \frac{0.268}{1 - Q_1}, \qquad \therefore Q_1 = \underline{0.772 \text{ m}^3 / \text{s}}.$$

Then: $H_J = 280 - 0.4 \times 0.772 - 7.7 \times 0.772^2 = \underline{275.1 \text{ m}},$
$Q_2 = \sqrt{(275.1 - 250) / 274.6} = \underline{0.302 \text{ m}^3 / \text{s}},$
$Q_3 = \sqrt{(300 - 275.1) / 686.5} = \underline{0.190 \text{ m}^3 / \text{s}},$
$Q_4 = \sqrt{(275.1 - 200) / 169.1} = \underline{0.666 \text{ m}^3 / \text{s}},$
$\therefore \Delta Q = 0.772 - 0.302 - 0.19 - 0.666 = -0.006. \quad \therefore$ ok.

11.23 Compute equivalent lengths and resistance coefficients using

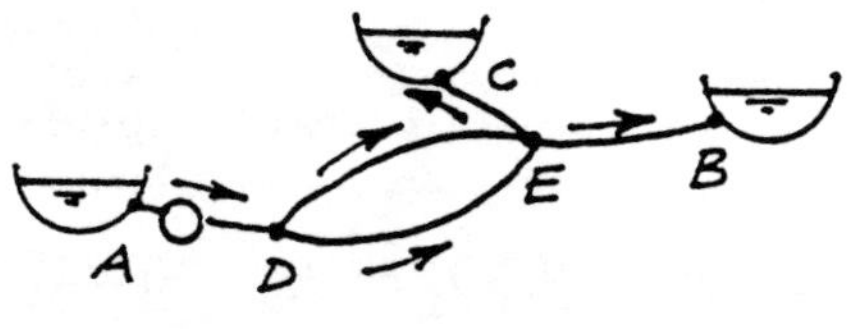

$$Le = \frac{D\Sigma K}{f} \text{ and } \overline{R} = \frac{8f(L + Le)}{g\pi^2 D^5}:$$

$$Le_1 = \frac{0.35 \times 2}{0.02} = 35, \quad \overline{R}_1 = \frac{8 \times 0.02 \times 135}{9.81 \times \pi^2 \times 0.35^5} = 42.5,$$

$$Le_2 = 0, \quad \overline{R}_2 = \frac{8 \times 0.02 \times 750}{9.81 \times \pi^2 \times 0.2^5} = 3873,$$

$$Le_3 = 0, \quad \overline{R}_3 = \frac{8 \times 0.02 \times 850}{9.81 \times \pi^2 \times 0.2^5} = 4390,$$

$$Le_4 = \frac{0.20 \times 2}{0.02} = 20, \quad \overline{R}_4 = \frac{8 \times 0.02 \times 520}{9.81 \times \pi^2 \times 0.2^5} = 2685,$$

$$Le_5 = \frac{0.25 \times 2}{0.02} = 25, \quad \overline{R}_5 = \frac{8 \times 0.02 \times 375}{9.81 \times \pi^2 \times 0.2^5} = 635.$$

Write energy eqn. from A to E:

$$\left(\frac{p}{\gamma}+z\right)_A + H_P = \overline{R}_1 Q_1^2 + \left(\frac{1}{\sqrt{\overline{R}_2}}+\frac{1}{\sqrt{\overline{R}_3}}\right)^{-2} Q_1^2 + H_E,$$

in which H_E = hyd. gr. line at loc. E. The following strategy is employed: Assume Q_1, compute H_E, then compute Q_4 and Q_5 using

$$Q_4 = \sqrt{\frac{H_E - H_C}{\overline{R}_4}} \text{ and } Q_5 = \sqrt{\frac{H_E - H_B}{\overline{R}_5}}, \text{ where } H_C = \left(\frac{p}{\gamma}+z\right)_C,\ H_B = \left(\frac{p}{\gamma}+z\right)_B.$$

Finally check continuity at E with the relation $\Delta Q = Q_1 - Q_4 - Q_5$.

Substitute known data into the above relations:

$$H_E = 0 + 60 - 10Q_1^2 - 42.5Q_1^2 - \left(\frac{1}{\sqrt{3873}}+\frac{1}{\sqrt{4390}}\right)^{-2} Q_1^2 = 60 - 1083Q_1^2,$$

$$Q_4 = \sqrt{(H_E - 50)/2685},\ Q_5 = \sqrt{(H_E - 48)/635}.$$

The results are tabulated below. The third estimate of Q_1 is found by interpolation.

Q_1 (m^3/s)	H_E (m)	Q_4 (m^3/s)	Q_5 (m^3/s)	ΔQ (m^3/s)
0.1	49.17	-0.01758	+0.04292	+0.07466
0.09	51.23	+0.02140	+0.07132	-0.00272
∴0.0904	51.15	+0.02070	+0.07043	-0.00073

Compute hydraulic grade line at D:

$$H_D = H_P - R_1 Q_1^2 = 60 - (10 + 42.5) \times 0.0904^2 = 59.57 \text{ m}.$$

$$\therefore Q_2 = \sqrt{\frac{H_D - H_E}{\overline{R}_2}} = \sqrt{\frac{59.57 - 51.15}{3873}} = 0.04663 \text{ m}^3/\text{s},$$

$$Q_3 = \sqrt{\frac{H_D - H_E}{\overline{R}_3}} = \sqrt{\frac{59.57 - 51.15}{4390}} = 0.04379 \text{ m}^3/\text{s}.$$

$\therefore Q_1 \cong \underline{90 \text{ L/s}},\ Q_2 \cong \underline{46 \text{ L/s}},$

$Q_3 \cong \underline{44 \text{ L/s}},\ Q_4 \cong \underline{20 \text{ L/s}},$ and $Q_5 \cong \underline{70 \text{ L/s}}.$

11.24 Compute the R-values for each pipe:

$$\text{Pipe 1:}\ \ Le = \frac{1 \times 0.333}{0.02} = 17 \text{ ft},\ \ R_1 = \frac{8 \times 0.02 \times 27}{32.2 \times \pi^2 \times 0.333^5} = 3$$

$$\text{Pipe 2:}\ \ Le = \frac{2 \times 0.333}{0.02} = 33 \text{ ft},\ \ R_2 = \frac{8 \times 0.02 \times 533}{32.2 \times \pi^2 \times 0.333^5} = 66$$

$$\text{Pipe 3:}\ \ Le = \frac{2 \times 0.333}{0.02} = 33 \text{ ft},\ \ R_3 = \frac{8 \times 0.02 \times 2{,}033}{32.2 \times \pi^2 \times 0.333^5} = 250$$

$$\text{Pipe 4:}\ \ Le = \frac{4 \times 0.333}{0.02} = 67 \text{ ft},\ \ R_4 = \frac{8 \times 0.02 \times 1{,}817}{32.2 \times \pi^2 \times 0.333^5} = 224$$

a) Compute the discharge and head loss in pipe 2:

$$Q_2 = 5{,}000 \frac{\text{gal}}{\text{h}} \times \frac{1 \text{ ft}^3}{7.48 \text{ gal}} \times \frac{1 \text{ hr}}{3600 \text{ sec}} = 0.186 \frac{\text{ft}^3}{\text{sec}},$$

$$\therefore (h_L)_2 = R_2 Q_2^2 = 66 \times 0.186^2 = \underline{2.28 \text{ ft}}$$

b) The flow path is pipe 1 – pipe 2 – pipe 3; there is no flow in pipe 4. The demand curve from A to B is

$$H_P = z_B - z_A + (h_L)_1 + (h_L)_2 + (h_L)_3 = 430 + (3 + 66 + 250)Q^2 = 430 + 319Q^2$$

A trial and error solution is employed: assume Q, compute H_P from the demand curve, and compare it with H_P from the pump characteristic curve.

Trial 1: $Q = 10{,}000$ gal/h $= 0.371$ ft³/sec,

H_P (demand) $= 430 + 319 \times 0.371^2 = 474$ ft, H_P (pump) $= 465$ ft

Trial 2: $Q = 9{,}000$ gal/h $= 0.334$ ft³/sec,

H_P (demand) $= 430 + 319 \times 0.334^2 = 467$ ft, H_P (pump) $= 466$ ft

∴ The discharge is approximately <u>0.33 ft³/sec, or 9,000 gal/h</u>.

c) $Q_2 = 11{,}000$ gal / min $= 0.409$ ft³ / sec, and from the pump curve, $H_P = 460$ ft. Write the energy equation from A to C:

$$H_C = H_P - (R_1 + R_2)Q_2^2 = 460 - 69 \times 0.409^2 = 449 \text{ ft}$$

The energy equation from C to D is $H_C = z_D + R_4 Q_4^2$, where H_C is the hydraulic grade line at location C. Therefore the discharge in pipe 4 is

$$Q_4 = \sqrt{\frac{H_C - z_D}{R_4}} = \sqrt{\frac{449 - 445}{224}} = 0.134 \text{ ft}^3/\text{sec}, \text{ or } \underline{3{,}600 \text{ gal/h}}.$$

11.25E a) The discharge through the pump is $Q = 12{,}000/(7.48 \times 3600) = 0.446 \text{ ft}^3/\text{sec}$, and from the characteristic curve the pump head is $H_P = 454$ ft. The efficiency is $\eta_P = 0.86$. Hence the required power is

$$\dot{W}_P = \frac{\gamma Q H_P}{\eta_P} = \frac{62.4 \times 0.446 \times 454}{0.86} = \underline{1.47 \times 10^4 \ \frac{\text{ft-lb}}{\text{sec}}, \text{ or } 26.7 \text{ hp}}$$

b) The energy equation from A to C is $H_P = z_C + p_C/\gamma - z_A + (h_L)_1 + (h_L)_2$.

$$\frac{p_C}{\gamma} = 454 - 300 - (3+66) \times 0.446^2 = 140 \text{ ft}, \quad \text{and } p_C = \frac{62.4 \times 140}{144} = \underline{61 \text{ lb/in}^2}$$

c) Write the energy equation from C to D and solve for the pressure at location D: $p_C/\gamma + z_C = p_D/\gamma + z_D$. (Since there is no flow in pipe 4, the energy equation reduces to the law of hydrostatics.) From part (b), the pressure at location C is 61 lb/in^2.

$$\therefore p_D = 61 + 62.4 \times \frac{300 - 445}{144} = \underline{-1.83 \text{ lb/in}^2}$$

11.26E a) From the pump curve, for $H_P = 460$ ft, $Q = 11{,}000$ gal/h, or 0.409 ft^3/sec. Write the energy equation from A to C and evaluate the hydraulic grade line:

$$H_C = H_P - (h_L)_1 - (h_L)_2 = 460 - 69 \times 0.409^2 = 449 \text{ ft}$$

Thus the discharge in pipe 3, between location C and location B is

$$Q_3 = \sqrt{\frac{H_C - H_B}{R_3}} = \sqrt{\frac{449 - 430}{250}} = \underline{0.276 \text{ ft}^3/\text{sec, or } 7{,}400 \text{ gal/h}}$$

b) The flow path is pipe 1 – pipe 2 – pipe 4; there is no flow in pipe 3. Write the energy equation from location A to location D:

$$H_P = z_D - z_A + (h_L)_1 + (h_L)_2 + (h_L)_4 = 445 + (3 + 66 + 224)Q^2 = 445 + 293Q^2$$

A trial and error solution is employed: assume Q, compute H_P from the demand curve, and compare it with H_P from the pump characteristic curve.

Trial 1: $Q = 10{,}000$ gal/h $= 0.371$ ft^3/sec
H_P (demand) $= 445 + 293 \times 0.371^2 = 485$ ft
H_P (pump) $= 465$ ft

Trial 2: $Q = 7{,}500$ gal/h = 0.278 ft³/sec
H_P (demand) = 445 + 293 × 0.278² = 468 ft
H_P (pump) = 475 ft

Trial 3: $Q = 8{,}000$ gal/h = 0.297 ft³/sec
H_P (demand) = 445 + 293 × 0.297² = 471 ft
H_P (pump) = 472 ft

Hence, the head across the pump is $\underline{H_P \approx 470 \text{ ft}}$.

11.27 a) Unknowns: Q_1, Q_2, Q_3, Q_4, Q_5 and H_J.

Eqns: $H_A = R_1Q_1^2 + H_J,\ H_J = R_2Q_2^2 + H_B,$
$H_J = R_3Q_3^2 + H_B,\ H_J = R_4Q_4^2 + H_B,$
$H_J = R_5Q_5^2 + H_C,\ Q_1 = Q_2 + Q_3 + Q_4 + Q_5.$

Reduce to one equation in the one unknown H_J:

$$H_J - H_B = R_2Q_2^2 = R_3Q_3^2 = R_4Q_4^2, \qquad \therefore Q_3 = Q_2\sqrt{\frac{R_2}{R_3}} \text{ and } Q_4 = Q_2\sqrt{\frac{R_2}{R_4}}.$$

Furthermore, $$Q_1 = \sqrt{\frac{H_A - H_J}{R_1}},\ Q_2 = \sqrt{\frac{H_J - H_B}{R_2}},\ Q_5 = \sqrt{\frac{H_J - H_C}{R_5}}.$$

Substitute into continuity relation:

$$\therefore \sqrt{\frac{H_A - H_J}{R_1}} = \sqrt{\frac{H_J - H_B}{R_2}}\left(1 + \sqrt{\frac{R_2}{R_4}} + \sqrt{\frac{R_2}{R_4}}\right) + \sqrt{\frac{H_J - H_C}{R_5}}$$

b)

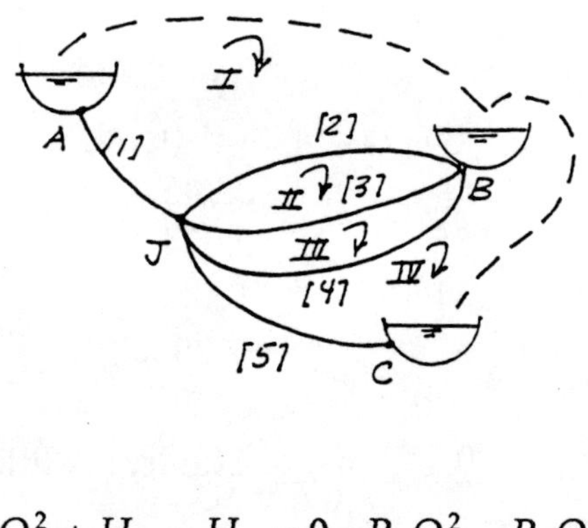

$$-R_1Q_1^2 - R_2Q_2^2 + H_A - H_B = 0,\ R_2Q_2^2 - R_3Q_3^2 = 0,$$
$$R_3Q_3^2 - R_4Q_4^2 = 0,\ R_4Q_4^2 - R_5Q_5^2 - H_C + H_B = 0$$

11.28 $\Delta Q_{\mathrm{I}} = \dfrac{-(\pm W_1 \pm W_2) - 10}{G_1 + G_2}$, $\Delta Q_{\mathrm{II}} = \dfrac{-(\pm W_2 \pm W_3) - 15}{G_2 + G_3}$,

$\pm W_i = \overline{R} Q_i |Q_i|$, $G_i = 2\overline{R}|Q_i|$.

Loop	Pipe	Q	W	G	Q	W	G	Q
I	1	-3.5	-24.50	14.00	-2.46	-12.10	9.84	-2.30
	2	0	0	0	+0.36	+0.26	1.44	+0.43
			Σ 24.50	14.00		Σ -11.84	11.28	

$$\Delta Q_{\mathrm{I}} = \frac{+24.50 - 10}{14.00} = +1.04 \qquad \Delta Q_{\mathrm{I}} = \frac{+11.84 - 10}{11.28} = +0.16$$

Loop	Pipe	Q	W	G	Q	W	G	Q
II	2	0	0	0	-0.36	-0.26	1.44	-0.43
	3	-3.5	-24.5	14.00	-2.82	-15.90	11.28	+2.73
			Σ -24.50	14.00		Σ -16.16	12.72	

$$\Delta Q_{\mathrm{II}} = \frac{+24.50 - 15}{14.00} = +0.68 \qquad \Delta Q_{\mathrm{II}} = \frac{+16.16 - 15}{12.72} = +0.09$$

$H_{\text{Junction}} = 50 - \overline{R}_1 Q_1^2 = 50 - 2 \times 2.3^2 = \underline{39.4}$

$\therefore \underline{Q_1 = 2.30}$ (into J), $\underline{Q_2 = 0.43}$ (into J), $\underline{Q_3 = 2.73}$ (out of J)

11.29 Consider the flow correction in a clockwise sense through the system:

$$\Delta Q = \frac{-(W_1 - H_P + W_2) - \Delta H}{G_1 + G_2} = \frac{-30Q_1|Q_1| + 20.4/Q_1 - 20Q_2|Q_2| - 20}{60|Q_1| + 20.4/Q_1^2 + 40|Q_2|}$$

Q_1 (m^3/s)	Q_2 (m^3/s)	ΔQ (m^3/s)
1	0.5	-0.345
0.655	0.155	-0.024
0.631	0.131	+0.000155
0.632	0.132	$+7.89 \times 10^{-9}$

Therefore, $\underline{Q_1 \approx 0.63 \text{ m}^3/\text{s}}$ and $\underline{Q_2 \approx 0.113 \text{ m}^3/\text{s}}$. The hydraulic grade line at location J is: $H_J = 5 + H_P - \overline{R}_1 Q_1^2 = 5 + \dfrac{20.4}{0.632} - 30 \times 0.632^2 = \underline{25.3 \text{ m}}$

11.30 First, continuity is satisfied as shown in the figure:

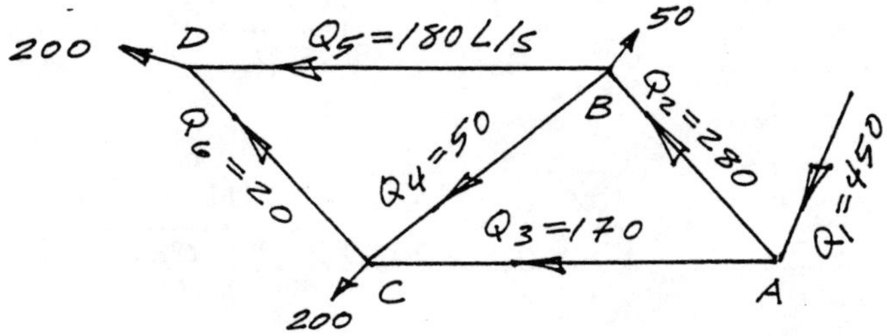

a) The hydraulic grade lines at the nodes (i.e., the nodal piezometric heads) can now be computed:

$$H_A = 100 - \overline{R}_1 Q_1^2 = 100 - 20 \times 0.45^2 = \underline{95.95 \text{ m}},$$
$$H_B = H_A - \overline{R}_2 Q_2^2 = 95.95 - 51 \times 0.28^2 = \underline{91.95 \text{ m}},$$
$$H_C = H_A - \overline{R}_3 Q_3^2 = 95.95 - 280 \times 0.17^2 = \underline{87.86 \text{ m}},$$
$$H_D = H_B - \overline{R}_5 Q_5^2 = 91.95 - 310 \times 0.18^2 = \underline{81.91 \text{ m}}.$$

b) $p_A = \gamma(H_A - z_A) = 9810(95.95 - 10) = \underline{8.43 \times 10^5 \text{ Pa}},$

$p_B = \gamma(H_B - z_B) = 9810(91.95 - 20) = \underline{7.06 \times 10^5 \text{ Pa}},$

$p_C = \gamma(H_C - z_C) = 9810(87.86 - 5) = \underline{8.13 \times 10^5 \text{ Pa}},$

$p_D = \gamma(H_D - z_D) = 9810(81.91 - 0) = \underline{8.04 \times 10^5 \text{ Pa}}.$

11.31E Compute equivalent lengths and resistance coefficients using

$$Le = \frac{D\Sigma k}{f} \quad \& \quad \overline{R} \frac{8f(L+Le)}{g\pi^2 D^5}:$$

$$Le_1 = 0, \qquad \overline{R}_1 = \frac{8 \times 0.015 \times 800}{32.2 \times \pi^2 \times (8/12)^5} = 2.29,$$

$$Le_2 = \frac{2 \times 3/12}{0.02} = 25, \qquad \overline{R}_2 = \frac{8 \times 0.02 \times 625}{32.2 \times \pi^2 \times (3/12)^5} = 322.2,$$

$$Le_3 = \frac{2 \times 3/12}{0.02} = 25, \qquad \overline{R}_3 = \frac{8 \times 0.02 \times 675}{32.2 \times \pi^2 \times (3/12)^5} = 348.0,$$

$$Le_4 = \frac{3 \times 3/12}{0.025} = 30, \qquad \overline{R}_4 = \frac{8 \times 0.025 \times 455}{32.2 \times \pi^2 \times (3/12)^5} = 293.2,$$

$$Le_5 = \frac{3 \times 4/12}{0.015} = 67, \qquad \overline{R}_5 = \frac{8 \times 0.015 \times 1067}{32.2 \times \pi^2 \times (4/12)^5} = 97.9.$$

Solve for single unknown H_J using eqn. derived in part (a), Pbm. 11.27:

$$\sqrt{\frac{H_A - H_J}{\overline{R}_1}} = \sqrt{\frac{H_J - H_B}{\overline{R}_2}}\left(1 + \sqrt{\frac{\overline{R}_2}{\overline{R}_3}} + \sqrt{\frac{\overline{R}_2}{\overline{R}_4}}\right) + \sqrt{\frac{H_J - H_C}{\overline{R}_5}},$$

$$\sqrt{\frac{650 - H_J}{2.29}} = \sqrt{\frac{H_J - 575}{322.2}}\left(1 + \sqrt{\frac{322.2}{348.0}} + \sqrt{\frac{322.2}{293.2}}\right) + \sqrt{\frac{H_J - 180}{97.9}},$$ which reduces to

$$F(H_J) = \sqrt{650 - H_J} - 0.2538\sqrt{H_J - 575} - 0.1529\sqrt{H_J - 180}.$$

Take first derivative and employ Newton's method of solution:

$$F'(H_J) = -\frac{0.5}{\sqrt{650 - H_J}} - \frac{0.1269}{\sqrt{H_J - 575}} - \frac{0.07645}{\sqrt{H_J - 180}}.$$

H_J	F	F'	$\Delta H_J = -F / F'$
600	2.669	-0.09982	+26.74
626.74	-0.2345	-0.1249	-1.877
624.86	-0.003064	-0.1213	-0.025
624.84	-0.0006379	-0.1213	-0.0026

$\therefore H_J \cong \underline{624.8 \text{ ft}}$,

$$Q_1 = \sqrt{\frac{H_A - H_J}{\overline{R}_1}} = \sqrt{\frac{650 - 624.8}{2.29}} = \underline{3.32 \text{ ft}^3 / \text{sec}},$$

$$Q_2 = \sqrt{\frac{H_J - H_B}{\overline{R}_2}} = \sqrt{\frac{624.8 - 575}{322.2}} = \underline{0.39 \text{ ft}^3 / \text{sec}},$$

$$Q_3 = \sqrt{\frac{H_J - H_B}{\overline{R}_3}} = \sqrt{\frac{624.8 - 575}{348.0}} = \underline{0.38 \text{ ft}^3 / \text{sec}},$$

$$Q_4 = \sqrt{\frac{H_J - H_B}{\overline{R}_4}} = \sqrt{\frac{624.8 - 575}{293.2}} = \underline{0.41 \text{ ft}^3 / \text{sec}},$$

$$Q_5 = \sqrt{\frac{H_J - H_C}{\overline{R}_5}} = \sqrt{\frac{624.8 - 180}{97.9}} = \underline{2.13 \text{ ft}^3 / \text{sec}}.$$

11.32

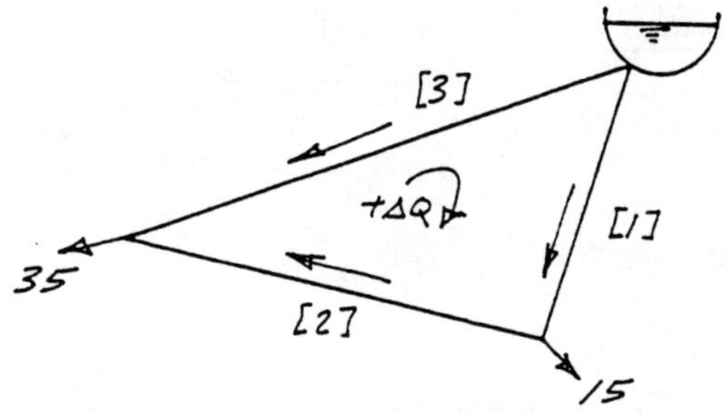

Assume flow directions as shown. Then: $\Delta Q = \dfrac{-\left(\overline{R}_1 Q_1^2 + \overline{R}_2 Q_2^2 - \overline{R}_3 Q_3^2\right)}{2\left(\overline{R}_1 Q_1 + \overline{R}_2 Q_2 + \overline{R}_3 Q_3\right)}$

Initial flow assumption: $Q_1 = 25,\ Q_2 = 10,\ Q_3 = 25.$

1st iteration: $\Delta Q = \dfrac{-\left(3 \times 25^2 + 5 \times 10^2 - 2 \times 25^2\right)}{2(3 \times 25 + 5 \times 10 + 2 \times 25)} = -3.21,$

$\therefore Q_1 = 25 - 3.21 = 21.79,\ Q_2 = 10 - 3.21 = 6.79,\qquad Q_3 = 25 + 3.21 = 28.21.$

2nd iteration: $\Delta Q = \dfrac{-\left(3 \times 21.79^2 + 5 \times 6.79^2 - 2 \times 28.21^2\right)}{2(3 \times 21.79 + 5 \times 6.79 + 2 \times 28.21)} = -0.20,$

$\therefore Q_1 = 21.79 - 0.20 = 21.59,\ Q_2 = 6.79 - 0.20 = 6.59,\quad Q_3 = 28.21 + 0.20 = 28.41.$

3rd iteration: $\Delta Q = \dfrac{-\left(3 \times 21.59^2 + 5 \times 6.59^2 - 2 \times 28.41^2\right)}{2(3 \times 21.59 + 5 \times 6.59 + 2 \times 28.41)} = -0.0041,$

$\therefore \underline{Q_1 \cong 21.6},\ \underline{Q_2 \cong 6.6},\ \underline{Q_3 \cong 28.4}.$

11.33

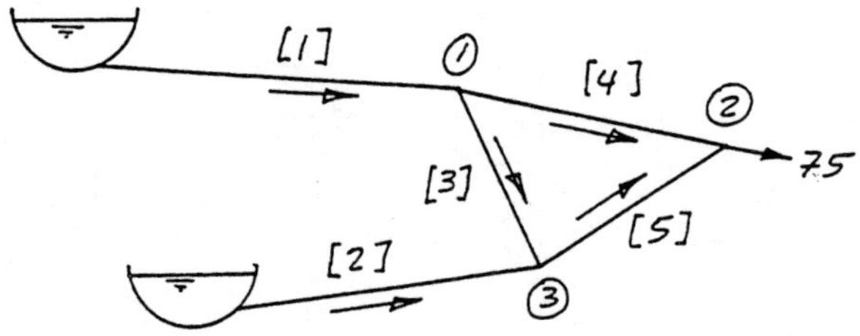


```
2000 REM ***** PROBLEM 11.33 *****
2010 DATA 5,3,0,SI,DW,1.E-6,1,1000,1000,.001
2020 DATA 1,500,300,0.15,0,40
2030 DATA 2,600,250,0.15,0,35
2040 DATA 3,50,150,0.15,5,5
2050 DATA 4,200,250,0.15,2,35
2060 DATA 5,200,300,0.15,2,40
2070 DATA 15,1,1,1,4,1,-5
2080 DATA 13,3,1,3,-2
2090 DATA 0,3,4,-5,-3
```

TRIAL NO.	RELATIVE FLOW CHANGE	AVERAGE FLOW CHANGE
1	7.408E-01	6.056E+01
2	1.948E-01	1.410E+01
3	2.829E-03	2.048E-01
4	9.658E-04	6.999E-02

PIPE NO.	FLOW RATE
1	138.529
2	-63.529
3	41.793
4	96.735
5	-21.735

NODE NO.	HYD. GRADE LINE
1	9.245
2	5.938
3	5.862

1.34 Hardy Cross: The unknown flow adjustment for each defined loop (ΔQ) is solved independently of the remaining loops in the network.
Linear: The unknown flow for each pipe in the network (Q) is solved simultaneously.
Single node: The unknown piez. head adjustment at each junction (ΔH) is solved independently of the piz. heads at the remaining junctions.
Simultaneous node: The unknown piez. head at all junctions of network (H) are solved simultaneously.

11.35 a) $H_B + H_P = H_A + (\overline{R}_1 + \overline{R}_2)Q^2$,

$$10 + 100 - 826Q^2 = 35 + (5000 + 300)Q^2,$$

$$6126\,Q^2 = 75, \qquad \therefore Q = \underline{0.111\ \text{m}^3/\text{s}}.$$

b)

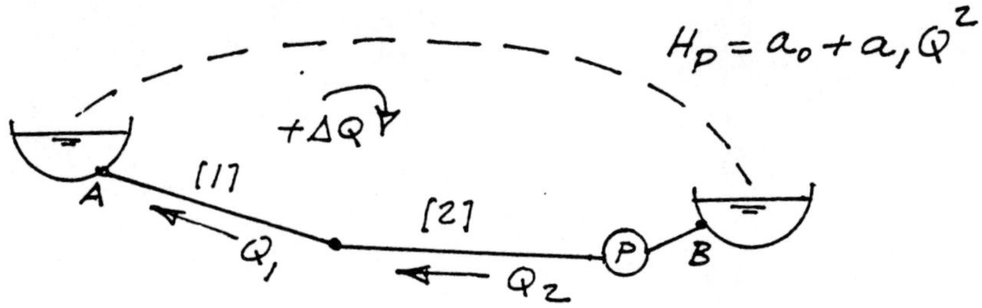

$$\Delta Q = \frac{-(\overline{R}_1 Q_1^2 + \overline{R}_2 Q_2^2 - H_P) - (H_A - H_B)}{2\overline{R}_1 Q_1 + 2\overline{R}_2 Q_2 - 2a_1 Q_2}$$

Let $Q_1 = Q_2 = Q$, and substituting in known data:

$$\Delta Q = \frac{-(5000 + 300)Q^2 + 100 - 826Q^2 - 25}{2(5000 + 300 + 826)Q} = \frac{-6126Q^2 + 75}{12252Q}$$

Iteration	Q	ΔQ
1	0.20	-0.0694
2	0.1306	-0.0184
3	0.1122	-0.00152
4	0.1107	-0.0000104

$\therefore Q \cong \underline{0.111\ \text{m}^3/\text{s}}.$

11.36

```
2000 REM ***** PROBLEM 11.36 *****
2010 DATA 5,2,0,SI,DW,1.E-6,1,1000,1000,.001
2020 DATA 1,200,100,.1,2,6
2030 DATA 2,150,50,.1,30,1
2040 DATA 3,500,100,.1,0,5
2050 DATA 4,35,50,.1,35,1
2060 DATA 5,120,100,.1,2,4
2070 DATA 125,1,1,1,3
2080 DATA 7,2,1,2
2090 DATA 10,3,1,3,5
2100 DATA 9,3,1,3,4
```

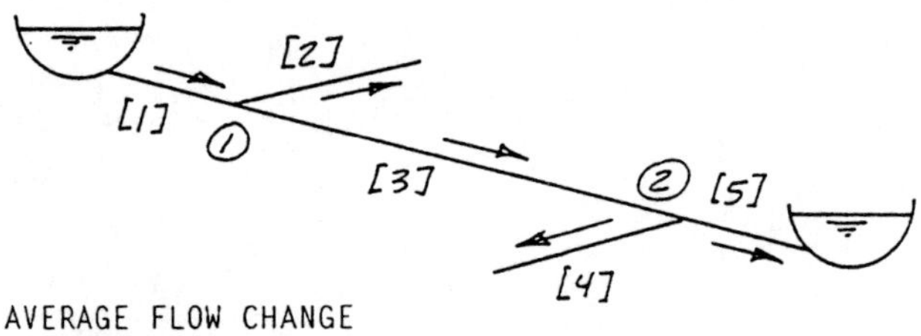

TRIAL NO.	RELATIVE FLOW CHANGE	AVERAGE FLOW CHANGE
1	4.016E-01	2.172E+00
2	3.844E-02	2.025E-01
3	7.737E-03	4.066E-02
4	1.303E-03	6.846E-03
5	5.124E-04	2.693E-03

PIPE NO.	FLOW RATE
1	9.287
2	1.584
3	7.702
4	0.557
5	7.146

NODE NO.	HYD. GRADE LINE
1	121.721
2	116.235

11.37

```
2000 REM ***** PROBLEM 11.37 *****
2010 DATA 5,2,1,SI,DW,1.E-6,1,1000,1000,.001
2020 DATA 1,200,100,.1,2,6
2030 DATA 2,150,50,.1,30,1
2040 DATA 3,500,100,.1,0,5
2050 DATA 4,35,50,.1,35,1
2060 DATA 5,120,100,.1,2,4
2065 DATA 1,UP,10
2070 DATA 125,1,1,1,3
2080 DATA 7,2,1,2
2090 DATA 10,3,1,3,5
2100 DATA 9,3,1,3,4
```

TRIAL NO.	RELATIVE FLOW CHANGE	AVERAGE FLOW CHANGE
1	7.382E-01	9.585E+00
2	6.599E-02	8.359E-01
3	2.580E-02	3.295E-01
4	1.344E-02	1.724E-01
5	7.571E-03	9.744E-02
6	4.406E-03	5.680E-02
7	2.599E-03	3.353E-02
8	1.543E-03	1.993E-02
9	9.200E-04	1.188E-02

PIPE NO.	FLOW RATE
1	23.123
2	4.792
3	18.331
4	2.589
5	15.742

NODE NO.	HYD. GRADE LINE
1	149.876
2	120.719

11.38 a) Let R_j be the resistance coefficient for a single tube in the condenser. Then $Q = \sum_{j=1}^{N} \sqrt{W/R_j} = N\sqrt{W/R_j}$. Solving for W, we have $W = \frac{R_j}{N^2}Q^2 = R_2 Q^2$.

Hence, $R_2 = \frac{R_j}{N^2}$.

b) $H_P = z_B - z_A + (R_1 + R_2 + R_3)Q^2$ $\qquad$ H_P and Q are the two unknowns.

c) $$\Delta Q = \frac{-(R_1 + R_2 + R_{3)})|Q|Q + (a_0 + a_1 Q + a_2 Q^2 + a_3 Q^3) + z_A - z_B}{2(R_1 + R_2 + R_{3)})|Q| - (a_1 + 2a_2 Q + 3a_3 Q^2)}$$

d) $R_1 = \frac{8f}{g\pi^2}\frac{L_1}{D_1^5} = 0.00516$, $R_2 = \frac{8f}{g\pi^2}\frac{L_2}{N^2 D_2^5} = 2.538$, $R_3 = \frac{8f}{g\pi^2}\frac{L_3}{D_3^5} = 0.0103$

Using the equation in part (c), $Q = 2.061$ m³/s after 5 iterations, beginning with an initial estimate $Q = 3$ m³/s.

Iteration	Q	ΔQ
1	3	-0.611
2	2.389	-0.277
3	2.113	-0.050
4	2.062	-0.0014
5	2.061	-5.1×10^{-13}

e)

$$H_C = z_A + \left[a_0 + a_1 Q + a_2 Q^2 + a_3 Q^{32}\right] - R_1 Q^2$$
$$= 2 + \left[30.4 - 31.8 \times 2.06 + 18.6 \times 2.06^2 - 4.0 \times 2.06^3\right] - 0.00516 \times 0.60^2$$
$$= 10.83 \text{ m}$$

$p_C = \gamma(H_C - z_C) = 9810 \times (10.83 - 6) = \underline{47380 \text{ Pa, or } 47.4 \text{ kPa}}$

$H_C' = H_C - R_2 Q^2 = 10.83 - 2.538 \times 2.06^2 = 0.060$ m

$p_C' = \gamma(H_C' - z_C) = 9810 \times (0.060 - 6) = \underline{-5830 \text{ Pa, or } -58.3 \text{ kPa}}$

11.39E

```
2000 REM ***** PROBLEM 11.39E *****
2010 DATA 12,8,0,EN,HW,0,1,1,12,.001
2020 DATA 1,1500,10,120,0,3
2030 DATA 2,400,10,120,0,2
2040 DATA 3,600,10,120,0,1.6
2050 DATA 4,800,8,120,0,.6
2060 DATA 5,700,8,120,0,1.5
2070 DATA 6,1000,10,120,0,2.5
2080 DATA 7,750,10,120,0,1.5
2090 DATA 8,450,10,120,0,2.5
2100 DATA 9,550,8,120,0,1.1
2110 DATA 10,800,10,120,0,1.6
2120 DATA 11,800,10,120,0,1
2130 DATA 12,1450,10,120,0,5
2140 DATA 275.4,1,1,1,2,1,3,1,4,1,9,1,-8,1,-7,1,6
2150 DATA 0,4,2,-10,-8,-11
2160 DATA 0,4,3,4,9,10
2170 DATA 0,5,8,-9,-5,-6,7
2180 DATA 10,4,1,11,-7,-12
```

TRIAL NO.	RELATIVE FLOW CHANGE	AVERAGE FLOW CHANGE
1	4.343E-01	7.274E-01
2	1.594E-01	2.645E-01
3	4.357E-02	7.143E-02
4	1.259E-02	2.056E-02
5	2.823E-03	4.608E-03
6	5.792E-04	9.453E-04

PIPE NO.	FLOW RATE
1	4.294
2	3.076
3	0.892
4	-0.108
5	0.631
6	1.631
7	1.075
8	2.292
9	-0.477
10	-0.184
11	1.218
12	3.706

NODE NO.	HYD. GRADE LINE
1	239.337
2	234.151
3	233.366
4	233.428
5	234.097
6	237.479
7	238.865
8	234.860

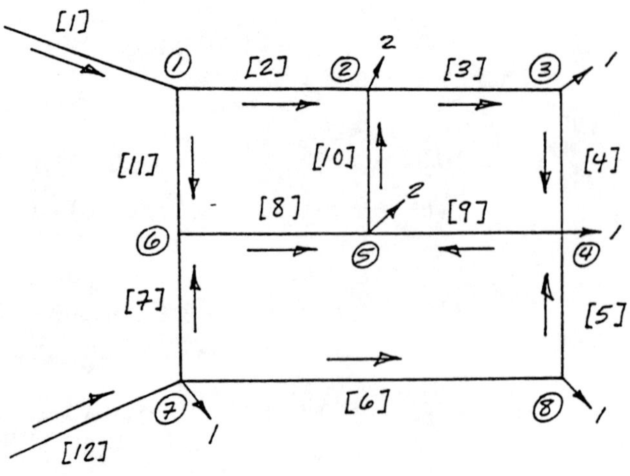

11.40

```
2000 REM ***** PROBLEM 11.40 *****
2010 DATA 14,8,1,SI,HW,0,1,1000,1000,.001
2020 DATA 1,610,350,100,8,250
2030 DATA 2,914,350,100,0,150
2040 DATA 3,760,350,100,0,40
2050 DATA 4,610,350,100,0,15
2060 DATA 5,610,350,100,10,45
2070 DATA 6,457,300,100,0,70
2080 DATA 7,610,350,100,0,140
2090 DATA 8,914,350,100,0,110
2100 DATA 9,760,350,100,0,250
2110 DATA 10,610,350,100,0,100
2120 DATA 11,30,200,100,0,55
2130 DATA 12,61,150,100,10,35
2140 DATA 13,1500,400,100,5,500
2150 DATA 14,975,300,100,0,80
2160 DATA 13,HQ,166,132,600,18,1000
2170 DATA 10,1,13,1,1,1,2,1,3,1,4,2,-4,-7,1,6,2,-6,-8
2180 DATA 0,4,1,10,-8,-9
2190 DATA 0,4,2,3,-7,-10
2200 DATA 0,4,7,4,-5,-6
2210 DATA -27,5,13,1,2,3,11
2220 DATA -31,3,13,9,14
2230 DATA -31,5,13,1,10,6,-12
```

TRIAL NO.	RELATIVE FLOW CHANGE	AVERAGE FLOW CHANGE
1	9.813E-02	1.373E+01
2	5.488E-02	7.649E+00
3	8.972E-03	1.243E+00
4	6.442E-03	8.885E-01
5	4.075E-03	5.609E-01
6	2.805E-03	3.856E-01
7	1.916E-03	2.632E-01
8	1.362E-03	1.870E-01
9	1.005E-03	1.379E-01
10	7.517E-04	1.031E-01

PIPE NO.	FLOW RATE
1	268.799
2	146.587
3	36.587
4	3.436
5	56.564
6	93.731
7	137.686
8	109.205
9	254.040
10	122.212
11	60.836
12	22.833
13	522.839
14	84.836

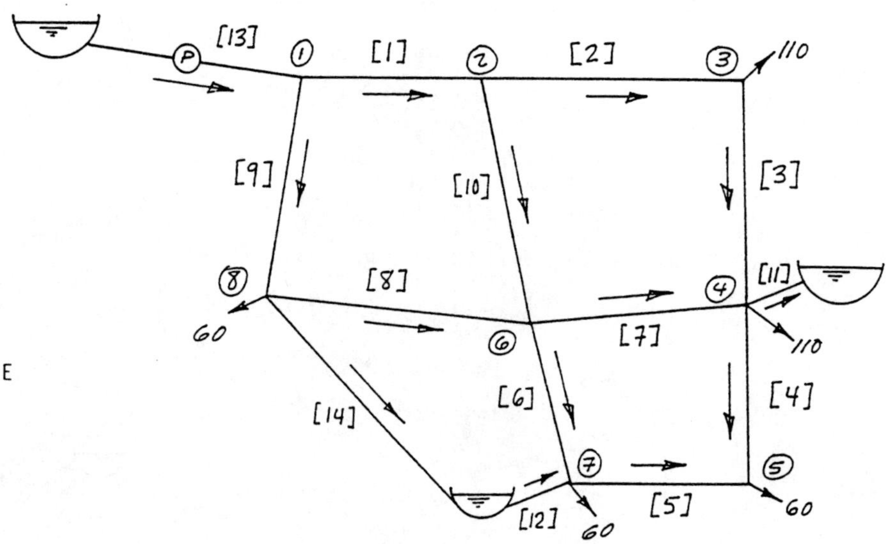

NODE NO.	HYD. GRADE LINE
1	69.254
2	47.453
3	38.378
4	37.801
5	37.795
6	43.194
7	38.994
8	48.454

11.41E This is a design problem, and use of the HARDY CROSS PIPE NETWORK ANALYSIS program (Fig. E.4) is recommended. One possible solution is shown below; it was determined by trial. The useful power concept for pumping was employed in the program.

```
2000 REM ***** PROBLEM11.41E *****
2010 DATA 11,9,1,EN,DW,1E-5,1,449,12,.001
2020 DATA 1,550,6,.00085,0,78
2030 DATA 2,625,6,.00085,0,2
2040 DATA 3,1000,6,.00085,0,1
2050 DATA 4,725,6,.00085,0,51
2060 DATA 5,725,6,.00085,0,151
2070 DATA 6,800,6,.00085,0,226
2080 DATA 7,675,6,.00085,0,101
2090 DATA 8,950,6,.00085,0,153
2100 DATA 9,725,6,.00085,0,128
2110 DATA 10,475,8,.00085,0,480
2120 DATA 11,825,6,.00085,0,1
2130 DATA 10,UP,10
2140 DATA 500,1,10,1,6,1,5,1,4,1,3,1,-11,2,11,2,1,-1,1,-9
2150 DATA 0,6,8,9,1,-2,-11,-7
2160 DATA 0,6,7,11,-3,-4,-5,-6
```

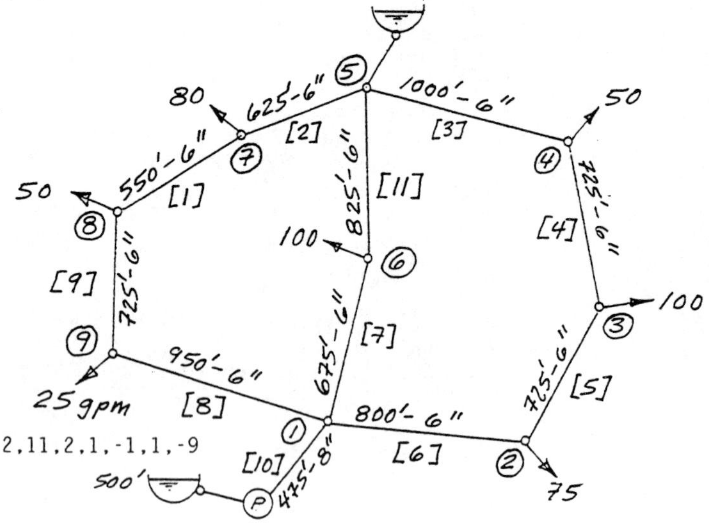

TRIAL NO.	RELATIVE FLOW CHANGE	AVERAGE FLOW CHANGE
1	3.716E-01	4.628E+01
2	6.661E-02	8.302E+00
3	2.152E-02	2.691E+00
4	6.558E-03	8.209E-01
5	1.952E-03	2.444E-01
6	5.846E-04	7.321E-02

PIPE NO.	FLOW RATE
1	55.196
2	24.804
3	-53.737
4	-3.737
5	96.263
6	171.263
7	178.541
8	130.196
9	105.196
10	480.000
11	78.541

NODE NO.	HYD. GRADE LINE
1	580.749
2	578.989
3	578.426
4	578.428
5	578.702
6	579.146
7	578.658
8	578.816
9	579.476

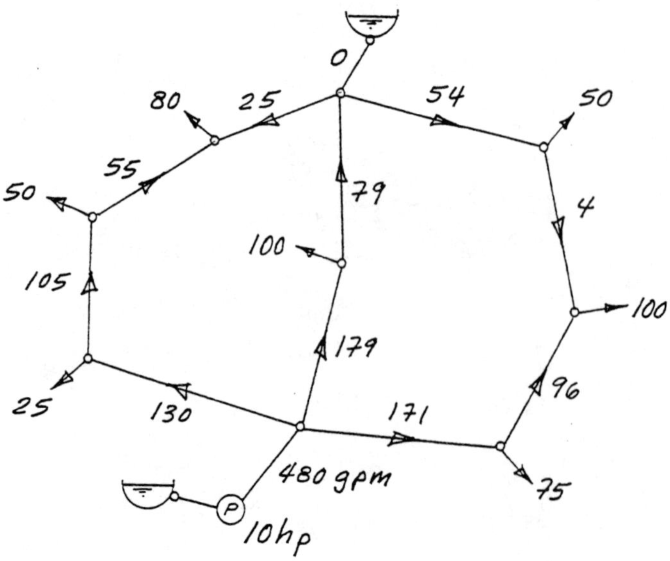

11.42E Using the HARDY CROSS PIPE NETWORK ANALYSIS program (Fig. E.4), the system can be designed by treating several groups of greens/fairways independently, each group branching from a main feeder pipe. To simplify the analysis, the four sprinklers on each green can be treated as a single sprinkler, and any sprinkler can be handled either using a known flow demand, or acting as a valve emptying into a reservoir (use a large *K*-value). Since all sprinklers are not run simultaneously, a watering strategy must be developed.

11.43

$$V_{ss} = \sqrt{\frac{2g(H_1 - H_3)}{\left(\frac{fL}{D} + K\right)}} = \sqrt{\frac{2 \times 9.81 \times 3}{\left(\frac{0.025 \times 2500}{0.10} + 0.15\right)}} = 0.307 \text{ m/s}$$

$$V = 0.99V_{ss} = 0.99 \times 0.307 = 0.304 \text{ m/s}$$

$$t = \frac{V_{ss}L}{2g(H_1 - H_3)} \ln\left[\frac{(V_{ss} + V)(V_{ss} - V_0)}{(V_{ss} - V)(V_{ss} + V_0)}\right]$$

$$= \frac{0.307 \times 2500}{2 \times 9.81 \times 3} \ln\left[\frac{(0.307 + 0.304)(0.307 - 0)}{(0.307 - 0.304)(0.307 + 0)}\right] = \underline{69.0 \text{ s}}$$

11.44

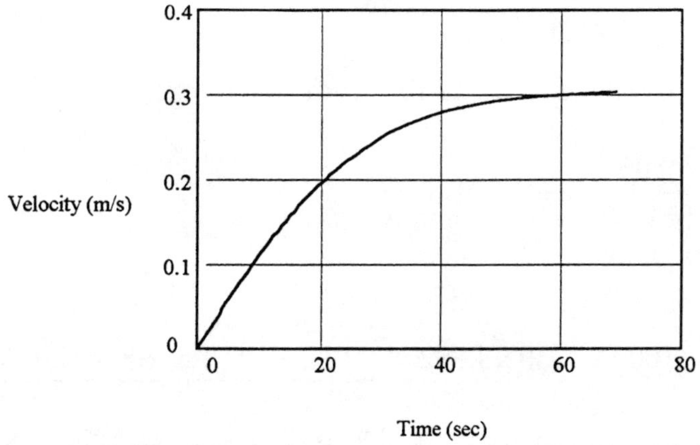

11.45 First compute the initial velocity V_0, with H_1-H_3 = 8 m, and K_0 = 275:

$$V_0 = \sqrt{\frac{2 \times 9.81 \times 8}{\frac{0.015 \times 800}{0.05} + 275}} = 0.552 \text{ m/s}$$

The final steady-state velocity is, with K_{ss} = 5:

$$V_{SS} = \sqrt{\frac{2 \times 9.81 \times 8}{\frac{0.015 \times 800}{0.05} + 5}} = 0.800 \text{ m/s}$$

The steady-state discharge is $Q = 0.7854 \times 0.05^2 \times 0.800 = \underline{0.00157 \text{ m}^3/\text{s}}$, and the time to reach 95% of that value is

$$V = 0.95V_{ss} = 0.95 \times 0.800 = 0.760 \text{ m/s}$$

$$\therefore t = \frac{0.800 \times 800}{2 \times 9.81 \times 8} \ln\left[\frac{(0.800 + 0.760)(0.800 - 0.552)}{(0.800 - 0.760)(0.800 + 0.552)}\right] = \underline{8.03 \text{ s}}$$

11.46 Compute the acoustic wave speed:

$$a = \sqrt{\frac{B}{\rho\left(1 + \frac{DB}{eE}\right)}} = \sqrt{\frac{220 \times 10^7}{1000\left(1 + \frac{200 \times 220 \times 10^7}{12 \times 150 \times 10^9}\right)}} = 1330 \text{ m/s}$$

a) Wave travel time from valve to reservoir is $t = \frac{L}{a} = \frac{800}{1330} = \underline{0.602 \text{ s}}$.

b) Pressure change at valve due to doubling of discharge:

$$\Delta V = \frac{Q}{\frac{\pi}{4}D^2} = \frac{0.05}{0.7854 \times 0.2^2} = 1.592 \text{ m/s}$$

$$\therefore \Delta p = -\rho a \Delta V = -1000 \times 1330 \times 1.592 = \underline{-2.12 \times 10^6 \text{ Pa, or } -2120 \text{ kPa.}}$$

Note the large pressure reduction due to the water hammer effect. The original pressure at the valve must be sufficiently large so that cavitation will not occur. Cavitation at the valve could be avoided by opening the valve slowly.

c) Pressure change at valve due to halving of the discharge:

$$\Delta V = -\frac{1}{2}\times\frac{Q}{\frac{\pi}{4}D^2} = -\frac{1}{2\times 0.7854\times 0.2^2}0.05 = -0.796 \text{ m/s}$$

$$\therefore \Delta \text{p} = \text{-}1000\times 1330\times(\text{-}0.796) = \underline{1.06\times 10^6 \text{ Pa, or } 1060 \text{ kPa.}}$$

11.47 a) Compute the acoustic wave speed:

$$a = \sqrt{\frac{B}{\rho\left(1+\frac{DB}{eE}\right)}} = \sqrt{\frac{1.05\times 10^9}{680\times\left(1+\frac{50\times 1.05\times 10^9}{2.5\times 70\times 10^9}\right)}} = \underline{1090 \text{ m/s.}}$$

b) The change in velocity due to rapid valve closure is $\Delta V = -V_{SS} = -0.800 \text{ m/s}$ (Problem 11.45). Hence the pressure rise is

$$\Delta p = -\rho a \Delta V = -680\times 1090\times(-0.800) = \underline{5.93\times 10^5 \text{ Pa, or } 593 \text{ kPa.}}$$

c) From Problem 11.45, $\Delta H = 8$ m, $f = 0.015$. Then the initial steady-state pressure at the valve is

$$p_{ss} = \rho\left(g\Delta H - \frac{fL}{D}\frac{V_{ss}^2}{2}\right) = 680\left(9.81\times 8 - \frac{0.015\times 800}{0.05}\times\frac{0.8^2}{2}\right) = 1120 \text{ Pa}$$

Hence the instantaneous pressure is

$$p = p_{ss} + \Delta p = 1120 + 5.93\times 10^5 = \underline{5.94\times 10^5 \text{ Pa, or } 594 \text{ kPa.}}$$

The instantaneous pressure due to rapid valve closure is about 590 kPa, which is over two times the allowable pressure in the pipe. It is possible that the pipe could rupture. This could be avoided by closing the valve more slowly, thereby reducing the maximum pressure rise due to water hammer activity.

11.48 Compute the acoustic wave speed:

$$a = \sqrt{\frac{1.5\times 10^9}{0.9\times 1000\times\left(1+\frac{0.5\times 1.5}{0.010\times 200}\right)}} = 1100 \text{ m/s}$$

a) The allowable changes in velocity and discharge due to pressure constraint are:

$$\Delta V = \frac{-\Delta p}{\rho a} = \frac{-600 \times 10^3}{900 \times 1100} = -0.606 \text{ m/s},$$

$$\Delta Q = \frac{\pi}{4} D^2 \Delta V = \frac{\pi}{4} \times 0.5^2 \times (-0.606) = \underline{-0.119 \text{ m}^3/\text{s}}$$

Hence the tolerable flow rate decrease is $\underline{(1\text{-}0.119/0.5) \times 100 = 76\%}$.

b) Wave travel time from the downstream to the upstream end of pipe:

$$t = \frac{L}{a} = \frac{4000}{1100} = \underline{3.64 \text{ s}}$$

11.48E Compute the acoustic wave speed:

$$a = \sqrt{\frac{217 \times 10^3 \times 144}{0.9 \times 1.94 \times \left(1 + \dfrac{20 \times 217 \times 10^3}{0.40 \times 29 \times 10^6}\right)}} = 3610 \text{ ft/sec}$$

a) The allowable changes in velocity and discharge due to pressure constraint are:

$$\Delta V = \frac{-\Delta p}{\rho a} = \frac{-90 \times 144}{0.9 \times 1.94 \times 3610} = -2.06 \text{ ft/s},$$

$$\Delta Q = \frac{\pi}{4} D^2 \Delta V = \frac{\pi}{4} \times \left(\frac{20}{12}\right)^2 \times (-2.06) = \underline{-4.49 \text{ ft}^3/\text{sec}}$$

Hence the tolerable flow rate decrease is $\underline{(1\text{-}4.49/20) \times 100 = 77\%}$.

b) Wave travel time from the downstream to the upstream end of pipe:

$$t = \frac{L}{a} = \frac{13000}{3610} = \underline{3.60 \text{ s}}$$

TWELVE

Turbomachinery

12.1 a) $r_1 = 0.15$ m, $V_1 = 3$ m / s, $\alpha_1 = 90°$, $r_2 = 0.3$ m, $V_2 = 6$ m / s, $\alpha_2 = 30°$.

$V_{t_1} = 0$, $V_{t_2} = V_2 \cos\alpha_2 = 6\cos 30° = 5.20$ m / s,

$T = \rho Q\left(r_2 V_{t_2} - \dot{r}_1 V_{t_1}\right)$ (assume pump)

$= 1000 \times 0.057(0.3 \times 5.2 - 0) = \underline{88.9 \text{ N}\cdot\text{m}}$, $\therefore$ machine is a <u>pump</u>.

$\dot{W} = \omega T = 25 \times 88.9 = \underline{2220 \text{ W}}$,

$H = \dot{W} / \gamma Q = 2220 / (9810 \times 0.057) = \underline{3.97 \text{ m}}$.

b) $r_1 = 0.15$ m, $v_1 = 5.3$ m / s, $\beta_1 = 45°$, $r_2 = 0.3$ m, $v_2 = 5.3$ m / s, $\beta_2 = 45°$.

$u_1 = \omega r_1 = 25 \times 0.15 = 3.75$ m / s, $V_{t_1} = u_1 - v_1 \cos\beta_1 = 3.75 - 5.3\cos 45° \cong 0$,

$u_2 = \omega r_2 = 25 \times 0.3 = 7.5$ m / s,

$V_{t_2} = u_2 - v_2 \cos\beta_2 = 7.5 - 5.3\cos 45° = 3.75$ m / s,

$T = \rho Q\left(r_2 V_{t_2} - r_1 V_{t_1}\right)$ (assume pump)

$= 1000 \times 0.057(0.3 \times 3.75 - 0) = \underline{64.1 \text{ N}\cdot\text{m}}$, $\therefore$ machine is a <u>pump</u>.

$\dot{W} = \omega T = 25 \times 64.1 = \underline{1600 \text{ W}}$,

$H = \dot{W} / \gamma Q = 1600 / (9810 \times 0.057) = \underline{2.86 \text{ m}}$.

c) $r_1 = 0.3$ m, $V_1 = 3.66$ m / s, $\alpha_1 = 80°$, $r_2 = 0.15$ m, $V_2 = 6.1$ m / s, $\alpha_2 = 30°$.

$V_{t_1} = V_1 \cos\alpha_1 = 3.66\cos 80° = 0.64$ m / s,

$V_{t_2} = V_2 \cos\alpha_2 = 6.1\cos 30° = 5.28$ m / s,

$T = \rho Q\left(r_1 V_{t_1} - r_2 V_{t_2}\right)$ (assume turbine)

$= 1000 \times 0.057(0.3 \times 0.64 - 0.15 \times 5.28) = \underline{-34.2 \text{ N}\cdot\text{m}}$, $\therefore$ machine is a <u>pump</u>.

$\dot{W} = \omega T = 25 \times 34.2 = \underline{855 \text{ W}}$,

$H = \dot{W} / \gamma Q = 855 / (9810 \times 0.057) = \underline{1.53 \text{ m}}$.

d) $r_1 = 0.15$ m, $v_1 = 3$ m / s, $\beta_1 = 90°$, $r_2 = 0.3$ m, $v_2 = 8.7$ m / s, $\beta_2 = 30°$.

$u_1 = \omega r_1 = 25 \times 0.15 = 3.75 \text{ m / s} = V_{t_1}$, $u_2 = \omega r_2 = 25 \times 0.3 = 7.5$ m / s,

$V_{t_2} = u_2 - v_2 \cos\beta_2 = 7.5 - 8.7\cos 30° \cong 0$,

$T = \rho Q\left(r_2 V_{t_2} - r_1 V_{t_1}\right)$ (assume pump)

$= 1000 \times 0.057(0 - 0.15 \times 3.75) = \underline{-32.1 \text{ N}\cdot\text{m}}$, $\therefore$ machine is a <u>turbine</u>.

$\dot{W} = \omega T = 25 \times 32.1 = \underline{802 \text{ W}},$

$H = \dot{W} / \gamma Q = 802 / (9810 \times 0.057) = \underline{1.43 \text{ m}}.$

e) $r_1 = 0.3 \text{ m}, V_1 = 6 \text{ m/s}, \alpha_1 = 30°, \; r_2 = 0.15 \text{ m}, V_2 = 3 \text{ m/s}, \alpha_2 = 80°.$

$V_{t_1} = V_1 \cos\alpha_1 = 6\cos 30° = 5.20 \text{ m/s}, \; V_{t_2} = V_2 \cos\alpha_2 = 3\cos 80° = 0.52 \text{ m/s}.$

$T = \rho Q\left(r_1 V_{t_1} - r_2 V_{t_2}\right)$ (assume turbine)

$= 1000 \times 0.057(0.3 \times 5.2 - 0.15 \times 0.52) = \underline{84.5 \text{ N} \cdot \text{m}},$ $\therefore$ machine is a $\underline{\text{turbine}}$.

$\dot{W} = \omega T = 25 \times 84.5 = \underline{2110 \text{ W}},$

$H = \dot{W} / \gamma Q = 2110 / (9810 \times 0.057) = \underline{3.77 \text{ m}}.$

f) $r_1 = 0.3 \text{ m}, v_1 = 3 \text{ m/s}, \beta_1 = 90°, \; r_2 = 0.15 \text{ m}, v_2 = 4.33 \text{ m/s}, \beta_2 = 30°.$

$u_1 = \omega r_1 = 25 \times 0.3 = 7.5 \text{ m/s} = V_{t_1}, \; u_2 = \omega r_2 = 25 \times 0.15 = 3.75 \text{ m/s},$

$V_{t_2} = u_2 - v_2 \cos\beta_2 = 3.75 - 4.33\cos 30° \cong 0,$

$T = \rho Q\left(r_1 V_{t_1} - r_2 V_{t_2}\right)$ (assume turbine)

$= 1000 \times 0.057(0.3 \times 7.5 - 0) = \underline{128 \text{ N} \cdot \text{m}},$ $\therefore$ machine is a $\underline{\text{turbine}}$.

$\dot{W} = \omega T = 25 \times 128 = \underline{3200 \text{ W}},$

$H = \dot{W} / \gamma Q = 3200 / (9810 \times 0.057) = \underline{5.72 \text{ m}}.$

12.2

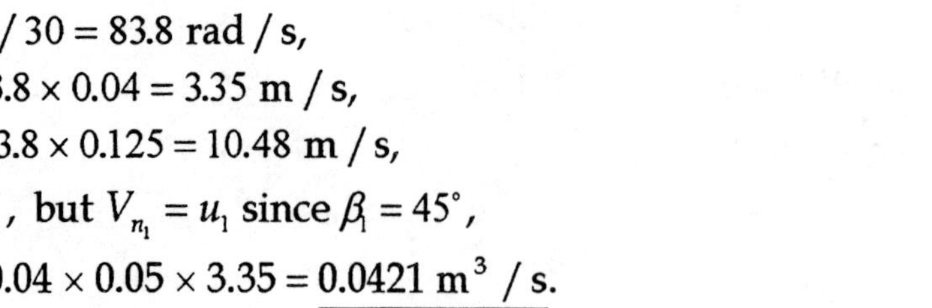

$\omega = 800 \times \pi / 30 = 83.8 \text{ rad/s},$

$u_1 = \omega r_1 = 83.8 \times 0.04 = 3.35 \text{ m/s},$

$u_2 = \omega r_2 = 83.8 \times 0.125 = 10.48 \text{ m/s},$

$Q = 2\pi r_1 b_1 V_{n_1},$ but $V_{n_1} = u_1$ since $\beta_1 = 45°,$

$\therefore Q = 2\pi \times 0.04 \times 0.05 \times 3.35 = \underline{0.0421 \text{ m}^3/\text{s}}.$

$$V_{n_2} = \frac{Q}{2\pi r_2 b_2} = \frac{0.0421}{2\pi \times 0.125 \times 0.025} = 2.14 \text{ m/s},$$

$$V_{t_2} = u_2 - \frac{V_{n_2}}{\tan\beta_2} = 10.48 - \frac{2.14}{\tan 30°} = 6.77 \text{ m/s},$$

$V_{t_1} = 0$ ($\alpha_1 = 90°$ under ideal conditions).

$$\therefore T = \rho Q\left(r_2 V_{t_2} - r_1 V_{t_1}\right) = 1000 \times 0.0421(0.125 \times 6.77 - 0) = \underline{35.6 \text{ N} \cdot \text{m}}.$$

$$\dot{W}_P = \omega T = 83.8 \times 35.6 = \underline{2980 \text{ W}},$$

$$H_t = \omega T / \gamma Q = 2980 / (9810 \times 0.0421) = \underline{7.22 \text{ m}}.$$

12.2E $\omega = 800 \times \pi / 30 = 83.8 \text{ rad/sec},$

$$u_1 = \omega r_1 = 83.8 \times \frac{1.5}{12} = 10.48 \text{ ft/sec},$$

$$u_2 = \omega r_2 = 83.8 \times \frac{5}{12} = 34.92 \text{ ft/sec},$$

$Q = 2\pi r_1 b_1 V_{n_1}$, but $V_{n_1} = u_1$ since $\beta_1 = 45^\circ$,

$$\therefore Q = 2\pi \times \frac{1.5}{12} \times \frac{2}{12} \times 10.48 = \underline{1.372 \text{ ft}^3/\text{sec}}.$$

$$V_{n_2} = \frac{Q}{2\pi r_2 b_2} = \frac{1.372}{2\pi \times \frac{5}{12} \times \frac{1}{12}} = 6.29 \text{ ft/sec},$$

$$V_{t_2} = u_2 - \frac{V_{n_2}}{\tan \beta_2} = 34.92 - \frac{6.29}{\tan 30^\circ} = 24.03 \text{ ft/sec},$$

$V_{t_1} = 0$ ($\alpha_1 = 90^\circ$ under ideal conditions).

$$\therefore T = \rho Q\left(r_2 V_{t_2} - r_1 V_{t_1}\right) = 1.94 \times 1.372 \times \left(\frac{5}{12} \times 24.03 - 0\right) = \underline{26.6 \text{ lb-ft}},$$

$$\dot{W}_P = \omega T = 83.8 \times 26.6 = \underline{2230 \text{ ft-lb/sec}}, \text{ or } \dot{W}_P = 2230/550 = \underline{4.05 \text{ horsepower}},$$

$$H_t = \omega T / \gamma Q = 2230 / (62.4 \times 1.372) = \underline{26.0 \text{ ft}}.$$

12.3

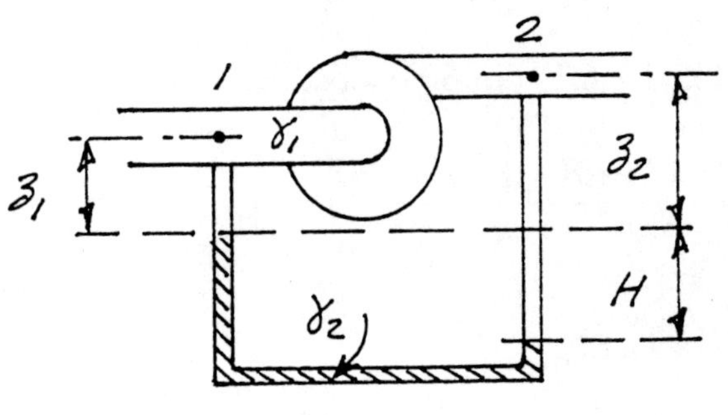

$D_1 = 200$ mm, $D_2 = 150$ mm,
$S_1 = 0.81$, $S_2 = 13.6$,
$H = 600$ mm, $Q = 115$ L/s.

Manometer relation: $p_1 + \gamma_1 z_1 + \gamma_2 H - \gamma_1 H - \gamma_1 z_2 = p_2$,

$$\frac{p_2 - p_1}{\gamma_1} + z_2 - z_1 = \left(\frac{\gamma_2}{\gamma_1} - 1\right)H = \left(\frac{S_2}{S_1} - 1\right)H = \left(\frac{13.6}{0.81} - 1\right)0.6 = 9.47 \text{ m}$$

Energy eqn. from loc. 1 to loc. 2: $\dfrac{p_1}{\gamma_1}+\dfrac{Q^2}{2gA_1^2}+z_1+H_P=\dfrac{p_2}{\gamma_1}+\dfrac{Q^2}{2gA_2^2}+z_2,$

$$\therefore H_P=\frac{Q^2}{2g}\left(\frac{1}{A_2^2}-\frac{1}{A_1^2}\right)+\frac{p_2-p_1}{\gamma_1}+z_2-z_1$$

$$=\frac{0.115^2}{2\times 9.81}\left[\frac{1}{(\pi/4)^2\times 0.15^4}-\frac{1}{(\pi/4)^2\times 0.2^4}\right]+9.47=\underline{10.95\text{ m}}.$$

12.4 $\omega=2000\times\dfrac{\pi}{30}=209\text{ rad/s},$

$$V_{n_2}=\frac{Q}{2\pi r_2 b_2}=\frac{0.0075}{2\pi\times 0.06\times 0.01}=1.99\text{ m/s},$$

$$u_2=\omega r_2=209\times 0.06=12.54\text{ m/s},$$

$$\therefore H_t=\frac{u_2}{g}(u_2-V_{n_2}\cot\beta_2)=\frac{12.54}{9.81}(12.54-1.99\cot 60^\circ)=\underline{14.6\text{ m}}.$$

$$\dot{W}=\gamma QH_t=9810\times 0.8\times 0.0075\times 14.6=\underline{859\text{ W}}.$$

12.4E $Q=2\text{ gal/sec}\times\dfrac{1\text{ ft}^3}{7.48\text{ gal}}=0.267\text{ ft}^3\text{/sec},$

$$\omega=2000\times\frac{\pi}{30}=209\text{ rad/sec},$$

$$V_{n_2}=\frac{Q}{2\pi r_2 b_2}=\frac{0.267}{2\pi\times\frac{2.5}{12}\times\frac{0.4}{12}}=6.12\text{ ft/sec},$$

$$u_2=\omega r_2=209\times\frac{2.5}{12}=43.54\text{ ft/sec},$$

$$\therefore H_t=\frac{u_2}{g}(u_2-V_{n_2}\cot\beta_2)=\frac{43.54}{32.2}(43.54-6.12\cot 60^\circ)=\underline{54.1\text{ ft}}.$$

$$\dot{W}=\gamma QH_t=62.4\times 0.8\times 0.267\times 54.1=\underline{721\text{ ft-lb/sec}},\text{ or}$$

$$\dot{W}=721/550=\underline{1.31\text{ horsepower}}.$$

12.5 $r_0=0.285\text{ m},\ r_i=0.135\text{ m},$

$$V_n=\frac{Q}{\pi(r_0^2-r_i^2)}=\frac{0.57}{\pi(0.285^2-0.135^2)}=2.88\text{ m/s},$$

$$u\cong\omega r_{\text{avg}}=1500\times\frac{\pi}{30}\left(\frac{0.285+0.135}{2}\right)=33.0\text{ m/s}.$$

Use the theoretical head relation

$$H_t = \frac{u^2}{g} - \frac{uV_n}{g}(\cot\alpha_1 + \cot\beta_2);\ \text{substitute in known data and solve for } \beta_2:$$

$$2.85 = \frac{33^2}{9.81} - \frac{33 \times 2.88}{9.81}(\cot 60^\circ - \cot\beta_2) = 111.0 - 9.69(0.5774 - \cot\beta_2),\ \text{or}$$

$\cot\beta_2 = 10.58, \quad \therefore \beta_2 = \underline{5.4^\circ}.$

12.6 Compute loss in suction pipe:

$$h_L = \left(f\frac{L}{D} + \Sigma K\right)\frac{Q^2}{2gA^2}$$

$$= \left(0.015 \times \frac{11}{0.1} + 2 \times 0.19 + 0.8\right)\frac{0.05^2}{2 \times 9.81 \times \left(\frac{\pi}{4}\right)^2 \times 0.1^4} = 5.85\text{ m.}$$

Water at 20°C: $\gamma = 9792\text{ N/m}^3$, $p_v = 2340$ Pa. Substitute known data into NPSH relation, solving for Δz:

$$\therefore \Delta z = \frac{p_{atm} - p_v}{\gamma} - h_L - \text{NPSH} = \frac{101 \times 10^3 - 2340}{9792} - 5.85 - 3 = \underline{1.23\text{ m}}.$$

12.7 a)

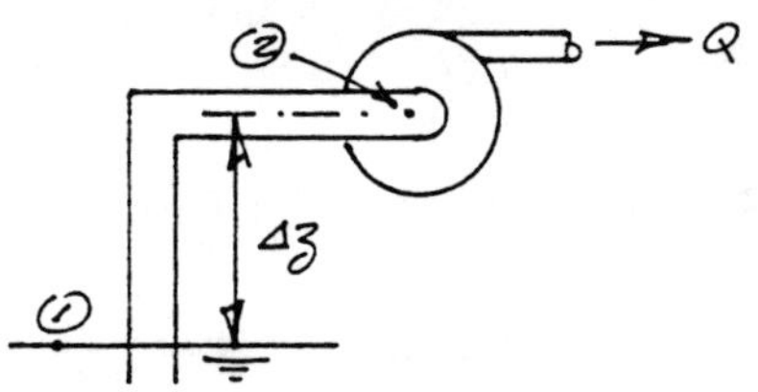

Neglecting losses, write energy eqn. from loc. 1 to loc. 2 to determine the magnitude of Δz.

For water at 25°C, $\gamma = 9780\text{ N/m}^3$ and $p_v = 3290$ Pa abs.

$$\Delta z = \frac{p_1 - p_2}{\gamma} - \frac{Q^2}{2gA_2^2} = \frac{0 - (-68.5 \times 10^3)}{9780} - \frac{0.05^2}{2 \times 9.81 \times \left(\frac{\pi}{4}\right)^2 \times 0.14} = \underline{4.94\text{ m}}.$$

$$\therefore \text{NPSH} = \frac{p_{atm} - p_v}{\gamma} - \Delta z = \frac{101 \times 10^3 - 3290}{9780} - 4.94 = \underline{5.05\text{ m}},\ \text{and}$$

$\sigma = \text{NPSH}/H_p = 5.05/35 = \underline{0.14}.$

b) If $p_{atm} = 83$ kPa, then

$$\Delta z = \frac{p_{atm} - p_V}{\gamma} - \text{NPSH} = \frac{83 \times 10^3 - 3290}{9780} - 5.04 = \underline{3.11\text{ m}}.$$

$\therefore$ Change in elev. is 4.94 - 3.11 = $\underline{1.83\text{ m lower}}$.

12.7E See Pbm. 12.7 for development. For water at 80°F, $\gamma = 62.1 \text{ lb/ft}^3$ and $p_v = 0.51 \text{ lb/in}^2$.

a) $$\Delta z = \frac{0-(-9.9\times 144)}{62.1} - \frac{(13/7.48)^2}{2\times 32.2\times\left(\frac{\pi}{4}\right)^2\times\left(\frac{4}{12}\right)^4} = \underline{16.8 \text{ ft.}}$$

$$\therefore \text{NPSH} = \frac{(14.7 - 0.51)\times 144}{62.2} - 16.8 = \underline{16.1 \text{ ft}}, \text{ and } \sigma = 16.1/115 = \underline{0.14}.$$

b) $$\Delta z = \frac{(12-0.51)\times 144}{62.2} - 16.1 = \underline{10.5 \text{ ft.}}$$

$\therefore$ Change in elev. is 16.8 - 10.5 = $\underline{6.3 \text{ ft lower}}$.

12.8 $$Q_1 = \frac{\omega_1}{\omega_2}Q_2 = \frac{N_1}{N_2}Q_2 = \frac{970}{1200}\times 0.8 = 0.65 \text{ m}^3/\text{s},$$

$\therefore$ from Fig. 12.9, $H_1 \cong 11.5$ m and $\dot{W}_1 \cong 91$ kW.

$$\therefore H_2 = \left(\frac{\omega_2}{\omega_1}\right)^2 H_1 = \left(\frac{1200}{970}\right)^2 \times 11.5 = \underline{17.6 \text{ m}},$$

$$\therefore \dot{W}_2 = \left(\frac{\omega_2}{\omega_1}\right)^3 \dot{W}_1 = \left(\frac{1200}{970}\right)^3 \times 91 = \underline{172 \text{ kW}}.$$

12.9 From Fig. 12.9, at best efficiency, $\text{NPSH}_1 \cong 5$ m. Hence for the new conditions

$$1200 \text{ rpm}, 0.8 \text{ m}^3/\text{s}), \ \text{NPSH}_2 = \left(\frac{\omega_2}{\omega_1}\right)^2 \text{NPSH}_1 = \left(\frac{1200}{970}\right)^2 \times 5 = 7.65 \text{ m}.$$

For water at 50°C, $\gamma = 9693 \text{ N/m}^3$ and $p_v = 12.3$ kPa.

$$\therefore \Delta z = \frac{p_{\text{atm}} - p_v}{\gamma} - \text{NPSH} = \frac{101\times 10^3 - 12.3\times 10^3}{9693} - 7.65 = \underline{1.51 \text{ m}}.$$

12.10

$$C_Q = \frac{Q}{\omega D^3} = \frac{Q/3600}{304\times 0.205^3} = 1.061\times 10^{-4} Q \qquad (Q \text{ in m}^3/\text{h})$$

$$C_H = \frac{gH}{\omega^2 D^2} = \frac{9.81H}{304^2\times 0.205^2} = 2.526\times 10^{-3} H \quad (H \text{ in m})$$

Tabulate C_Q and C_H using selected values of Q and H from Fig. 12.6:

Q (m³/h)	$C_Q \times 10^3$	H (m)	$C_H \times 10$
0	0	54	1.36
50	0.53	53	1.34
100	1.06	52	1.31
150	1.59	50	1.26
200	2.12	47	1.19
250	2.65	41	1.04
300	3.18	33	0.83

The dimensionless curve shown in Fig. 12.12 is for the 240-mm impeller. Since the impellers are not the same (240 mm versus 205 mm), dynamic similitude does not exist, and thus the curves are not the same.

12.11 Given: $N_1 = 2900$ rpm, $N_2 = 3300$ rpm, $Q_2 = 200$ m²/h. To find H_2, we must first find the corresponding Q_1 and H_1.

$$\frac{H_2}{H_1} = \left(\frac{\omega_2}{\omega_1}\right)^2 = \left(\frac{3300}{2900}\right)^2 = 1.295\,, \quad \frac{Q_2}{Q_1} = \frac{\omega_2}{\omega_1} = \frac{3300}{2900} = 1.138$$

$$\therefore Q_1 = \frac{Q_2}{1.138} = \frac{200}{1.138} = 176 \text{ m}^3/\text{h}$$

From Fig. 12.6, $H_1 \approx 58$ m. $\therefore \underline{H_2 = 1.295 \times 58 = 75.1 \text{ m}}$.

12.12 Compute the specific speed: $\Omega_P = \dfrac{\omega\sqrt{Q}}{(gH_P)^{3/4}} = \dfrac{1800 \times \frac{\pi}{30} \times \sqrt{0.15}}{(9.81 \times 22)^{3/4}} = \underline{1.30}$,

hence use a mixed flow pump. As an alternate, since Ω_p is close to unity, a radial flow pump could be employed.

12.13 Compute the specific speed: $\Omega_P = \dfrac{\omega\sqrt{Q}}{(gH_P)^{3/4}} = \dfrac{2000 \times \frac{\pi}{30} \times \sqrt{0.17}}{(9.81 \times 104)^{3/4}} = \underline{0.033}$,

hence use a radial flow pump.

12.14 Fig. 12.13: At $\eta = 0.75$ (best eff.), $C_Q \cong 0.048$, $C_H \cong 0.018$, $C_{\dot{W}} \cong 0.0011$, $C_{\text{NPSH}} \cong 0.023$.

$$\omega = 750 \times \frac{\pi}{30} = 78.5 \text{ rad/s},$$

(a) $$D = \left(\frac{Q}{C_Q \omega}\right)^{1/3} = \left(\frac{1.25}{0.049 \times 78.5}\right)^{1/3} = \underline{0.688 \text{ m}},$$

$$H = \frac{C_H \omega^2 D^2}{g} = \frac{0.018 \times 78.5^2 \times 0.688^2}{9.81} = \underline{5.35 \text{ m}},$$

$$H_{\text{NPSH}} = \frac{C_{\text{NPSH}} \omega^2 D^2}{g} = \frac{0.023 \times 78.5^2 \times 0.688^2}{9.81} = \underline{6.84 \text{ m}},$$

$$\dot{W} = C_{\dot{W}} \rho \omega^3 D^5 = 0.0011 \times 1000 \times 78.5^3 \times 0.688^5 = \underline{8.20 \times 10^4 \text{ W}}, \text{ or } \underline{82 \text{ kW}}$$

(b) $$Q = \omega D^3 C_Q = 78.5 \times 0.688^3 C_Q = 25.56 C_Q$$

$$H = \frac{\omega^2 D^2}{g} C_H = (78.5^2 \times 0.688^2 / 9.81) C_H = 297.3 C_H$$

$$\dot{W} = \rho \omega^3 D^5 \times 10^{-3} C_{\dot{W}} = 1000 \times 78.5^3 \times 0.688^5 \times 10^{-3} C_{\dot{W}} = 7.457 \times 10^4 C_{\dot{W}}$$

C_Q	C_H	$C_{\dot{W}}$	$Q(\text{m}^3/\text{s})$	$H(\text{m})$	$\dot{W}(\text{kW})$
0.015	0.036	—	0.38	10.7	—
0.02	0.031	—	0.51	9.22	—
0.03	0.024	—	0.77	7.14	—
0.04	0.022	0.0013	1.02	6.54	97
0.049	0.018	0.0011	1.25	5.35	82
0.055	0.006	0.0006	1.41	1.78	45

12.14E See Pbm. 12.14 for development.

a) $$D = \left(\frac{45}{0.049 \times 78.5}\right)^{1/3} = \underline{2.27 \text{ ft}},$$

$$H = \frac{0.018 \times 78.5^2 \times 2.27^2}{32.2} = \underline{17.8 \text{ ft}},$$

$$H_{\text{NPSH}} = \frac{0.023 \times 78.5^2 \times 2.27^2}{32.2} = \underline{22.7 \text{ ft}},$$

$$\dot{W} = 0.0011 \times 1.94 \times 78.5^3 \times 2.27^5 = \underline{6.22 \times 10^4 \text{ ft-lb/sec}},$$

or $\dot{W} = 6.22 \times 10^4 / 550 = \underline{113 \text{ horsepower}}.$

b) $Q = 78.5 \times 2.27^3 C_Q = 918.2 C_Q$

$H = (78.5^2 \times 2.27^2 / 32.2) C_H = 986.1 C_H$

$\dot{W} = (1.94 \times 78.5^3 \times 2.27^5 / 550) C_{\dot{W}} = 1.028 \times 10^5 C_{\dot{W}}$

C_Q	$Q(\text{ft}^3/\text{sec})$	$H(\text{ft})$	$\dot{W}(\text{h}_\text{p})$
0.015	13.8	35.5	—
0.02	18.4	30.6	—
0.03	27.6	23.7	—
0.04	36.7	21.7	134
0.049	45.0	17.8	113
0.055	50.5	5.9	62

12.15 $N_1 = 400 \text{ rpm}, \; Q_1 = 0.085 \text{ m}^3/\text{s}, \; H_1 = 7.6 \text{ m},$

$N_2 = 400 \text{ rpm}, \; D_2/D_1 = 2/3, \; g_2/g_1 = 1/2$

$$Q_2 = \frac{N_2}{N_1}\left(\frac{D_2}{D_1}\right)^3 Q_1 = \frac{400}{400}\left(\frac{2}{3}\right)^3 \times 0.085 = \underline{0.0252 \text{ m}^3/\text{s}},$$

$$H_2 = \frac{g_1}{g_2}\left(\frac{N_2}{N_1}\right)^2\left(\frac{D_2}{D_1}\right)^2 H_1 = 2\left(\frac{400}{400}\right)^2\left(\frac{2}{3}\right)^2 \times 7.6 = \underline{6.76 \text{ m}},$$

$$\eta_2 = 1 - \left(\frac{D_1}{D_2}\right)^{1/4}(1-\eta_1) = 1 - \left(\frac{3}{2}\right)^{1/4}(1-0.7) = 0.67,$$

$$\therefore \dot{W}_2 = \rho g_2 Q_2 H_2 / \eta_2 = 1000 \times \frac{9.81}{2} \times 0.0252 \times 6.76 / 0.67 = \underline{1247 \text{ W}}.$$

12.16 $\omega = 600 \times \dfrac{\pi}{30} = 62.8 \text{ rad/s}, \quad Q = 22.7/60 = 0.378 \text{ m}^3/\text{s},$

$$\Omega_P = \frac{\omega\sqrt{Q}}{(gH_P)^{3/4}} = \frac{62.8\sqrt{0.378}}{(9.81 \times 19.5)^{3/4}} = \underline{0.751}, \quad \therefore \text{Use a } \underline{\text{radial flow pump}}.$$

12.17 $H_1 = 19.5 \text{ m}, \; Q_1 = 0.378 \text{ m}^3/\text{s}, \; N_1 = 600 \text{ rpm},$

$\dot{W}_1 = \gamma Q_1 H_1 / \eta = 9810 \times 0.378 \times 19.5 / 0.7 = 103 \times 10^3 \text{ W}, \quad H_2 = 30.5 \text{ m}.$

$$\therefore N_2 = \sqrt{\frac{H_2}{H_1}}\left(\frac{D_2}{D_1}\right) N_1 = \sqrt{\frac{30.5}{19.5}}(1)600 = \underline{750 \text{ rpm}},$$

$$Q_2 = \frac{N_2}{N_1}\left(\frac{D_2}{D_1}\right)^3 Q_1 = \frac{750}{600}(1)^3 0.378 = \underline{0.472 \text{ m}^3/\text{s}},$$

$$\dot{W}_2 = \left(\frac{N_2}{N_1}\right)^3 \left(\frac{D_2}{D_1}\right)^5 \dot{W}_1 = \left(\frac{750}{600}\right)^3 (1)^5\, 103 \times 10^3 = \underline{201 \times 10^3 \text{ W}}$$

12.18 $\frac{H_2}{H_1} = \left(\frac{\omega_2}{\omega_1}\right)^2 \left(\frac{D_2}{D_1}\right)^2 = 2, \quad \frac{Q_2}{Q_1} = \left(\frac{\omega_2}{\omega_1}\right)\left(\frac{D_2}{D_1}\right)^3 = 2,$

$$\therefore \left(\frac{\omega_2}{\omega_1}\right)^2 \left(\frac{D_2}{D_1}\right)^2 = \left(\frac{\omega_2}{\omega_1}\right)\left(\frac{D_2}{D_1}\right)^3, \quad \text{or} \quad \frac{\omega_2}{\omega_1} = \frac{D_2}{D_1}.$$

$$\therefore \left(\frac{\omega_2}{\omega_1}\right)^4 = 2, \quad \omega_2 = \sqrt[4]{2}\,\omega_1 = \underline{1.19\,\omega_1}, \quad \text{and } D_2 = \sqrt[4]{2}\,D_1 = \underline{1.19\,D_1}.$$

12.19 $\omega = 1400 \times \frac{\pi}{30} = 146.6 \text{ rad / s},$

$$\Omega_P = \frac{\omega\sqrt{Q}}{(gH_P)^{3/4}} = \frac{146.6\sqrt{0.182}}{(9.81 \times 60.6)^{3/4}} = 0.52, \quad \therefore \text{use } \underline{\text{radial flow pump}}.$$

Fig. 12.12: $\eta_P \cong 0.76$ (best eff), $C_Q \cong 0.0165$, $C_H = 0.125$.

$\omega D^3 = Q / C_Q = 0.182 / 0.0165 = 11.03,$

$\omega D = \sqrt{gH_P / C_H} = \sqrt{9.81 \times 60.6 / 0.125} = 68.96,$

$\therefore D = \sqrt{11.03 / (\omega D)} = \sqrt{11.03 / 68.96} = \underline{0.40 \text{ m}},$

and $\omega = 68.96 / D = 68.96 / 0.40 = \underline{172.4 \text{ rad / s}}$, or $N = 172.4 \times 30 / \pi = \underline{1646 \text{ rpm}}$.

12.19E $\omega = 1400 \times \frac{\pi}{30} = 146.6 \text{ rad / sec}, \quad Q = 2900 / 449 = 6.46 \text{ ft}^3 \text{ / sec},$

$$\Omega_P = \frac{\omega\sqrt{Q}}{(gH_P)^{3/4}} = \frac{146.6\sqrt{6.46}}{(32.2 \times 200)^{3/4}} = 0.52, \quad \therefore \text{use } \underline{\text{radial flow pump}}.$$

$\omega D^3 = Q / C_Q = 6.46 / 0.0165 = 391.5,$

$\omega D = \sqrt{gH_P / C_H} = \sqrt{32.2 \times 200 / 0.125} = 227,$

$\therefore D = \sqrt{391.5 / (\omega D)} = \sqrt{391.5 / 227} = \underline{1.31 \text{ ft}},$

and $\omega = 227 / D = 227 / 1.31 = \underline{173.3 \text{ rad / sec}}$, or $N = 173.3 \times 30 / \pi = \underline{1655 \text{ rpm}}$.

12.20 Assume a pump speed $N = 2000$ rpm, or $\omega = 2000 \times \dfrac{\pi}{30} = 209$ rad / s.

$$H_P = \dot{W}_f / (\gamma Q) = 200 \times 10^3 / (8830 \times 0.66) = 34.3 \text{ m},$$

$$\therefore \Omega_P = \frac{\omega\sqrt{Q}}{(gH_P)^{3/4}} = \frac{209\sqrt{0.66}}{(9.81 \times 34.3)^{3/4}} = 2.16.$$

The specific speed suggets a mixed-flow pump. However, if $N = 1000$ rpm, a radial-flow pump may be appropriate. Consider both possibilities.

Mixed flow: from Fig. 12.14, at best η_t: $C_{\dot{W}} \cong 0.0117,\ C_Q \cong 0.148,\ C_H \cong 0.067.$

Use $C_Q = Q/(\omega D^3)$ and $C_H = gH_P/(\omega^2 D^2)$; combining and solving for D and ω,

$$D = \sqrt{\frac{Q/C_Q}{\sqrt{gH_P/C_H}}} \quad \text{and} \quad \omega = \frac{Q}{C_Q D^3}. \qquad \therefore D = \sqrt{\frac{0.66/0.148}{\sqrt{9.81 \times 34.3/0.067}}} = \underline{0.251 \text{ m}},$$

$$\omega = 0.66/(0.148 \times 0.251^3) = \underline{282 \text{ rad/s}}, \quad \text{or } N = 282 \times 30/\pi = \underline{2674 \text{ rpm}}.$$

$$\rho = \gamma / g = 8830/9.81 = 900 \text{ kg/m}^3,$$

$$\therefore \dot{W}_P = C_{\dot{W}}\rho\omega^3 D^5 = 0.0117 \times 900 \times 282^3 \times 0.251^5 = \underline{2.35 \times 10^5 \text{ W}}, \text{ or } \underline{235 \text{ kW}}.$$

Radial flow: from Fig. 12.12, at best η_t: $C_{\dot{W}} \cong 0.027,\ C_Q \cong 0.0165,\ C_H \cong 0.125.$

$$\therefore D = \sqrt{\frac{0.66/0.0165}{\sqrt{9.81 \times 34.3/0.125}}} = \underline{0.878 \text{ m}},$$

$$\omega = 0.66/(0.0165 \times 0.878^3) = \underline{59.1 \text{ rad/s}}, \quad \text{or } N = 59.1 \times 30/\pi = \underline{564 \text{ rpm}}.$$

$$\dot{W}_P = 0.0027 \times 900 \times 59.1^3 \times 0.878^5 = \underline{2.62 \times 10^5 \text{ W}}, \quad \text{or } \underline{262 \text{ kW}}.$$

Hence, a mixed-flow pump is preferred.

12.21 Compute required pump head using energy eqn.:

$$H_P = \Delta z + \left(f\frac{L}{D} + \Sigma K\right)\frac{Q^2}{2gA^2}$$

$$= 30 - 15 + \left(0.019 \times \frac{500}{0.75} + 3\right)\frac{1^2}{2 \times 9.81 \times \left(\frac{\pi}{4}\right)^2 \times 0.75^4} = 19.1 \text{ m},$$

$$\omega = 600 \times \pi / 30 = 62.8 \text{ rad/s}, \qquad \Omega_P = \frac{\omega\sqrt{Q}}{(gH_P)^{3/4}} = \frac{62.8\sqrt{1}}{(9.81 \times 19.1)^{3/4}} = 1.24.$$

$\therefore$ Use either a mixed-flow or radial-flow pump.

12.22 Compute required pump head using energy eqn.:

$$H_P = \Delta z + \left(\Sigma K + \frac{fL}{D} \right) \frac{Q^2}{2gA^2}$$

$$= 122 + \left(1.5 + \frac{0.01 \times 61}{0.075} \right) \frac{0.0142^2}{2 \times 9.81 \times \left(\frac{\pi}{4} \right)^2 \times 0.075^4} = 17.3 \text{ m.}$$

From Fig. 12.12: At best efficiency ($\eta_P \cong 0.76$), $C_Q \cong 0.0165$ and $C_H \cong 0.125$. Use $C_Q = Q/(\omega D^3)$ and $C_H = gH_P/(\omega^2 D^2)$ by combining and solving for D and ω:

$$\therefore D = \sqrt{\frac{Q/C_Q}{\sqrt{gH_P/C_H}}} = \sqrt{\frac{0.0142/0.0165}{\sqrt{9.81 \times 17.3/0.125}}} = \underline{0.153 \text{ m}},$$

$$\omega = Q/(C_Q D^3) = 0.0142/(0.0165 \times 0.153^3) = \underline{240 \text{ rad/s}},$$

or $N = 240 \times 30/\pi = \underline{2290 \text{ rpm}}$.

$$\dot{W}_P = \gamma Q H_P / \eta_P = 9810 \times 0.85 \times 0.0142 \times 17.3/0.76 = \underline{2695 \text{ W}}.$$

Expected pump efficiency would be lower because of smaller scale, and since oil has a higher viscosity than water.

12.23

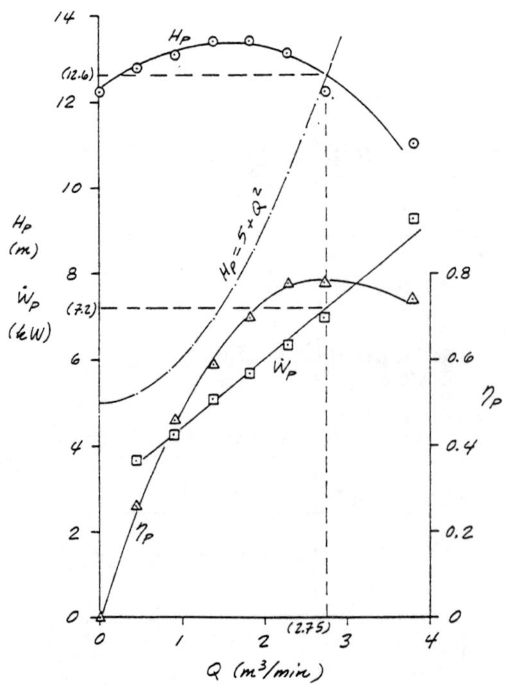

The intersection of the system demand curve with the head-discharge curve yields $\underline{Q = 2.75 \text{ m}^3/\text{min}}$, $\underline{H_P = 12.6 \text{ m}}$, $\underline{\dot{W}_P = 7.2 \text{ kW}}$.

12.24 $N_1 = 1350$ rpm, $Q_1 = 2.75\ \text{m}^3/\text{min}$, $H_1 = 12.6$ m,
$\dot{W}_1 = 7.2$ kW, $N_2 = 1200$ rpm, $D_2 = D_1$.

$$\therefore Q_2 = Q_1\left(\frac{N_2}{N_1}\right)\left(\frac{D_2}{D_1}\right)^3 = 2.75\left(\frac{1200}{1350}\right) = \underline{2.44\ \text{m}^3/\text{min}},$$

$$H_2 = H_1\left(\frac{N_2}{N_1}\right)^2\left(\frac{D_2}{D_1}\right)^2 = 12.6\left(\frac{1200}{1350}\right)^2 = \underline{9.96\ \text{m}},$$

$$\dot{W}_2 = \dot{W}_1\left(\frac{N_2}{N_1}\right)^3\left(\frac{D_2}{D_1}\right)^5 = 7.2\left(\frac{1200}{1350}\right)^3 = \underline{5.06\ \text{kW}}.$$

Efficiency will remain approximately the same.

12.25

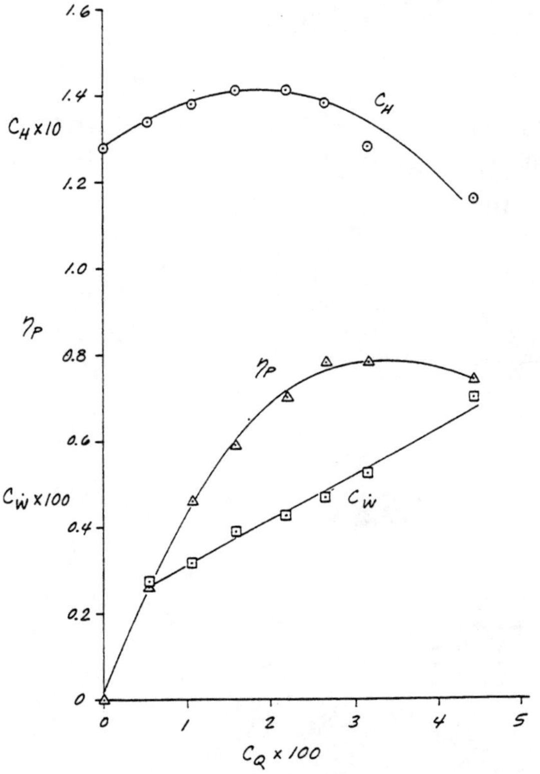

$\omega = 1350 \times \pi / 30 = 141.4$ rad/s,

$C_Q = (\omega D^3)^{-1} Q = (141.4 \times 0.216^3)^{-1} Q = \underline{0.702Q}$ (Q in m^3/s)

$C_H = (\omega D)^{-2} g H_P = (141.4 \times 0.216)^{-2} H_P = \underline{0.0105 H_P}$ (H_P in m)

$C_{\dot{W}} = (\rho\omega^3 D^5)^{-1} \dot{W}_P = (1000 \times 141.4^3 \times 0.216^5)\dot{W}_P = \underline{7.53 \times 10^{-7}\dot{W}_P}$ ($\dot{W}_P$ in W)

At best η_P, $C_Q \cong 0.032$, $C_H \cong 0.134$, $\therefore \Omega_P = C_Q^{1/2} C_H^{-3/4} = (0.032/2)^{1/2}(0.134)^{-3/4} = \underline{0.57}$

(Note: Use $C_Q/2$ in place of C_Q because of double entry.)

12.26 Compute system demand:

$$H_P = \left(f\frac{L}{D} + \Sigma K\right)\frac{V^2}{2g} = \left(0.01 \times \frac{14}{0.3} + 4 \times 0.1\right)\frac{3^2}{2 \times 9.81} = \underline{0.40 \text{ m}},$$

$$Q = VA = 3 \times \frac{\pi}{4} \times 0.3^2 = \underline{0.212 \text{ m}^2}, \quad \omega = 300 \times \pi / 30 = \underline{31.4 \text{ rad/s}}.$$

$$\therefore \Omega_P = \frac{\omega\sqrt{Q}}{(gH_P)^{3/4}} = \frac{31.4\sqrt{0.212}}{(9.81 \times 0.4)^{3/4}} = \underline{5.19}, \quad \therefore \text{Axial pump is appropriate.}$$

12.27 From Fig. 12.13, at best η_P, $C_Q \cong 0.049$, and $C_H \cong 0.019$.

$$\therefore \omega = \frac{Q}{C_Q D^3} = \frac{0.212}{0.049 \times 0.3^3} = \underline{160 \text{ rad/s}}, \text{ or } \quad N = 160 \times 30/\pi = \underline{1530 \text{ rpm}},$$

$$H_P = \frac{C_H \omega^2 D^2}{g} = \frac{0.019 \times 160^2 \times 0.3^2}{9.81} = \underline{4.46 \text{ m}}.$$

Since the head rise is too large for the demand, the pump is not well suited for the desired application. In an attempt to reduce H_P, try operating the pump at the extreme end of the curve, i.e., $C_Q \cong 0.055$ and $C_H \cong 0.005$. Then:

$$\omega = \frac{0.212}{0.055 \times 0.3^3} = \underline{143 \text{ rad/s}} \quad (N = 1370 \text{ rpm}), \text{ and}$$

$$H_P = \frac{0.005 \times 1143^2 \times 0.3^2}{9.81} = \underline{0.94 \text{ m}}.$$

12.28 a)

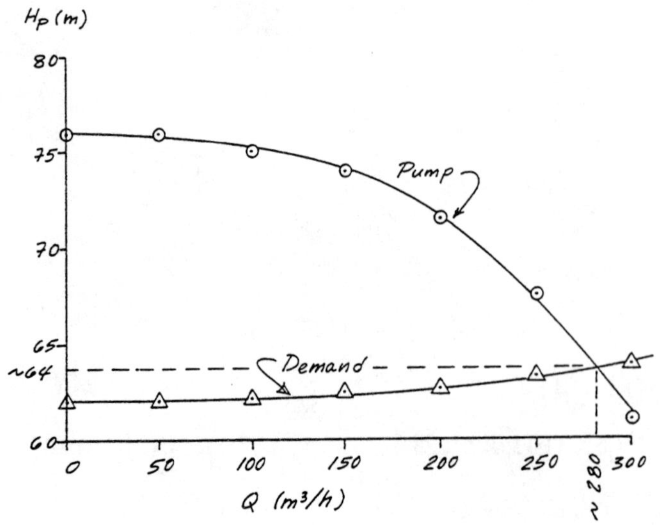

The intersection of the pump curve with the demand curve yields $H_P \cong \underline{64 \text{ m}}$ and $Q \cong \underline{280 \text{ m}^3/\text{h}}$. $\therefore$ From Fig. 12.6, $\dot{W}_P \cong \underline{64 \text{ kW}}$ and NPSH $\cong$ $\underline{8.3 \text{ m}}$.

(b)

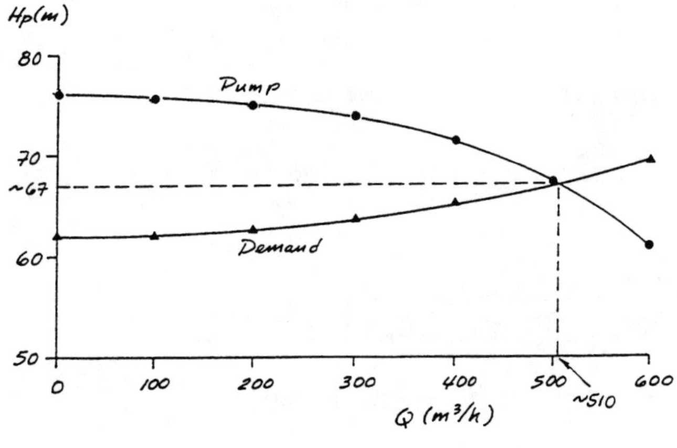

The intersection of the two curves gives $H_P \cong 67$ m and $Q \cong \underline{510\text{m}^3/\text{h}}$.
$\therefore$ From Fig. 12.6; with $\underline{Q = 255 \text{ m}^3/\text{h}}$ for one pump, $\dot{W}_P \cong \underline{2 \times 60 = 120 \text{ kW}}$, and $\underline{\text{NPSH} \cong 7.6 \text{ m}}$.

12.29 $Q = 600/3600 = 0.167 \text{ m}^3/\text{s}$, $H_P = \Delta p/(\rho g) = 1000 \times 10^3/(9.81 \times 607) = 168$ m, $\omega = 2500 \times \pi/30 = 262$ rad/s.

$$\Omega_P = \frac{\omega\sqrt{Q}}{(gH_P)^{3/4}} = \frac{262\sqrt{0.167}}{(9.81 \times 168)^{3/4}} = \underline{0.41}, \quad \therefore \text{use a } \underline{\text{radial-flow pump}}.$$

From Fig. 12.12, at best efficiency, $C_Q \cong 0.0165$ and $C_H \cong 0.125$,

$$\therefore D = \left(\frac{Q}{C_Q\omega}\right)^{1/3} = \left(\frac{0.167}{0.0165 \times 262}\right)^{1/3} = \underline{0.338 \text{ m}},$$

$$H = \frac{C_H\omega^2D^2}{g} = \frac{0.125 \times 262^2 \times 0.338^2}{9.81} = \underline{100 \text{ m}}.$$

$\therefore$ Since required pumping head is 168 m, $\underline{\text{use two stages}}$.

12.30 Use energy eqn. to establish system demand:

$$f \cong 1.325\left[\ln 0.27\frac{e}{D}\right]^{-2} = 1.325\left[\ln 0.27 \times \frac{0.25}{450}\right]^{-2} = 0.017,$$

$$H_P = \Delta z + f\frac{L}{D}\frac{Q^2}{2gA^2} = 195 + \frac{0.017 \times 5000}{2 \times 9.81 \times (\pi/4)^2 \times 0.45^5}Q^2 = 195 + 381Q^2.$$

From Fig. 12.6, at best η_P, $Q \cong 260/3600 = \underline{0.072 \text{ m}^3/\text{s}}$, and $H_P \cong 66$ m. Assume three pumps in series, so that $H_P = 3 \times 66 = 198$ m. Then the demand head is $H_P = 195 + 381 \times 0.072^2 = \underline{197 \text{ m}}$. Hence three pumps in series are appropriate. The required power is

$$\dot{W}_P = \gamma Q H_P / \eta_P = 9810 \times 0.86 \times 0.072 \times 198 / 0.75 = \underline{1.60 \times 10^5 \text{ W}},$$

or $\dot{W}_P = \underline{160 \text{ kW}}$.

12.30E Use energy eqn. to establish system demand:

$$f \cong 1.325\left[\ln 0.27\frac{e}{D}\right]^{-2} = 1.325\left[\ln 0.27 \times \frac{0.00085}{1.5}\right]^{-2} = 0.017,$$

$$H_P = \Delta z + f\frac{L}{D}\frac{Q^2}{2gA^2} = 640 + \frac{0.017 \times 3 \times 5280}{2 \times 32.2 \times (\pi/4)^2 \times 1.5^5}Q^2 = 640 + 0.893Q^2.$$

From Fig. 12.6, at best η_P, $Q \cong \frac{1100}{449} = \underline{2.45 \text{ ft}^3/\text{sec}}$, and $H_P \cong 215$ ft. Assume three pumps in series, so that $H_P = 3 \times 215 = 645$ ft. Then the demand head is $H_P = 640 + 0.893 \times 2.45^2 = \underline{645 \text{ ft}}$. Hence three pumps in series are appropriate. The required power is

$$\dot{W}_P = \gamma Q H_P / \eta_P = 62.4 \times 0.86 \times 2.45 \times 645 / 0.75 = \underline{1.13 \times 10^5 \ \frac{\text{ft - lb}}{\text{sec}}},$$

or $\dot{W}_P = 1.13 \times 10^5 / 550 = 206$ hp.

12.31E Given:

$$Q = \frac{500}{7.48 \times 60} = 1.114 \text{ ft}^3/\text{s}, \quad p_2 = 80 \times 144 = 11520 \text{ lb/ft}^2,$$

$$L = 150 \text{ ft}, \ \text{D} = \frac{10}{12} = 0.833 \text{ ft}, \ \Delta z = 75 \text{ ft, and } \text{V}_2 = \frac{1.114}{0.7854 \times 0.833^2} = 2.044 \text{ ft/s}$$

The required pumping head is

$$H_P = \frac{p_2}{\gamma} + \Delta z + \left(1 + \frac{fL}{D}\right)\frac{V^2}{2g} = \frac{11520}{62.4} + 75 + \left(1 + \frac{0.02 \times 150}{0.833}\right)\frac{2.044^2}{2 \times 32.2} = 259.9 \text{ ft}$$

The performance data for the mixed flow pump (Fig. 12.12) is
$C_Q = 0.0165, \quad C_H = 0.124, \quad \eta = 0.75$. Determine the speed and head delivered by a one-stage impeller:

$$d = \frac{8}{12} = 0.667 \text{ ft}, \quad \omega = \frac{Q}{C_Q d^3} = \frac{1.114}{0.0165 \times 0.667^3} = 228 \text{ rad/s}$$

$$\therefore N = \frac{30\omega}{\pi} = \underline{2180 \text{ rpm}}, \text{ and } H = \frac{C_H \omega^2 d^2}{g} = \frac{0.124 \times 228^2 \times 0.667^2}{32.2} = 89 \text{ ft..}$$

Number of stages required: $\frac{H_P}{H} = \frac{260}{89} = 2.92$ (use three stages)

Power requirement: $\dot{W}_P = \frac{62.4 \times 1.114 \times 3 \times 89}{0.75} = \underline{2.48 \times 10^4 \text{ ft-lb/sec, or } 45 \text{ hp}}.$

12.32 a) For water at 80°C, $p_v = 46.4 \times 10^3$ Pa, and $\rho = 9553$ kg/m³. Write the energy equation from the inlet (section i) to the location of cavitation in the pump:

$$\frac{p_i}{\gamma} + \frac{V_i^2}{2g} = \frac{p_v}{\gamma} + NPSH,$$

$$\therefore NPSH = \frac{p_i - p_v}{\gamma} + \frac{V_i^2}{2g} = \frac{(83 - 46.4) \times 10^3}{9533} + \frac{6^2}{2 \times 9.81} = \underline{5.67 \text{ m}}.$$

b) $NPSH_1 = 5.67$ m, $N_1 = 2400$ rpm, $N_2 = 1000$ rpm, $D_2/D_1 = 4$.

$$\therefore NPSH_2 = NPSH_1 \left(\frac{N_2}{N_1}\right)^2 \left(\frac{D_2}{D_1}\right)^2 = 5.67 \left(\frac{1000}{2400}\right)^2 (4)^2 = \underline{15.8 \text{ m}}.$$

12.33 Compute H_P and ω:

$$H_P = z_2 - z_1 + \left(\Sigma K + \frac{fL}{D}\right) \frac{Q^2}{2gA^2}$$

$$= 23 - 20 + \left(1.5 + \frac{0.02 \times 60}{0.75}\right) \frac{1.25^2}{2 \times 9.81 \times (0.7854 \times 0.75^2)^2} = 4.265 \text{ m}$$

$$\omega = 600 \times \frac{\pi}{30} = 62.8 \text{ rad/s}$$

(a) $\Omega_P = \frac{\omega \sqrt{Q}}{(gH_P)^{3/4}} = \frac{62.8\sqrt{1.25}}{(9.81 \times 4.265)^{3/4}} = 4.27$. Hence use an axial pump.

b) From Fig. 12.13, $\eta_P \approx 0.75$.

$$\therefore \dot{W}_P = \frac{\gamma Q H_P}{\eta_P} = \frac{9810 \times 1.25 \times 4.265}{0.75} = 6.97 \times 10^4 \text{ W, or approx. } \underline{70 \text{ kW}}.$$

12.34 Given:

$L = 200$ m, $D = 0.05$ m, $\Delta z = z_2 - z_1 = -3$ m, $V = 3$ m / s, $\nu = 6 \times 10^{-7}$ m^2 / s,

$\rho = 1593$ kg / m^3, $p_v = 86.2 \times 10^3$ Pa, $p_a = 101 \times 10^3$ Pa.

Compute the pump head:

$$\text{Re} = \frac{3 \times 0.05}{6 \times 10^{-7}} = 2.5 \times 10^5,\ f = 1.325\left[\ln\left(5.74\text{Re}^{-0.9}\right)\right]^{-2} = 1.325\left[\ln 5.74 \times (2.5 \times 10^5)^{-0.9}\right]^{-2} = 0.015,$$

$$H_P = \Delta z + \left(1 + \frac{fL}{D}\right)\frac{V^2}{2g} = -3 + \left(1 + \frac{0.015 \times 200}{0.05}\right)\frac{3^2}{2 \times 9.81} = 25.0 \text{ m}.$$

a) Choose a radial-flow pump. Use Fig. 12.12 to select the size and speed:

$C_H = 0.124,\ C_Q = 0.0165,\ \eta = 0.75,$

$Q = 3 \times \frac{\pi}{4} \times 0.05^2 = 0.00589$ m^2,

$$\therefore D = \sqrt[4]{\frac{C_H Q^2}{C_Q^2 g H_P}} = \sqrt[4]{\frac{0.124 \times 0.00589^2}{0.0165^2 \times 9.81 \times 25.0}} = \underline{0.090 \text{ m}}$$

$$\therefore \omega = \frac{Q}{C_Q D^3} = \frac{0.00589}{0.0165 \times 0.090^3} = 490 \text{ rad / s, or } N = 490 \times \frac{30}{\pi} = \underline{4680 \text{ rpm}}.$$

b) Available net positive suction head:

$$NPSH = \frac{p_a - p_v}{\rho g} - \Delta z = \frac{101 \times 10^3 - 86.2 \times 10^3}{1593 \times 9.81} + 3 = \underline{3.95 \text{ m}}.$$

c)

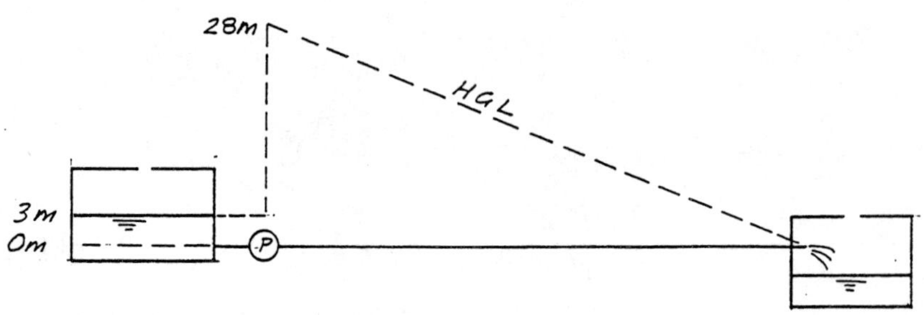

12.35 Given: $D = 1.5$ m, $\rho = 1.2\ \text{kg/m}^3$, $U = 20$ m/s, $A = 2\text{m}^2$, $N = 2500$ rpm.

Compute speed, discharge, and C_Q:

$$\omega = 2500 \times \frac{\pi}{30} = 262\ \text{rad/s},\quad Q = UA = 20 \times 2 = 40\ \text{m}^3/\text{s},$$

$$C_Q = \frac{Q}{\omega D^3} = \frac{40}{262 \times 1.5^3} = 0.045$$

Hence, from Fig. 12.13, $C_W = 0.0012$, and $C_H = 0.02$.

(a) Evaluate the power:

$$\dot{W} = C_{\dot{W}} \rho \omega^3 D^5 = 0.0012 \times 1.2 \times 262^3 \times 1.5^5 = 1.97 \times 10^5\ \text{W, or } \underline{197\ \text{kW}}.$$

(b) Compute the head rise across the fan, and the corresponding pressure rise:

$$H_P = \frac{C_H \omega^2 D^2}{g} = \frac{0.02 \times 262^2 \times 1.5^2}{9.81} = 315\ \text{m},$$

$$\Delta p = \rho g H_P = 1.2 \times 9.81 \times 315 = \underline{3710\ \text{Pa}}.$$

12.36 $\omega = 120 \times \pi / 30 = 12.6\ \text{rad/s},$

$$\alpha_1 = \cot^{-1}(2\pi r_1^2 b_1 \omega / Q + \cot\beta_1)$$

$$= \cot^{-1}(2\pi \times 4.5^2 \times 0.85 \times 12.6/150 + \cot 75^\circ) = \underline{6.1^\circ}.$$

$$V_{t_1} = u_1 + V_{n_1}\cot\beta_1 = \omega r_1 + \frac{Q}{2\pi r_1 b_1}\cot\beta_1$$

$$= 12.6 \times 4.5 + \frac{150}{2\pi \times 4.5 \times 0.85}\cot 75^\circ = 58.37\ \text{m/s},$$

$$V_{t_2} = u_2 + V_{n_2}\cot\beta_2 = \omega r_2 + \frac{Q}{2\pi r_2 b_2}\cot\beta_2$$

$$= 12.6 \times 2.5 + \frac{150}{2\pi \times 2.5 \times 0.85}\cot 100^\circ = 29.52\ \text{m/s},$$

$$\therefore T = \rho Q(r_1 V_{t_1} - r_2 V_{t_2}) = 1000 \times 150(4.5 \times 58.37 - 2.5 \times 29.52) = \underline{2.83 \times 10^7\ \text{N}\cdot\text{m}}.$$

$$\dot{W}_T = \omega T = 12.6 \times 2.83 \times 10^7 = \underline{3.57 \times 10^8\ \text{W}},\ \text{or } \underline{357\ \text{MW}}.$$

Under ideal conditions $\eta_T = 1$, and $\dot{W}_T = \dot{W}_f$, hence

$$H_T = \dot{W}_T / \gamma Q = 3.57 \times 10^8 / (9810 \times 150) = \underline{243\ \text{m}}.$$

12.37 $\dfrac{D_1}{D_2}=\dfrac{1}{5}$, $\dot{W}_1 = 3$ kW, $N_1 = 360$ rpm, $H_1 = 1.8$ m, $H_2 = 5.8$ m.

From the similarity rules $\dfrac{\dot{W}_2}{\dot{W}_1}=\left(\dfrac{N_2}{N_1}\right)^3\left(\dfrac{D_2}{D_1}\right)^5$ and $\dfrac{H_2}{H_1}=\left(\dfrac{N_2}{N_1}\right)^2\left(\dfrac{D_2}{D_1}\right)^2$.

Eliminate $(N_2/N1)$ and solve for $\dot{W}_2$. From the second relation,

$\dfrac{N_2}{N_1}=\left(\dfrac{H_2}{H_1}\right)^{1/2}\left(\dfrac{D_1}{D_2}\right)$, and substituting into the first relation,

$$\dot{W}_2=\dot{W}_1\left(\frac{H_2}{H_1}\right)^{3/2}\left(\frac{D_2}{D_1}\right)^2=3\left(\frac{5.8}{1.8}\right)^{3/2}(5)^2=\underline{434\text{ kW}}.$$

Subsequently N_2 is determined:

$$N_2=N_1\left(\frac{H_2}{H_1}\right)^{1/2}\left(\frac{D_1}{D_2}\right)=360\left(\frac{5.8}{1.8}\right)^{1/2}\left(\frac{1}{5}\right)=\underline{129\text{ rpm}}.$$

12.38 $N_2 = 240$ rpm, $\dot{W}_2 = 2200$ kW, $H_2 = 50$ m, $D_2 = 1$ m, $\dot{W}_1 = 9$ kW, $H_1 = 7.6$ m.

From the similarity rules, $\dfrac{\dot{W}_2}{\dot{W}_1}=\left(\dfrac{N_2}{N_1}\right)^3\left(\dfrac{D_2}{D_1}\right)^5$ and $\dfrac{H_2}{H_1}=\left(\dfrac{N_2}{N_1}\right)^2\left(\dfrac{D_2}{D_1}\right)^2$.

Substitute second eqn. into the first to eliminate (N_2/N_1), and solve for D_1:

$$\therefore D_1=D_2\left(\frac{\dot{W}_1}{\dot{W}_2}\right)^{1/2}\left(\frac{H_2}{H_1}\right)^{3/4}=1\left(\frac{2}{2200}\right)^{1/2}\left(\frac{50}{7.6}\right)^{3/4}=\underline{0.263\text{ m}}$$

$$N_2=N_1\left(\frac{H_1}{H_2}\right)^{1/2}\left(\frac{D_2}{D_1}\right)=240\left(\frac{7.6}{50}\right)^{1/2}\left(\frac{1}{0.263}\right)=\underline{356\text{ rpm}}.$$

12.38E See Pbm. 12.38 for development. $N_2 = 240$ rpm, $\dot{W}_2 = 2200$ kW, $D_2 = 3$ ft, $\dot{W}_1 = 9$kW, $H_1 = 25$ ft.

$$\therefore D_1=D_2\left(\frac{\dot{W}_1}{\dot{W}_2}\right)^{1/2}\left(\frac{H_2}{H_1}\right)^{3/4}=3\left(\frac{2}{2200}\right)^{1/2}\left(\frac{150}{25}\right)^{3/4}=\underline{0.736\text{ ft}},$$

$$N_2=N_1\left(\frac{H_1}{H_2}\right)^{1/2}\left(\frac{D_2}{D_1}\right)=240\left(\frac{25}{150}\right)^{1/2}\left(\frac{3}{0.736}\right)=\underline{399\text{ rpm}}.$$

12.39 From Fig. 12.24, $\eta_T = 0.91$, $C_H = 0.23$, and $C_{\dot{W}} = 0.027$. Use definitions of dimensionless coefficients to determine D and ω:

$$\omega = \left(\frac{gH_T}{C_H D^2}\right)^{1/2} = \left(\frac{\dot{W}_T}{C_{\dot{W}} \rho D^5}\right)^{1/3}, \text{ and } \quad D = \left(\frac{\dot{W}_T}{C_{\dot{W}} \rho}\right)^{1/2} \left(\frac{C_H}{gH_T}\right)^{3/4}$$

From given data compute $\dot{W}_T$: $\dot{W}_T = \gamma Q H_T \eta_T = 9810 \times 3 \times 80 \times 0.91 = 2.14 \times 10^6$ W,

$$\therefore D = \left(\frac{2.14 \times 10^6}{0.027 \times 1000}\right)^{1/2} \times \left(\frac{0.23}{9.81 \times 80}\right)^{3/4} = \underline{0.631 \text{ m}},$$

$$\text{and } \omega = \left(\frac{9.81 \times 80}{0.23 \times 0.631^2}\right)^{1/2} = \underline{92.6 \text{ rad/s}}, \text{ or } N = 92.6 \times \frac{30}{\pi} = \underline{884 \text{ rpm}}.$$

12.40 Write energy eqn. from upper reservoir (loc. 1) to surge tank (loc. 2) and solve for Q:

$$Q = \left[2gA^2 \frac{D}{fL}(z_1 - z_2)\right]^{1/2}$$

$$= \left[\frac{2 \times 9.81 \times (\pi/4)^2 \times 0.85^5}{0.025 \times 2000}(650 - 648.5)\right]^{1/2} = \underline{0.401 \text{ m}^3/\text{s}}.$$

Apply energy eqn. from loc. 2 to lower reservoir (loc. 3) and determine H_T:

$$H_T = z_2 - z_3 - \left(f\frac{L}{D} + K_v\right)\frac{Q^2}{2gA^2}$$

$$= 648.5 - 595 - \left(\frac{0.02 \times 100}{0.85} + 1\right)\frac{0.401^2}{2 \times 9.81 \times \left(\frac{\pi}{4}\right)^2 \times 0.85^5} = \underline{53.4 \text{ m}}.$$

$\therefore \dot{W}_T = \gamma Q H_T \eta_T = 9810 \times 0.401 \times 53.4 \times 0.9 = \underline{1.89 \times 10^5 \text{ W}}$, or $\underline{189 \text{ kW}}$.

∴ From Fig. 12.32, use a <u>Francis turbine</u>.

A representative value of the specific speed is 2 (Fig. 12.20):

$$\therefore \omega = \frac{\Omega_T (gH_T)^{5/4}}{(\dot{W}_T/\rho)^{1/2}} = \frac{2(9.81 \times 53.3)^{5/4}}{(2.67 \times 10^5/1000)^{1/2}} = \underline{306 \text{ rad/s}},$$

or $N = 306 \times 30/\pi = \underline{2920 \text{ rpm}}$.

12.41 Prototype: $N_1 = 420$ rpm, $H_1 = 3$ m, $Q_1 = 0.312\ \text{m}^3/\text{s}$, $\eta_1 = 0.9$;

Model: $N_2 = 2000$ rpm; $\dfrac{D_2}{D_1} = \dfrac{1}{6}$.

$\dot{W}_1 = \gamma Q_1 H_1 \eta_1 = 9810 \times 0.312 \times 3 \times 0.9 = 8260$ W.

Use similarity rules to compute H_2, Q_2, $\dot{W}_2$ and η_2:

$$\therefore H_2 = H_1\left(\frac{N_2}{N_1}\right)^2\left(\frac{D_2}{D_1}\right)^2 = 3\left(\frac{2000}{420}\right)^2\left(\frac{1}{6}\right)^2 = \underline{1.89\ \text{m}},$$

$$Q_2 = Q_1\left(\frac{N_2}{N_1}\right)\left(\frac{D_2}{D_1}\right)^3 = 0.312\left(\frac{2000}{420}\right)\left(\frac{1}{6}\right)^3 = \underline{0.0688\ \text{m}^3/\text{s}},$$

$$\dot{W}_2 = \dot{W}_1\left(\frac{N_2}{N_1}\right)^3\left(\frac{D_2}{D_1}\right)^5 = 8260\left(\frac{2000}{420}\right)^3\left(\frac{1}{6}\right)^5 = \underline{115\ \text{W}},$$

$$\eta_2 = 1 - (1-\eta_1)\left(\frac{D_1}{D_2}\right)^{1/4} = 1 - (1-0.9)(6)^{1/4} = \underline{0.84}.$$

12.42 $\omega = 200 \times \pi/30 = 20.94$ rad/s, and from Fig. 12.29 at best efficiency ($\eta_T = 0.8$), $\phi \approx 0.42$, $c_v = 0.94$.

$$V_1 = c_v\sqrt{2gH_T} = 0.94\sqrt{2\times 9.81\times 120} = 45.6\ \text{m/s},$$

$$\therefore Q = \frac{\dot{W}_T}{\gamma H_T \eta_T} = \frac{4.5\times 10^6}{9810\times 120\times 0.8} = \underline{4.78\ \text{m}^3/\text{s}}$$ (this is the discharge from all of the jets).

Determine the wheel radius r: $\quad r = \dfrac{\phi\sqrt{2gH_T}}{\omega} = \dfrac{0.42\sqrt{2\times 9.81\times 120}}{20.94} = 0.973$ m,

Hence the diameter of the wheel is $2r = 2 \times 0.973 = \underline{1.95\ \text{m}}$.

Compute diameter of one jet: $\quad D_j = 2r/8 = 1.95/8 = \underline{0.244\ \text{m}}$, or $\underline{244\ \text{mm}}$.

Let N_j = no. of jets. Then each jet has a discharge of Q/N_j and an area $\dfrac{Q/N_j}{V_1} = \dfrac{\pi}{4}D_j^2$. $\therefore$ Solving for N_j: $N_j = \dfrac{Q}{V_1}\dfrac{1}{\left(\frac{\pi}{4}D_j^2\right)} = \dfrac{4.78}{45.6\times\left(\frac{\pi}{4}\right)\times 0.244^2} = \underline{2.24}$,

$\therefore$ Use $\underline{\text{three}}$ jets. $\quad \Omega_T = \dfrac{\omega(\dot{W}_T/\rho)^{1/2}}{(gH_T)^{5/4}} = \dfrac{20.94(4.5\times 10^6/1000)^{1/2}}{(9.81\times 120)^{5/4}} = \underline{0.204}.$

12.43

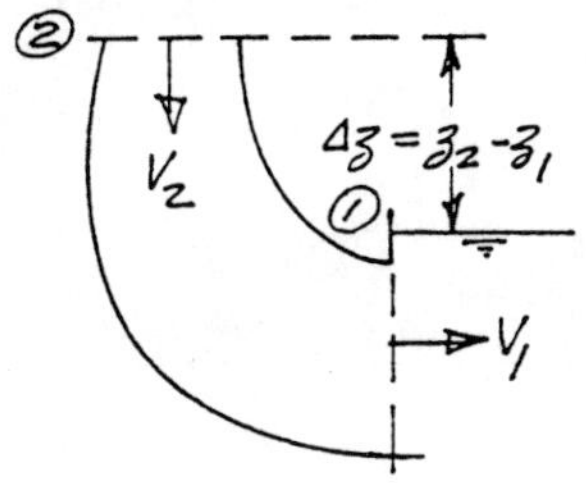

$$V_1 = \frac{Q}{\pi r_1^2} = \frac{85}{\pi \times 5^2} = 1.08 \text{ m/s},$$

$$V_2 = \frac{Q}{\pi r_2^2} = \frac{85}{\pi \times 2.5^2} = 4.33 \text{ m/s}.$$

Assume $p_{\text{atm}} = p_1 = 101$ kPa

a) Write energy eqn. from loc. 2 to loc. 1 and solve for p_2 (absolute):

$$p_2 = p_1 + \frac{\rho}{2}\left(V_1^2 - V_2^2\right) - \gamma \Delta z$$

$$= 101 \times 10^3 + \frac{998}{2}(1.08^2 - 4.33^2) - 9800 \times 2.5 = \underline{6.77 \times 10^4 \text{ Pa}}, \text{ or } \underline{68 \text{ kPa}}.$$

b) $$\Delta z = \frac{p_{\text{atm}} - p_v}{\gamma} - \sigma H_T$$

$$= \frac{101 \times 10^3 - 2340}{9800} - 0.14 \times 31.8 = \underline{5.62 \text{ m}}.$$

12.44E $Q = 73{,}530/6 = 12{,}260 \text{ ft}^3/\text{sec}$ (one unit),

$$H_T = \frac{\dot{W}_T}{\gamma Q \eta_T} = \frac{427{,}300 \times 550}{62.4 \times 12\,260 \times 0.85} = 361.4 \text{ ft}$$

Write energy eqn. from upper reservoir (loc. 1) to lake (loc. 2):

$$z_1 = \left(f\frac{L}{D} + \Sigma K\right)\frac{Q^2}{2gA^2} + H_T + z_2,$$

$$\therefore 1030 = \left(\frac{0.01 \times 1300}{D} + 0.5\right)\frac{12{,}260^2}{2 \times 32.2 \times \left(\frac{\pi}{4}\right)^2 D^4} + 361.4 + 660,$$

which reduces to $8.6 - \dfrac{4.919 \times 10^7}{D^5} - \dfrac{1.892 \times 10^6}{D^4} = 0.$ $\therefore$ Solving, $\underline{D \cong 25.8 \text{ ft}}$

$\dot{W}_T = 437{,}000 \times 0.746 \cong 326{,}000$ kW, $H_T = 361.4 \times 0.3048 \cong 110$ m,

$\therefore$ from Fig. 12.32, a Francis <u>or pump/turbine</u> unit is indicated.

12.45 From Fig. 12.24, $\eta_T = 0.91$, $C_Q = 0.13$, $C_H = 0.23$.

$\omega = 480 \times \pi / 30 = 50.3 \text{ rad/s}$,

$$\therefore D = \left(\frac{gH_T}{C_H \omega^2} \right)^{1/2} = \left(\frac{9.81 \times 9.5}{0.23 \times 50.3^2} \right)^{1/2} = \underline{0.400 \text{ m}},$$

$$Q = C_Q \omega D^3 = 0.13 \times 50.3 \times 0.4^3 = \underline{0.418 \text{ m}^3/\text{s}},$$

$$\dot{W}_T = \gamma Q H_T \eta_T = 9.81 \times 0.418 \times 9.5 \times 0.9 = \underline{3.55 \times 10^4 \text{ W}}, \text{ or } \underline{35.5 \text{ kW}}.$$

12.46 a) Let H be the total head and Q the discharge delivered to the turbine; then

$$H_T = 0.95H = 0.95 \times 305 = 289.8 \text{ m, and}$$

$$Q = \frac{\dot{W}_T}{\gamma H_T \eta_T} = \frac{10.4 \times 10^6}{9810 \times 289.8 \times 0.85} = 4.30 \text{ m}^3/\text{s}.$$

Write energy eqn. from reservoir to turbine outlet:

$$H = H_T + \left(f\frac{L}{D} + \Sigma K \right) \frac{Q^2}{2gA^2},$$

$$\therefore 305 = 289.8 + \left(\frac{0.02 \times 3000}{D} + 2 \right) \frac{4.3^2}{2 \times 9.81 \times (\pi/4)^2 D^4},$$

which reduces to $15.2 - \frac{91.67}{D^5} - \frac{3.06}{D^4} = 0.$ $\therefore$ Solving $\underline{D = 1.45 \text{ m}}$.

b) Compute jet velocity: $V_1 = c_v \sqrt{2gH_T} = 0.98\sqrt{2 \times 9.81 \times 289.8} = 73.9 \text{ m/s}$

The flow through one nozzle is $Q/4$, and the jet area is $\pi D_j^2 / 4$. Hence

$$\frac{\pi D_j^2}{4} = \frac{Q/4}{V_1} = \frac{4.3/4}{73.9} = 0.0146 \text{ m}^2,$$

$$\therefore D_j = \sqrt{\frac{4 \times 0.0146}{\pi}} = \underline{0.136 \text{ m}}.$$

12.47 Determine power available from <u>each</u> turbine:

$$\dot{W}_T = \rho\left[\frac{\Omega_T (gH_T)^{5/4}}{\omega}\right]^2 = 1000\left[\frac{2.42\times(9.81\times3.7)^{5/4}}{50\times\pi/30}\right]^2 = 1.69\times10^6 \text{ W}$$

The total power developed by all turbines is

$$\gamma Q H_T \eta_T = 9810\times282\times3.7\times0.9 = 9.21\times10^6 \text{ W}$$

Hence, the required number of units is 9.21/1.69 = 5.4, ∴ <u>use six turbines</u>.

12.48 $\dot{W}_T = 1000\left[\frac{4.15\times(9.81\times3.7)^{5/4}}{50\times\pi/30}\right]^2 = 4.99\times10^6 \text{ W (one unit)}.$

Total power developed is 9.21×10^6 W. Hence, required number of units is $9.21/4.99 = 1.8$, ∴ <u>use two turbines</u>.

12.49 a)

$$Q = \frac{1200}{60\times1000} = 0.02 \text{ m}^3/\text{s},$$

$$H_T = z_1 - z_2 - \left(\frac{fL}{D}+\Sigma K\right)\frac{Q^2}{2gA^2}$$

$$= 70-47-\left(\frac{0.02\times105}{0.10}+2\right)\frac{0.02^2}{2\times9.81\times(0.7854\times0.10^2)^2} = 15.4 \text{ m},$$

$$\dot{W}_f = \gamma Q H_T = 9810\times0.02\times15.4 = \underline{3020 \text{ W}}.$$

b) From Fig. 12.24, $C_H = 0.23$, $C_{\dot{W}} = 0.027$, $C_Q = 0.132$, and $\eta_T = 0.91$.

$$\omega D^3 = \frac{Q}{C_Q} = \frac{0.02}{0.132} = 0.152, \text{ or } \omega^2 D^6 = 0.0231$$

$$\omega^2 D^2 = \frac{gH_T}{C_H} = \frac{9.81\times15.4}{0.23} = 657, \qquad \therefore D = \sqrt[4]{\frac{0.0231}{657}} = \underline{0.077 \text{ m}},$$

$$\therefore \omega = \frac{0.152}{D^3} = \frac{0.152}{0.077^3} = 333 \text{ rad/s, or } N = 333\times\frac{30}{\pi} = \underline{3180 \text{ rpm}}.$$

c)

$$\eta_2 = 1-(1-\eta_1)\left(\frac{D_1}{D_2}\right)^{1/4} = 1-(1-0.91)\left(\frac{1000}{77}\right)^{1/4} = \underline{0.83}$$

$$\dot{W}_T = \dot{W}_f \eta = 3020\times0.83 = \underline{2510 \text{ W}}$$

12.50 a)

$$H_T = z_1 - z_2 - \frac{fL}{D}\frac{Q^2}{2gA^2}$$

$$= 915 - 892 - \frac{0.015 \times 350 \times 0.25^2}{0.3 \times 2 \times 9.81 \times (0.7854 \times 0.3^2)^2} = 11.8 \text{ m},$$

$$\dot{W}_T = \gamma Q H_T \eta_T = 9810 \times 0.25 \times 11.8 \times 0.85 = 2.46 \times 10^4 \text{ W}, \text{ or } \underline{24.6 \text{ kW}}.$$

b) Compute the specific speed of the turbine:

$$\omega = N\frac{\pi}{30} = 1200 \times \frac{\pi}{30} = 126 \text{ rad/s},$$

$$\therefore \Omega_T = \frac{\omega\sqrt{\dot{W}_T/\rho}}{(gH_T)^{5/4}} = \frac{126\sqrt{2.46 \times 10^4/1000}}{(9.81 \times 11.8)^{5/4}} = 1.65$$

Hence, from Fig. 12.20, a Francis turbine is appropriate.

c) From Fig. 12.24, the turbine with $\Omega_T = 1.063$ is chosen: $C_H = 0.23$, $C_Q = 0.13$, and $\eta_T = 0.91$.

$$D = \sqrt[4]{\frac{C_H Q^2}{C_Q^2 g H_T}} = \sqrt[4]{\frac{0.23 \times 0.25^2}{0.13^2 \times 9.81 \times 11.8}} = 0.293 \text{ m, or approximately } \underline{0.30 \text{ m}};$$

$$\omega = \frac{Q}{C_Q D^3} = \frac{0.25}{0.13 \times 0.30^3} = 71.2 \text{ rad/s, or } N = 71.2 \times \frac{30}{\pi} = \underline{680 \text{ rpm}};$$

$$\dot{W}_T = \gamma Q H_T \eta_T = 9810 \times 0.25 \times 11.8 \times 0.91 = 2.63 \times 10^4 \text{ W, or } \underline{26.3 \text{ kW}}.$$

Calculate a new specific speed based on the final design data:

$$\Omega_T = \frac{71.2 \times \sqrt{2.63 \times 10^4/1000}}{(9.81 \times 11.8)^{5/4}} = \underline{0.96}.$$ This value is acceptable according to Fig. 12.20.

THIRTEEN

Measurements in Fluid Mechanics

13.1 A reading of 4 cm provides a vertical measurement of $4\sin 20^\circ = 1.368$ cm.
$\therefore p = .01368 \times (9810 \times 13.6) = \underline{1825\ \text{Pa}}.$

13.2 a) $\dfrac{V^2}{2} \times \rho = p = \dfrac{\gamma_{Hg} h}{100}; \quad \therefore \dfrac{V^2}{2} \times 1.22 = (9810 \times 13.6)\dfrac{h}{100}$,

$\therefore V = 46.8\sqrt{h}$ and $\underline{C = 46.8}$.

b) $\dfrac{V^2}{2} \times (.8217 \times 1.22) = (9810 \times 13.6)\dfrac{h}{100}. \quad \therefore V = 51.6\sqrt{h}$, and $\underline{C = 51.6}$.

13.3 $\dfrac{1}{2}\rho V^2 = \gamma h. \quad \therefore h = \dfrac{1}{2} \times 1.22 \times 8^2 \,/\, 9810 = .00398$ m or $\underline{3.98\ \text{mm}}$.
The reading is too small for accurate measurements.

13.4 $Q = \dfrac{V\!\!\!\!-}{\Delta t} = \dfrac{2 \times 10^{-3}}{10 \times 60} = \underline{3.33 \times 10^{-6}\ \text{m}^3/\text{s}}.$

$\dot{m} = \rho A V = \rho Q = 1000 \times 3.33 \times 10^{-6} = \underline{0.00333\ \text{kg/s}}.$

$V = \dfrac{Q}{A} = \dfrac{3.33 \times 10^{-6}}{\pi \times .001^2} = \underline{1.06\ \text{m/s}}. \quad \therefore \text{Re} = \dfrac{1.06 \times .002}{10^{-6}} = 2120.$

The flow may be laminar or turbulent depending on the smoothness of the pipe walls, the entry construction, the flow fluctuation level, and the length of the pipe (is it short and entry flow only, is it fully developed?).

13.4E $Q = \dfrac{V\!\!\!\!-}{\Delta t} = \dfrac{.5 \times \dfrac{1}{7.481}}{10 \times 60} = \underline{1.114 \times 10^{-4}\ \text{ft}^3/\text{sec}}.$

$\dot{m} = \rho Q = 1.94 \times 1.114 = \underline{2.16 \times 10^{-4}\ \text{slug/sec}}.$

$V = \dfrac{Q}{A} = \dfrac{1.114 \times 10^{-4}}{\pi \times .05^2 / 144} = \underline{2.04\ \text{fps}}. \quad \therefore \text{Re} = \dfrac{2.04 \times .1/12}{10^{-5}} = \underline{1702}.$

The flow is laminar.

13.5 $Q = [.5 \times 2.6 + .5 \times 6.65 + 1(8.65 + 9.5 + 9.9 + 10)] \times 100 \times 10^{-4} \times 2 = \underline{0.854 \text{ m}^3/\text{s}}.$

$V = \frac{Q}{A} = \frac{.854}{.1 \times 1} = \underline{8.54 \text{ m/s}}.$

13.6 $Q = [\pi \times 1^2 \times 10 + \pi(2^2 - 1^2) \times 9.9 + \pi(3^2 - 2^2) \times 9.5 + \pi(4^2 - 3^2) \times 8.65 +$

$\pi(4.5^2 - 4^2) \times 6.65 + \pi(5^2 - 4.5^2) \times 2.6] \times 10^{-4} = \underline{0.0592 \text{ m}^3/\text{s}}.$

$V = \frac{Q}{A} = \frac{.0592}{\pi \times .05^2} = \underline{7.54 \text{ m/s}}.$

13.7 $p + \gamma h = \gamma_{Hg} h.\quad V_1 = \frac{.09}{\pi \times .06^2} = 7.96 \text{ m/s}.\quad V_0 = 31.8 \text{ m/s}.\quad \frac{D_0}{D} = 0.5.$

$\text{Re} = \frac{VD}{\nu} = \frac{7.96 \times .12}{10^{-6}} = 9.55 \times 10^5.\quad \therefore K = 1.02.\quad \therefore \frac{Q}{A_0} = V_0 = K\sqrt{2g(h_1 - h_2)}.$

$\frac{31.8^2}{1.02^2} = 2g\frac{p_1 - p_2}{\gamma}.\quad \therefore p_1 - p_2 = \frac{31.8^2}{1.02^2} \times \frac{9810}{2 \times 9.81} = 486\,000 \text{ Pa}.$

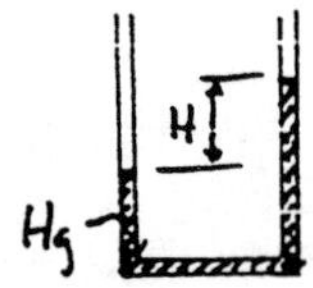

Manometer: $p_1 + \gamma H = p_2 + \gamma_{Hg} H.\quad \therefore \frac{p_1 - p_2}{\gamma_{Hg} - \gamma} = H.$

$\therefore H = \frac{486\,000}{9810(13.6 - 1)} = \underline{3.93 \text{ m}}.$

13.8 See the sketch above: $\frac{p_1 - p_2}{\gamma_{Hg} - \gamma} = H.\quad \therefore p_1 - p_2 = 9810(13.6 - 1) \times .12 = 14\,830 \text{ Pa}.$

$\therefore h_1 - h_2 = \frac{p_1 - p_2}{9810} = \frac{14\,830}{9810} = 1.512 \text{ m}.\quad \frac{D_0}{D} = \frac{15}{24} = 0.625.$

a) Assume $\text{Re} = 10^5.\quad \therefore K = .68.\quad \therefore Q = .68\pi \times .075^2\sqrt{2 \times 9.81 \times 1.512} = 0.0654 \text{ m}^3/\text{s}.$

Check: $V = \frac{Q}{A} = 1.45 \text{ m/s}.\quad \text{Re} = \frac{1.45 \times .24}{10^{-6}} = 3.5 \times 10^5.$

$\therefore K = .67$ and $\underline{Q = 0.064 \text{ m}^3/\text{s}}.$

b) Assume $\text{Re} = 10^5.\quad \therefore K = 1.05.\quad \therefore Q = 1.05\pi \times .075^2\sqrt{2 \times 9.81 \times 1.512} = \underline{0.101 \text{ m}^3/\text{s}}.$

Check: $V = \frac{Q}{A} = 2.23 \text{ m/s}.\quad \text{Re} = \frac{2.23 \times .24}{10^{-6}} = 5.4 \times 10^5.\quad \therefore$ OK.

13.9 $p_1 + 9810 \times .12 = p_2 + 13.6 \times 9810 \times .12. \quad \therefore p_1 - p_2 = 14\,830 \text{ Pa.} \quad \frac{D_0}{D} = \frac{6}{12} = 0.5$

$$h_1 - h_2 = \frac{14\,830}{9810} = 1.512 \text{ m.}$$

a) $\text{Re} = 10^5. \quad \therefore K = 1.0. \quad \therefore Q = 1.0\pi \times .03^2 \sqrt{2 \times 9.81 \times 1.512} = \underline{0.0154 \text{ m}^3/\text{s.}}$

Check: $V = \frac{Q}{A} = \frac{.0154}{\pi \times .06^2} = 1.36 \text{ m/s.} \quad \therefore \text{Re} = \frac{1.36 \times .12}{.661 \times 10^{-6}} = 2.5 \times 10^5.$

$K = 1.01. \quad \therefore \text{OK.}$

b) Assume $\text{Re} = 10^5. \quad \therefore K = .63. \quad \therefore Q = .63\pi \times .03^2 \sqrt{2 \times 9.81 \times 1.512} = \underline{0.0097 \text{ m}^3/\text{s.}}$

Check: $V = \frac{Q}{A} = \frac{.0097}{\pi \times .06^2} = 0.858 \text{ m/s.} \quad \therefore \text{Re} = \frac{.858 \times .12}{.661 \times 10^{-6}} = 1.6 \times 10^5. \quad \therefore \text{OK.}$

13.10 $Q = KA\sqrt{\frac{R\Delta p}{D\rho}}$ where $K = 1 - \frac{6.5}{\sqrt{\text{Re}}}. \quad \text{Re} = \frac{VD}{\nu} = \frac{Q4}{\pi D \nu}.$

a) Assume $\text{Re} = 10^5$. Then $K = .98. \quad Q = .98\pi \times .05^2 \sqrt{\frac{.2 \times 80\,000}{.1 \times 1000}} = 0.0974 \text{ m}^3/\text{s.}$

Check $V = \frac{Q}{A} = \frac{.0974}{\pi \times .05^2} = 12.4 \text{ m/s.} \quad \therefore \text{Re} = \frac{12.4 \times .1}{10^{-6}} = 1.2 \times 10^6. \quad \therefore K = .99 \quad \therefore \text{OK}$

b) Assume $\text{Re} = 10^6$. Then $K = .99. \qquad \therefore \underline{Q = 0.098 \text{ m}^3/\text{s.}}$

$$Q = .99\pi \times .025^2 \sqrt{\frac{.2 \times 80\,000}{.05 \times 1000}} = \underline{0.035 \text{ m}^3/\text{s.}}$$

Check: $\text{Re} = \frac{.035 \times 4}{\pi \times .05 \times 10^{-6}} = 9 \times 10^5. \quad \therefore \text{OK.}$

c) Assume $\text{Re} = 10^6$. Then $K = .99. \quad Q = .99\pi \times \left(\frac{2}{12}\right)^2 \sqrt{\frac{(8/12) \times 10 \times 144}{(4/12) \times 1.94}} = \underline{3.33 \text{ cfs.}}$

Check: $\text{Re} = \frac{3.33 \times 4}{\pi \times 4/12 \times 10^{-5}} = 1.3 \times 10^6. \quad \therefore \text{OK.}$

d) Assume $\text{Re} = 10^6$. Then $K = .99. \quad Q = .99\pi \times \left(\frac{1}{12}\right)^2 \sqrt{\frac{(8/12) \times 10 \times 144}{(2/12) \times 1.94}} = \underline{1.18 \text{ cfs.}}$

Check: $\text{Re} = \frac{1.18 \times 4}{\pi \times 2/12 \times 10^{-5}} = 9 \times 10^5. \quad \therefore \text{OK.}$

13.11 a) $v_1(\rho, p_T, p_1) = \sqrt{2(102-95)\times10^3/680} = 4.537\ \text{m/s},$

$v_1(\rho+\delta\rho, p_T, p_1) = \sqrt{2(102-95)\times10^3/730} = 4.379\ \text{m/s},$

$v_1(\rho, p_T+\delta p_T, p_1) = \sqrt{2(103-95)\times10^3/680} = 4.581\ \text{m/s},$

$v_1(\rho, p_T, p_1+\delta p_1) = \sqrt{2(102-96)\times10^3/680} = 4.201\ \text{m/s},$

$\therefore\ \delta v_1 = \sqrt{(4.379-4.537)^2+(4.851-4.537)^2+(4.201-4.537)^2} = \underline{0.486\ \text{m/s}}$

b) $v_1(\rho, p_T - p_1) = 4.537\ \text{m/s}$

$v_1(\rho+\delta\rho, p_T - p_1) = 4.379\ \text{m/s}$

$v_1[\rho, p_T - p_1 + \delta(p_T - p_1)] = 4.851\ \text{m/s}$

$\therefore \delta V_1 = \sqrt{(4.379-4.537)^2-(4.851-4.537)^2} = \underline{0.3515\ \text{m/s}}$

c) Arrangement b) is preferable, since δv_1 is smaller (one less reading, resulting in one less uncertainty measurement).

13.12 $A_0 = \frac{\pi}{4}\times\left(\frac{33.3}{1000}\right)^2 = 8.709\times10^{-4}\text{m}^2,\ \beta = 33.3/54 = 0.617$

$$K = \frac{Q}{A_0\sqrt{2g(S-1)\Delta h}} \quad (\Delta h \text{ in meters of mercury})$$

$$= \frac{Q}{8.709\times10^{-4}\sqrt{2\times9.81\times(13.6-1)\Delta h}} = 73.03\frac{Q}{\sqrt{\Delta h}} = K(Q,\Delta h)$$

$$\delta K = \sqrt{[K(Q,\Delta h)-K(Q+\delta Q,\Delta h)]^2+[K(Q,\Delta h)-K(Q,\Delta h+\delta(\Delta h))]^2}$$

No.	$K(Q,\Delta h)$	$K(Q+\delta Q,\Delta h)$	$K(Q,\Delta h+\delta(\Delta h))$	δK
1	1.076	1.147	1.002	0.1026
2	1.024	1.093	0.9805	0.0816
3	1.017	1.082	0.9862	0.0719
4	1.026	1.095	1.001	0.0734
5	1.031	1.100	1.012	0.0716
6	1.041	1.112	1.025	0.0728
7	1.027	1.096	1.012	0.0706
8	1.025	1.084	1.013	0.0602
9	1.005	1.063	0.9938	0.0591
10	1.025	1.083	1.015	0.0589
11	1.041	1.099	1.032	0.0587
12	1.027	1.085	1.018	0.0587
13	1.043	1.103	1.036	0.0604
14	1.040	1.109	1.033	0.0694
15	0.9935	1.050	0.9864	0.0569

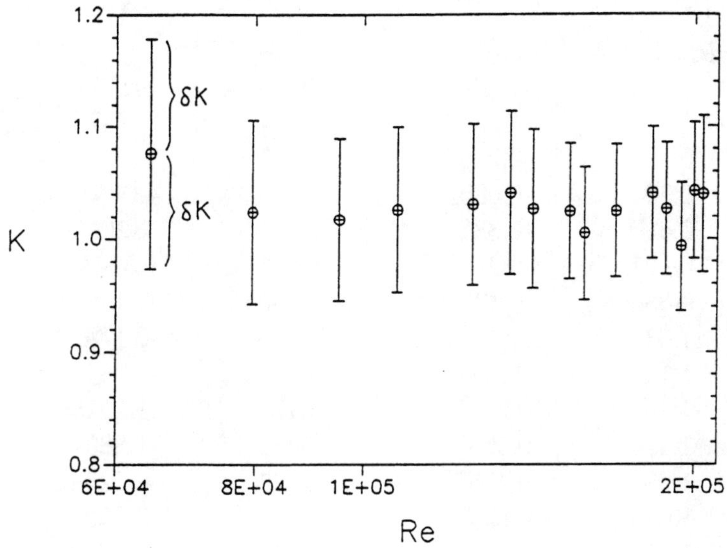

Compare these data with Fig. 13.10, curve labeled "Venturi meters and nozzles," $\beta = 0.6$.

13.13 From Example 13.1:

No.	$\delta K / K$
1	0.1173
2	0.0859
3	0.0789
4	0.0703
5	0.0633
6	0.0628
7	0.0622
8	0.0614
9	0.0507
10	0.0507
11	0.0455
12	0.0449
13	0.0455
14	0.0448
15	0.0441

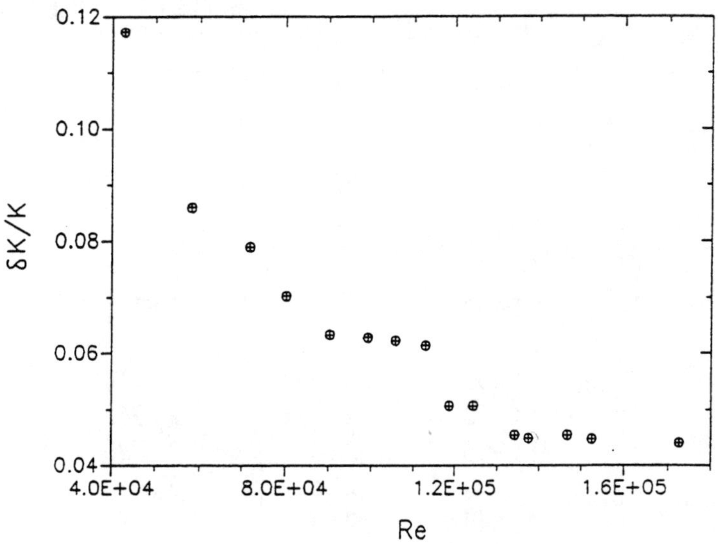

Relative uncertainty $\delta K / K$ decreases with increasing Re.

13.14

i	$(S-1)\Delta h$ (m water)	$x_i^{(1)}$	$y_i^{(2)}$	$x_i y_i$	x_i^2
1	0.164	-1.8079	-6.3890	11.5507	3.2685
2	0.277	-1.2837	-6.1754	7.9274	1.6479
3	0.403	-0.9088	-5.9955	5.4487	0.8259
4	0.504	-0.6852	-5.8746	4.0253	0.4695
5	0.680	-0.3857	-5.7199	2.2062	0.1488
6	0.781	-0.2472	-5.6408	1.3944	0.06111
7	0.882	-0.1256	-5.5940	0.7026	0.01578
8	1.033	0.03247	-5.5165	-0.1791	0.001054
9	1.147	0.1371	-5.4846	-0.7519	0.01880
10	1.260	0.2311	-5.4171	-1.2519	0.05341
11	1.424	0.3535	-5.3412	-1.8881	0.1250
12	1.550	0.4383	-5.3124	-2.3284	0.1921
13	1.688	0.5235	-5.2533	-2.7501	0.2741
14	1.764	0.5676	-5.2344	-2.9710	0.3222
15	1.764	0.5676	-5.2805	-2.9972	0.3222
	Σ	-2.5929	-84.2292	+18.1376	+7.7464

[(1)] $x_i = \ln[(S-1)\Delta h]$, $S = 13.6$, Δh in meters mercury, [(2)] $y_i = \ln Q$, Q in m^3/s

$$\therefore m = \frac{18.1376 - (-2.5929)(-84.2292)/15}{7.7464 - (2.5929)^2/15} = \underline{0.4902},$$

$$b = \frac{-84.2292 - 0.4902(-2.5929)}{15} = -5.5305, \qquad \therefore C = \exp(-5.5305) = \underline{0.00396}.$$

From Pbm. 13.11, $A_0 = 8.709 \times 10^{-4}\,\text{m}^2$ and $K_{\text{avg}} = 1.029$. Hence in Eq. 13.3.8, $K_{\text{avg}} A_0 \sqrt{2g} = 1.029 \times (8.709 \times 10^{-4})\sqrt{2 \times 9.81} = \underline{0.00397}$, and the exponent is $\underline{0.500}$.

13.15E

i	$x_i = \ln Y$	$y_i = \ln Q$	$x_i y_i$	x_i^2
1	-0.9493	-1.9733	1.8733	0.9012
2	-0.9729	-2.0402	1.9849	0.9465
3	-0.9835	-2.0557	2.0218	0.9673
4	-1.0300	-2.1716	2.2367	1.0609
5	-1.0906	-2.3403	2.5523	1.1894
6	-1.1394	-2.3677	2.6978	1.2982
7	-1.1648	-2.5145	2.9289	1.3568
8	-1.1842	-2.5498	3.0195	1.4023
9	-1.1973	-2.5889	3.0997	1.4335
10	-1.2140	-2.6636	3.2336	1.4738
11	-1.3356	-2.9604	3.9539	1.7838
12	-1.4106	-3.1350	4.4222	1.9898
13	-1.4355	-3.1749	4.5576	2.0549
14	-1.5799	-3.4451	5.4429	2.4961
Σ	-16.6876	-35.9810	+44.0251	+20.3545

$$\therefore m = \frac{44.0251 - (-16.6876)(-35.9810)/14}{20.3545 - (16.6876)^2/14} = \underline{2.4533},$$

$$b = \frac{-35.9810 - 2.4533(-16.6876)}{14} = 0.3542,$$

$$\therefore C = \exp(0.3542) = \underline{1.425} \text{ and } \underline{Q = 1.425\ Y^{2.45}}$$

Eq. 10.4.27, use $\theta = 60^\circ$, $C_d \cong 0.58$; then

$$Q = 0.58 \times \frac{8}{15}\sqrt{2 \times 32.2}\ \tan\left(\frac{60^\circ}{2}\right) Y^{5/2} = \underline{1.433 Y^{5/2}}$$

FOURTEEN

Environmental Fluid Mechanics

14.1 $q=-K\frac{dT}{dx}=-0.86\times\frac{15\text{-}25}{0.005}=\underline{1720\ \text{W}/\text{m}^2}$

14.2 a) The temperature changes only in the *x*-direction

b) $q=-K\frac{dT}{dx}=\frac{T_2-T_1}{L}$

c) $q=-K\frac{dT}{dx}=-35\ \times\frac{50\text{-}100}{0.4}=\underline{4.3\ \times\ 10^3\ \text{W/m}^2}$

d) $q=-K\left(i\frac{\partial T}{\partial x}+j\frac{\partial T}{\partial y}+k\frac{\partial T}{\partial z}\right)$

$(q=-K\,\nabla T=-K\ \text{grad}\ T\quad \text{where}\ \ \nabla=i\frac{\partial}{\partial x}+j\frac{\partial}{\partial y}+k\frac{\partial}{\partial z})$

14.3 a) $J=-D\frac{dC}{dx}=-0.142\times\frac{1}{10^4}\times\frac{0\text{-}35}{0.01}=\underline{4.97\times10^{-2}\ \text{mol}/\text{m}^2-\text{s}}$

b) $M=J\times A\times MW\times t=4.97\times10^{-2}\times3.14\times\frac{1}{10^4}\times44.01\times2\times3600=\underline{4.9\ g}$,

or $\underline{4.9\times10^{-3}\ \text{kg}}$.

14.4

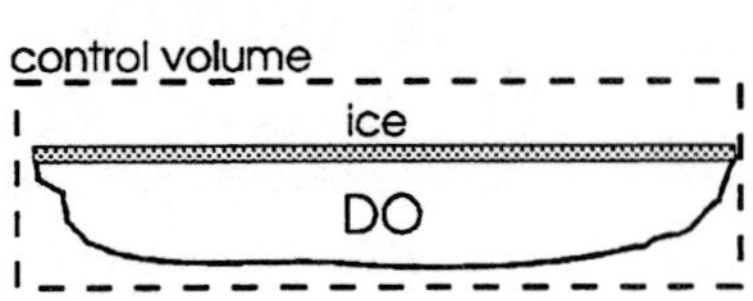

The control volume is shown in the accompanying figure. The control volume is constant, and there is no DO flux at an inlet and an outlet locations. Therefore Eq. 14.3.1 reduces to $r\,\forall=\frac{dC}{dt}\,\forall$. Substituting given expression for r we can rearrange the above equation in the manner $\int_{C(0)}^{C(t)} \frac{dC}{C} = -k\int_0^t dt$. The solution is $C(t) = C_0 \exp(-k\,t)$ where C_0 is the initial concentration.

a) Substitute known values of k and C_o into the solution to find the concentration after 30 days:

$$C(t) = C_o \exp(-kt) = 12.4\exp\left(-0.1 \times \frac{1}{86400} \times 30 \times 86400\right) = \underline{0.62\,\text{mg}/\text{L}}$$

b) The required time follows from the solution of the above equation:

$$t = -\frac{1}{k}\ln\left(\frac{C(t)}{C_o}\right) = -\frac{1}{0.1 \times \frac{1}{86400}}\ln\left(\frac{3}{12.4}\right) \approx \underline{14\,\text{days}}$$

14.5

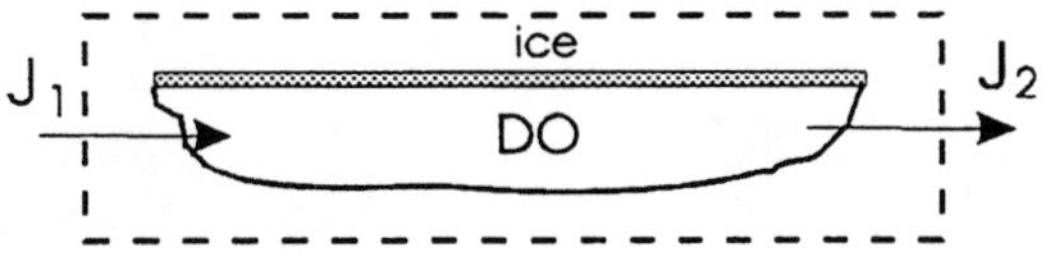

The control volume is shown in the accompanying figure. The control volume is constant, and there is constant flux of DO at an inlet and an outlet locations. Therefore Eq. 14.3.1 reduces to $r\,\forall = \frac{dC}{dt}\forall + J_2 - J_1$. Substituting in the given expression for r, and noting that $J_1 = Q\,C_s$ and $J_2 = Q\,C$, we can rearrange the equation in the manner $\frac{dC}{dt} + \left(\frac{Q}{\forall} + k\right) C = \frac{Q\,C_s}{\forall}$. This differential equation is subjected to the constant concentration initial condition: at $t = 0$, $C = C_o$. The solution is $C(t) = C_e + (C_o - C_e)\exp\left[-\left(\frac{Q}{\forall} + k\right) t\right]$ where $C_e = \frac{Q\,C_s}{Q + k\forall}$. As $t \to \infty$, the equation reduces to $C = C_e$, and the pond concentration approaches an equilibrium concentration.

a) Substituting known values of $Q = 2$ L/s, $k = 0.1$ day^{-1}, $C_o = 12.4$ mg/l, $C_s = 13.8$ mg/l, $\forall = 10^4$ m^3, and $t = 30$ days:

$$C_e = \frac{2 \times \frac{1}{10^3} \times 13.8}{2 \times \frac{1}{10^3} + 0.1 \times \frac{1}{86400} \times 10^4} = \underline{2.0\,\text{g}/\text{m}^3}, \text{ or } \underline{2.0\,\text{mg}/\text{L}} \text{ , and}$$

$$C = 2.0 + (12.4 - 2.0)\exp\left[-\left(\frac{2\times\frac{1}{10^3}}{10^4} + 0.1\times\frac{1}{86400}\right)\times 30\times 86400\right] = \underline{2.3\ \mathrm{g/m^3}}.$$

b) The required time follows from the solution of the above equation:

$$t = -\frac{1}{\frac{Q}{\forall}+k}\ln\left(\frac{C\text{-}C_e}{C_o - C_e}\right) = -\frac{1}{\frac{2\times\frac{1}{10^3}}{10^4} + 0.1\times\frac{1}{86400}}\ln\left(\frac{3-2}{12.4-2}\right) \approx \underline{20\ \text{days}}.$$

Notice also that the DO concentration is never lower than the equilibrium concentration $C_e = 2.0$ mg/L.

14.6

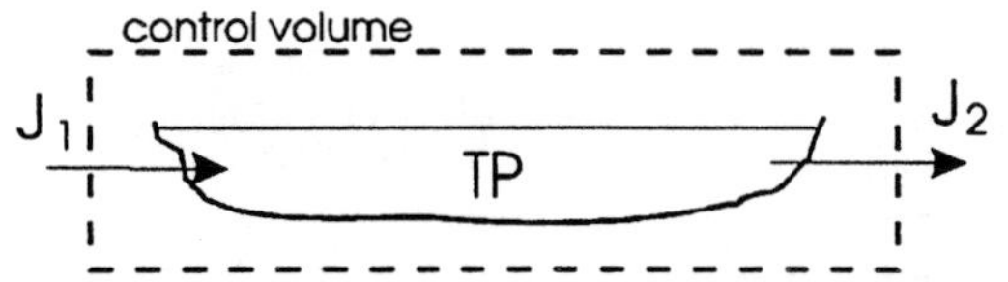

The control volume is shown in the accompanying figure. The control volume is constant, and there is constant flux of total phosphorus (TP) at the inlet and outlet locations. The governing equation reduces to the same form as given in Problem 14.5. The solution is $C(t) = C_e + (C_o - C_e)\exp\left[-\left(\frac{Q}{\forall}+k\right)t\right]$, where

$C_e = \frac{J_1}{Q+k\forall}$. We can estimate the initial concentration C_o from a mass balance prior to the sewage diversion:

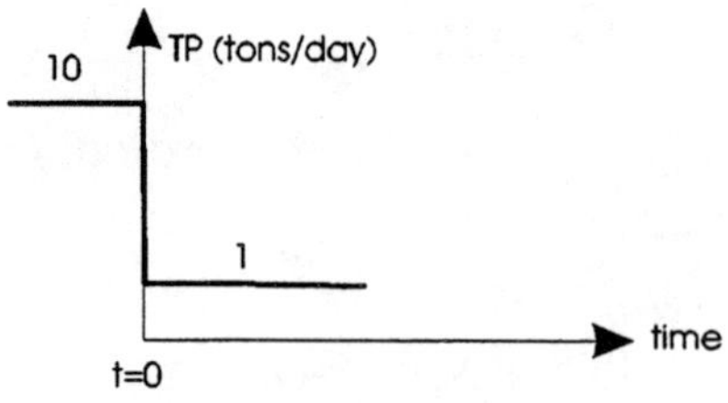

The TP is discharged for a long time, that is $\frac{d}{dt}(\ ) = 0$, prior to the sewage diversion. Therefore the steady-state mass balance becomes $\left(\frac{Q}{\forall}+k\right)C = \frac{Q\,C_s}{\forall}$.

Substituting the expression for the input load $J_1 = Q\,C_s$, the initial concentration C_o is $C_o = \dfrac{J_1}{Q + k\forall}$.

a) The conservative case implies that $k = 0$. The concentration of TP in a lake over several years is given by $C(t) = C_e + (C_o - C_e)\exp\left[-\left(\frac{Q}{\forall}\right) t\right]$, where

$$C_e = \frac{J_1}{Q} = \frac{1 \times 10^6 \times \frac{1}{86400}}{500} = 0.023\ \text{g}/\text{m}^3,$$ or $0.023\ \text{mg}/\text{L}$. The initial concentration, that is, the concentration prior to the sewage diversion, is

$$C_o = \frac{J_1}{Q} = \frac{10 \times 10^6 \times \frac{1}{86400}}{500} = 0.23\ \text{g}/\text{m}^3.$$ The TP concentration after one year is given by $C(t) = 0.023 + (0.23 - 0.023)\exp\left[-\left(\frac{500 \times 365 \times 86400}{380 \times 10^8}\right)1\right] = 0.16\ \text{g}/\text{m}^3$, or 0.16 mg/L. The TP concentration over several years is given in the figure below.

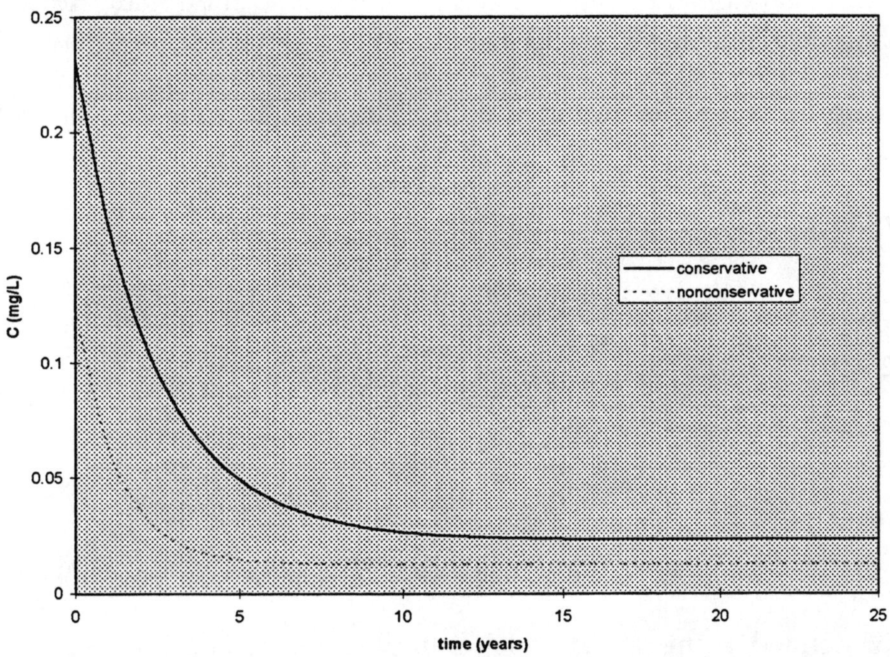

b) The nonconservative case implies that $k \neq 0$. The concentration of TP in the lake over several years is given by $C(t) = C_e + (C_o - C_e)\exp\left[-\left(\frac{Q}{\forall} + k\right) t\right]$, where $C_e = \dfrac{J_1}{Q + k\mathcal{V}} = \dfrac{1 \times 10^6 \times \dfrac{1}{86400}}{500 + 0.001 \times \dfrac{1}{86400} \times 380 \times 10^8} = 0.012\,\mathrm{g/m^3}$, or 0.012 mg/L.

The initial concentration is

$$C_o = \frac{J_1}{Q + k\mathcal{V}} = \frac{10 \times 10^6 \times \dfrac{1}{86400}}{500 + 0.001 \times \dfrac{1}{86400} \times 380 \times 10^8} = 0.12\,\mathrm{g/m^3}.$$

The TP concentration after one year is given by

$$C(t) = 0.012 + (0.12 - 0.012)\,\exp\left[-\left(\frac{500 \times 365 \times 86400}{380 \times 10^8} + 0.001 \times 365\right)1\right]$$

$$= \underline{0.06\,\mathrm{g/m^3},\ \text{or}\ 0.06\,\mathrm{mg/L}}$$

The TP concentration over several years is given in the above figure.

14.7 The scales of motion in a turbulent flow can range from the Batchelor scale η_B, to the computational domain scale L. The size of the computational grid should be smaller than the Batchelor scale. In a volume of size L^3 we should have $n = \left(\dfrac{L}{\eta_B}\right)^3$ where n is the total number of grid points. The scale at which gradients of the contaminant disappear is $\eta_B = \left(\dfrac{D^2\,\nu}{\varepsilon}\right)^{1/4}$, where $D = 8.2 \times 10^{-10}$ m²/s, $\varepsilon = 5 \times 10^{-6}\ \mathrm{m^2/s^3}$, and $\nu = 1.141 \times 10^{-6}\ \mathrm{m^2/s}$ (Table B.1, T=15°C). Hence,

$\eta_B = \left(\dfrac{(8.2 \times 10^{-10})^2\, 1.141 \times 10^{-6}}{5 \times 10^{-6}}\right)^{1/4} = 1.98 \times 10^{-5}$ m, and the total number of grid points is $n = \left(\dfrac{50}{1.98 \times 10^{-5}}\right)^3 = \underline{1.6 \times 10^{19}}$.

14.8 a) Power is defined as the rate at which energy is transferred into heat:

$$P = m\,c_p \frac{\Delta T}{\Delta t} = 20 \times 4180 \times \frac{2}{30 \times 60} = \underline{92.9\,\mathrm{W}}.$$

b) The rate of energy dissipation per unit mass is $\varepsilon = \dfrac{P}{m} = \dfrac{92.9}{20} = 4.64\,\mathrm{m^2/s^3}$,

$$\therefore\ \eta_K = \left(\frac{\nu^3}{\varepsilon}\right)^{1/4} = \left(\frac{(1.1\times10^{-6})^3}{4.64}\right)^{1/4} = \underline{2.3\times10^{-5}\ \mathrm{m}}$$

c) $$\eta_B = \left(\frac{D^2\,\nu}{\varepsilon}\right)^{1/4} = \left(\frac{(7.2\times10^{-10})^2\times1.1\times10^{-6}}{4.64}\right)^{1/4} = \underline{5.9\times10^{-7}\ \mathrm{m}}$$

14.9 The flux of NTA that results from molecular diffusion is

$$J = -D\,\frac{d\overline{C}}{dz} = -6.5\times10^{-10}\,\frac{(2.1\text{-}2.6)\times10^{-6}}{2} = \underline{1.6\times10^{-16}\ \mathrm{mol/m^2-s}},$$

and the flux of NTA that results from turbulent diffusion is

$$J = \overline{w'\,C'} = -D_{tz}\,\frac{d\overline{C}}{dz} = -1.2\times10^{-4}\,\frac{(2.1\text{-}2.6)\times10^{-6}}{2} = \underline{3\times10^{-11}\ \mathrm{mol/m^2\text{-}s}}.$$

14.10 a) The turbulent heat flux that results from turbulent diffusion is

$$q = r\,c_p\,\overline{w'\,T'} = -K_{tz}\,\frac{d\overline{T}}{dz} = -r\,c_p a_{tz}\,\frac{d\overline{T}}{dz}$$

where K_{tz} is the turbulent thermal conductivity. The density of air at 10°C is 1.247 kg/m³ (Table B.2). Thus

$$q = \rho\,c_p\,\overline{w'\,T'} = -\rho c_p a_{tz}\,\frac{d\overline{T}}{dz} = -1.247\times1003\times10\times0.02 = \underline{-2.5\times10^2\ \mathrm{W/m^2}}$$

b) The air temperature increases with height. The heat is transferred from high to low temperature, therefore the flux is negative.

14.11 a) Assume that the H_e concentration changes only in the x-direction. The mass balance over the control volume $A\Delta x$, in the absence of any chemical reaction between H_e and N_2, is given by $\dfrac{\partial}{\partial t}(\overline{C}\,A\Delta x) = J\,A|_x - J\,A|_{x+Dx}$. Dividing by

$A\Delta x$ and taking the limit as $\Delta x \to 0$ yields $\frac{\partial \overline{C}}{\partial t} = -\frac{\partial J}{\partial x}$. The oscillating grid does not generate the mean velocity along the x-axis in the chamber. The turbulent flux of H_e along the x-axis is given by Eq. 14.4.11. Thus the above equation reduces to $\frac{\partial \overline{C}}{\partial t} = D_t \frac{\partial^2 \overline{C}}{\partial x^2}$, where $\overline{C}$ is the time averaged H_e concentration in the chamber.

b) The initial concentration of H_e is $\overline{C} = 0$ in the chamber with N_2. Thus, the initial condition for positive x is $t = 0,\ x > 0,\ \overline{C} = 0$. The helium will diffuse from high to low concentration in the positive x-direction after the removal of the barrier. The concentration of H_e at $x = \infty$ is zero because it is too far from the H_2-N_2 interface to be present at the end of the long chamber. The concentration C_o is kept constant at $x = 0$ in the large, well-mixed tank. Therefore, the boundary conditions are:

$$t \geq 0,\ x = 0,\ \overline{C} = \overline{C}_o$$
$$x = \infty,\ \overline{C} = 0$$

c) The solution of the diffusion equation subject to the above boundary conditions is given in Table 14.3: $\frac{\overline{C}}{\overline{C}_o} = 1 - \operatorname{erf}\left(\frac{x}{2\sqrt{D_t\, t}}\right)$, where erf is the error function. The error function is available in tabulated form in Table 14.3; it can be related to cumulative normal probability distribution. If one plots the above equation with $\frac{\overline{C}}{\overline{C}_o}$ as an ordinate on an arithmetic-probability scale, and $\frac{x}{\sqrt{t}}$ as an abscissa on a linear scale, it becomes a straight line. Helium concentration measurements are plotted in the figure below. Assuming D_t is a constant, one can utilize this plot to estimate the turbulent diffusion coefficient. Assume that there is a point on the abscissa at which $\frac{x}{\sqrt{t}} = \frac{\sqrt{D_t}}{2}$. Substituting this in the above equation results in

$\frac{\overline{C}}{\overline{C}_o} = 1 - \operatorname{erf}\left(\frac{\sqrt{D_t}}{2 \times 2\sqrt{D_t}}\right) = 1 - \operatorname{erf}\left(\frac{1}{4}\right) = 1 - 0.27 = 0.73$. Thus, we can read the value of $\frac{x}{\sqrt{t}}$ that corresponds to $\frac{\overline{C}}{\overline{C}_o} = 0.73$ from the plot. This allows us to calculate the turbulent diffusion coefficient. Hence, $\frac{x}{\sqrt{t}} = 1.54 = \frac{\sqrt{D_t}}{2}$,

and $D_t = 9.49$ cm^2/s.

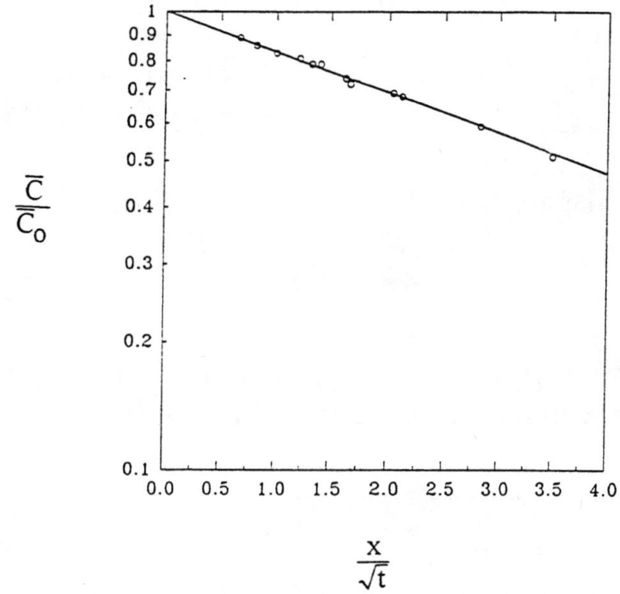

14.12 We assume that the acrolein herbicide (AH) is well-mixed over the channel cross-sectional area at the point of discharge. The mass balance of AH along the x-axis is described by Eq. 14.4.19. Substituting in the given expression for r in a uniform flow with constant dispersion coefficient reduces Eq. 14.4.19 to $\frac{\partial C}{\partial t} + U\frac{\partial C}{\partial x} = K\frac{\partial^2 C}{\partial x^2} - k\,C$. The AH is injected continuously at a constant input rate over the whole cross-section of the channel during a time period $\Delta t = 5$ hr. The input is "turned off" after time Δt. This is a "rectangular" input concentration condition. The solution of the above differential equation subject to the above condition is given in Table 14.3:

$$C(x,t) = \frac{C_o}{2}\exp\left(-\frac{k\,x}{U}\right)\left[G(t)\,\mathrm{erfc}\left(\frac{x\text{-}U\,t\left(1+\phi\right)}{\sqrt{4\,K\,t}}\right)\text{-}G\left(t\text{-}\Delta t\right)\mathrm{erfc}\left(\frac{x\text{-}U\left(t\text{-}\Delta t\right)\left(1+\phi\right)}{\sqrt{4\,K\left(t\text{-}\Delta t\right)}}\right)\right],$$

where $\phi = \frac{2\,k\,K}{U^2}$, $G(t - \Delta t) = 1$ when $t - \Delta t > 0$, and $G(t - \Delta t) = 0$ when $t - \Delta t < 0$.

a) The initial AH concentration at the point of discharge is

$$C_o = \frac{M}{Q\,\Delta t} = \frac{360 \times 10^3}{20 \times 1 \times 3600} = \underline{5\,\mathrm{mg/L}}$$

b) The longitudinal dispersion coefficient is needed in order to estimate AH concentration downstream from the discharge point. A formulation for the longitudinal dispersion coefficient is given in Table 14.4: $K = \alpha_k \dfrac{U^2 b^2}{K_y}$, where

$\alpha_k = 0.0066$, $K_y = U_* H$, $U_* = \sqrt{g H S} = \sqrt{9.81 \times 1 \times 0.00018} = 0.042$ m/s,
$K_y = 0.6 \times 0.042 \times 1 = 0.025$ m²/s, $U = Q/A = 20/20 = 1$ m/s, and $b = 20$m.

Hence the dispersion coefficient is

$$K = \alpha_k \frac{U^2 b^2}{K_y} = 0.0066 \times \frac{1 \times 20^2}{0.025} = \underline{105.6 \,\mathrm{m}^2/\mathrm{s}}.$$

The estimated AH concentration versus time at $x = 1$ km downstream from the discharge point is given in the following figure.

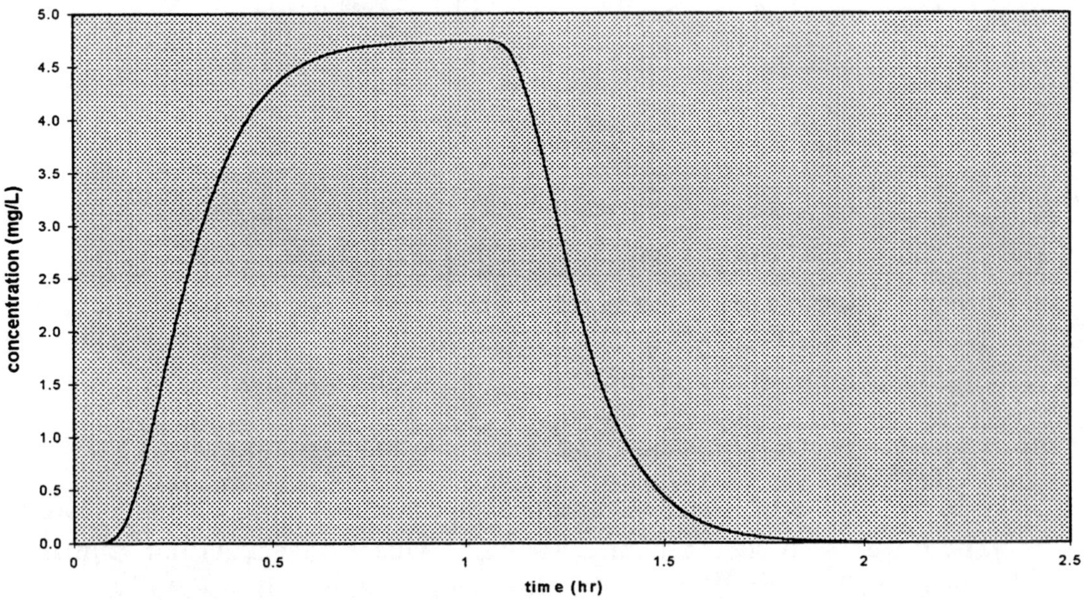

14.13 The mass balance of substance along the channel is described by Eq. 14.4.19. Transport that results only from advection for the nonconservative substance in a steady-state flow reduces Eq. 14.4.19 to $U \dfrac{dC}{dx} = r$. The measurements at two points suggest that the substance concentration decreases along the river. Therefore assume that $r = -kC$, where the k is the rate constant. The above differential equation is subject to the constant concentration boundary condition $C = C_o$ at $x = 0$. Integrating the above differential equation yields the solution

$$C = C_o \exp\left(-\frac{k}{U}x\right).$$

The concentration at the discharge point can be estimated knowing the concentration at the measurement location A:

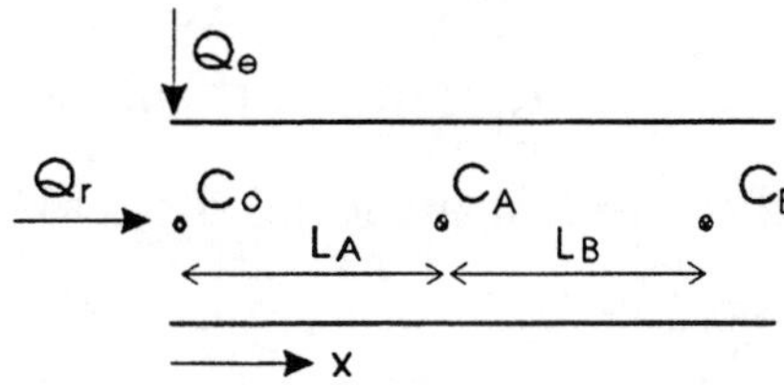

$$C_o = \frac{C_A}{\exp\left(-\frac{k}{U}L_A\right)}$$

The rate constant k can be estimated from the measurements between sections A and B: $k = \frac{U(\ln C_A - \ln C_B)}{L_B}$. The concentration measurements at those two points along the river yield $k = \frac{2\ (\ln 2 - \ln 1.5)}{2 \times 10^3} = 2.9 \times 10^{-4}\ \mathrm{s}^{-1}$. The concentration at the discharge point is $C_o = \frac{2}{\exp\left(-\frac{2.9 \times 10^{-4}}{2} \times 10^3\right)} = 2.3\ \mathrm{mg/L}$. We can estimate the effluent concentration at the discharge point from a simple weighted mass balance at $x = 0$: $C_e = \frac{Q_o C_o - Q_r C_r}{Q_e} = \frac{21 \times 2.3 - 20 \times 1}{1} = 28.3\ \mathrm{mg/L}$

<u>The company's claim is not true. The effluent concentration is 28.3 mg/L.</u>

14.14 a) The change with time of heat inside the control volume, $A\Delta x$, is equal to the algebraic sum of the heat fluxes across all boundaries of the control volume and the algebraic sum of heat sources within the control volume. Assume that the water temperature changes only in the x-direction. Thus, the temperature T is a cross-sectional average; lateral and vertical deviations from the average are considered to be small. The heat balance over the control volume is given by $\frac{\partial}{\partial t}(\rho c T A\Delta x) = q A\big|_x - q A\big|_{x+Dx} + SA\Delta x$. Dividing by $A\Delta x$ and taking the limit as $\Delta x \to 0$ yields $\frac{\partial}{\partial t}(\rho c T) = -\frac{\partial q}{\partial x} + S$. The total heat flux that results from advection and dispersion is $q = \rho c_v U T - \rho c_v E \frac{\partial T}{\partial x}$, where U is the cross-sectional average velocity, and E is the longitudinal dispersion coefficient for heat. The control volume sink term is the heat loss at the water surface. The overall surface heat transfer coefficient is defined per unit surface-contact area, A_s. The overall volumetric heat transfer coefficient is $kA_s/V = k/H$. Thus, the sink term is $S = -\frac{k}{H}(T - T_e)$. Combining the heat balance equation, sink term, and the total heat flux yields the one-dimensional dispersion model for heat transport: $\frac{\partial T}{\partial x} + U\frac{\partial T}{\partial x} = E\frac{\partial^2 T}{\partial x^2} - \frac{k}{\rho c_v H}(T - T_e)$.

b) For the steady-state and advective flow the above equation becomes $\frac{dT}{dx} = -\frac{k}{\rho c_v HU}(T - T_e)$. This differential equation is subject to the constant temperature boundary condition, $T = T_o$ at $x = 0$. The solution is

$$T = T_e + (T_o - T_e)\exp\left(-\frac{k}{\rho c_v HU}x\right).$$

c) At a long distance downstream from the discharge point ($x \to \infty$) the above equation reduces to $T = T_e$. Therefore, the river temperature approaches the equilibrium temperature. We can estimate the initial temperature from a simple weighted heat balance in the vicinity of the diffuser:

$$T_o = \frac{Q_r T_r + Q_{ef} T_{ef}}{Q_r + Q_{ef}} = \frac{20 \times 20 + 5 \times 45}{20 + 5} = 25\,°\text{C}.$$

d) The water density at 25 °C is 996.9 kg/m³ (Table B.1). The distance downstream of the diffuser where white bass cannot survive follows from the solution of the heat equation:

$$x = -U\frac{\rho c_v H}{k}\ln\left(\frac{T - T_e}{T_o - T_e}\right)$$

$$= -0.5 \times \frac{996.9 \times 4180 \times 1}{100}\ln\left(\frac{23 - 19}{25 - 19}\right) \approx 8448 \text{ m} \approx 8.5\ km$$

Thus the white bass will not survive 8.5 km beyond the effluent discharge along the river.

14.15 a) The length of the advective zone is given by Eq. 14.5.1: $L_x = \alpha_x \frac{U L_t^2}{K_y}$. The transverse dispersion coefficient is $K_y = 0.6HU_*$, where the mean shear stress velocity $U_* = \sqrt{gHS} = \sqrt{9.81 \times 1 \times 0.00015} = 0.038 \text{ m/s}$. Then $K_y = 0.6 \times 1 \times 0.038 = 0.023$ m²/s. We assume $\alpha_x = 0.3$, and $L_t = 0.7\, b = 0.7 \times 50 = 35$ m. The mean flow velocity $U = Q/A = 30/50 = 0.6$ m/s. Hence the length of the advective zone is $L_x = 0.3 \times \frac{0.6 \times 35^2}{0.023} \approx \underline{9.6 \text{ km}}$.

b) The required time $t = \frac{L_x}{U} = \frac{9.6}{0.6} \approx \underline{4.4 \text{ hr}}$.

14.16 a) The vertical turbulent diffusion coefficient is estimated from Eq. 14.5.8. The measured water temperature is used in place of a substance concentration. Thus Eq. 14.5.8 reads

$$D_{tz} = -\frac{\dfrac{\partial}{\partial t}\displaystyle\int_{z}^{z_{max}} T(z)\, A(z)\, dz}{\dfrac{\partial T(z)}{\partial z} A(z)}$$

where $T(z)$ is the measured water temperature. A numerical approximation of the above numerator can be carried out on a spreadsheet.

Step 1: Approximate the numerator of the equation (see Example 14.10) with $\frac{1}{\Delta t}\sum_{i=1}^{n} \Delta T_{ti}\, \Delta z_i\, A_i$.

Layer (m)	A_i (10^6 m²)	T_{1i} (°C)	T_{2i} (°C)	ΔT_{12i} (°C)	$\Delta T_{12i} A_i \Delta z$ (°C m³ 10^6)	$\frac{1}{\Delta t}\sum \Delta T_{12i} A_i \Delta z$ (°C m³/d 10^4)
12-13	0.69	9.5	10.02	0.52	0.359	3.512
13-14	0.60	9.17	9.55	0.38	0.228	2.017
14-15	0.53	9.05	9.25	0.20	0.106	1.077
15-16	0.43	8.97	9.10	0.13	0.056	0.625
16-17	0.31	8.90	9.00	0.10	0.031	0.392
17-18	0.21	8.82	8.95	0.13	0.027	0.262
18-19	0.13	8.75	8.93	0.18	0.023	0.150
19-20	0.06	8.7	8.92	0.22	0.013	0.054

Step 2: Approximate the denominator of the above equation. The temperature gradients are estimated using central differences.

Depth (m)	A (10^6 m^2)	$\left.\frac{\Delta T}{\Delta z}\right\|_1$ (°C /m)	$\left.\frac{\Delta T}{\Delta z}\right\|_2$ (°C /m)	$\left.\frac{\Delta T}{\Delta z}\right\|_{1\text{-}2}$ (°C /m)	$\left.\frac{\Delta T}{\Delta z}\right\|_{1\text{-}2} A$ (°C m 10^6)
12	0.72	-0.50	-0.55	-0.52	-0.374
13	0.66	-0.32	-0.47	-0.39	-0.257
14	0.55	-0.12	-0.30	-0.21	-0.115
15	0.52	-0.07	-0.15	-0.11	-0.057
16	0.35	-0.07	-0.10	-0.08	-0.028
17	0.28	-0.07	-0.05	-0.06	-0.017
18	0.15	-0.07	-0.01	-0.04	-0.006
19	0.12	-0.05	-0.01	-0.03	-0.004

Step 3: Compute the turbulent diffusion coefficient.

Depth (m)	12	13	14	15	16	17	18	19
D_{tz} (cm^2/s)	0.011	0.009	0.011	0.013	0.016	0.018	0.029	0.016

b) The stability frequency is estimated with Eq. 14.5.7. Water density profiles are estimated from the water temperature profiles using the expression given in the problem. The reference water density ρ_0 is selected to be the one that corresponds to the highest temperature in a water temperature profile, ρ_{01} = 997.7327 kg/m³ (21.85 °C), ρ_{02} = 996.797 kg/m³ (25.65 °C). Numerical approximations can be performed on a spreadsheet. The following results are obtained:

Depth (m)	T_1 (°C)	T_2 (°C)	ρ_1 (kg/m³)	ρ_2 (kg/m³)	ρ_{1e} (kg/m³)	ρ_{2e} (kg/m³)	$\left.\frac{\Delta\rho_{1e}}{\Delta z}\right\|_1$ (kg/m⁴)	$\left.\frac{\Delta\rho_{2e}}{\Delta z}\right\|_2$ (kg/m⁴)	N_1 (s^{-1})	N_2 (s^{-1})	N^2_{12} 10^{-4} (s^{-2})
11	10.25	10.85	999.635	999.576	1.902	2.778					
12	9.75	10.30	999.681	999.630	1.948	2.832	0.0440	0.0524	0.0208	0.0227	4.7089
13	9.25	9.75	999.723	999.681	1.990	2.883	0.0272	0.0422	0.0163	0.0204	3.3856
14	9.10	9.35	999.735	999.715	2.002	2.917	0.0099	0.0252	0.0099	0.0157	1.6384
15	9.00	9.15	999.743	999.731	2.010	2.933	0.0059	0.0121	0.0076	0.0109	0.8464
16	8.95	9.05	999.747	999.739	2.014	2.941	0.0058	0.0079	0.0075	0.0088	0.6724
17	8.85	8.95	999.754	999.747	2.021	2.949	0.0057	0.0039	0.0075	0.0062	0.4624
18	8.80	8.95	999.758	999.747	2.025	2.949	0.0056	0.0012	0.0074	0.0034	0.2916
19	8.7	8.92	999.765	999.749	2.033	2.951	0.0037	0.0012	0.0060	0.0034	0.2209
20	8.7	8.92	999.765	999.749	2.033	2.951					

In the table, T_1 is the water temperature profile on day 1, T_2 is the water temperature profile on day 25, ρ_1 is the water density profile on day 1, ρ_2 is the water density profile on day 25, $\rho_{1e} = (\rho_1 - \rho_{01})$ is the density above ρ_{01}, $\rho_{2e} = (\rho_2 - \rho_{02})$ is the density above ρ_{02}, and N^2_{12} is the square of the average stability frequency between N_1 and N_2.

c) The turbulent diffusion coefficient versus square of stability frequency N_{12}^2 is plotted below. The estimates, denoted by symbols, refer to the depth interval between 12 and 19 m.

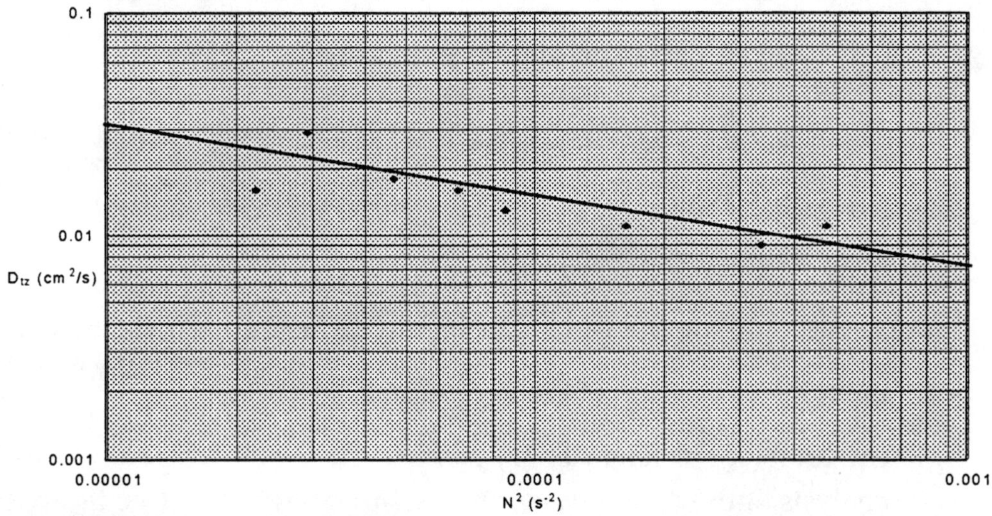

d) The vertical turbulent diffusion coefficient decreases with increasing stability frequency.

14.17 The total amount of dye in the channel is given by $M = \int_{-\infty}^{+\infty} C\,A\,dx$. Since we are considering a conservative tracer, the mass of the dye should remain constant in the channel. Normalizing the one-dimensional diffusion Eq. 14.3.6 with the total mass of the dye, we obtain $\frac{\partial P}{\partial t} = D\frac{\partial^2 P}{\partial x^2}$. Note that the total amount of dye in the channel is $\int_{-\infty}^{\infty} P(x)\,dx = 1.0$.

a) The coordinate of the centroid of the diffusing dye is by definition $x_c = \int_{-\infty}^{\infty} x\,P(x)\,dx$. Thus multiplying the above equation by x and integrating from $-\infty$ to ∞ yields $\frac{\partial}{\partial t}\int_{-\infty}^{\infty} x\,P(x)\,dx = D\int_{-\infty}^{\infty} x\frac{\partial^2 P}{\partial x^2}dx$. The integral on the left-hand side is clearly the definition of the centroid of the dye. Integrating the right-hand side by parts yields $\frac{d x_c}{d t} = D\left(x\frac{\partial P}{\partial x}\Big|_{-\infty}^{\infty} - \int_{-\infty}^{\infty} \frac{\partial P}{\partial x} dx \right) = 0$. Both expressions on the right-hand side reduce to zero at the upper and lower limits. Thus, the coordinate of the centroid does not change over the time.

b) The variance of the diffusing dye is defined as $\sigma^2 = \int_{-\infty}^{\infty} x^2\, P(x)\, dx$. Multiplying the normalized diffusion equation by x^2, integrating twice by parts, and using the fact that the total amount of dye in the channel is equal to 1.0, yields $\frac{d\,\sigma^2}{d\,t} = 2\,D$. This expression is integrated as follows:

$\int_0^t d\,\sigma^2 = 2\,D\int_0^t dt\ ;\sigma^2(t) = \sigma_0^2 + 2\,D\,t$. Using $\sigma(0)=0$ so that $\sigma_0^2 = 0$, the variance is $\sigma^2(t) = 2\,D\,t$.

14.18 a) We can estimate the vertical turbulent diffusion coefficient from the measured time rate of increase of the variance of the SF_6: $D_{tz} = \frac{1}{2}\frac{d\sigma_t{}^2}{d\,t}$. Integrating this equation (see problem 14.17) yields $\sigma^2(t) = \sigma_0^2 + 2\,D_{tz}\,t$. A least-squares analysis of the data provide the best-fit straight line (see figure below). The result is $\sigma^2(t) = \sigma_0^2 + 2\,D_{tz}\,t$, that is, $\sigma^2(t) =$ const + $0.634t$. Thus, the turbulent diffusion coefficient is $D_{tz} = \frac{0.634}{2} = 0317\ \mathrm{m}^2/\mathrm{d}$ = $\underline{3.6 \times 10^{-2}\ \mathrm{cm}^2/\mathrm{s}}$

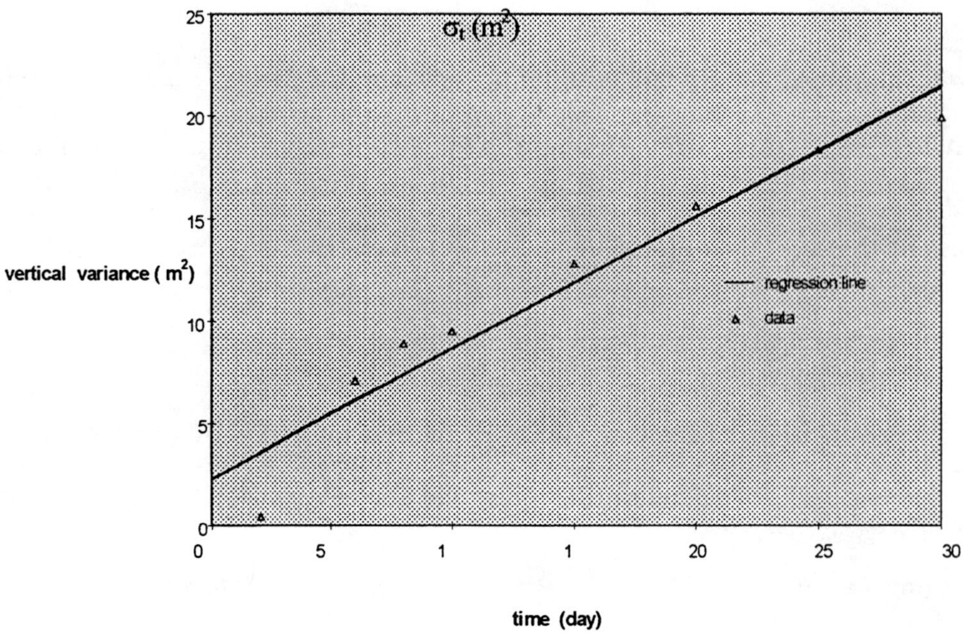

b) The tracer is injected at a depth of 17 m along a 3 km distance. The injection should be along the line of constant water temperature and not along the constant depth below the lake surface. In our case the tracer is injected along a line of constant depth. This initial tracer adjustment is reflected in the finite variance quantity at $t = 0$.

FOURTEEN

Environmental Fluid Mechanics

Corrections to 1st Printing

1. Page 682, line 3 from the top:

READS $\frac{dJ}{dx} - k_s C = 0$ SHOULD READ $\frac{dJ}{dx} + k_s C = 0$

---

2. Page 682, line 8 from the top:

READS $-D\frac{d^2C}{dx^2} - k_s C = 0$ SHOULD READ $-D\frac{d^2C}{dx^2} + k_s C = 0$

---

3. Page 682, line 8 from the bottom:

READS $\frac{C}{C_o} = e^{-\left(k_s/D\right)x}$ SHOULD READ $\frac{C}{C_o} = e^{-\sqrt{\left(k_s/D\right)}\,x}$

---

4. Page 682, line 3 from the bottom:

READS $\ln C = -\frac{k_s}{D}x + \ln C_o$ where $\frac{k_s}{D} = 0.002$

SHOULD READ $\ln C = -\sqrt{\frac{k_s}{D}}x + \ln C_o$ where $\frac{k_s}{D} = 0.002$

5. Page 701, Example 14.8:

READS:

Mississippi River at station 2

	May 11													May 12		
t (hours)	20	20.5	22	24	25.5	27	29	31	33	35	37	39	41	43	45	47
C (mg/L)	0.00	0.18	0.94	12.9	19.0	8.5	2.5	1.4	0.93	0.71	0.51	0.40	0.29	0.21	0.16	0.13

t (hours)	49	51	53	55	57
C (mg/L)	0.00	0.18	0.94	12.9	19.0

SHOULD READ:

Mississippi River at station 2

	May 11				May 12											
t (hours)	20	20.5	22	24	25.5	27	29	31	33	35	37	39	41	43	45	47
C (mg/L)	0.00	0.2	2.8	5.0	8.8	8.1	7.4	6.0	4.4	3.0	2.0	1.45	1.0	0.75	0.56	0.42

	May 13				
t (hours)	49	51	53	55	57
C (mg/L)	0.3 2	0.25	0.19	0.15	0.11

---

6. Page 708, Problem 14.6, Figure 14.6:

READS

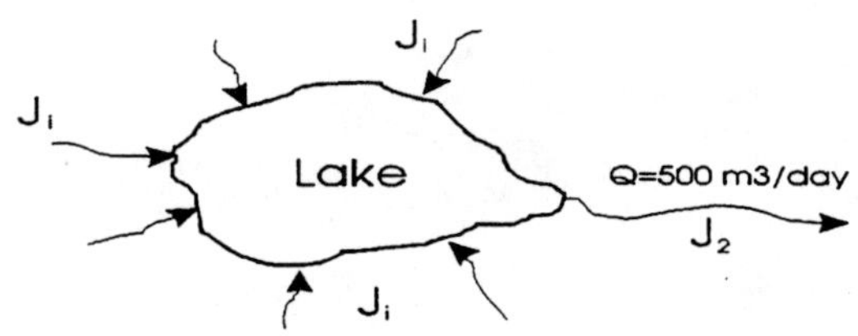

SHOULD READ

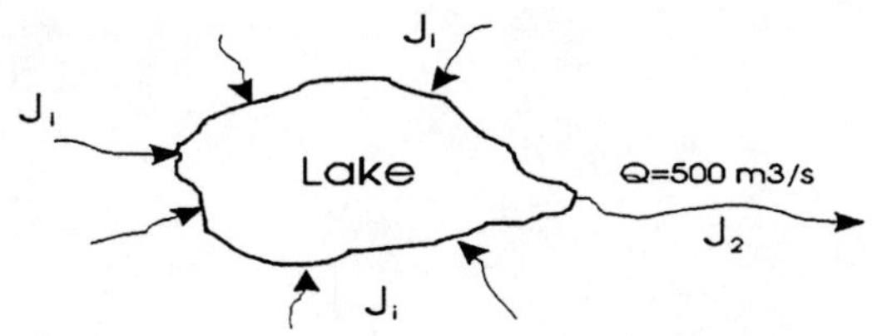

7. Page 709, Problem 14.13, Figure P14.13:

READS:

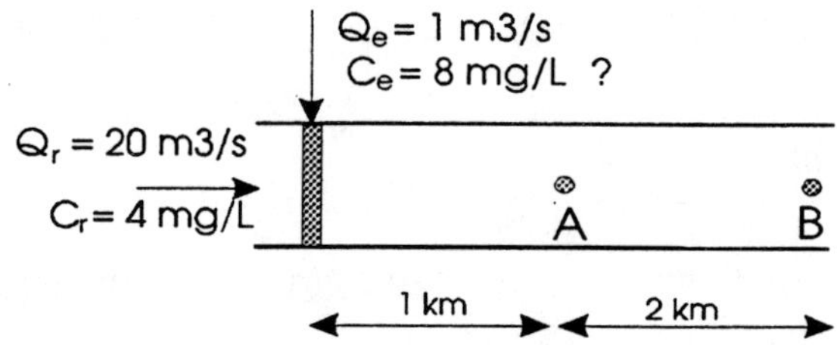

SHOULD READ:

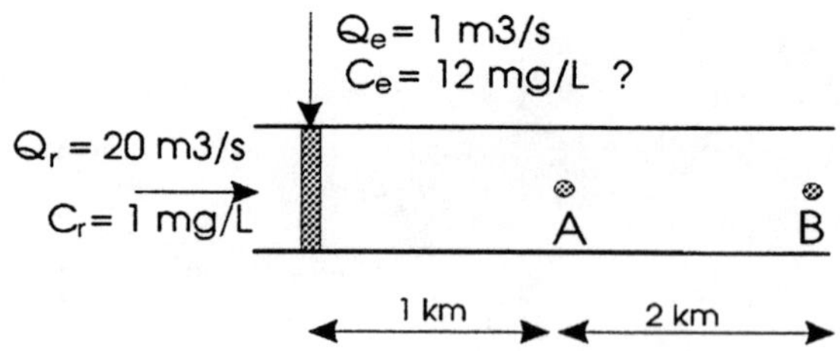

---

8. Page 710, Problem 14.14, line 2 from the bottom:

READS $k = 100\ \mathrm{J/m^2 \cdot min \cdot {}^\circ C}$

SHOULD READ $k = 100\ \mathrm{J/m^2 \cdot s \cdot {}^\circ C}$

Solutions Manual
Mechanics of Fluids 2/e
Merle C. Potter, David C. Wiggert , Midhat Hondzo
Prentice-Hall, Inc.

YOU SHOULD CAREFULLY READ THE TERMS AND CONDITIONS BEFORE USING THE DISKETTE PACKAGE. USING THIS DISKETTE PACKAGE INDICATES YOUR ACCEPTANCE OF THESE TERMS AND CONDITIONS.

Prentice-Hall, Inc. provides this program and licenses its use. You assume responsibility for the selection of the program to achieve your intended results, and for the installation, use, and results obtained from the program. This license extends only to use of the program in the United States or countries in which the program is marketed by authorized distributors.

LICENSE GRANT
You hereby accept a nonexclusive, nontransferable, permanent license to install and use the program ON A SINGLE COMPUTER at any given time. You may copy the program solely for backup or archival purposes in support of your use of the program on the single computer. You may not modify, translate, disassemble, decompile, or reverse engineer the program, in whole or in part.

TERM
The License is effective until terminated. Prentice-Hall, Inc. reserves the right to terminate this License automatically if any provision of the License is violated. You may terminate the License at any time. To terminate this License, you must return the program, including documentation, along with a written warranty stating that all copies in your possession have been returned or destroyed.

LIMITED WARRANTY
THE PROGRAM IS PROVIDED "AS IS" WITHOUT WARRANTY OF ANY KIND, EITHER EXPRESSED OR IMPLIED, INCLUDING, BUT NOT LIMITED TO, THE IMPLIED WARRANTIES OR MERCHANTABILITY AND FITNESS FOR A PARTICULAR PURPOSE. THE ENTIRE RISK AS TO THE QUALITY AND PERFORMANCE OF THE PROGRAM IS WITH YOU. SHOULD THE PROGRAM PROVE DEFECTIVE, YOU (AND NOT PRENTICE-HALL, INC. OR ANY AUTHORIZED DEALER) ASSUME THE ENTIRE COST OF ALL NECESSARY SERVICING, REPAIR, OR CORRECTION. NO ORAL OR WRITTEN INFORMATION OR ADVICE GIVEN BY PRENTICE-HALL, INC., ITS DEALERS, DISTRIBUTORS, OR AGENTS SHALL CREATE A WARRANTY OR INCREASE THE SCOPE OF THIS WARRANTY.

SOME STATES DO NOT ALLOW THE EXCLUSION OF IMPLIED WARRANTIES, SO THE ABOVE EXCLUSION MAY NOT APPLY TO YOU. THIS WARRANTY GIVES YOU SPECIFIC LEGAL RIGHTS AND YOU MAY ALSO HAVE OTHER LEGAL RIGHTS THAT VARY FROM STATE TO STATE.

Prentice-Hall, Inc. does not warrant that the functions contained in the program will meet your requirements or that the operation of the program will be uninterrupted or error-free.

However, Prentice-Hall, Inc. warrants the diskette(s) on which the program is furnished to be free from defects in material and workmanship under normal use for a period of ninety (90) days from the date of delivery to you as evidenced by a copy of your receipt.

The program should not be relied on as the sole basis to solve a problem whose incorrect solution could result in injury to person or property. If the program is employed in such a manner, it is at the user's own risk and Prentice-Hall, Inc. explicitly disclaims all liability for such misuse.

LIMITATION OF REMEDIES
Prentice-Hall, Inc.'s entire liability and your exclusive remedy shall be:
1. the replacement of any diskette not meeting Prentice-Hall, Inc.'s "LIMITED WARRANTY" and that is returned to Prentice-Hall, or
2. if Prentice-Hall is unable to deliver a replacement diskette that is free of defects in materials or workmanship, you may terminate this agreement by returning the program.

IN NO EVENT WILL PRENTICE-HALL, INC. BE LIABLE TO YOU FOR ANY DAMAGES, INCLUDING ANY LOST PROFITS, LOST SAVINGS, OR OTHER INCIDENTAL OR CONSEQUENTIAL DAMAGES ARISING OUT OF THE USE OR INABILITY TO USE SUCH PROGRAM EVEN IF PRENTICE-HALL, INC. OR AN AUTHORIZED DISTRIBUTOR HAS BEEN ADVISED OF THE POSSIBILITY OF SUCH DAMAGES, OR FOR ANY CLAIM BY ANY OTHER PARTY.

SOME STATES DO NOT ALLOW FOR THE LIMITATION OR EXCLUSION OF LIABILITY FOR INCIDENTAL OR CONSEQUENTIAL DAMAGES, SO THE ABOVE LIMITATION OR EXCLUSION MAY NOT APPLY TO YOU.

GENERAL
You may not sublicense, assign, or transfer the license of the program. Any attempt to sublicense, assign or transfer any of the rights, duties, or obligations hereunder is void.

This Agreement will be governed by the laws of the State of New York.

Should you have any questions concerning this Agreement, you may contact Prentice-Hall, Inc. by writing to:
Director of New Media
Higher Education Division
Prentice-Hall, Inc.
1 Lake Street
Upper Saddle River, NJ 07458

Should you have any questions concerning technical support, you may write to:
New Media Production
Higher Education Division
Prentice-Hall, Inc.
1 Lake Street
Upper Saddle River, NJ 07458

YOU ACKNOWLEDGE THAT YOU HAVE READ THIS AGREEMENT, UNDERSTAND IT, AND AGREE TO BE BOUND BY ITS TERMS AND CONDITIONS. YOU FURTHER AGREE THAT IT IS THE COMPLETE AND EXCLUSIVE STATEMENT OF THE AGREEMENT BETWEEN US THAT SUPERSEDES ANY PROPOSAL OR PRIOR AGREEMENT, ORAL OR WRITTEN, AND ANY OTHER COMMUNICATIONS BETWEEN US RELATING TO THE SUBJECT MATTER OF THIS AGREEMENT.